The Life Cycles
of
Extratropical Cyclones

The Life Cycles
of
Extratropical Cyclones

Edited by

Melvyn A. Shapiro
National Center for Atmospheric Research

Sigbjørn Grønås
Geophysical Institute, University of Bergen

American Meteorological Society
Boston
1999

ISBN 1-878220-35-7

Published by the American Meteorological Society
45 Beacon Street, Boston, Massachusetts 02108 USA

Printed in the United States of America
by the Canterbury Press, Rome, New York

Contents

We Remember Bergen: A Photographic Recollection of Our Friends

Preface

This monograph contains expanded versions of the invited papers presented at the International Symposium on the Life Cycles of Extratropical Cyclones, held in Bergen, Norway, 27 June–1 July 1994. The symposium coincided with the 75th anniversary of the introduction of Jack Bjerknes's frontal-cyclone model presented in his seminal article, "On the Structure of Moving Cyclones." The event was attended by approximately 300 scientists and students from around the world and included 207 papers and poster presentations, of which 17 were invited lectures. The symposium provided a state-of-the-science account of advances in the research and forecasting of extratropical cyclones. The symposium was organized by Sigbjørn Grønås of the Geophysical Institute of the University of Bergen, Norway, and Melvyn Shapiro of the National Oceanic and Atmospheric Administration (NOAA)/Environmental Technology Laboratory, Boulder, Colorado, USA. Preprints of the invited and submitted presentations were published in three volumes and are available through the Geophysical Institute of the University of Bergen. A special issue of *Tellus* (1995), **47A**, 525 pp., was dedicated to the publication of 27 reviewed papers from the symposium.

This monograph should be of interest to historians of meteorology, researchers, and forecasters. It contains material appropriate for teaching courses in advanced undergraduate and graduate meteorology. The chapter bibliographies provide a valuable source for key references on many aspects of extratropical cyclones.

The symposium was hosted by the University of Bergen and cosponsored and generously supported by societies, universities, government organizations, private companies, and endowments from Norway and the United States. Norwegian support was provided by the Norwegian Geophysical Society; University of Bergen; Norwegian Meteorological Institute; Research Council of Norway; Norwegian Department of Church, Education and Research; Meltzers Høyskolefond; O. Kalvi og Knut Kalvi's Almennyttige Fond; Intel Supercomputer Systems; Cray Research; Silicon Graphics Inc.; Aanderaa Instruments; and Vesta Foriksring. United States support was provided by the NOAA/Environmental Research Laboratories; American Meteorological Society (AMS); National Science Foundation, Office of Naval Research; University Corporation for Atmospheric Research; and the National Center for Atmospheric Research.

The first nine chapters of the monograph present a historical overview of extratropical cyclone research and forecasting from the early 18th century into the mid-20th century. The first chapter presents Vilhelm Bjerknes's 1904 paper, which outlined a rational approach to weather forecasting through the synthesis of classical hydrodynamics and thermodynamics with meteorological observations. Bjerknes envisioned weather forecasting as a problem in mechanics and physics in which the dynamical equations for atmospheric motion were integrated through numerical methods. Arnt Eliassen highlights the early studies of Vilhelm Bjerknes and their connection to Jack Bjerknes's Bergen school cyclone model. Hans Volkert discusses observations, theories, and conceptual models prior to 1920 that reaffirm the international scope of the scientific milieu that sowed the seeds for the subsequent Norwegian conceptual models of extratropical cyclones. Science historian Robert Marc Friedman chronicles the political, societal, and economic factors that contributed to the development of scientific thought leading to the capstone of the Bergen school contributions: The Life Cycles of Extratropical Cyclones and the Polar Front Theory of Atmospheric Circulation, published in 1922 by Jack Bjerknes and Halvor Solberg. Chester Newton and Harriet Rodebush Newton review the advances in American meteorology before 1919 and how the Bergen school polar-front and cyclone concepts came to be adopted into daily weather forecasting by the U.S. Weather Bureau. Peter Lynch critiques the numerical methods applied by L. F. Richardson in his historic attempt at dynamical weather forecasting. Lynch repeats Richardson's forecast, with more successful results, using modern computational initialization methods. The next chapter presents a transcript of a tape-recorded informal lecture by Erik Palmén delivered in Helsinki, Finland, in 1979. Palmén spoke about the contributions of the Bergen school and the period when he worked with Jack Bjerknes on the analysis of the northern European serial thermograph sounding experiment, reported on in their classical 1930s papers on the three-dimensional structure of extratropical cyclones. Brian Hoskins recaps the career of Reginald Sutcliffe and his contributions to the formulation to quasigeostrophic theory and its application to the theoretical understanding of cyclone development. The historical chapters are concluded with a commemorative photograph album, compiled by Melvyn Shapiro, which portrays some of the key contributors to the advancement of meteorology.

The succeeding chapters present an overview of contemporary research on the theory, observations, analysis, diagnosis, and prediction of extratropical cyclones. The sequence of presentations transcends the planetary-scale to mesoscale scales of motion. Eero Holopainen presents an overview of recent observational studies of extratropical cyclone climatology and their relation to planetary-scale waves. Isaac Held explores selected aspects of planetary wave dynamics and their interactions with smaller scales. Brian Farrell presents a historical review of advances in the theory of cyclone development and introduces a contemporary approach toward a generalized theory of baroclinic development. Adrian Simmons discusses numerical simulations of idealized and actual cyclone life cycles with an emphasis on the larger scales of motion, including downstream and upstream baroclinic development, and recent results from predictions produced at the European Centre for Medium-Range Weather Forecasts (ECMWF). Melvyn Shapiro and collaborators present a planetary-scale to mesoscale perspective of cyclone life cycles which aspires to build a conceptual bridge between theoretically idealized cyclone life cycles and those observed in nature. Lance Bosart provides an overview of observed cyclone and anticyclone life cycles and discusses current and future directions in extratropical cyclone and anticyclone research. Huw Davies reviews the theoretical studies of frontogenesis within the framework of quasi- and semigeostrophic dynamics. Daniel Keyser reviews classical and contemporary perspectives on the representation and diagnosis of frontal circulations in two and three dimensions. Keith Browning documents observed mesoscale aspects of extratropical cyclones, which include frontal fracture, air streams, dry intrusions, and mesoscale substructures associated with surface cold fronts. Alan Thorpe addresses the dynamics of extratropical cyclones associated with mesoscale structures and includes discussions of potential vorticity "thinking," frontal-wave cyclones, and the role of conditional symmetric instability in the formation of frontal rain bands. Thor Erik Nordeng presents an overview of numerical simulations of mesoscale substructures and physical processes within extratropical cyclones, addressing topics such as frontal structures and dynamics, surface friction and fluxes of heat and moisture, latent heating, and nonclassical mesoscale cyclone development. Louis Uccellini, Paul Kochin, and Joseph Sienkiewicz describe forecasting advances at the U.S. National Meteorological Center, with an emphasis on the performance of models and forecasters in predicting oceanic cyclogenesis. Michael McIntyre discusses, among other topics, his "vision of the future" for the potential for humans to interact usefully and efficiently with computer-based systems in advancing the accuracy of numerical weather predictions. The monograph concludes with a collection of photographs taken at the Bergen Symposium by Carlye Calvin and Nadine Lindzen (lead photographers), with additional contributions provided by David Schultz, Howie Bluestein, and Melvyn Shapiro.

The editors acknowledge and thank the authors and reviewers for their contributions to the monograph. We express our appreciation to NOAA's Environmental Research Laboratories (ERL), James Rasmussen, Director; ERL/Environmental Technology Laboratory, Stephen Clifford, Director; and the Faculty of Natural Sciences, University of Bergen, Kjell Saelen, Director, for providing the resources essential for carrying the monograph to completion. We acknowledge and give special thanks to Sandra Rush for her exemplary contribution as technical editor of the monograph and for her preparation of the camera-ready versions of the chapters for publication. Our appreciation to Keith Seitter of the American Meteorological Society for his encouragement of this project and for his preparation of the monograph subject index. Special acknowledgments are due to Carlye Calvin for preparing the Bergen symposium photo album, and to Frank Cleveland for his assistance in obtaining many of the photographs appearing in the historical photo album, and to Paul Neiman for his help in constructing the historical photo album.

Melvyn A. Shapiro   Sigbjørn Grønås
Boulder Colorado, USA   Bergen, Norway

The Problem of Weather Forecasting as a Problem in Mechanics and Physics

VILHELM BJERKNES

If it is true, as every scientist believes, that subsequent atmospheric states develop from the preceding ones according to physical law, then it is apparent that the necessary and sufficient conditions for the rational solution of forecasting problems are the following:

1. A sufficiently accurate knowledge of the state of the atmosphere at the initial time.
2. A sufficiently accurate knowledge of the laws according to which one state of the atmosphere develops from another.

1

The determination of the state of the atmosphere at the initial time is the task of observational meteorology This problem has not yet been solved to the extent that is necessary for rational forecasting. There are two major gaps in the observations. The first one is that only land stations participate in the daily programs of the weather services. Over the seas, which cover four-fifths of the earth's surface and must therefore exert an overwhelming influence, no observations are made for the purposes of current weather analysis. Furthermore, the observations that are used in current analysis are made only at the surface of the earth and all data pertaining to the state of the higher layers of the atmosphere are missing.

But we already have the technical means that will enable us to fill these two gaps. With the help of wireless telegraphy, we will be able to include among the reporting stations the ships moving in fixed routes. And to judge by the great forward steps that have been made in recent years in the techniques of upper air soundings, it will be possible to obtain

daily observations of the higher atmospheric layers not only from fixed land positions but also from traveling stations on the sea.

We can hope, therefore, that the time will soon come when either as a daily routine, or for certain designated days, a complete diagnosis of the state of the atmosphere will be available. The first condition for putting forecasting on a rational basis will then be satisfied.

2

The second problem then arises as to whether we know, with sufficient accuracy, the laws according to which one state of the atmosphere develops out of another.

The atmospheric processes are of a mixed mechanical and physical nature. Each one of these processes can be expressed in one or more mathematical equations according to mechanical or physical principles. We have sufficient knowledge of the laws according to which the atmosphere develops if we can set up as many independent equations as there are unknown quantities. From a meteorological point of view, the state of the atmosphere is specified, at an arbitrary time, if we can determine for that time at each point, the velocity, density, pressure, temperature, and humidity of the air. The velocity, as a vector, is given by three scalar quantities, the three velocity components, and one must therefore deal with seven unknown quantities.

To compute these quantities, we can set up the following equations:

1. The three hydrodynamical equations of motion. These are differential relations among the three velocity components, the density, and pressure.
2. The continuity equation, which expresses the principle of the conservation of mass during motion. This equation is again a differential relation between the velocity components and the density.

---

*Das Problem der Wettervorhersage, betrachtet vom Standpunkte der Mechanik und der Physik. Meteorologische Zeitschrift, January 1904, pp. 1-7. English translation by Y. Mintz, Los Angeles, 1954.

3. The equation of state of atmospheric air, which is a relation in finite form among the density, pressure, temperature, and humidity of a given mass of air.
4. The two laws of thermodynamics, which allow us to set up two differential relations giving the rates of change of energy and entropy during the changes of state that are taking place. These equations introduce no new unknowns into the problem, as the energy and entropy are expressed by the same variables that appear in the equation of state and connect the changes of these quantities with other quantities considered as known. These other quantities are, first, the work done by the mass of air, which is determined by the same variables that appear in the dynamical equations; and second, the amount of heat given up or received by the mass of air, which is determined by the physical data on incoming and outgoing radiation and on conduction where the air is in contact with the ground.

It should be emphasized that a basic simplification of the problem can be achieved if there is no condensation or evaporation of water, so that the water vapor of the air can be considered as a constant constituent. Then the problem will have one variable less, and one of the equations, the one that comes from the second law of thermodynamics, can be eliminated. On the other had, if we had to deal with several variable constituents of the atmosphere, then the second law of thermodynamics would give a new equation for each new constituent.

For the computation of the normally occurring seven variables, we can set up seven independent equations. So that, as we now see the problem, we must conclude that we do have sufficient knowledge of the laws of atmospheric processes upon which a rational weather forecasting system can be based. But it must be admitted that we could have overlooked important factors on account of the incompleteness of our knowledge. The interference of cosmic effects of an unknown kind may be imagined. Furthermore, the large-scale atmospheric phenomena are accompanied by a long train of subsidiary effects, for example those of an electrical or optical nature, and the question is to what extent such subsidiary effects could react in a significant way on the course of the atmospheric processes. These reactions exist, of course: for instance, the rainbow modifies the distribution of incoming radiant energy from the sun, and electrical potentials influence the condensation processes. But until now there is no evidence that processes of this kind react upon the large-scale atmospheric processes in any significant way. Yet in any case, the scientific method is to start with the simplest problem that can be formulated, which is the problem, posed above, of seven variables and seven equations.

3

Of the seven equations, only one, the equation of state, has a finite form. The other six are partial differential equations. Of the seven unknowns, one can be eliminated with the aid of the equation of state, and the problem then becomes the integration of a system of six partial differential equations with six unknowns, and with the utilization of initial conditions as given by the observations of the initial state of the atmosphere.

An exact analytical integration of the system of equations is out of the question. Even the computation of the motion of three mass-points, which influence each other according to a law as simple as that of Newton, exceeds the limits of today's mathematical analysis. Naturally there is no hope of understanding the motion of all the points of the atmosphere, which have far more complicated reactions upon one another. Moreover, the exact analytical solution, even if we could write it down, would not give the result that we need. For to be practical and useful, the solution has to have a readily seen, synoptic form and has to omit the countless details that would appear in every exact solution. The prognosis need only deal, therefore, with averages over sizeable distances and time intervals; for example, from degree of meridian to degree of meridian and from hour to hour, but not from millimeter to millimeter or from second to second.

We therefore forgo any thought of analytical methods of integration and instead pose the problem of weather prediction in the following practical form:

Based upon the observations that have been made, the initial state of the atmosphere is represented by a number of charts which give the distribution of the seven variables from level to level in the atmosphere. With these charts as the starting point, new charts of a similar kind are to be drawn that represent the new state from hour to hour.

For the solution of the problem in this form, graphical or mixed graphical and numerical methods are appropriate, which methods must be derived either from the partial differential equations, or from the dynamical-physical principles that are the basis of these equations. There is no reason to doubt, beforehand, that these methods can be worked out. Everything will depend upon whether we can successfully divide, in a suitable way, a total problem of insurmountable difficulty into a number of partial problems of which none is too difficult.

4

To accomplish this division into partial problems, we have to draw upon the general principle that is the basis of the infinitesimal calculus of several variables. For purposes of computation, one can replace the simultaneous variation of several variables with the sequential variation of single variables or groups of variables. If one goes to the limit of infinitesimal intervals, one arrives at the exact methods of the infinitesimal calculus. If one uses finite intervals, one arrives at the approximation methods of finite difference computations and mechanical quadrature, which we must use here.

These principles cannot be used blindly, however, because the practical usefulness of the method will depend on the natural grouping of the variables so that one gets comprehensible partial problems, well-defined in mathematical and physical respects. Above all, the first division will be basic. It must follow a natural line of division in the overall problem.

One such natural line of division may be indicated. It follows the boundary-line between the specifically dynamic and the specifically physical processes out of which the atmospheric processes are composed. The division along this boundary-line divides the overall problem into pure hydrodynamic and pure thermodynamic partial problems.

The link which ties the hydrodynamic and the thermodynamic problems together is very easy to cut, so easy indeed that the theoretical hydrodynamicists have fully used it to avoid every serious contact with meteorology; for the connecting link is the equation of state. If we suppose that temperature and moisture do not enter into this equation, then we come to a "supplementary" equation, used ordinarily by the hydrodynamicists, which is a relation only between density and pressure. Thereby one is led to the study of fluid motions under such circumstances that each explicit consideration of the thermodynamic processes automatically falls away.

Instead of making the temperature and the moisture disappear entirely from the equation of state, we can regard them, for short time intervals, as given quantities, with values derived from the observations or from the preceding calculations. When the dynamical problem for that time interval is solved, then one computes afterwards new values of temperature and moisture according to purely thermodynamical methods. These are regarded as given quantities when one solves the hydrodynamic problem for the next time interval, and so on.

5

This, then, is the general principle for the first subdivision of the main problem. In the practical solution of this problem, there are several different ways in which the separation may be done, according to the manner in which one introduces the hypotheses about temperature and moisture. But there is no need to go into more detail in a general discussion of this kind.

The next major question will be, however, whether the hydrodynamic and the thermodynamic partial problems can be individually solved in a sufficiently simple way.

We consider, first, the hydrodynamic problem, which is the principal one, because the dynamic equations are the true prognostic equations. Only through them is time introduced as an independent variable in the problem. The thermodynamic equations do not contain time.

The hydrodynamic problem lends itself well to graphical solutions. Instead of computing with the three dynamic equations, one can execute simple parallelogram constructions for a suitable number of selected points, with graphical or visual interpolation for the regions in between. The main difficulty will lie in the restriction on the motion, which follows from the equation of continuity and the boundary conditions. But the test of whether the continuity equation is satisfied or not can also be performed by graphical methods, and in doing so, one can take into account the topography of the earth, performing the construction on charts that represent this topography in the usual way.

One will not encounter great mathematical difficulties in the solution of the hydrodynamic partial problems. However, there is a serious gap in our knowledge of the factors that we must take into account, as we have a very incomplete knowledge of the frictional stress in the atmosphere. True friction depends upon velocity differences in the infinitesimally small, but meteorologists are forced to deal with the average movements of large masses of air. One cannot, therefore, apply the frictional terms of the hydrodynamical equations by using the coefficients of friction found in laboratory experiments, but one must draw upon empirical results about the effective resistance opposing the motion of large masses of air. However, we already have sufficient data of this kind to make the first attempts in the computational prediction of air movements, and these attempts will create, in time, the necessary corrections and completions.

The thermodynamical partial problem can be considered to be much simpler, in mathematical respect, than the hydrodynamical. From the solution to the hydrodynamical problem, one obtains the work done by the air masses during their displacement. Knowing this work, and knowing the amounts of heat introduced during the time interval by incoming radiation and given up by outgoing radiation, one computes the new distribution of temperature and moisture according to known thermodynamic principles. These computations will not be more difficult in mathematical respects than similar computations in laboratory experiments where masses of air are at rest in a closed space. We have extensive pioneering work also in the investigations of Hertz, V. Bezold, and others.

As in the hydrodynamical problem, the main difficulty will be the lack of knowledge of the different factors with which the computations are to be carried out. The estimates of the amount of heat that the air masses receive as the difference between incoming and outgoing radiation, and the estimates of the amount of water that evaporates from the surface of the ocean or that condenses into clouds and falls as rain, will be very uncertain in the beginning. However, we have sufficient knowledge for a trial performance of the first computation, and through continued work will gradually find more exact values of the constants that relate to the different countries and oceans, to different heights in the atmosphere, to different weather situations, to different amounts of cloudiness, and so forth.

6

It is certain that we will not encounter insurmountable mathematical difficulties in following through with these methods.

After the graphical techniques have been worked out and the necessary tabular aids have been assembled, the individual operations will probably be easy to execute. The number of individual operations does not have to be excessively large. The number will depend upon the length of the time interval for which the dynamic partial problem is to be solved. The shorter one chooses the time interval, the more involved the work becomes, but so also does the result become more exact; the longer one chooses the time interval, the faster one arrives at the goal, but at the cost of accuracy. A final decision about the best time interval to use can be determined only by experience. Even if striving for high accuracy, a one-hour interval should suffice. For air masses will only exceptionally travel longer distances in an hour than a degree of longitude, and only exceptionally will their path appreciably change curvature during that time. Thereby the conditions are fulfilled under which we can carry out simple parallelogram constructions with straight line sections. If one gains sufficient experience and learns to utilize instinct and visual estimates, one would probably be able to work easily with much larger time intervals, such as six hours. For a 24-hour weather forecast, one would then carry out the hydrodynamic construction four times, and four times compute the thermodynamic corrections of temperature and moisture.

It may be possible some day, perhaps, to utilize a method of this kind as the basis for a daily practical weather service.

But however that may be, the fundamental scientific study of atmospheric processes sooner or later has to follow a method based upon the laws of mechanics and physics. And thereby we will arrive, necessarily, at a method of the kind outlined here.

If this is admitted, the general plan for dynamical-meteorological research is given.

The main task of observational meteorology will be to obtain regular simultaneous observations of all parts of the atmosphere, at the surface of the earth and aloft, over land and over sea.

The first task of the theoretical meteorologist will be to work out, on the basis of these observations, the best possible overall picture of the physical and dynamical state of the atmosphere at the time of these observations. And this representation must have such a form that will enable it to serve as the starting point for weather prediction by rational dynamical-physical methods.

Even the first preliminary task is a sizeable one. For it is of course much more difficult to represent the state of the atmosphere at all elevations than only at sea level as it is now done. In addition, our direct observations of the higher layers of air will always be very limited. One must therefore use each observation from the higher levels to the utmost. From the directly observable quantities one has to compute to the greatest extent all accessible data about the non-observable ones. In doing this, one has to utilize the physical relationships between the quantities. Even to construct a coherent picture of the total state of the atmosphere out of scattered observations, one has to use, to a large extent, dynamical-physical methods.

The second, and most important, task of theoretical meteorology will finally be to construct, with this representation of the state of the atmosphere as the starting point, the representation of the future states, either according to the methods outlined here, or by methods of a similar kind. The comparison of the predicted fields with those that are given afterwards by the observations will reveal the general accuracy of the method, and at the same time will provide empirical knowledge of better values of the constants, as well as hints on the improvement of the method. On later occasions, I shall return to the various principal points of this program.

Vilhelm Bjerknes's Early Studies of Atmospheric Motions and Their Connection with the Cyclone Model of the Bergen School

ARNT ELIASSEN

Institute of Geophysics, University of Oslo, Oslo, Norway

The frontal cyclone model that was put forward by the Bergen team at the end of World War I has, with only minor adjustments, survived all these years as an effective tool in weather analysis and forecasting. The Bergen school concepts have also been an inspiration and a challenge to theoreticians and have contributed to a fruitful development of atmospheric dynamics.

The first version of the frontal cyclone model was presented by Jacob (Jack) Bjerknes (1919) in his celebrated paper entitled "On the structure of moving cyclones." The model was further refined and extended to include the typical life cycle of frontal cyclones by Jack Bjerknes and his teammates Halvor Solberg and Tor Bergeron. However, it was Jack's father, the classical theoretical physicist Vilhelm Bjerknes, who put these young scientists to work and gave them the research tools that led them to their results.

Vilhelm Bjerknes's influence upon the development of the Bergen school has been documented in two excellent books, which in recent years have enriched the literature on the history of meteorology, namely: *The Thermal Theory of Cyclones* by Gisela Kutzbach (1979) and *Appropriating the Weather* by Robert Marc Friedman (1989). I do not have much to add, but I shall give some personal views and take a closer look at some of Vilhelm Bjerknes's early meteorological studies.

It was in 1897, while he was working on the hypothesis of his father, Carl Anton Bjerknes, on a possible hydrodynamic explanation of electrostatic and magnetic forces, that Vilhelm Bjerknes stumbled across the circulation theorem that bears his name; it expresses how the circulation of a material closed curve is changed by baroclinicity. At the time this was in seeming contradiction to the established theory of inviscid fluids that led to the theorem of Helmholtz on vortex conservation and Kelvin's equivalent theorem on conservation of circulation. Clearly, however, these conservation

theorems cannot be generally valid, since they are refuted by everyday phenomena such as the air motion in a bonfire, or the motion of water in a heated kettle. In view of this, it is quite astonishing that the Bjerknes circulation theorem was not expressed in the literature much earlier. Bjerknes noted that the conservation theorems were based on a very special fluid model in which the density is for all particles the same function of pressure, so that baroclinicity is precluded. By relaxing the restriction on the density so that baroclinicity became possible, Bjerknes (1898) could turn Kelvin's conservation theorem into a theorem stating how the circulation of a material closed curve changes with time.

In his paper, Bjerknes also derived a second circulation theorem, in which velocity was replaced by momentum density. This theorem lacks the simplicity of the first circulation theorem, however, and is hardly of much use except in a simple approximate form; but then it might just as well be considered as an approximate form of the first circulation theorem. Most writers have ignored the second circulation theorem, and I believe rightfully so. Bjerknes's own motivation for presenting the two versions of the circulation theorem seems to have been that his theory of analogy between electrostatic and hydrodynamic phenomena required such a dual representation of motion, in terms of velocity on the one hand and momentum density on the other.

Circulating currents and baroclinic solenoids are plentiful in the atmosphere and the oceans, and Bjerknes realized that his circulation theorem must therefore have important applications in these media. This realization made him change his field of scientific activity from physics to geophysics. Robert Friedman (1989) has shown from Bjerknes's letters that he was reluctant at first, and took up geophysics only after much encouragement and persuasion from his Swedish colleagues. In his autobiographic article at the end of *Physikalische Hydrodynamik* (Bjerknes et al. 1933), Bjerknes credits his Swedish student and assistant Johan W. Sandström for the decision to take up geophysics. He writes:

> I would not have concerned myself any further with the alien sciences of meteorology and oceanography, had not . . . my student at Stockholm Högskola, J. W. Sandström been prepared to undertake related practical work in this direction and work out the numerical and graphical tools for the practical applications.

Soon Bjerknes devoted his full time and energy to these "alien" geophysical sciences. It was clear to him that the first law of thermodynamics must be invoked to deal with the changing temperature and density of air and seawater. Of course, he was not the first to apply thermodynamics in geophysics, but he was probably the first to combine the equations of thermodynamics and hydrodynamics systematically into a complete set of governing equations for the atmosphere. In a paper (V. Bjerknes 1904) entitled "The problem of weather forecasting from the view point of mechanics and physics," he proposed that weather prediction should be treated as an initial value problem of mathematical physics. He argued that the science of meteorology should have as its principal objective to determine future states of the atmosphere by integrating the governing differential equations, starting from an observed initial state.

With this as an ultimate goal for meteorological research, certain preparatory tasks present themselves. The observation network, which at the time was quite inadequate, would have to be extended and improved. Diagnostic methods must be developed for deriving the spatial distribution of the state variables from the observations. And finally, methods for integrating the equations must be found.

Bjerknes envisaged representing the fields of the state variables by maps and carrying out the integration process by a combination of graphical and numerical methods. He expressed his confidence in the feasibility of such a procedure. He proposed to split the total problem into a hydrodynamic and thermodynamic part, and wrote:

> One will not encounter great mathematical difficulties in the solution of the hydrodynamic partial problems. . . . The thermodynamical partial problem can be considered to be much simpler, in mathematical respect, than the hydrodynamical. . . . It is certain that we will not encounter unsurmountable mathematical difficulties in following through with these methods

Today we know, in hindsight, that Bjerknes grossly underestimated the difficulties, in particular if he wanted to integrate fast enough to beat the weather itself, at a time when electronic computers were not even dreamt of. On the other hand, his unwarranted optimism undoubtedly did much more to promote the meteorological science than a sober pessimistic attitude would have done.

Curiously, Bjerknes's paper contains a technical blunder. To close the set of equations when humidity is included among the dependent variables, he proposed to use the second law of thermodynamics, instead of the continuity equation for the water substance. But this is a minor imperfection in a paper whose important message was that meteorology should be treated as an exact science, based on the laws of physics.

In 1905 Bjerknes presented his ideas about weather prediction before a U.S. audience in Washington, D.C., and got an enthusiastic response. The lecture resulted in a yearly grant from the Carnegie Institution of Washington, which enabled Bjerknes to employ research assistants throughout the rest of his active life, up to World War II. The money from the Carnegie Institution could hardly have found a better use; a dozen of Scandinavia's most prominent geophysicists were recruited and educated through this arrangement.

With the financial support from Carnegie, Bjerknes now embarked upon a major geophysical research project, aiming at nothing less than to realize his research plan from 1904. He began by attacking the problems of diagnosis. As his first Carnegie assistant and collaborator, he employed Sandström, who had already worked with him for some years. Their results are presented in *Dynamic Meteorology and Hydrography, Part I: Statics*, by V. Bjerknes and J. W. Sandström (Bjerknes and Sandström 1910). According to V. Bjerknes (1910), the manuscript had been submitted in 1907, but was delayed in press for three years.

Statics is a book of 146 pages and many tables. It opens with an account of absolute units, in particular the bar and millibar, which took Bjerknes many years to get introduced instead of centimeters or inches of mercury. It is explained why geopotential should be used as vertical coordinate instead of geometrical height. It is set forth that, excepting cases of violent motions, the hydrostatic equation may be assumed to hold along every vertical in the atmosphere .

Two methods of representation of the three dimensional pressure field are discussed: the obvious one, to draw isobars in several level surfaces, and a new invention: contour lines that define the topography of several isobaric surfaces. The virtues of these two methods are compared, and the conclusion is reached that the topography of isobaric surfaces is to be preferred because it requires less calculation and is quicker. Bjerknes published this result in 1910. It is a curious fact that most countries ignored it and did not introduce isobaric surface maps in their upper air analyses until about 30 years later. A notable exception is the German Weather Service which introduced isobaric contour maps from the beginning of their aerological service before World War I.

Another point in favor of the isobaric surface maps is that the relation between geostrophic wind and the slope of the isobaric surface is the same at all levels. This was probably not noticed by Bjerknes and Sandström, since they did not mention it.

To construct the contour maps for a selection of standard isobaric surfaces, Bjerknes and Sandström in their *Statics* paper recommended drawing first the 1000 mb contour map, and to this add consecutive thickness maps by graphical addition. To obtain the thickness values for standard isobaric layers from ascents measuring corresponding values of pressure, temperature, and relative humidity, they invented a graphical method: from a plot of the virtual temperature versus the logarithm of pressure, the area mean of the virtual temperature in the standard isobaric layers could be estimated by eye with sufficient accuracy (Fig. 1). These methods, in various versions, have been in extensive use in the world's weather services, in particular for many decades after World War II, until electronic computers took over.

To illustrate the methods, Bjerknes and Sandström carried through the analysis on two examples: one over the eastern United States on the morning of 23 September 1898, with seven kite ascents reaching 900 mb; and the other over Europe on the morning of 7 November 1901, with five self-recording balloon ascents, which reached 300 mb; in addition they used temperatures extrapolated from the ground. Of course, these upper air observations were utterly deficient; the authors emphasized, however, that the intention was not to produce an accurate analysis, but to demonstrate the analysis methods. Some of their maps for Europe are reproduced in Fig. 2. They are probably the first upper air isobaric contour maps ever made.

In 1907 Vilhelm Bjerknes left Stockholm and returned to Norway as professor at Christiania (Oslo). Together with two new Carnegie assistants, the Norwegian students Theodor Hesselberg and Olaf Devik, he started a comprehensive study of kinematic analysis. The results of their investigations are reported in *Dynamic Meteorology and Hydrography, Part II: Kinematics*, by V. Bjerknes, Th. Hesselberg, and O. Devik (Bjerknes et al. 1911).

The authors pointed out that the two-dimensional wind field should not be represented merely as a number of arrows on the map, but should be given a continuous representation amenable to graphic algebra and differentiation. They proposed two such representations, as streamlines and isotachs, or as isogons and isotachs, and they showed how the fields of two-dimensional divergence and vorticity can be derived by graphical differen-

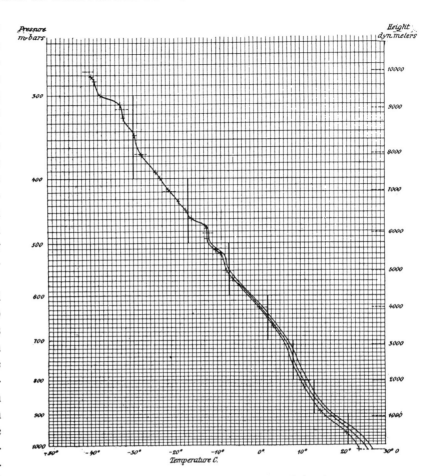

FIG. 1. Plot of temperature, saturated virtual temperature, and virtual temperature versus logarithm of pressure, showing the evaluation of the thickness of decibar layers (Bjerknes and Sandström 1910).

tiation from either of these representations. Isogon analysis had previously been used in maps of the magnetic declination; Sandström (1909) proposed to use this method also in maps of the velocity field.

The authors of *Kinematics* were aware that the vertical velocity of synoptic scale cannot easily be observed, and proposed instead that it should be inferred from kinematical principles: at the ground from the wind multiplied by the slope of the earth's surface, and at higher elevations an additional contribution due to convergence of horizontal mass flux.

As an illustration, Bjerknes and his co-workers carried out such analyses for the ground level over the United States. Figures 3 and 4 show, respectively, their isogon-isotach and streamline-isotach analyses for 0800 EST 28 November 1905. The analyses show streamlines spiraling toward a cyclonic center in southern Minnesota. Moreover, the streamline picture reveals at least two lines of confluence where the streamlines come together.

This streamline map is now famous. In a slightly simplified version, Bjerknes (1910) showed it at a lecture in London, and it has also been reproduced by both Kutzbach and Friedman in their books as an illustration of Bjerknes's early interest in confluence lines. And with good reason—the confluence lines are undoubtedly the seeds of the Bergen school fronts.

Shown in *Kinematics* is also the distribution of horizontal mass flux convergence, corresponding to the increase of vertical mass flux with height.

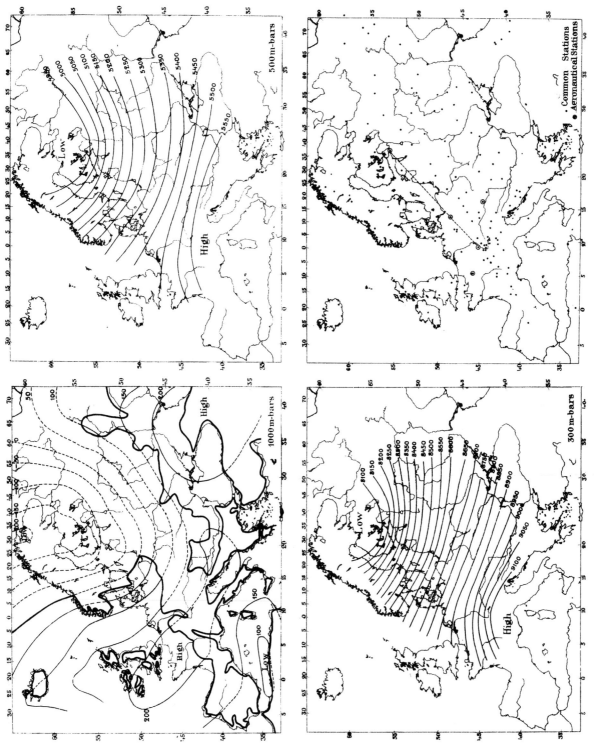

FIG. 2. Contour maps of the 1000, 500, and 300 mb surfaces, and map showing Europe's five aerological stations and ground stations. Contour interval: 50 dyn. meters (Bjerknes and Sandström 1910).

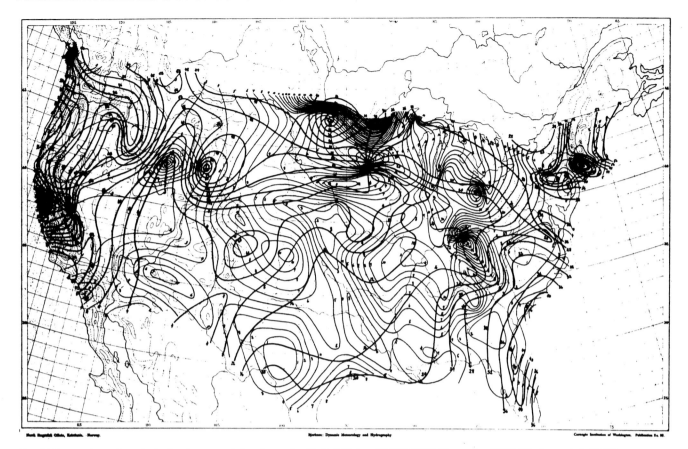

FIG. 3. Isogon-isotach analysis for the ground level over the United States, 0800 EST 28 November 1905 (Bjerknes et al. 1911).

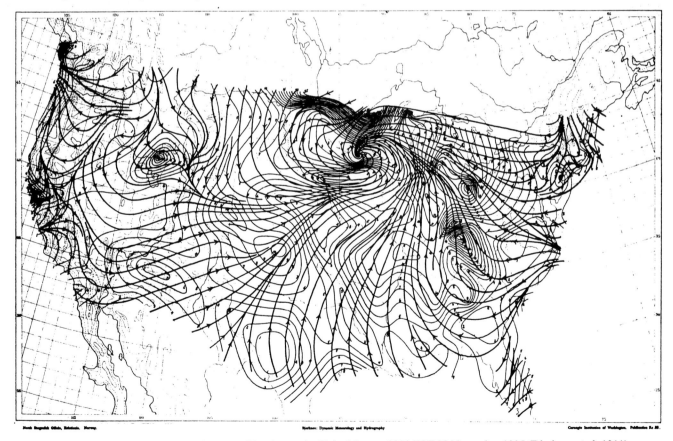

FIG. 4. Streamline-isotach analysis for the ground level over the United States, 0800 EST 28 November 1905 (Bjerknes et al. 1911).

FIG. 5. Vertical mass flux divergence $\delta\omega/\delta p$ (unit $10^{-5}s^{-1}$) superimposed upon the streamlines at the ground level (Fig. 4) (adapted from Bjerknes et al. 1911).

When superposed upon the streamline picture, it reveals that the confluence lines are located in areas of horizontal mass flux convergence and ascending motion (Fig. 5).

On the other hand, no analysis of ground temperature is given, and there is no discussion of the temperature field in relation to the confluence lines. There is, however, an analysis of the 1000/900 mb thickness, which by itself roughly fits the picture of a frontal cyclone. But the band of thickness contrast does not coincide too well with the confluence lines, as will be seen in Fig. 6, in which the thickness lines are superimposed on the streamlines. But then we must expect the thickness analysis to be very inaccurate, as it was mostly based on extrapolated ground temperatures.

Figure 7 shows the surface pressure analysis and fronts from Historical Weather Maps, Daily Synoptic Series. The position of Bjerknes's confluence lines from 1910 (Fig. 5) agree roughly with the fronts of the analysis made after World War II.

In *Kinematics*, Bjerknes and his coworkers come very close to presenting the quasi-static equations in pressure coordinates. On page 62 they give the geostrophic wind formula in terms of the isobaric slope, probably for the first time. This shows that they knew how to express the horizontal pressure gradient force in pressure coordinates. What is more remarkable, on page 145 they derive the continuity equation in pressure coordinates. I realized this only a few months ago, and felt quite embarrassed for not having noticed it earlier; I had thought until then that the first derivation was due to the Belgian meteorologist O. Godart, as stated in a memorandum by Sutcliffe and Godart (1942).

The derivation on pages 143–145 in *Kinematics* is not so easy to follow, but the subsequent application to a summer situation over Europe (0700 GMT 25 July 1907) makes the matter clear. Here the upward mass flux through each of the isobaric surfaces, 1000, 900, 800, 700 and 600 mb, are shown on maps obtained by adding up the convergences of the pressure-integrated wind in each of the underlying isobaric layers. This is a direct application of the continuity equation in pressure coordinates, integrated over the isobaric layers. Note that Bjerknes's vertical mass flux, or specific momentum as he calls it, contains a hidden factor g due to the use of "dynamic height," and is thus identical with our "omega," the material rate of change of pressure, with opposite sign.

Dynamic Meteorology and Hydrography was a pioneering work, which contained a wealth of new valuable analysis techniques. The three Carnegie assistants who were co-authors undoubtedly gave significant contributions. They all

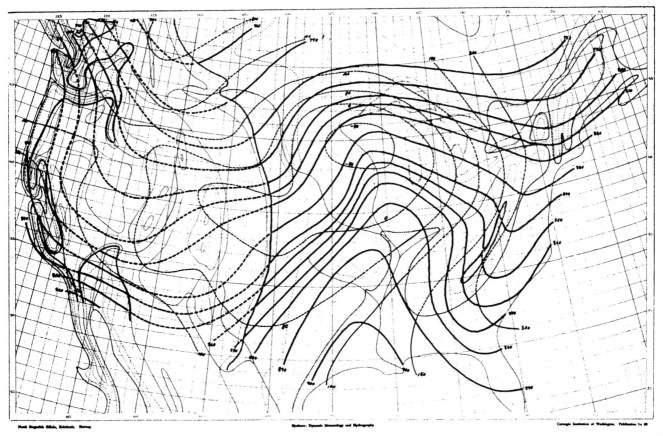

FIG. 6. The 1000/900 mb thickness map (interval 10 dynamic meters ~ 3K) superimposed upon the streamlines at the ground level (Fig. 4) (adapted from Bjerknes et al. 1911).

became prominent geophysicists: Sandström was later director of the Swedish Meteorological-Hydrological Institute, Devik became a leading hydrologist in Norway, and Hesselberg was director of the Norwegian Meteorological Institute from 1915 to 1955 and president of the International Meteorological Organization from 1937 to 1947.

In 1913 Bjerknes moved to Leipzig in Germany, where a chair and the directorship of a new Geophysical Institute had been offered to him. In his inauguration lecture (Bjerknes 1914), he repeated his plan from 1904 to predict the state of the atmosphere by integration of the governing equations. However, many years of occupation with the diagnostic problems had taught him to express himself more carefully about the feasibility of the integration procedure than he had done in his 1904 paper. He said (Bjerknes 1914):

> The problem is of huge dimensions. Its solution can only be the result of long development. An individual investigator will not advance very far, even with his greatest exertions. However, I am convinced that it is not too soon to consider this problem as the objective of our researches. One does not always aim only at what one expects soon to attain. The effort to steer straight toward a distant, possibly unattainable point serves, nevertheless, to fix one's course. So, in the present case, the far-distant goal will give an invaluable plan of work and research.

Bjerknes realized that even if a feasible integration method was found, the chances of being able to carry out such a huge amount of calculation and graphical operations faster than the weather itself was slim indeed. But to this problem he had an answer. He said in his lecture:

> What satisfaction is there in being able to calculate tomorrow's weather if it takes us a year to do it?
> To this I can only reply: I hardly hope to advance even so far as this. I shall be more than happy if I can carry on the work so far that I am able to predict the weather from day to day after many years of calculation. If only the calculation shall agree with the facts, the scientific victory will be won. Meteorology would then have become an exact science, a true physics of the atmosphere. When that point is reached, *then* practical results will soon develop.

The Geophysical Institute in Leipzig had a promising start with a scientific staff that included Robert Wenger, an able German meteorologist, and the two Carnegie assistants

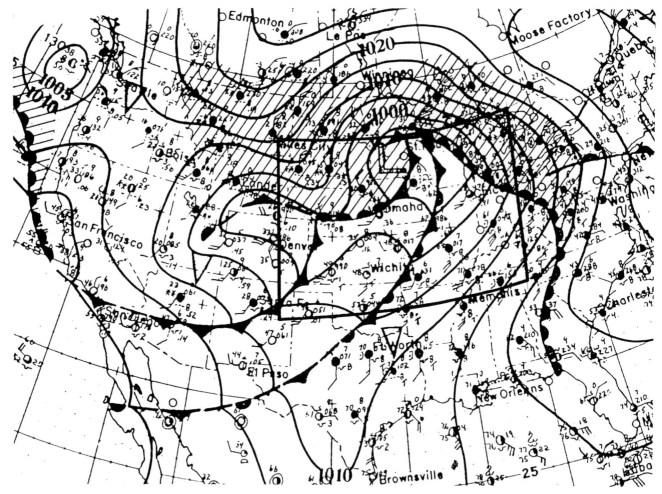

FIG. 7. Fronts and isobars for 1300 GMT 28 November 1905 (from Historical Weather Maps. Daily Synoptic Series, U.N. Joint Meteorological Committee) superimposed on the streamline pattern (Fig. 4). Inner box shows the area of the streamline and mass flux analysis of Figure 5 (Bjerknes et al. 1911).

Th. Hesselberg and Harald U. Sverdrup; the latter was for many years head of Scripps Institution of Oceanography in California.

The Leipzig institute started to publish a series of sea level and aerological isobaric contour analyses over Europe, with the intention that they should serve as initial states in prognosis experiments. One of these analyses, for 0700 GMT 20 May 1910, was actually used for such an experiment; Lewis Richardson (1922) took it as the initial state for his worked example in his famous book. But Bjerknes himself never started any serious experiments toward numerical or graphical prognosis. Thus the planned Part III of *Dynamic Meteorology and Hydrography*, which should have treated the prognosis problem, never appeared.

Bjerknes has given two reasons for this. In the first place, the Leipzig institute was gradually paralyzed by World War I. Bjerknes writes in his "Bibliographie mit historischen Erläuterungen" in Bjerknes et al. (1933):

> In particular, through series of consecutive diagnoses which were as complete as possible, we hoped to be able to clarify the processes in the atmosphere in order to apply the methods of dynamic-thermodynamic prognosis. . . . In this comprehensively organized work, we had just reached far enough so that everything seemed to progress well, and we were ready to extend the work in breadth and depth in order to be able to seriously attack the main problems, when the war made every extension of the work impossible. The assistants at the institute were withdrawn; five out of ten doctor candidates fell [in the war] . . . the work was paralysed. (Translated from German).

Secondly, Bjerknes stated (in Hergesell and Bjerknes 1922) that he had considered his Leipzig analyses too inaccurate to serve as initial state in prognostic calculations, because the observations on which they were based were much too scattered.

We know now, of course, that even with the best data and computer assistance, neither Bjerknes nor Richardson could have obtained a realistic integration result from the primitive equations without additional knowledge, in particular about methods of initialization and the Courant-Friedrichs-Lewy criterion of numerical stability .

Instead, at Bjerknes's Leipzig institute some interesting studies were made on air flow and turbulent friction. The study of confluence lines was resumed by the German doctoral student Herbert Petzold, who investigated confluence lines associated with the invasion of cold air. But Petzold fell at Verdun in 1916, and the study of confluence (and difluence) lines was taken over by Vilhelm Bjerknes's son Jack, who wrote a short paper about the subject (J. Bjerknes 1917). The following quotation is taken from that paper (translated):

> The real convergence lines are associated with ascending motion which, depending upon the air humidity, may result in cloud formation and rain. One can find many confirmations of these rules by comparing the wind maps in "Veröffentlichungen des Geophysikalischen Instituts" with the corresponding maps of cloud cover and precipitation. Depending upon their strength, one observes that the convergence lines are accompanied by bandformed regions of overcast, ordinary rain, or showers and frontal storms (Frontgewitter). In the two latter cases they are often immediate precursors of cold air intrusion. The advancing cold air pushes the warm [air] ahead, and a short distance ahead of the front of the cold air a convergence line is formed, which is parallel to.this and which represents the location of the strongest ascent of the warm air.

In my opinion, this paper shows that Jack Bjerknes was already in Leipzig close to contriving the concept of a front. This must have been a good starting point for the Bergen school.

Vilhelm Bjerknes's influence on the progress of meteorology was only in part due to his own scientific contributions. Just as important was his rare ability to attract and inspire bright young students. Wherever he was, in Stockholm, Oslo, Leipzig, or Bergen, he created a fertile milieu around himself. Moreover, he did not hide his light under a bushel, but shared his accomplishments not only with the scientific community, but also with politicians and the public as well. This attitude brought him support to carry through with his research projects. An intellectual type, Vilhelm Bjerknes was distinctively visual; whenever possible, he would give geo-metrical interpretations to his equations. Jule Charney once said that he thought Vilhelm Bjerknes in trying to understand atmospheric dynamics would imagine that he himself was an air particle, trying to decide where to go.

For several years after the war, when Vilhelm Bjerknes was in his eighties, he still came every day to his office in the Astrophysics building, which also housed the geophysicists at the University of Oslo. He was an ardent listener to seminars and thus gave his encouragement and moral support to the research activity around him.

References

Bjerknes, J., 1917: Uber die Fortbewegung der Konvergenz - und Divergenzlinien. *Meteor., Z.* **34**, 345–349.

———,1919: On the structure of moving cyclones. *Geof. Publ. 1* No. 2, 1–8.

Bjerknes, V., 1898: Uber einen hydrodynamis chen Fundamentalsatz und seine Anwendung besonders auf die Mechanik der Atmosphäre und des Weltmeeres. *K. Svenska Vetensk.-Akad . Handl.*, **31**, No. 4, 33 p.

———,1904: Das Problem der Wettervorhersage, betrachtet vom Standpunkte der Mechanik und der Physik. *Meteor. Z.*, **21**, 1-7. (Weather Forecasting as a Problem in Mechanics and Physics. English translation by Y. Mintz, mimeographed, Los Angeles 1954).

———,1910: Synoptical representation of atmospheric motions. *Quart. J. Roy. Meteor. Soc.*, **36**, 167–286.

———,1913: *Die Meteorologie als Exakte Wissenschaft.* Fr. Vieweg & Sohn, Braunschweig, 16 pp.

———,1914: Meteorology as an exact science. *Mon. Wea. Rev.,* 11–14.

———, and Sandström, 1910: *Dynamic Meteorology and Hydrography. Part I, Statics.* Carnegie Institution, Washington, D.C., 146 pp., tables

———, Th. Hesselberg, and O. Devik, 1912: *Dynamic Meteorology and Hydrography. Part II, Kinematics.* Publ. 58, Carnegie Institution, Washington, D.C., 175 pp., 60 plates (German Edition, Fr. Vieweg & Schn, Braunschweig, 1912).

———, J. Bjerknes, H. Solberg and T. Bergeron, 1933: *Physikalische Hydrodynamik.* Springer, Berlin, 797 pp.

Friedman, R. M., 1989: *Appropriating the Weather.* Cornell University Press, Ithaca, N.Y., 251 pp.

Hergesell, H., and V. Bjerknes, 1922: Robert Wenger. *Beitr. z. Phys. d. fr. Atm.,* X, pp. I-VIII.

Kutzbach, G., 1979: *The Thermal Theory of Cyclones.* American Meteorological Society, Boston, Mass., 255 pp.

Richardson, L. F. 1922: *Weather Prediction by Numerical Process.* Cambridge University Press, Cambridge, 236 pp.

Sandström, J. W., 1909: Über die Bewegung der Flüssigkeiten. *Ann. d. Hydrogr. u. Marit. Meteor.,* 244–254.

Sutcliffe, R. C., and O. Godart, 1942: Isobaric analysis Meteor. Office, London, S D T M No 50

Components of the Norwegian Cyclone Model: Observations and Theoretical Ideas in Europe Prior to 1920

HANS VOLKERT

Institut für Physik der Atmosphäre, DLR-Oberpfaffenhofen, Weßling, Germany

1. Introduction

The publication of a scientific article is occasionally used to mark, in retrospect, the beginning of a new era for a scientific discipline. For meteorology, Jacob Bjerknes's article of 1919, "On the structure of moving cyclones" (referred to as JB19 in the following), provides such a landmark, as it first introduced the model of the *ideal cyclone*, which greatly influenced research and practical weather forecasting for many years to come. The achievement made by the Bergen school of meteorology at the end of World War I can, it is thought, be esteemed especially well if related observations and theoretical considerations, published before 1920 are recalled.

At the beginning of JB19, J. Bjerknes gave the following "characteristic traits" of the surface flow of a moving cyclone (Fig. 1): (a) a spiraling inflow toward the cyclone center; (b) two lines of convergence (*steering line* and *squall line*, which were later called warm and cold fronts, respectively[1]); and (c) the pronounced warm sector to the south of the cyclone center, signifying a pronounced asymmetry in the temperature distribution. In this chapter, we look for evidence as to what extent these components of the Norwegian cyclone model were individually described in the European literature during the period 1880 to 1920.

The Bergen school's visualization of their concepts through skillfully crafted conceptual sketches and later also by detailed weather maps appears to be one of the foundations of its success. This was in contrast to the general practice for fewer figures to appear in journal articles before 1920 than nowadays. But figures have been published earlier and some of these, which are to date probably not widely known, are reproduced here. Thus, we can obtain a quite direct impression of the way extratropical cyclones and their components were viewed in Europe during the four decades around the turn of the century.

A very thorough historical investigation of the much wider topic, the thermal theory of cyclones, was carried out by Gisela Kutzbach (1979; K79 in the following). It includes 49 historical diagrams, and mentions, in its later chapters, the majority of the works quoted here. However, most of the figures reproduced here were not included. A special feature of Kutzbach's book is its appendix, which contains short biographies of scientists who had made important contributions during the early phase of cyclone research. We also make some biographical remarks in footnotes to stimulate interest in the acting persons.

This chapter is organized as follows: Section 2 deals with investigations regarding cyclones as a whole, some of them indicating an asymmetric temperature field. Section 3 presents observations, a laboratory experiment, and theoretical ideas regarding squall lines as examples of discontinuities in the atmosphere. Section 4 mentions descriptions of pronounced warm sectors. Section 5 briefly outlines the reception and tradition of these early studies, together with the Norwegian cyclone model after 1920. Finally, conclusions are drawn concerning the value of such a retrospective investigation.

2. Asymmetric Cyclones

The terms *depression, barometric minimum*, or *cyclone* have been used for larger-scale regions of lower surface pressure and significant weather since the early nineteenth century. Mostly thermodynamic concepts were used in the numerous attempts to explain the internal structure and the development and progression of cyclones (K79).

---

1. The now familiar terms were introduced by Bjerknes and Solberg (1921), where the change in nomenclature is briefly commented in a footnote (on p. 25).

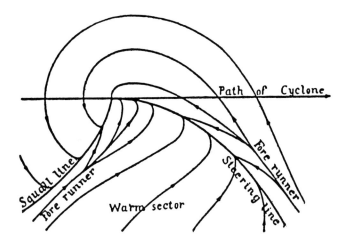

FIG. 1. First appearance of the Norwegian cyclone model and its flow structure. The components are an asymmetric cyclone moving from left to right, a sector of warm air in the south, and two separation lines ahead of and behind the warm sector (JB19, Fig. 1).

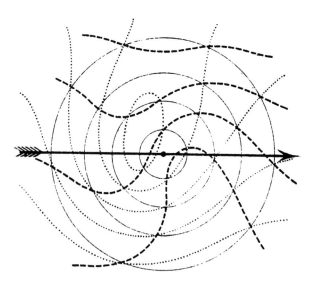

FIG. 2. Interrelationship among idealized circular surface isobars (solid lines), cirrus-level isobars (dotted lines), and mean isotherms of the air mass between (dashed lines) (Köppen 1882a, Fig. 2a on plate 23).

There have also been attempts to link the effects of dynamics and thermodynamics in a consistent manner. Köppen[2] (1882a), when reviewing empirical rules put forward by Ley (1872), introduced a consistent conceptual model, which related a circular surface low, a slightly westward shifted upper-level trough, and a highly asymmetric mean temperature distribution of the intermediate layer (Fig. 2). The temperature field was obtained by graphical subtraction, that is, by joining points of equal pressure differences. The upper level pressure field was constructed from observations of cirrus cloud motion provided by Ley. This method of *indirect aerology*[3] was later used during the early years of the Bergen school. Köppen's schematic and its background were also discussed in the widely read German textbook on meteorology by Sprung (1885).

The meteorological services established in many European countries during the second half of the last century thoroughly investigated cases of strong depressions passing over their areas of interest. A very revealing example is the case study by Shaw[4] et al. (1903), when "the British Isles were visited by a storm of unusual severity." They stressed

the value of time series from self-recording instruments (see Fig. 3) as a supplement to synoptic charts. They discussed the coincidence of pressure fall and temperature rise (during the night!) at stations south of the storm's center (Valencia, Southport, and Stonyhurst) and contrasted it with the development at places which were "under the path of the storm" (Glasgow and Aberdeen). In their concluding paragraph, Shaw et al. (1903) mentioned the role of the "strong and warm Southerly wind . . . independent of any local action within the storm area" for the development of this particular event. As an aside we note Shaw's remark in the discussion that "he had no wish to add another to the theories of storms—there are too many of these already," but he wanted to increase the knowledge of the facts and to deal with their classification.

In 1906, Shaw and Lempfert published the pioneering monograph, *The Life History of Surface Air Currents*, in which they applied their concept of air trajectories to show that cyclones comprise air masses from very different source regions. One of the many diagrams is reproduced in Fig. 4, depicting a mature cyclone over the Atlantic with a polar airstream in its rear and a subtropical one ahead of it. The methodology, originally developed for more mesoscale studies over the British Isles, had been applied to a "historic" data set, which covered 13 months in 1882 and 1883, in an attempt to describe the complete *life history* of surface air currents in

---

2. Wladimir Peter Köppen (1846–1940) was appointed director of the department of storm warnings and forecasting in the newly founded Deutsche Seewarte (literally, German sea observatory) in Hamburg in 1875, and worked as chief scientist and director of the research department of this institution from 1879 to 1919. Köppen is best known for his later work regarding climatology, for example, for his classification of climates, although he also made important contributions to synoptic and mesoscale meteorology during the earlier years of his career (cf. K79, 237–238).

3. The term *aerology* was suggested by Köppen before the Commission on Scientific Aeronautics in 1906 to distinguish the study of the free atmosphere, throughout its vertical extent, from investigations confined to the atmospheric layer near the earth's surface.

---

4. Sir William Napier Shaw (1854–1945) was director of the Meteorological Office in England from 1905 to 1920, knighted in 1915, and was first professor of meteorology at Imperial College until 1924. He introduced trained scientific staff at the Meteorological Office, emphasized studies of the physics of the atmosphere, and produced the four volume *Manual of Meteorology* (cf. K79, 243–244).

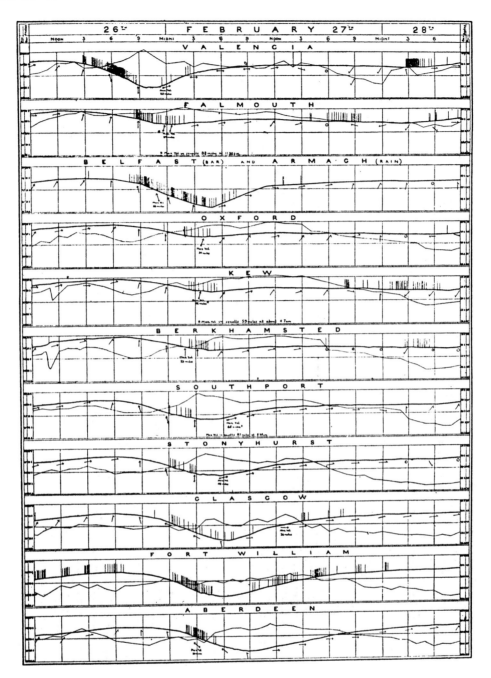

FIG. 3. The passage of a strong depression over the British Isles documented by time series of pressure (thick line), temperature (thin line), horizontal wind, and rainfall (short vertical lines on pressure trace; each line marks 0.01 inch = 0.25 mm) for the period 0900 GMT 26 February 1903 until 0900 GMT 28 February 1903 at eleven stations. The horizontal line in the middle of each series represents 29 inch Hg (980 hPa) and 40°F (4.4°C) with ranges of 1 inch Hg (33.8 hPa) and 10°F (5.6 K) above and below; 3-hour periods are separated by vertical lines (Shaw et al. 1903).

the vicinity of cyclones. In their conclusions, the authors admitted uncertainties in the database and suggested that "observations must be extended on the one hand by enlarging the area and on the other by including measurements made in the upper air."

Techniques of obtaining upper air data from ascents of manned and unmanned balloons and of kites began to be developed before the turn of the century (cf. the very recent

account by Hoinka (1997)). Soon coordinated ascents from several European stations were organized. A paper by Hergesell[5] (1900) collected results of these international

5. Hugo Hergesell (1859–1938) was founding president of the International Aerological Commission from 1896 to 1914, then director of the aerological observatory in Lindenberg. He developed instruments, improved observation procedures for the free atmosphere, and encouraged close cooperations between meteorology and aeronautics.

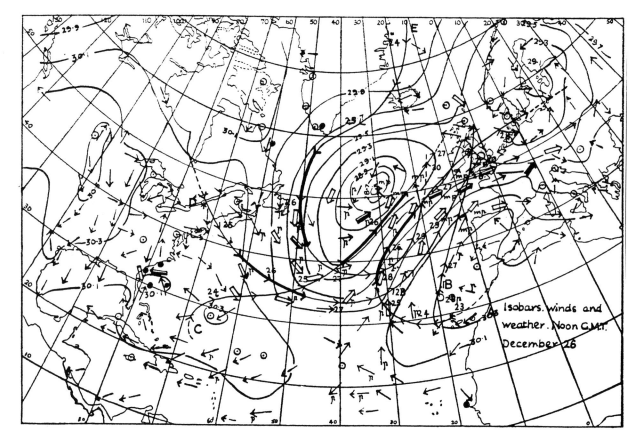

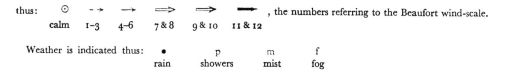

NOTES.—In the Synoptic Charts on Plates XXIV, XXV and XXVI direction and force of wind have been indicated

thus: ⊙   - →   →   ⇒   ⇒   ➤ , the numbers referring to the Beaufort wind-scale.
        calm   1–3   4–6   7 & 8   9 & 10   11 & 12

Weather is indicated thus: •   p   m   f
                              rain   showers   mist   fog

FIG. 4. Synoptic surface chart of 1200 GMT 26 December 1882 with superimposed trajectories for the period 23–30 December. The synoptic information is in light gray (blue in the original print) and contains pressure (increment 0.2 inch Hg = 6.8 hPa), wind and weather (see legend). The trajectories are in black (labeled A-E); small arrows along them mark 12-hourly intervals (the day numbers are given at the noon positions); the 24-hour period centered on the synoptic map time is thickened (Shaw and Lempfert 1906, Fig. 2 of plate xxiv).

ascents. Besides many technical details it contained approximated upper air charts such as the distribution of isotherms at a height of 10 km for 13 May 1897 (Fig. 5). Today's practice to construct upper air charts on constant pressure surfaces had not yet been adopted. The cold core over central Europe was put in relation to the extended cold period at the ground, which was classified as a good example of the quasi-regular weather anomaly in mid-May known as *Eisheiligen* in German.

3. Squall Lines

Elongated, narrow regions of rapid or quasi-discontinuous change in meteorological parameters were investigated throughout the last century. Most frequently, meteorologists described passages "of a sudden strong wind or turbulent storm" (dictionary definition of *squall*), which were found to progress over extended areas as organized lines separating a warmer air mass from an advancing colder one (cf. K79, Section 6.7).

In a very detailed account in two parts, Köppen (1882b) documented the event of 9 August 1881. The bulk of this study dealt with the description of the impact that this fierce cold front had in various places on its way across Germany. An account was also given of how the data collection was gradually extended after Köppen had realized that the event was not just a local thunderstorm, as he had first suspected. The pressure and temperature analyses during the peak phase are of particular interest (Fig. 6). They show a distinct mesoscale trough-ridge system; an isolated area of low pressure to the east of 12°E and south of 48°N, probably due to Föhn in the Alpine lee; and a narrow band of distinct

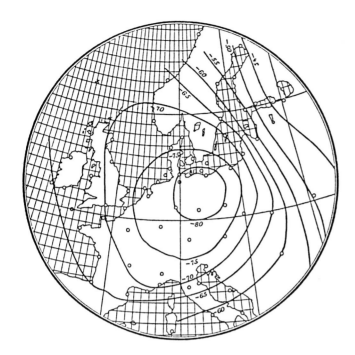

FIG. 5. Isotherms at a height of 10 km (lowest value in the center: −80°C; increment: 5 K) constructed from simultaneous balloon ascents on 13 May 1897 (Hergesell 1900, plate i).

temperature contrast all the way from the Baltic coast (at the northern edge of the panel) to the Black Forest (in the southwest), which was partly collocated with storm force winds. All these features, incidentally, are remarkably similar to recent analyses of a cold front observed during the Fronts Experiment 1987 (see Volkert et al. 1992).

The few other diagrams in Köppen's monograph include barograms from eight stations—some with distinct pressure jumps—and qualitative isochrons of the leading edges of thunderstorms associated with the squall line. Köppen mentioned as mechanisms conducive for the strong temperature contrast: airflow from different source regions ahead of and behind the front, strong insolation in the east, and evaporative cooling in areas experiencing heavy rain and even hail in the west.

Ten years later, Durand-Gréville[6] (1894) systematically investigated thunderstorms and their relation to squall lines, many of which progressed from France into the eastern parts of Europe. Hourly isochrons and a mesoscale pressure analysis (Fig. 7) document the "particularly severe squall line of 27

August 1890." The general conclusions of the investigation comprised, inter alia: (1) squall lines can be detected by the bulged form of the isobars, (2) significant weather is triggered by the highly disturbed flow field, and (3) there are only a few depressions without an area of squalls where precipitation is likely to be produced. He suggested a combination of self-registering instruments in the western parts of France with the telegraph transmission of data to the east as a tool for timely warnings.

Beginning in 1895, Margules[7] used a mesoscale network of four barographs around Vienna in an attempt to relate high winds at the central location with the pressure gradient in its neighborhood, especially for thunderstorm situations. Although the number of stations proved to be insufficient for general conclusions, Margules (1897) discussed, among other cases, the pressure traces during the passage of a "devastating storm of short duration" on 26 August 1896 (Fig. 8). Westerly surface winds of 90 km h^{-1} coincided with the onset of a steep pressure rise (about 10 hPa over 2 h) in Vienna. Comparisons with stations farther to the west and from Sonnblick mountain observatory (3106 m) revealed that the cold (i.e., denser) air advanced eastbound north of the Alps, but was confined to a layer below the Alpine crest height. Special mention was made of the quasi-instantaneous (i.e., within less than 12 min) pressure jump of about 10 hPa in Gmunden, situated at the Alpine baseline about 200 km to the west of Vienna. At the end, Margules considered his work as meager compared to Köppen's study. During the following years, he thoroughly investigated temperature steps in thermograms from mountain and nearby valley stations for three-dimensional analyses of the progression of air masses. This observational evidence eventually served as the background for Margules's celebrated theoretical studies on the energetics of storms and surfaces of discontinuity (cf. K79, Section 6.6). We note that highly resolved pressure traces continue to be relevant for the investigation of frontal modification by the Alps (e.g., Hoinka et al. 1990; Volkert et al. 1991).

Line-squalls constituted a topic of detailed research at the Meteorological Office in England during the first decade of this century. Lempfert[8] (1906) and Lempfert and Corless

---

6. Emile-Alix Durand-Gréville (1838–1914) was a French writer and independent scientist who translated works by novelists such as Turgenev into French and taught for a long time in St. Petersburg before returning to Paris in 1872. From 1890 onward he carried out highly resolved analyses of squall lines, thunderstorms, and hail storms. He contributed mathematical, physical, and climatological articles to several encyclopedic works, was in charge of all meteorological aspects for the publisher of the *Grande Encyclopédie*, and served as vice-president of the Société Astronomique.

7. Max Margules (1856–1920), a physicist and chemist of outstanding calibre, but of rather introverted and partly eccentric personality, worked at the Zentralanstalt für Meteorologie in Vienna from 1882 to 1906. After the turn of the century he published the first thorough theoretical analyses of atmospheric energy processes using thought experiments derived from his earlier observational studies (cf. K79, 239–240).

8. Rudolf Gustav Karl Lempfert (1875–1957) was a university assistant to Shaw in Cambridge (1898–1900) and held various positions at the Meteorological Office between 1902 and 1938, ranging from personal scientific assistant of the director (Shaw) to assistant director. He drew all charts of the 1906 memoir on surface air currents and later thoroughly investigated line-squalls (cf. K79, 238–239).

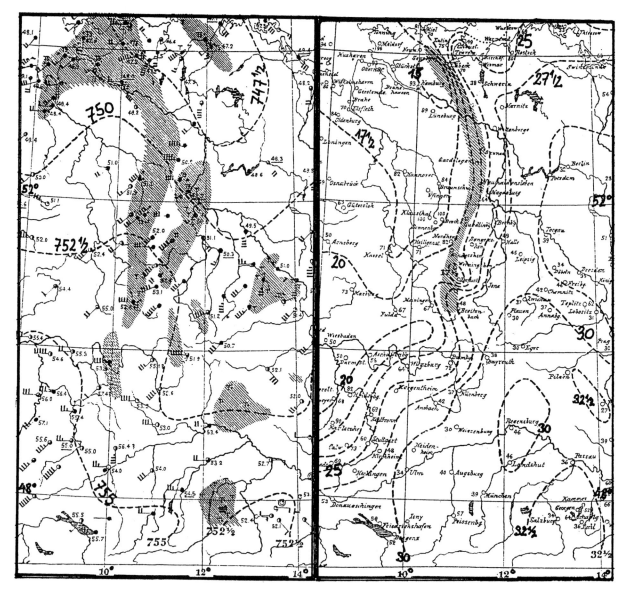

FIG. 6. Mesoscale surface charts over central Germany for 9 August 1881, 2 p.m. The left chart shows station reports of wind (long and short flags, respectively, for single and double Beaufort grades), cloud cover (octa), sea level pressure (mm Hg above 700), pressure analysis with 2.5 mm Hg (3.3 hPa) increment, and areas with thunderstorms (hatched). The right chart displays station reports of relative humidity, a temperature analysis (reduced to sea level with a 2.5 K interval), and area of stormforce winds (hatched) (Köppen 1882b; central portions of plates 20 and 21).

(1910) presented several detailed case studies containing a number of synoptic weather maps, isochron charts, and careful mesoscale analyses of the pressure field. Of special interest is the practice of identifying the squall line with a fault in the pressure analysis (Fig. 9) to achieve consistency with the jumps in the barograph traces at fixed stations. Lempfert (1906) stated that "line-squalls tend to arrange themselves with regard to depressions, and to rotate round their centres, like the spokes of a wheel." Lempfert and Corless (1910) described four cases and culminated in a discussion of "the wind vector changes" across the discontinuities. Conceptual vertical sections (see K79, Fig. 46) depicted the flow in a fixed frame of reference and in one steadily moving with an idealized backward-sloping "linear

front," clearly indicating "upward motion in front of the line and downward motion behind it."

At approximately that same time, the intrusion of a denser fluid under a lighter one was investigated in the laboratory by Schmidt[9] (1911) with explicit reference to

9. Wilhelm Schmidt (1883–1936) was a research scientist at the Zentralanstalt für Meteorologie in Vienna from 1905 to 1919, then professor of physics of the earth in Vienna and, from 1930 until his sudden death, successor of F.M. Exner as director of the Zentralanstalt. He was renowned for his laboratory experiments with gravity currents, and he constructed a variograph to register atmospheric pressure differences at squall lines. Around 1920 he coined the term "Austausch" for the description of a whole range of geophysical exchange processes for which coefficients can be determined empirically.

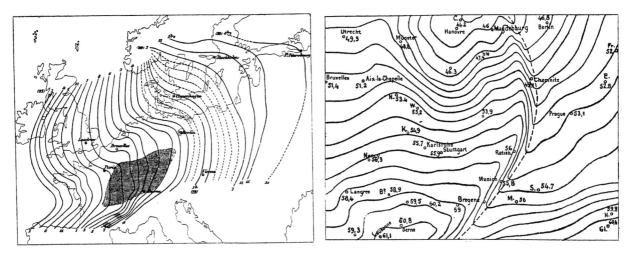

FIG. 7. The squall line of 27 August 1890. Left: hourly isochrons of pressure jump (longer intervals in the east) and areas with thunderstorms (hatched); locations are from west to east: London, Paris, Brussels, Berne, Copenhagen, Berlin, Vienna, Stockholm, St. Petersburg. Right: pressure observations (reduced, mm Hg above 700) and analysis over central Europe for 27 August 1890, 9 pm (increment: 1 mm Hg = 1.33 hPa) (Durand-Greville 1894, Figs. 1 and 2).

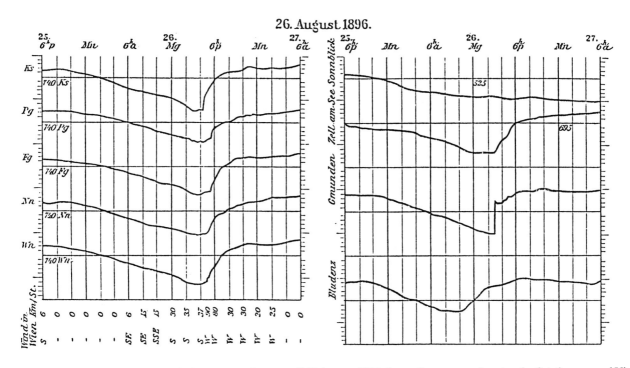

FIG. 8. Surface pressure traces for the period 25 August, 6 p.m., until 27 August 1896, 6 a.m., for a mesoscale network of stations around Vienna (left) and other stations in the west of Austria (right). Vertical lines every two hours; Mn: midnight; Mg: noon; tick marks every 1 mm Hg (= 1.33 hPa); horizontal lines designate reference pressure of 740 mm Hg, except for Zell am See (695 mm Hg) and Sonnblick (525 mm Hg). Station codes: Ks - Krems, Pg - Bratislava, Fg - Feldsberg, Nn - Neunkirchen, Wn - Vienna. Wind speed (km h^{-1}) and direction for Vienna are given every 2 hours under the left diagram (Margules 1897, Fig. 5).

squall lines. Such a *gravity current* was used as a prototype for the interaction of warm and cold air masses. Schmidt worked with an elongated, two chambered tank bounded by glass walls. Salt water was used as the denser fluid. Inclination angles varying between 0.5° and 35° resulted in different flow speeds once the separation between the chambers was lifted. Time exposure (2 s) photographs of illuminated sus-

pended particles (sawdust) were produced to infer the instantaneous velocity distribution (Fig. 10). Characteristic was the elevated head of the denser fluid with some turbulent motion above and recirculation behind it. Schmidt translated the results of his experiments to temperature differences and propagation speeds of squall lines and found good agreement with Köppen's case. At the same time he regretted the

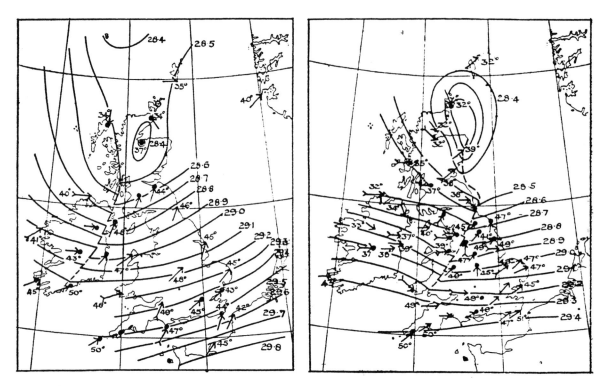

FIG. 9. Synoptic surface charts with wind and temperature (°F) observations as well as a mesoscale pressure analysis (increment: 0.1 inch Hg = 3.4 hPa) with a line-squall (dashed) extending southward from a surface low for 1800 GMT 19 February 1907 (left) and 2100 GMT 19 February 1907 (right) (Lempfert and Corless 1910, Figs. 7 and 8).

insufficiency of observational data to evaluate the analogy between laboratory and atmosphere in a more rigorous fashion, without mentioning the detailed case studies by Lempfert.

Köppen (1914) reviewed his earlier studies, the work of Durand-Gréville and Schmidt, and he presented observations of the severe squall line that passed over the German Bight on 9 September 1913 where it caused the crash of the airship L1. Quickly alternating upward and downward motions of considerable strength forced the airship down to 100 m above the sea, up to 1400 m and down onto the sea. When it touched the water with some 20 m s^{-1}, it broke apart and sank after a short while. The close proximity of updrafts and downdrafts was said to have been directly observed earlier and was considered characteristic for squalls connected with thunderstorms.

During the second half of the last century, theoretical considerations based on physical principles were gradually applied to meteorological problems, beginning with *thermodynamics*, for example, the effects of variable temperature and moisture in the atmosphere (typically at rest); and then also leading to *dynamics*, that is, the role of various forces for the movement of the atmospheric fluid. The latter topic included the investigation of different air masses separated by a surface of discontinuity and the stability of such an arrangement subject to small perturbations.

The renowned physician and physicist Hermann von Helmholtz[10] (1888) was probably the first to consider math-

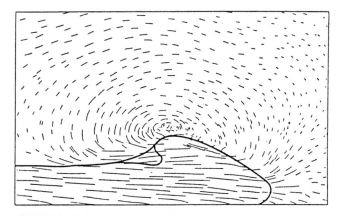

FIG. 10. Instantaneous velocity distribution during the intrusion of dense fluid (thin lines show the movement during 2 sec). Flow direction is from left to right. The approximate outline of the "head" is marked by the bold line (Schmidt 1911, Fig. 2).

10. Hermann Ludwig Ferdinand von Helmholtz (1821–1894) was the most influential scientist during the late nineteenth century in Germany as was his contemporary and friend Lord Kelvin in England. He received his basic training as an army surgeon, and later was professor of physiology. From 1871 he was professor of physics at a newly created institute in Berlin. He gave a mathematical treatment of the energy conservation principle in 1847. In fluid dynamics he investigated discontinuous motions and the condition of dynamic equilibrium along surfaces of discontinuity (see K79, 235–236).

ematically the equilibrium condition of the surface between different air masses on the rotating globe, specifically of an idealized (axisymmetric and inviscid) atmosphere, by juxtaposing two homogeneous layers of different angular momentum (i.e., wind speed) and potential temperature (i.e., density). He found that a dynamically stable equilibrium can exist if the two layers are separated by an *inclined* surface of discontinuity. The derivation of the stability condition is followed by a detailed qualitative discussion including frictional effects and the role of typical cyclones and anticyclones for mixing processes at the postulated prototype discontinuity.

Later, Margules (1905) introduced thought experiments dealing with the available potential energy, which is stored in adjacent air masses of different mean temperatures and which may be converted to kinetic energy (of a storm) when the colder air mass moves under the warmer one. Explicit mention is made of the better known motion fields along squall lines and the lesser understood ones within complete cyclones, when a warmer air mass is often followed by a cold air outbreak in the rear. Interestingly, JB19 quoted this passage as supporting evidence for his new conceptual model, although Margules concluded his paper with the frank remark: "The source of the storms lies, as far as I can see, entirely in the potential energy.... The horizontal pressure distribution appears as a kind of transmission within the storm's motion, by which a fraction of the air mass can aquire high velocities. . . . This leads to problems which cannot be solved by considering solely energetics."

One year later, Margules (1906) transformed Helmholtz's frontal-equilibrium condition to a constantly rotating Cartesian coordinate system. He wrote in the introduction: "I am trying to derive Helmholtz's equation in a way, that may well offer easier applications to meteorological problems." The simple model of a steady-state front involved an inclination angle α, which can be determined as a function of the velocity and temperature differences across the front. This well-known textbook formula was also quoted in JB19 (but erroneously from the 1905 paper) during the discussion of the squall line (cold front) together with Schmidt's laboratory study. Margules checked his idealized model using distinctly differing registrations of temperature and wind of stations not far from each other (Vienna and Bratislava). He also considered a temperature gradient zone in relation to vertical changes in wind speed rather than a sharp discontinuity. Finally, he followed Helmholtz's discussion on mixing processes acting on separated air masses or fluids and introduced simple thought experiments, for which he derived formulas for the energy conversion through mixing.

4. Warm Sector

Asymmetries in the temperature distribution within cyclones were hinted at in the literature before 1920. We have mentioned Köppen (1882a), Shaw et al. (1903), and Margules (1905) as examples. Detailed case studies of extended regions of anomalous warm air progressing over the whole of northern Asia were available by 1911.

Ficker[11] (1911) analyzed 11 episodes of about 10 days each from the years 1898 to 1902 when *warm waves* (as he called them) had traveled all the way from 40°E to at least 110°E. He confined his analyses mostly to isochron charts of the daily positions of the leading edge of the wave of warm air, but also presented some synoptic temperature analyses (Fig. 11). A massive warm sector spanned as much as 20 degrees of longitude over Siberia and was bordered by a zone of warming ahead of it and cooling behind it. Ficker also gave a conceptual scheme (Fig. 12) as a summary of this study and its companion (Ficker 1910), which dealt with the progression of *cold waves*. The scheme emphasized the thermal structure and did not give a characteristic length scale. But if one associates the bounding isotherm with what the Bergen school later called the *polar front*, similarities become obvious.

In Vienna, an intermediate position between the theoretician Margules and the data analyst Ficker was occupied by Exner[12]. For many of the present meteorological community, the factor to convert temperature to potential temperature (Exner's function) appears to be the only association with his name, although he was among the leading figures of European meteorology of his day, besides, for example, N. Shaw and V. Bjerknes.

Exner's textbook on dynamical meteorology (completed 1915, published 1917) was to a large extent based on concepts first developed by Margules. In chapter 74 on "the genesis of cold air outbreaks," for instance, a Margulean type front was introduced separating a cap of cold polar air, which

---

11. Heinrich von Ficker (1881–1957) was a research assistant at the Zentralanstalt in Vienna in 1906, held university positions in Innsbruck and Graz from 1909 to 1914, was a professor in Berlin from 1923 to 1937, and director of the Zentralanstalt as successor of W. Schmidt from 1937 to 1953. Fisker made many balloon ascents in Föhn flows for his dissertation, scrutinized the large body of surface data from Russia, and explored the interactions of troposphere and stratosphere in the development of cyclones (see K79, p. 233).

12. Felix Maria Exner (1876-1930) worked at the Zentralanstalt in Vienna from 1900 to 1910 with a leave of one year (1904–1905) for a world tour with extended stays in India and the United States. He was a professor in Innsbruck from 1910 to 1917 and then director of the Zentralanstalt and professor of physics of the earth in Vienna. He pioneered numerical weather forecasting by evaluating the surface pressure tendency for a geostrophically balanced flow with a climatologically determined thermal forcing before 1908, published a textbook on dynamical meteorology in 1917, undertook rotating tank experiments to study analogies of tornados as well as the general circulation with embedded cyclones, and attempted to explain cyclogenesis by the interaction between the westerlies and cold air outbreaks from polar regions (see Shields 1995, 1–3).

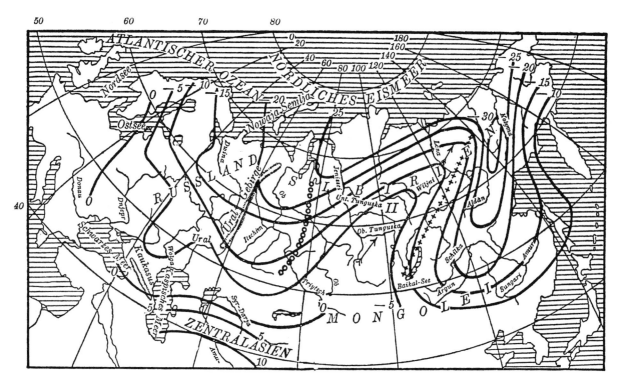

FIG. 11. Surface isotherms on 3 December 1901. Along the line ++++, quick warming is in progress; along the line ○○○○, fast cooling. Labels in °C; interval: 5 K. (First published by Ficker (1911, Fig. 11), taken from Exner (1917, Fig. 56)).

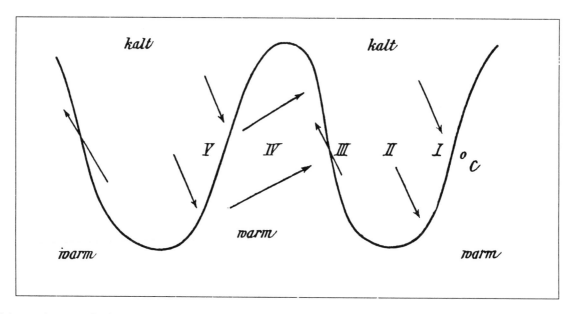

FIG. 12. Scheme of warm and cold waves over Asia depicting a bounding isotherm (wavy line) between cold air to the north and warm air to the south as well as predominant wind directions. The roman numerals designate different states at different places at the same time as well as the sequence of events at a location, say C, when the complete system passes from west to east. I: onset of cold wave with NW winds; II: lowest temperatures; III: gradual warming with winds from S to SE; IV: extremely warm SW winds; V: as I; phases III to V resemble the passage of a depression (Ficker 1911, Fig. 13).

is in steady westward motion relative to the warmer air aloft and farther south (Fig. 13). This may be viewed as a cross section through the *polar front*, a term introduced by the Bergen school some years later. Exner investigated how the frontal inclination α changes when the cold air gets retarded, for example, by frictional processes. He found a shallowing of the front equivalent to a gain of potential energy, which in part might be responsible for the strong winds observed during outbreaks of cold air toward the south.

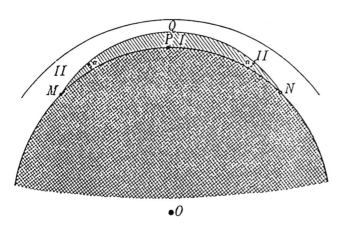

FIG. 13. Polar cap of a cold air mass (I) underneath a warm air mass (II). *O*, center of the earth; *P*, north pole; *M O P N*, span of the solid earth (Exner 1917, Fig. 58).

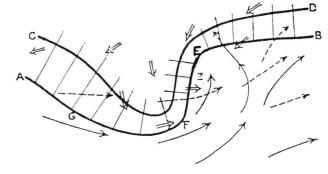

FIG. 14. Form and position of the separation surface between cold and warm air in a cold-tongue region. Full single arrows: warm airflow at the surface; dashed arrows: warm airflow aloft; double arrows: cold surface airflow (Exner 1920, Fig. 6).

One year after JB19 had been published, V. Bjerknes organized a meteorologists' conference in Bergen to better acquaint colleagues from abroad with his recent findings. For political reasons it took place in two sittings (Friedman 1989, 196). The German-speaking scientists gathered during the second part, when Exner (1920) presented his view of cold and warm airflow, in a lecture that Davies (1990) called "seminal but comprehensively unacknowledged." Having started with the general circulation including a Margulean type of polar front, and having presented observational evidence of cold air outbreaks, Exner developed a "hypothetical picture of the shape and position of the boundary between cold and warm air in the region of a cold-air-tongue, which progressed eastwards" (Fig. 14). The section EB of the surface front line AB was identified with the steering line (warm front) and EF with the squall line (cold front). CD gave the front position at a higher level; so the cold front was considered to be steeper than the warm front. It was noted that the scheme described best the development of shallow (i.e., only tropospheric) cyclones, which occurred often over North America and could be seen as early stages of mature cyclones, which extended into the stratosphere and were more frequent in the Atlantic sector. In a footnote, Exner stated that he became acquainted with the Norwegian cyclone model and the new *polar front theory* at the conference after he had written down his lecture (dated Vienna, 7 July 1920). He found the correspondence between both concepts encouraging and mentioned possible extensions of the theory of asymmetric cyclones.

5. Reception in the Secondary Literature

The preceding sections document the availability prior to 1920 of a variety of studies that dealt with components of the Norwegian cyclone model (first published in JB19). Today it

appears to be impossible—and somewhat irrelevant—to exactly reconstruct which of these studies were known to J. Bjerknes, his father, and their co-workers in Bergen.[13] A comparison of all the papers quoted in the preceding sections with JB19 reveals the special quality of the latter: on only a few pages a quite elaborate conceptual model was introduced ("I have been led to some general results concerning the structure of cyclones"), which had been inspired by detailed analyses from meteorological archives and a daily weather forecasting practice, and which outstandingly combined earlier findings. This section attempts to back this view by looking at some pieces of secondary literature about the Norwegian cyclone model and its evolution.

The immediate impact exerted by Norwegian cyclone model and the polar front theory, not only on practical aspects of weather analysis but also on the scientific meteorological community, must have been considerable. Invited by the editors of *Meteorologische Zeitschrift* (one of them Exner), Ficker (1923) compiled a detailed review article about the "polarfront and the lifecycle of cyclones." In essence, he congratulated V. Bjerknes and collaborators for the introduction of a new compact scheme with clear and memorable diagrams and for coining short and characteristic names for the various phenomena. Ficker considered the description of the life cycle of cyclones as the most novel finding of the

---

13. A detailed account of the situation in Bergen during the summer of 1918, when J. Bjerknes wrote JB19, is given by Friedman (1989, chapter 6). In his (mainly personal) bibliography with historical remarks, V. Bjerknes (1933, 783–787) mentions Ficker's (1910) and Helmhotz's (1888) papers as sources of inspiration for empirical data analyses and theoretical investigations, respectively, during the few years at Leipzig University (1913–1917). Note that the research program of this new institute could not be pursued continuously during the course of World War I. The move to Bergen in 1917 proved to provide "lucky circumstances, which made it possible to continue the work, though in a much different way, namely as practical rather than theoretical weather predication."

Bergen school, which also had implications for practical forecasting. On the other hand, he strongly disagreed with the claim that a radically new theory had been presented, and related the components of the Norwegian concept to earlier findings of others, especially from the Viennese school.

Textbooks for students may be used as indicators of how the development of concepts is generally seen in a scientific discipline. Raethjen (1953) presented a detailed monograph about extratropical cyclones and their dynamics some 30 years after the origin of the Norwegian cyclone model. He put strong emphasis on the historical development of concepts and compiled a detailed bibliography of 840 titles published between 1830 and 1950, including the studies of Köppen, Helmholtz, Margules, Ficker, Exner, Shaw, and Vilhelm and Jacob Bjerknes. In his conclusions, Raethjen emphasized two main cyclogenetic mechanisms: the thermal contrast between pole and equator, which can induce upper-level cyclogenesis in the jet stream regime, and gradient zones of latent or sensible heat at lower levels. Within frontal zones, both effects tend to be combined and most effective. He also noted that this kind of duality had already been advocated by the Viennese school of meteorology whereas the Bergen school had emphasized processes at lower levels.

More contemporary textbooks, such as Wallace and Hobbs (1978), tend to introduce just the Norwegian cyclone model as the standard—although idealized—conceptual model for extratropical cyclones. In their biographical footnotes about scientists who were claimed to have made major contributions to the atmospheric sciences, Jacob and Vilhelm Bjerknes, Helmholtz, Margules, and Shaw were mentioned.

There are also reviews that concentrate on the development of concepts from various points of view. Bergeron (1959), one of the Bergen school scientists, gave a wide and balanced historical review of the development of synoptic meteorology, putting into perspective the early case studies mentioned above in relation to the work of the Bergen school. One of his conclusions was that apparent rediscoveries were frequent in meteorology because of the complexity of the subject, an inefficiency in international bibliography and terminology and, very often, due to insufficient data to validate new hypotheses. He saw an essential difference between the case studies of, say, Köppen or Durand-Gréville and the Bergen school findings in the fact that the latter used "a rationally introduced concept on a routine basis on the daily synoptic charts."

The Bergen school period served as the terminus in Kutzbach's admirable book (K79), which discusses all primary sources on cyclone research between 1840 and 1920. Its conclusion states that "the polar front theory of cyclones is seen as an outstanding synthesis reconciling new insights and findings with important earlier results in meteorological theory" (p. 219), rather than representing a sharp break with older theories.

Schwerdtfeger (1981) criticized the indifferent attitude of the Bergen school scientists toward referencing older studies that had relevance to the cyclone model. Furthermore, he considered Bergeron's influence as essential for the acceptance of the new concepts in the synoptic practice in Europe, because, as Schwerdtfeger put it, Bergeron showed how to analyze the observations, whereas the Bergen papers (as JB19) mostly presented model sketches.

Friedman's (1989) scientific biography of V. Bjerknes discusses in depth the development of the cyclone and polar front models. Yet, both concepts feature as just two components of the elder Bjerknes's astonishingly dense curriculum vitae in an epoch of dramatic political, social, and technological change. Friedman's psychologizing attitude to rationalize in hindsight the sequence of events may be debatable, but he certainly presents a wealth of detailed background material, for example, from exchanges of correspondence.

A considerable amount of consciousness about the development of current research from quite old roots is apparent in the Atmospheric Science group at ETH Zurich. In the first chapter of his treatise on extratropical cyclogenesis, Schär (1989) sketched the development of ideas from the nineteenth century to the present time and juxtaposed the concepts of the Viennese and Bergen schools around 1920. Davies (1990) presented a masterpiece in historical condensation at the centenary of the meteorological institute in Innsbruck—150 years and nearly 100 citations on less than 10 pages of skillfully crafted text. He loosely grouped the theoretical concepts into three categories: (1) cyclogenesis attributed to an instability of a quasi two-dimensional sharp front to three-dimensional perturbations, (2) formation of lows linked to the passage of an upper-level trough over a band of low-level baroclinicity, and (3) cyclones regarded as outcome of a wave growing within a broad, but deep baroclinic zone with frontal features as some embroidery. The achievements of the Bergen school were considered as a milestone in the first group, "a masterful crystallisation of the prime observed features . . . and sandwiched in time between contributions of Exner." A more detailed account recently became available in Davies (1997).

6. Concluding Remarks

This chapter illustrates that the history of extratropical cyclone research in general and of the Bergen school in particular is well documented, perhaps better than for any other branch of the atmospheric sciences. The achievements made in Norway, starting with JB19, were recognized from the beginning and are highly valued to the present day. But there are also quite a number of important roots of the Norwegian cyclone model that deserve to be remembered and should be touched upon in courses of synoptic meteorology. Closer looks at the development of ideas regarding extratropical cyclones clearly reveal what it has in common with the subject itself: *Rich structures that are fascinating to explore.*

The question of why the Bergen ideas feature so prominently up to the present time cannot be answered in our context. Hints may be that especially V. Bjerknes devoted considerable energy to the widespread promotion of their concepts. Examples for this effort are a lecture before the congress of Scandinavian geophysicists and its double publication in *Monthly Weather Review* (1919) and *Meteorologische Zeitschrift* (1919), a letter to *Nature* (1920b) and *Monthly Weather Review* (1921), a lecture before the Royal Meteorological Society that was published in detail in its *Quarterly Journal* (1920a), and a keynote lecture in the physics section during the centenary conference of German scientists and physicians (1922 in Leipzig with about 10,000 participants; Kölzer, 1922). This was in contrast to other scientists, many of whom used academy transactions as their main medium for publication.

In more recent years, the Norwegian model has been mentioned as an example for a static conceptual model that led to an "unhealthy" situation for the present development of research (Hoskins 1983). The contributions in this volume certainly document how fit meteorology is at present, especially in the important field of extratropical cyclones and their life cycles. A good understanding of the development of various ideas and their interdependence is thought to be conducive to progress, especially if interchanges are achieved across political boundaries and language barriers. Overlooking longer periods of time will also create some modesty about the most recent achievements, as Köppen (1932) remarked with direct reference to both Bjerkneses when, at the age of 86 years, he reminisced:

> I happily acknowledge that meteorology has made great advances compared to the situation in . . . 1880, especially through V. and J. Bjerknes. . . . We also had seen air masses of different character within the general airflow, but we could not make much of our observations.

And he continued:

> For the present generation of researchers it is perhaps interesting to learn that an earlier generation tackled the same problems, partly with some success and partly without.

This was 66 years ago. It certainly is still interesting—and still true—today.

Acknowledgments

Expeditions into archives need assistance. It is my pleasure to thank I. Bernlochner (Oberpfaffenhofen), M. Dorninger (Vienna), G. Hartjenstein (Munich), D. Intes (Paris), and H. Lejenäs (Stockholm) for their help in obtaining copies of some of the referenced articles and biographical information. H. C. Davies (Zurich), M. A. Shapiro (Boulder), and A. J. Thorpe (Reading) provided valuable encouragement, and words of caution, during the early phase of this historical research. C. W. Newton (Boulder) kindly directed me around the punctuation and spelling pitfalls of American English.

References

The English translations of German titles are given in parentheses.

Bergeron, T., 1959: Methods in scientific weather analysis and forecasting: An outline in the history of ideas and hints at a program. *The Atmosphere and the Sea in Motion* (Rossby Memorial Volume), B. Bolin, Ed., Rockefeller Institute Press, New York, 440–474. (Reprinted in 1981 as Synoptic meteorology: A historical review in *Pure Appl. Geophys.*, **119**, 443–473.)

Bjerknes, J., 1919 (JB19): On the structure of moving cyclones. *Geofys. Publ.*, **1** (2), 1–8; and *Mon. Wea. Rev.*, **47**, 95–99.

——, and H. Solberg, 1921: Meteorological conditions for the formation of rain. *Geofys. Publ.*, **2** (3), 1–60 (plus 1 plate).

Bjerknes, V., 1919: Weather forecasting. *Mon. Wea. Rev.*, **47**, 90–95. (German version: Wettervorhersage. *Meteor. Z.*, **36**, 68–75.)

——, 1920a: On the structure of the atmosphere when rain is falling. *Quart. J. Roy. Meteor. Soc.*, **46**, 119–140.

——, 1920b and 1921: The meteorology of the temperate zone and the general atmospheric circulation. *Nature*, **105**, 522–524; and *Mon. Wea. Rev.*, **49**, 1–3.

——, 1933: Bibliographie mit historischen Anmerkungen. *Physikalische Hydrodynamik mit Anwendung auf die dynamische Meteorologie* (Bibliography with historical remarks; last chapter of *Physical Hydrodynamics with Applications to Dynamical Meteorology*), V. Bjerknes, J. Bjerknes, H. Solberg, and T. Bergeron, Springer, Berlin, 777–790.

Davies, H. C., 1990: The relationship of surface fronts and the upper-level flow to cyclogenesis—A historical resumé of the concepts. *100 Jahre Meteorologie Innsbruck* (*One Hundred Years of Innsbruck Meteorology*), Universität Innsbruck, ISNBN 3-900 259-14-3, 165–178.

——, 1997: Emergence of the mainstream cyclogenesis theories. *Meteor. Z.*, N.F. **6**, 261–274.

Durand-Gréville, E., 1894: Böen und Gewitter (Squall lines and thunderstorms; short version of monograph *Les grains et les orages, Ann. Bureau Centrale Météorol. de France*, 1892). *Meteor. Z.*, **11**, 312–314.

Exner, F. M., 1917: *Dynamische Meteorologie* (*Dynamical Meteorology*, 2d rev. ed., 1925). Teubner, Leipzig, x + 308 pp.

——, 1920: Anschauungen über kalte und warme Luftströmungen nahe der Erdoberfläche und ihre Rolle in den niedrigen Zyklonen (Perceptions about cold and warm airflows near the earth's surface and their role in shallow cyclones; Lecture presented at the meteorologists' conference in Bergen in August 1920). *Geogr. Annaler*, **3**, 225–236.

Ficker, H. von, 1910: Die Ausbreitung kalter Luft in Rußland und Nordasien (The spreading of cold air in Russia and northern Asia). *Sitz.-Ber. Wiener Akad. Wiss.*, **119**, Abt. IIa, 1769–1837.

——, 1911: Das Fortschreiten der Erwärmungen in Russland und Nordasien (The progression of warmings in Russia and northern Asia). *Sitz.-Ber. Wiener Akad. Wiss.*, **120**, Abt. IIa, 754–836.

——, 1923: Polarfront, Aufbau, Entstehung und Lebensgeschichte der Zyklonen (Polar front, structure, genesis and life cycle of cyclones). *Meteor. Z.*, **40**, 65–79.

Friedman, R. M., 1989: *Appropriating the Weather: Vilhelm Bjerknes and the Construction of a Modern Meteorology*. Cornell University Press, xx + 251 pp.

Helmholtz, H. von, 1888: Über atmosphärische Bewegungen I (On atmospheric motions, Part 1). *Wissenschaftliche Abhandlungen von H.v.Helmholtz*, Band 3, Barth, Leipzig (1895), 289–308.

Hergesell, H., 1900: Ergebnisse der internationalen Ballonfahrten (Results of international balloon ascents). *Meteor. Z.*, **17**, 1–28 (plus 3 plates).

Hoinka, K. P., 1997: The tropopause: discovery, definition and demarcation. *Meteor. Z.*, N.F. **6**, 281–303.

————, M. Hagen, H. Volkert, and D. Heimann, 1990: On the influence of the Alps on a cold front. *Tellus*, **42A**, 140–164.

Hoskins, B. J., 1983: Dynamical processes in the atmosphere and the use of models. *Quart. J. Roy. Meteor. Soc.*, **109**, 1–21.

Kölzer, J., 1922: Die Meteorologie auf der "Hundertjahrfeier Deutscher Naturforscher und Ärzte" (Meteorology at the centenary of German scientists and physicians). *Meteor. Z.*, **39**, 377–379.

Köppen, W., 1882a: Über den Einfluß der Temperaturverteilung auf die obere Luftströmungen und die Fortpflanzung der barometrischen Minima (On the influence of temperature distributions on the upper air currents and the progression of barometric minima). *Ann. Hydrogr. Maritim. Meteor.*, **10**, 657–666 (plus 1 plate).

————, 1882b: Der Gewittersturm vom 9. August 1881 (The thunderstorm of 9 August 1881). *Ann. Hydrogr. Maritim. Meteor.*, **10**, 595–619 (plus 3 plates) and 714–737 (plus 1 plate).

————, 1914: Über Böen, insbesondere die Böe vom 9. September 1913 (On squall lines, in particular the squall line of 9 September 1913). *Ann. Hydrogr. Maritim. Meteor.*, **42**, 303–320 (plus 1 plate).

————, 1932: Die Anfänge der deutschen Wettertelegraphie in den Jahren 1862–1880 (The beginnings of the German weather telegraphy during the years 1862 to 1880). *Beitr. Phys. fr. Atmos.*, **19**, 27–33.

Kutzbach, G., 1979 (K79): *The Thermal Theory of Cyclones*. American Meteorological Society, xiv + 255 pp.

Lempfert, R. G. K, 1906: The development and progress of the line-squall of February 8, 1906. *Quart. J. Roy. Meteor. Soc.*, **32**, 259–280.

————, and R. Corless, 1910: Line-squalls and associated phenomena. *Quart. J. Roy. Meteor. Soc.*, **36**, 135–164; discussion, 164–170.

Ley, C., 1872: *The Laws of the Winds Prevailing in Western Europe*. London, 164 pp.

Margules, M., 1897: Vergleichung der Barogramme von einigen Orten rings um Wien (Comparison of barogrammes from some places around Vienna). *Meteor. Z.*, **14**, 241–258.

————, 1905: Über die Energie der Stürme (On the energetics of storms). Anhang zum *Jahrbuch der Zentralanstalt f. Meteorologie 1903*, 1–26.

————, 1906: Über die Temperaturschichtung in stationär bewegter und in ruhender Luft (On the temperature stratification in constantly moving and in resting air). *Meteor. Z.*, **23 A**, 243–254.

Raethjen, P., 1953: *Dynamik der Zyklonen* (*Dynamics of Cyclones*). Akad. Verlagsgesellschaft, xii + 384 pp.

Schär, C. J., 1989: Dynamische Aspekte der aussertropischen Zyklogenese—Theorie und numerische Simulation im Limit der balancierten Strömungssysteme (Dynamical aspects of extratropical cyclogenesis—Theory and numerical simulation in the limit of balanced flow-systems). Ph.D. dissertation No. 8845, ETH Zürich, viii + 241 pp. [Available from Atmospheric Science, ETH Hönggerberg, CH-8093 Zürich, Switzerland.]

Schmidt, W., 1911: Zur Mechanik der Böen (On the mechanics of squall lines). *Meteor. Z.*, **28**, 355–362.

Schwerdtfeger, W., 1981: Comments on Tor Bergeron's contribution to synoptic meteorology. *Pure Appl. Geophys*, **119**, 501–509.

Shaw, W. N., and R. G. K. Lempfert, 1906: The life history of surface air currents. A study of the surface trajectories of moving air. Meteor. Office Memoir No. 174, reprinted in *Selected Meteorological papers of Sir Napier Shaw*, Macdonald, 1955, 15–131.

————, ————, and F. J. Brodie, 1903: The meteorological aspects of the storm of February 26–27, 1903. *Quart. J. Roy. Meteor. Soc.*, **29**, 233–258; discussion, 258–262.

Shields, L., 1995: Biographical note on F. M. Exner and English translation of his 1908 paper "Über eine erste Annäherung zur Vorausberechnung synoptischer Wetterkarten" (A first approach towards calculating synoptic forecast charts). Historical Note No. 1, Meteorological Service, Dublin 9, ISBN 0-9521232-1-5, vi + 31 pp. [Available from Meteorological Service, Library, Glasnevin Hill, Dublin, Ireland.]

Sprung, A., 1885: *Lehrbuch der Meteorologie* (*Textbook of Meteorology*). Hoffmann und Campe, 407 pp.

Volkert, H., L. Weickmann and A. Tafferner, 1991: The papal front of 3 May 1987: A remarkable example of frontogenesis near the Alps. *Quart. J. Roy. Meteor. Soc.*, **117**, 125–150.

————, M. Kurz, D. Majewski, T. Prenosil, and A. Tafferner, 1992: The front of 8 October 1987—Predictions of three mesoscale models. *Meteor. Atmos. Phys.*, **48**, 179–191.

Wallace, J. M., and P. V. Hobbs, 1978: *Atmospheric Science: An Introductory Survey*. Academic Press, xvii + 467 pp.

Constituting the Polar Front, 1919–1920

ROBERT MARC FRIEDMAN

Department of History, University of Oslo, Oslo, Norway

Between 1918 and 1924, a group of Scandinavian researchers under the leadership of Vilhelm Bjerknes (1862–1951) established a new conceptual foundation for atmospheric science. The formulation of the so-called Bergen meteorology has long been recognized as one of the principal turning points in the development of the modern science. For many meteorologists the Bergen school—with its air masses, polar fronts, and evolutionary cyclones—actually represents the birth of a comprehensive science of the weather. This chapter analyzes the genesis and early elaboration of one of these concepts— the polar front. As it first appeared in the Bergen school's work during the winter of 1919–1920, the polar front consisted of a single three-dimensional surface of discontinuity, stretching around the Northern Hemisphere, separating polar and tropical air masses (V. Bjerknes 1920a). Extratropical cyclones—the type of low-pressure system common in the midlatitudes—were understood to develop and evolve as wavelike disturbances along the polar front. The concept played a central role in the Bergen school's forecasting and theorizing during the 1920s and 1930s.

The Bergen meteorologists constituted the polar front as part of the process by which they devised an innovative system of weather forecasting. Earlier in the century, Bjerknes's fear of disciplinary isolation and his understanding of the importance of aviation for the advancement of meteorology originally led him to turn from a mechanically based mathematical physics to atmospheric science. Following a shift in emphasis during World War I toward practical forecasting, Bjerknes and his assistants developed in 1918 and 1919 an innovative cyclone model and new forecasting techniques. The special forecasting goals arising from the onset of commercial aviation, the rapid exchanges of weather data and predictions afforded by advances in wireless teleg-

raphy, and the new cyclone model combined to form a single perspective for meteorological discourse. Additional changes in predictive goals and communications possibilities, arising from expectations for regular transatlantic flights, led the Bergen meteorologists to broaden as well the conceptual basis for the forecasting system. This broadening—the "constituting of the polar front"—owed much both to the experience of World War I, which provided the image of the front along which two forces battled, and to the communications network that formed the technological basis for the proposed hemispheric forecasting system, which would utilize such fronts. The claims made for the polar front (some of them exaggerated) reflected the school's twin concerns to make weather predictions accurate enough for aviation and to put Norway in the center of the discipline of meteorology.

1. Bjerkneses Turn to Atmospheric Science

As a young scientist, Vilhelm Bjerknes had not considered studying the atmosphere. He began as an assistant to Heinrich Hertz in Bonn (1890–1891), and in 1895 he accepted an appointment as professor of mechanics and mathematical physics at Stockholm University. He first investigated electric waves, especially multiple resonance phenomena. He soon turned to James Clerk Maxwell's electromagnetic theory, in an effort to unify all physics by a mechanics based on contiguous forces. Specifically, he attempted to complete and extend his father's endeavors to find mathematical analogies between hydrodynamic and electromagnetic fields of force.

Whereas the elder Carl Anton Bjerknes had generally worked in isolation, Vilhelm intended to be part of and to shape the forefront of the physics discipline. However, he soon recognized that to pursue his life's goals of recasting mathematical physics so as to unify all the branches of physics might lead to his becoming peripheral to the advance of European physics. He gradually realized that the direction

---

This chapter is an abridged and edited version of Friedman (1982). A comprehensive history of the Bergen school's endeavors can be found in Friedman (1989): *Appropriating the Weather: Vilhelm Bjerknes and the Construction of a Modern Meteorology.*

and content of a particular science owes much to a structure of authority and power within the professional discipline. That is, neither the consensus by which research programs are selected nor the reception and acceptance of new findings proceed by rational methods alone. The research program of Bjerknes and his father, after some initial renown, suffered a gradual eclipse after 1900, as physicists increasingly focused on new research areas such as electron theory, radioactivity, and X-rays. Moreover, the effort that some German physicists made, prior to relativity and quantum theory, to replace the mechanical foundations of physics by an electromagnetic world view further led Bjerknes to believe that his endeavors had become irrelevant for the forefront of the discipline. Unable to control resources for influencing the discipline, such as journals, research institutions, and laboratories, and unable to develop a school of disciples, he could only watch passively from Stockholm.

Bjerknes's response to the major changes occurring in physics was decidedly conservative. Unwilling to abandon the mechanical world view in which both he and his father had heavily invested, he found he could preserve this mechanical physics by extending it to atmospheric and oceanic phenomena. (In fact, Bjerknes never abandoned his mechanical world view, and even in the 1920s and 1930s he urged physicists to take up the mechanical foundations of electromagnetism.) In his circulation theorems for fluids, in which density is a function of both pressure and temperature, he possessed a new tool for analyzing geophysical phenomena. (Unlike the classical circulation theorems of Helmholtz and Kelvin, in which vortex motions are conserved, Bjerknes's theorems (1898) allow for the formation and decay of circulation within fluids (V. Bjerknes 1898).) Bjerknes then regarded the atmosphere "exactly as a big laboratory for hydrodynamics" (from interview with Olaf Devik, assistant to Bjerknes, Oslo, January 1975).

The geophysical sciences commanded strong interest and support at both Stockholm and Oslo, where Bjerknes went in 1907. Yet at first Bjerknes considered the application of his hydrodynamics to the atmosphere and ocean primarily to be a side interest on which his younger collaborators (J. W. Sandström and V. W. Ekman) might devote time. He did not plan a major research program, nor work actively in meteorology until he recognized how little influence he could exert on European physics. At that point, however, he also began to recognize that meteorology would necessarily become a science of great significance as the commercial and military roles of aeronautics grew. Moreover, since both Norway and Sweden already possessed strong traditions in geophysics, he might be able to found a school that could direct the advance of these sciences and thus make the traditional scientific periphery into a center. Thereby Bjerknes would avoid his father's fate of scientific isolation and obscurity.

2. An Exact Physics of the Atmosphere

Bjerknes proposed establishing an exact physics of the atmosphere, which, once constituted, would allow rational predictions of the weather based on the laws of mechanics (V. Bjerknes 1904). Essentially, this ambitious project called for defining the state of the atmosphere at a given time and then using appropriate hydrodynamic and thermodynamic equations to precalculate future states of the atmosphere. Bjerknes claimed that ideally the state of the atmosphere would be defined by knowing the temperature, pressure, humidity, density, and wind velocity at each of many points of a grid covering a geographical area at the earth's surface and at various levels above the surface. Even if it took three months to calculate the change of the atmospheric state over a three-hour period, this would represent a major victory, for it would indicate that the physical laws underlying this change were known.

Developments in aviation and meteorology began to confirm Bjerknes's belief that the two would become interdependent. Enthusiasm for aeronautics promoted the advance of aerology, a subspecialty focusing on the investigation of the "free" atmosphere well above the earth's surface. By 1910 this field had emerged as the most dynamic growing point in meteorology, but because it required the technical support of a thriving enterprise in aviation, Bjerknes again found himself uncomfortably on the outside. Although he dreamed of making Norway a leading scientific center, he recognized that the only way to have his theories and methods accepted and developed more broadly was to move to Germany, which had become the leading center for aerology. The move took place in 1912, when Bjerknes was named director and professor of the geophysics institute established that year at Leipzig University. Behind the foundation of the institute and Bjerknes's appointment lay the desire to aid aeronautics by placing the ever-growing number of aerological observations on an improved scientific foundation. In Leipzig, Bjerknes received generous funds, spacious working conditions, and numerous students and assistants. By the time World War I brought work at the institute to a near halt in 1917, much progress had been made in Bjerknes's theoretical project. The war also changed the dynamics for the development of meteorology, however.

3. An Attempt at Practical Forecasting

The war forced Bjerknes to return to Norway from Leipzig in 1917. There he first planned to continue his long-term project to establish an exact physics of the atmosphere at the new geophysics institute at the Bergen Museum. Wartime, however, soon diverted his research activities: the needs of the expected air traffic and of taxed agricultural resources prompted him to turn to practical weather forecasting instead. Planning for a practical weather service to complement the existing Norske Meteorologiske Institutt in Oslo began when Bjerknes discussed with Minister of Defense Christian Holtfodt the provision of weather services for Norway's expanding air forces. Bjerknes submitted a plan modeled on the German field weather services. Following the entry of the United States into the war in 1917, grain exports to neutral Norway were stopped, and by the autumn severe food shortages appeared imminent. Consequently, the government established emergency laws and agencies to encourage the production and regulation of foodstuffs. In response, Bjerknes approached Prime Minister Gunnar Knudsen with a plan for expanding the existing weather services for eastern Norway and establishing a new service for western Norway. Approval came rapidly, for Knudsen's liberal Venstre party was already looking to science to overcome shortages of raw materials. Bjerknes placed his son, Jacob, in charge of forecasting for the west, and his other young assistant, Halvor Solberg, went to Oslo to head the expanded summer forecasting for eastern Norway. Bjerknes himself did not engage in the daily forecasting routines; instead, he devoted his energies to establishing the forecasting service, to devising new predictive methods, and to supervising his assistants in their chores.

During summer 1918, the attempt to develop rational forecasting methods for predicting rain focused on so-called lines of convergence in the horizontal windfield. These constructs, discovered in the Bjerknes group's earlier kinematics analyses of the wind field (see Eliassen, this volume) were known to be associated with rain. To compensate for foreign weather data no longer available during the war and to facilitate calculations of the movements of these lines of convergence, the group set up a considerably denser observation network than usual in forecasting at that time. They discovered quickly that the observations were still too sparse for the satisfactory use of prognostic equations; however, the data's quality did provide a new and unexpected means to predict rain. By carefully examining charts of wind flow based on precise wind measurements, Jacob Bjerknes surprisingly found twin lines of convergence, which, rather than simply being associated with cyclones, appeared to be one of their fundamental characteristics (Fig. 1). Furthermore, analysis of cyclones on the group's weather maps revealed an asymmetrical thermal structure consisting of a distinct "tongue" of warm air bounded by colder air. Similarly, the rain pattern in a cyclone, generally thought to be distributed somewhat symmetrically around the cyclone's central core, now appeared to be more closely associated with the lines of convergence composing the cyclone. Moreover, the line drawn tangent to the forward "steering" line where it meets the "squall line" in the cyclone's center showed the instantaneous direction of the cyclone and thereby pro-

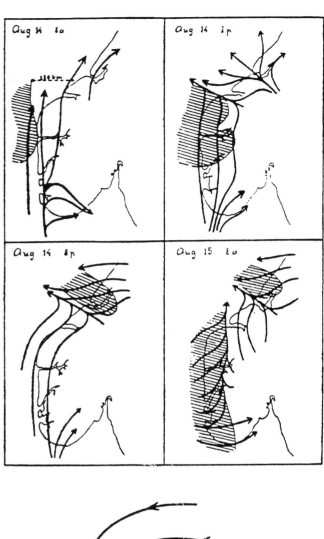

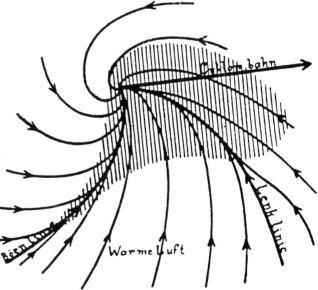

FIG. 1. Top: streamlines in the wind field near the surface showing lines of convergence and rain (shaded areas) along the west coast of Norway during the summer 1918 forecasting experiment. Bottom: initial Bergen cyclone model in horizontal cross section, showing two lines of convergence and the direction taken by the cyclone. The rain areas, here somewhat diffusely defined, were soon restricted to the lines of convergence (Bjerknes 1919; cf. Fig. 5 of Eliassen, this volume).

vided an empirical means for predicting a cyclone's movement.

In his first analysis, written during the fall, Jacob Bjerknes attempted to integrate his findings into the original hydrodynamics-based theoretical project, which led to some conceptual difficulties (J. Bjerknes 1919). His desire to interpret the discovery with the help of structures and models developed earlier from purely hydrodynamic considerations is understandable, given the original direction of the Bjerknes group. The article also has a short section describing the cloud and weather patterns associated with the new asymmetrical cyclone model. This section receives only cursory treatment, being overshadowed by the kinematic analyses of the wind flow. However, when during the next several months Vilhelm Bjerknes and his assistants accepted practical weather forecasting as a legitimate and critical part of their overall endeavors, and when they therefore attempted to establish systematic methods for practical weather prediction, these physical aspects of the new cyclone model assumed greater significance. In the process of becoming a basis for practical forecasting, Jacob's initial cyclone model was itself transformed: two-dimensional lines of convergence in the wind field were replaced by three-dimensional weather fronts.

4. Postwar Challenges and Opportunities

Even after the war ended and the immediate threat of famine had passed, Vilhelm Bjerknes and his assistants devoted their energies almost exclusively to devising new forecasting methods and atmospheric models. Bjerknes, the theoretician at heart, willingly and expediently prolonged his sojourn in practical matters. Forecasting practice, he learned, could react upon theory and, moreover, could help provide the social and material conditions for the general advancement of atmospheric science. Indeed, his son Jacob had in the course of the forecasting work "picked up" the cyclone's most "important secret." Recognizing that this "actual contact with practice [*praxis*]" resulted in a discovery that had implications both for theory and for practice, Bjerknes came to believe that close contact between theoretical dynamic meteorology and daily practical forecasting was essential for the progress of both. Practical methods ought to be based on a physical understanding of the atmosphere, and concepts and insight arising from the study of weather maps and of actual weather phenomena could provide direction and enrich theoretical inquiry.

Bjerknes now sought to connect a permanent forecasting service to his all-but-stillborn institute at the Bergen Museum. Forecasting not only promised to advance theory, it could also help develop a scientific meteorological profession in Norway. Bjerknes conceived of recruiting science students into the forecasting service. In order to function as forecasters, using methods based on physical principles,

these recruits would receive training in dynamic meteorology. Subsequently, they could pursue both theory and practice on the job. Bjerknes expected a permanent expansion of the national weather services in which numerous positions would be created because he, like most contemporary meteorologists, recognized the imminent advent of commercial aviation. New means of warfare, such as long-range artillery bombardments, gas attacks, and especially aerial combat, had already led to a mobilization and rapid expansion of meteorological services. Postwar expectations for rapid development of air transport similarly prompted meteorologists to promote the growth of their science, because new predictive methods and greater understanding of atmospheric phenomena would be required to insure safe operations of aerial routes.

During and immediately following the war, this perceived need and belief in meteorology led, in many lands, to the establishment of new professorships and university institutions, reorganization and expansion of weather forecasting services, and founding of professional societies for advancing meteorology. Subsequent to the reestablishment of international cooperation, which excluded scientists from the defeated Central Powers, meteorologists began to tackle the challenges posed by aviation. By early 1919 regularly scheduled flights had begun connecting London, Paris, and Brussels, and expectations had mounted that aerial networks would soon link the nations of the world. Both practical and theoretical aspects of atmospheric science gained a new sense of urgency and direction (*Scientific American* 1918).

5. Short-term Forecasting and Surfaces of Discontinuity

Wartime experience had taught meteorologists that to be effective for aviation, forecasts had to be much more geographically precise and detailed than traditional predictions and had to emphasize the short-term changes of weather conditions two to six hours in advance. In contrast to giving general vague weather outlooks for a broad region 12 to 24 hours in advance, as was the usual practice prior to the war, forecasters during the war had to attempt to answer such inquiries as, "Would the sky clear before midnight?" and "Would there be fog at such and such a place some 50 to 60 miles away at 2 a.m.?" Moreover, demand for such precise, detailed short-term forecasts necessitated changes in communication systems as well. Before the war, forecasters collected observations and distributed predictions by telegraph, which were then made available in the form of placards on buildings, newspapers, and a variety of signal flags and lanterns. In contrast, short-term forecasts for two to six hours in advance had to be transmitted directly to air fields as quickly as possible, by telephone and increasingly by wireless telegraphy. Thus commercial aviation operations required not only short-term forecasting but rapid communications.

When, during the winter of 1918–1919, the Bergen meteorologists began devising new systematic methods of weather forecasting, they had in mind the then apparent needs of aviation, particularly for the short-term predictions. They believed that by developing a cyclone model based on three-dimensional surfaces of discontinuity and integrated with other forecasting methods, they would be able to provide the detail and precision required for aviation. Although agriculture was a concern, they believed that satisfying aviation's needs would also fulfill those of agriculture. (Fishery did not become a significant concern for them until 1920.)

When the Bergen meteorologists changed their focus to practical forecasting, they dropped the mode of conceptualization based on kinematics and hydrodynamics and instead began developing models based on a physical atmosphere. Rather than two-dimensional lines of convergence in a horizontal wind field, they now spoke of three-dimensional surfaces of discontinuity—the physical boundaries between differing air masses (V. Bjerknes 1920b). These discontinuities, or fronts as they were later designated, constituted cyclones and in so doing provided the Bergen group with a means for comprehending the three-dimensional weather phenomena associated with these systems (Fig. 2). Not only could the meteorologists locate the occurrence of rain or snow associated with cyclones along these discontinuities, they could also specify the location of other weather phenomena critical for flight: cloud cover and sharp discontinuities in wind velocity, temperature, humidity, and air pressure. Along the forward discontinuity—called the steering surface (and later the warm front)—warm moist air rises up along a smooth incline over denser colder air. As the rising air gradually cools and condenses, a characteristic pattern of clouds and steady rain or snow forms. Meanwhile, along the squall surface (later termed the cold front) the cold air ploughs under the warm, rapidly lifting the lighter air, and thereby producing a narrow band of showers or thundershowers.

Successful use of these concepts in a forecasting practice geared for detailed and precise short-term predictions required rapid communications in order to collect large amounts of observational data and relay the forecasts for specific areas to those needing them; otherwise, the speed and accuracy of the new system of forecasting based on the concept of the surface of discontinuity would be meaningless. Why attempt to specify as precisely as possible the time and place for which a forecast is valid if the information cannot be commu-

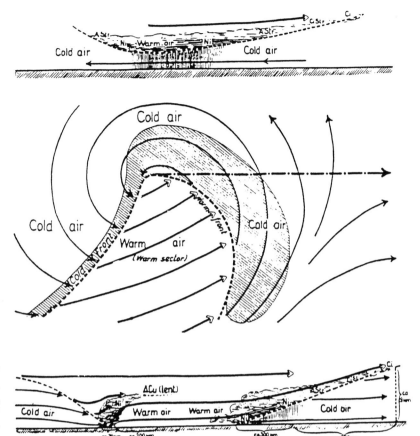

FIG. 2. The Bergen school's 1919 cyclone model (modified slightly; the "front" terminology was introduced after 1920). Top: vertical section north of cyclonic center. Middle: air motion and rain in a cyclone, horizontal projection. Bottom: vertical section south of cyclonic center (Bjerknes and Solberg 1921).

nicated to specific locations quickly enough to be of value? Bjerknes believed that although his group would have to rely on telephone and telegraph to begin with, wireless telegraphy would very soon be introduced at home and abroad.

In short, the Bergen meteorologists conceived of their 1919 cyclone model as part of a forecasting practice, and the model derived its initial meaning from this practice, sharing its goals and technological foundation. The surfaces of discontinuity also derived meaning from existing theory and physical laws (see conclusion) and could become objects for theoretical study, but their initial significance for Bjerknes and his assistants derived from their potential use in the detailed and precise short-term forecasts needed for aviation and agriculture.

6. Transatlantic Flight: Long-term Forecasting and Hemispheric Communications

By December 1919, when the polar front was first articulated, both the significance of aviation for and its challenge to meteorology had surpassed even Bjerknes's normally farsighted expectations. Recognizing that international laws and treaties must now accommodate aviation, delegates at the Paris Peace Treaty conference drafted a set of regulations to be adopted by individual nations desiring to join the projected League of Nations. This Convention for the Regulation of Aerial Navigation (13 October 1919) included an extensive set of proposals for providing aviation with special forecasts and observations; these had been worked out in part by meteorologists meeting in July and in early October. When the Paris Peace Treaty was ratified in November, the Norwegian meteorologists recognized that they would have to satisfy the international regulations for frequency and type of both observations and predictions. They spent the autumn working toward "a comprehensive plan for the development of a weather service for aerial transport." It was in this context that the Bergen meteorologists formulated the polar front.

Prominent during the fall of 1919 were two issues that proved crucial for the polar front's emergence: long-term forecasting, to complement the short-term, consisting of general weather outlooks three to four days in advance, and rapid hemispheric exchanges of weather information by wireless telegraphy. After two airplanes flew east from Newfoundland and an airship made the round trip from Great Britain to New York, "Atlantic fever" raged on both sides of the ocean: commentators predicted that fleets of Zeppelin airships would soon traverse the ocean regularly. Prior to and during these flights, extensive meteorological activities were coordinated to provide weather information for safeguarding the trips. Articles on the problem of forecasting for transatlantic flights appeared frequently in meteorological journals. The challenge could only be described as awesome. Transatlantic aviators had to contend with meteorological obstacles of a radically different order from anything previously confronting forecasters. Moving at high speeds with poor navigational equipment, often through seemingly boundless clouds and fog; being much more vulnerable than ships to sudden wind shifts, temperature and pressure changes, icing conditions, and severe weather; and having too limited a fuel capacity to allow for appreciable alterations of course, post–World War I aircraft could safely cross the ocean only when weather permitted. An aviator had to know in advance what weather conditions would confront him while crossing the Atlantic and had to follow changing conditions while in flight. Not surprisingly, from 1919 on, the weather over the North Atlantic and the problems entailed in assisting aviators crossing the ocean emerged as topical themes in meteorological literature and conferences. These issues dovetailed with the prior interests of the Bergen meteorologists.

Wireless communication, as previously noted, preoccupied Bjerknes from the start of the Bergen school's efforts to devise a new system of forecasting. To be truly effective as a forecasting tool, the new cyclone model had to be traced well beyond Norway's boundaries from observations collected by wireless communications. Further improvement would accrue if the observations made from ships in the Atlantic Ocean could also be incorporated. Once accomplished, a transatlantic wireless network could provide an "excellent foundation" for the expansion of weather forecasting; moreover, these reforms in communications could be justified by aviation's need for "an accurate, fast, and reliable international weather service." It looked like it might be possible to predict general weather outlooks several days in advance. Thus both the goals and the technological means for forecasting had broadened since the spring of 1919. Soon the conceptual foundation for forecasting was also broadened, as the empirical basis upon which the polar front was postulated was laid down.

7. The Polar Front as a Hemispheric Battleground

Following the cessation of the summer 1919 forecasting service Bjerknes and his assistants received authorization to issue, when necessary, storm warnings for the west coast during the fall and winter. A storm warning service, primarily for fishermen, had been instituted as early as 1909, but now the Bergen meteorologists hoped to provide greater accuracy by using their new cyclone model and forecasting methods. In turn, since the frequency with which cyclones approach Norway's west coast (and their intensity) increases dramatically during the fall, the Bergen meteorologists would have an excellent opportunity to refine their recent model and methods through daily analyses of the changing weather.

Recent changes in European and Norwegian weather forecasting operations also helped. Prior to the war, national weather services collected observations by means of telegraph once or twice daily. The advent of commercial aviation following the war led meteorologists to recommend an increase. During the summer of 1919, some nations began to take observations four times daily; Norway had initially increased them to three times daily. Extra sets of observations not only provided meteorologists with further raw data for aviators: in the case of the Bergen school meteorology, in which each map analysis was also an investigation and experiment, they also aided the understanding of the development and movement of weather systems. For example, a cyclone first appearing on a weather map when the morning observations were analyzed could, within 12 to 24 hours, pass beyond the map's boundaries, or develop in an unexpectedly dramatic fashion. Each extra observation period, therefore, provided further opportunity to study the evolution of weather systems and discover new phenomena.

One such discovery was that a mature cyclone can "give birth" to a new cyclone. At the tail end of one of the squall surfaces [cold fronts], extending out from a mature cyclone, a small wavelike pattern forms: this marks the development of the new cyclone. As frequent cyclones passed Norway's west coast during that fall, the meteorologists soon recognized also that not all these storms were autonomous entities: some were linked together—secondary cyclones growing out of and following the other. New and surprising though this discovery was, we should not mistakenly read it as a step inevitably leading to the conceptualization of the polar front. Not until mid-December of 1919, during an informal discussion among Bergen meteorologists, did the idea emerge that a single line, in fact a "battle line," stretched around the Northern Hemisphere. Previously the group had tried to conceptualize their new cyclone model (Fig. 2) in terms of a battle along the two surfaces of discontinuity (V. Bjerknes 1920b):

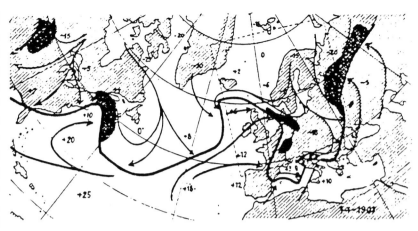

FIG. 3. "Polar front" demarcating polar and tropical air on map for 7 January 1907, as reanalyzed by Halvor Solberg in early 1920 (V. Bjerknes 1920a).

> We have before us a struggle between a warm and a cold air current. The warm is victorious to the east of the centre. Here it rises up over the cold, and approaches in this way a step towards its goal, the pole. The cold air, which is pressed hard, escapes to the west, in order suddenly to make a sharp turn towards the south, and attacks the warm air in the flank: it penetrates under it as a cold West wind

The team now extended this World War I battle image to hemispheric dimensions. Someone suggested that the polar air is the "enemy," initiating an attack toward the equator, while in response the warm equatorial air counterattacks with thrusts toward the pole. In between the two opposing types of air lies a battlefront that extends around the hemisphere, making the polar air's furthest advances—hence the name polar front. Just as battles during the war raged along the front, as one or the other army attempted to advance, atmospheric skirmishes were conceived of as occurring along the polar front; cyclones form on the front, representing struggles between polar and equatorial air, each attempting to advance into the other's territory. During the next several months, the Bergen group alternated use of the expression "polar front" with "battle line" [*kamplinje*] or "battlefront" [*kampfront*].

Once the concept of the polar front was suggested, Solberg looked for proof of its existence by reanalyzing a series of old weather charts that covered the north Atlantic Ocean and much of the North American continent. These reanalyzed charts "showed" the existence of a major portion of the single circumpolar line of discontinuity (Fig. 3). Yet these maps were based on observations from a sparse network of stations, and those from the oceanic areas were both scanty and poorly synchronized. The polar front, a narrow line of discontinuity on a map separating two different types of air, could not be "discovered," much less proven, from such charts. Even some members of the Bergen school recognized that this exercise merely depicted a hypothetical construct. Skeptical foreign meteorologists failed totally to be convinced: when during

conferences in Bergen and Leipzig in 1920 Jacob Bjerknes presented these maps and others on which he had drawn wavy polar fronts across the Atlantic Ocean, based on few observations, he left many observers agape and unconvinced.

Empirical evidence alone surely did not lead to the initial postulation of the polar front, a single surface of discontinuity. Rather, it was originally postulated at the conceptual level. Although the discovery that a secondary cyclone can form on the squall surface extending out from an older cyclone was a prerequisite for articulating the concept, it does not explain why Bjerknes and his group made the jump to hemispheric discontinuity, nor why this concept should become so important in their work.

8.   The Polar Front and the Forecasting Practice

Recalling that the Bergen meteorologists were actively devising a forecasting system that could satisfy the observational and predictive needs of aviation makes the development of the polar front understandable. The polar front was initially postulated when the battle metaphor was extended from denoting a condition along the surfaces of discontinuity within individual cyclones to evoking a single hemispheric "battle" front encompassing the individual skirmishes. First the Bergen school had believed that short-term forecasts could be provided by a forecasting practice based on these surfaces of discontinuity, which in turn were grounded in rapid local communications. Now this entire system, extrapolated to hemispheric dimensions, could also provide long-term forecasts. Belief that a hemispheric communications network would soon be implemented made it possible for the school to conceive of a hemispheric surface of discontinuity; indeed, the technology here seems to inform the concept. Since the cyclonic discontinuities were originally conceived of as integrated with the rapid communications system, the promise of a hemispheric network provided a rationale for expanding the notion of a surface of discontinuity to a similar scale, with the aid of the battle metaphor. A single discontinuity, hemispheric in scope, could provide a conceptual basis for the envisioned circumpolar weather service, and the needs and materials of the weather service simultaneously rendered the polar front a significant construct, because it could then be used to predict the weather.

During the winter of 1919–20, Solberg and Vilhelm Bjerknes further elaborated the polar front into a central concept for both forecasting practice and theoretical study. They attributed the formation of both cyclones and anticyclones to the motions of the polar front that followed attacks and counterattacks by polar and equatorial air. They regarded the polar front, whose undulations swept across most of the temperate zone latitude as the whole system moved from west to east, as a new means for comprehending the general circulation of the atmosphere and therefore for predicting the weather. Bjerknes and Solberg considered the meteorological events of the temperate zone to be "details" within the large-scale general circulation, which in turn were "correlated" with the motions of the polar front. To realize this predictive function, the polar front would have to be used in conjunction with a "circumpolar weather service." All initial references to the polar front are linked to discussions of a circumpolar weather service, and together they form a single focus of attention.

If a network of observation stations was established stretching around the hemisphere and linked together by appropriate communications facilities, it would then be possible to keep track of the polar front and its motions. A circumpolar weather service of this kind, based on the polar front, "would certainly be a great benefit to all occupations dependent upon the weather, such as agriculture, fishing, and shipping, and perhaps no less than a necessity for the realization of transoceanic air routes." Bjerknes claimed that predictions based on the polar front could provide precise short-term forecasts as well as general long-term ones. The polar front "can not fail to exert a considerable influence upon the methods of weather forecasting. . . . *An effective survey of this front all round the pole will form the rational basis of short-range as well as long-range weather forecasts*" (V. Bjerknes 1920c, p. 389). Furthermore, according to Bjerknes (1920a), "These two kinds of forecasts could be extended to all regions of the temperate zone—oceanic as well as continental."

The form and characteristics initially given the polar front helped determine the manner in which the daily study of it could provide improved forecasts. Cyclones were said to develop and move as part of the polar front; hence, the cyclonic precipitation, wind, and temperature structure specified in the original Bergen model could be forecast from a survey of the polar front. In this manner, short-term and "normal" (12–24 hours) forecasts could be secured. Although the mechanism by which cyclones evolve was not as

yet fully understood, the Bergen school conceived of a cyclone as a form of wave moving along the polar front. Along those portions of the polar front that do not constitute cyclones, fog occurs. Hence an idealized section of the polar front was represented by the diagram in Fig. 4 and described as follows (V. Bjerknes 1920a):

> Along the whole of this [polar] front line we have the conditions, especially the contrasts, from which atmospheric events originate—the strongest winds, the most violent shifts in wind, and the greatest contrasts in temperature and humidity. Along the whole of the line formation of fog, clouds and precipitation is going on, fog prevailing where the line is stationary, clouds and precipitation where it is moving.

Immediately apparent in this account is the reference to the weather phenomena most important for aviation, especially transatlantic flights. Strong winds, especially during a pronounced shift in wind direction, and sharp temperature and humidity contrasts over a short distance were well-known specific weather problems for airships (and airplanes). Although Bjerknes's remarks here imply this concern, his attribution of fog to the "stationary" portions of the polar front is most remarkable. All literature written at this time on the problem of transatlantic flight and "oceanic meteorology" discusses the hazards posed by the huge stretches of fog over the Atlantic Ocean. Fog could not be predicted reliably by the then prevalent forecasting methods. Bjerknes and his colleagues endowed their new construct with the ability to account for these fogs and, as a consequence, with the means to predict them.

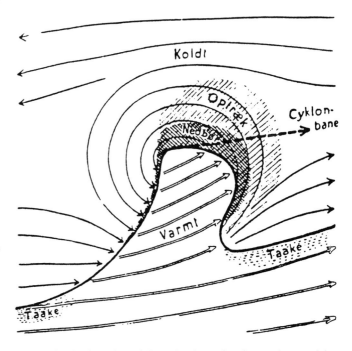

Fig. 4. Idealized section of the polar front showing cyclone and fog [*taake*] (V. Bjerknes 1920a).

In retrospect such claims can be seen as exaggerations. Yet it is precisely this almost cavalier manner in which Bjerknes and his assistants constructed the original polar front that reveals their motivations. Why should they definitively characterize the polar front by these regions of fog, for they offer no physical reason why fog must occur where the polar front is stationary? In fact, this characteristic was dropped shortly thereafter in subsequent articles on the polar front. Moreover, the initial idea of a single discontinuity stretching around the hemisphere proved theoretically clumsy, revealing hasty conceptualization. In short, the original description of the polar front reinforces the assertion that the Bergen meteorologists were consciously thinking of aviation's meteorological needs; that interest was in turn related to their thoughts on how best to mobilize the international discipline to accept and use their meteorology. Details of their plans to "sell" their methods and concepts support this suggestion, while also indicating the role of the polar front as a means to acquire a central position within the discipline.

9. The Polar Front as a Means to Acquire Authority

Bjerknes recognized the need to campaign actively for advancing both the new system of forecasting and the new ideas. From his earlier experiences he believed that to exert an influence on the major science centers he must acquire both the intellectual and material resources that the international discipline demanded and, not least, disciples who could form a sizable school of thought. From the start, Bjerknes began a campaign to interest foreign meteorologists in the Bergen school's new cyclone model and forecasting methods. He repeatedly emphasized the significance of this work for aviation. In Paris and London, he lectured on the embryonic Bergen meteorology, tried to recruit young scientists to Bergen, and, in order to gain a central influential position, captured the presidency of the international commission for aerology and a membership on the international commission for the application of meteorology to aerial navigation. By establishing a forecasting practice that claimed to satisfy the challenges facing the international discipline, Bjerknes and his eager young assistants could hope to make Bergen the new capital for meteorology. This goal in part shaped the way they postulated and developed the polar front concept. They believed that the polar front provided them with a necessary basis for erecting a new meteorology.

Bjerknes's plan for the circumpolar weather service consisted in uniting the various observation networks in the Northern Hemisphere nations and in supplementing these with a number of floating observation stations on ships and with several stations in the polar region erected for this specific purpose. This aspect of Bjerknes's plan did not differ greatly from other general plans for a hemispheric weather

network; still, it has one unique characteristic. Bjerknes calls for the establishment of "an international [weather] central" that would use the hemispheric data to survey the polar front and its motions on a daily basis and in turn transmit the results of its analyses of the polar front to various national weather services. The latter could then proceed with their own regional duties: only now they could prepare their forecasts with the addition of a "full knowledge of the general weather conditions of the whole northern hemisphere" and of their expected changes based on the study of the polar front. Thus, this information could provide the basis for predictions of "the general character of the weather for much more than four days ahead, and at the same time give to the short-range forecasts for a day or two a degree of precision hitherto by far not attained."

And where else but Bergen would the expertise exist to analyze the polar front? Even if it proved impossible to locate the "international weather central" in Bergen, the central would still have to be staffed with persons well trained in the Bergen methods. The Bergen's school's "microscopic analysis," which stressed a physical interpretation of each observation, had to be learned from someone possessing this skill. Therefore, to learn how to use this system of forecasting, with its claims to be able to provide short- and long-term predictions, meteorologists would have to come to Bergen, or invite Bergen scientists to their universities and weather-forecasting institutions.

Consequently, Bjerknes planned two conferences to be held in Bergen during the summer of 1920. Here he hoped to reveal to meteorologists of other lands the Bergen system of meteorology and the plan for a circumpolar weather service. After an encouraging conference with English meteorologists, Bjerknes intensified his efforts to spread the Bergen meteorology. Asserting that they should "strike while the iron is still hot," he asked Solberg to assist him in elaborating the mathematical basis for the new meteorology and in interesting others in their work. Bjerknes claimed that the conference was "very encouraging and if we can follow up our victory immediately, then perhaps in the course of couple of years we will organize the polar front meteorology." He contemplated sending Jacob to Germany, Holland, Paris, Brussels, and London in the fall and going himself to Washington after Christmas with Solberg to "teach" the new meteorology to Marvin, the head of the U.S. Weather Bureau. Bjerknes, in fact, refers to his assistants as "apostles" who must spread the new meteorology around the world, and this feeling was shared by other key members of the Bergen school.

The emerging Bergen school faced significant challenges also at home. By late 1920, a rapidly deteriorating economy threatened to undermine the forecasting system. Bjerknes and his collaborators, however, soon discovered a new source of political support in the Norwegian fisheries. In seeking to make predictions yet more reliable, detailed, and localized in time and space so that these could benefit fishery, the meteorologists could justify economic support for the forecasting system. Toward this goal they increased the number of observation periods and map analyses as well as enriched the types of observations collected. Swedish members of the Bergen school, Ernst Calwagen and Tor Bergeron, helped perfect "indirect aerology" in which fine-scale differences in cloud and weather observations could be read to infer three-dimensional atmospheric processes that were not readily accessible through direct measurements. Over the period 1919–1922 Bergeron used this method to articulate the occlusion process by which a cyclonic cold front overtakes the leading warm front, lifting the warm sector off the ground. Based on the collective contributions of all the members of the group, Jacob Bjerknes and Solberg published, in 1922, *Life Cycle of Cyclones and the Polar Front Theory of Atmospheric Circulation,* which can be considered the capstone of the early years of the Bergen school. Here the classic depiction of cyclonic evolution appears—from a slight wave on a preexisting polar front separating warm and cold air masses growing into a mature cyclone with a warm sector and attendant cold and warm fronts, to occlusion and decay as a cold vortex. Refinements in forecasting practice also allowed the Bergen scientists to replace the single polar front with a system of four polar fronts, each connected to one of four currents of air moving south from the polar regions. In this model for the atmospheric circulation of the Northern Hemisphere, each polar front extends from southwest to northeast, moving west to east. Cyclones originate on the southwest end of a polar front and propagate as waves along the surface of discontinuity, evolving through the various phases of cyclonic development. On any given polar front a cyclone "family" can develop, each member at a different stage of cyclonic development. This schema of polar fronts and cyclonic evolution became the basis in the ensuing years for forecasting efforts and for theoretical study.

Bjerknes's expectations for acceptance of the polar front proved in the short run to be too high. In spite of refinements in the concept, acceptance came slowly. Appropriately, it depended in part on the fate of aviation. Economic and technological difficulties postponed detailed plans for large-scale commercial transatlantic flying until the 1930s; only then did a working circumpolar weather service come into existence. Indeed, only with the anticipation of regularly scheduled transatlantic flights did the polar front receive widespread acceptance.

10. Further Reflections

In this discussion of the origin of the polar front, little has been said about theory. Although the Bergen meteorologists had set aside theoretical investigations in 1918 and 1919,

theory did play a role during this period: existing theory provided precedents and structural guidelines for constituting new models. Yet theory did not lead to or cause conceptual development. As early as 1915, Bjerknes and his assistants read various works on atmospheric discontinuities, either as entities in themselves or in relation to the problem of the energy of cyclones, but these had virtually no impact on the way they conceptualized atmospheric structures. Only when the group began developing a three-dimensional cyclone model during the winter of 1918–19 did they consider some of these works seriously, and then only to help resolve problems within their own project. Without direct measurements of the air above the ground, they used the results of Max Margules and Wilhelm Schmidt as guidelines for conceiving three-dimensional structures. Margules's formulas for the slope of a surface of discontinuity between two differing air masses and Schmidt's laboratory models of squall lines provided preliminary depictions of how the surfaces of discontinuity constituting the Bergen cyclone model might look. Equally significant, these and other findings were used to support the existence of three-dimensional discontinuities in the atmosphere. Moreover, the Bergen models assumed the validity of atmospheric thermodynamics, which had been fruitfully developed during the previous fifty years (Kutzbach 1979). Yet none of these earlier contributions inspired the Bergen school to pursue similar problems. They were engaged instead with developing atmospheric models that could serve as the conceptual foundation for innovative forecasting methods. These models could become the object for theory afterwards.

In the case of the polar front, forecasting and related disciplinary concerns, not theoretical problems, provided the impetus and rationale for development. Theory, so to speak, was already built into the 1919 cyclone model: the surfaces of discontinuity had sufficient validity in their relation to theory (Margules) and in their reproducibility in daily forecasting practice. Thus, when the Bergen meteorologists contemplated extending the forecasting service to include long-term predictions based on a hemispheric communications system, they felt confident they could, with the assistance of their military metaphor, extend the surface of discontinuity to hemispheric proportions. Once they had postulated it, the Bergen meteorologists concretized the polar front by devising analytical techniques for interpreting various weather data as a system of signs for detecting the discontinuity. In this manner, the polar front could be reproduced by weather forecasters who had learned these techniques. Similarly, after the spring of 1920, V. Bjerknes began to integrate the polar front into a hydrodynamic theory of atmospheric wave motions so that the concept could be reproduced also on a theoretical plane. Theoretical study followed rather than caused conceptualization.

True, the idea of atmospheric "struggles" and

discontinuities had occurred earlier in meteorological literature, but these were purely speculative and were rarely taken seriously. When the Bergen school had recourse to Helmholtz's thoughts on this manner, they sought to legitimize their bold new concept by linking it with his name. On more sober reflection, Bjerknes rightly noted that Helmholtz's ideas are not even valid precursors, since they are a muddle of speculations concerning phenomena of very different scales from that of the polar front. Rather than evolving from existing ideas, or forming a logical and necessary step in the development of a theory of general atmospheric circulation, the polar front emerged and possessed a specific meaning within a context of problems entailing forecasting goals and practices. And these problems in turn played a causal role in that conceptual development because of their significance for the then rapidly growing professional discipline in which Bjerknes and his assistants hoped to gain positions of authority.

References

Bjerknes, J., 1919: On the structure of moving cyclones. *Geofys. Publ.*, 1(2) (also in *Mon. Wea. Rev.*, **47**, 95–99).
———, and H. Solberg, 1921: Meterological conditions for the formation of rain.
Bjerknes, V., 1898: Über die Bildung von Cirkulationsbewegungen und Wibbeln in reibungslosen Flüssigkeiten. *Skrifter udgit af Videnskabsselskabet i Christiania. Mathematisk-naturvidenskabelig Klass*, No. 5.
———, 1904: Das Problem der Wettervorhersage, betrachtet vom Standpunkte der Mechanik und der Physik. *Meteor. Z.*, **21**, 1–7.
———, 1919: Veirforutsigelse: Foredrag ved Geofysikermøtet i Göteborg 28 Aug. 1918. *Naturen*, **43**, 3–16.
———, 1920a: The meteorology of the temperate zone and the general circulation of the atmosphere. *Nature*, **105**, 522–524 (also in *Mon. Wea. Rev.*, **49**, 1–3).
———, 1920b: The structure of the atmosphere when rain is falling. *Quart. J. Roy. Meteor. Soc.*, **46**, 119–130.
Scientific American, 27 April 1918: Commercial aviation in Norway. **118**, 396.

The Bergen School Concepts Come to America

CHESTER W. NEWTON

National Center for Atmospheric Research[1], Boulder, Colorado

HARRIET RODEBUSH NEWTON

Boulder, Colorado

1. Introduction

From the outset of the polar front theory, its originators carried on a campaign to make the concept known in the United States and to recommend improvements in observations that would enable polar front analyses. At the time, there was already a long tradition of research, but weather forecasting was an empirical "art" little influenced by scientific principles, being guided instead by an accumulation of rules based on experience. Thus the seeds of the polar front theory were cast upon a field that had been both fertilized by scientific progress and made sterile by adherence to tradition. Consequently, its reformation entailed both major changes in the weather services and the establishment of meteorology curricula in universities.

The polar front and air mass concepts were suited to these purposes, since they furnished a coherent framework for the analysis of observations and a unified view of the processes and evolution of weather systems. Although these ideas found early advocacy by some individuals, two decades passed before polar front methods were adopted into general practice by the U.S. Weather Bureau. This change was forced by outside pressures, notably to meet the exacting needs of the developing aviation industry. Private foundations played an essential role through support of extragovernmental committees, research and education, and a model airway weather service.

In order to examine the influence of the Bergen school upon meteorology in America, we are concerned with several questions: What was the state of knowledge at the time polar front thinking was introduced? How were the ideas spread? What reception was accorded these ideas? What measures were taken to adopt them into forecasting practice in the weather services? What were the further Scandinavian influences on cyclone research in the United States?

2. American Meteorology Before 1919

In the United States, as in Europe, there had been roughly a century of scientific studies of cyclones prior to the introduction of the polar front theory. A thorough account and interpretation of the evolution of cyclone theories during this period is presented in *The Thermal Theory of Cyclones: A History of Meteorological Thought in the Nineteenth Century*, by Gisela Kutzbach (1979; hereafter cited as K with page numbers). As she observes in the preface, "The discovery of the laws of thermodynamics during the 1840's and 1850's produced profound changes in many fields of Nineteenth Century science . . ." and they "also played an essential role in the emergence of modern meteorology." During the same period, the establishment of limited observational networks enabled synoptic analyses that, eventually coming together with theoretical principles, led the way toward sound concepts of the nature of cyclones.

The course had been generally positive but erratic, as put by Bergeron (1959a) in a critical epignosis: "At every stage of the development of a scientist or a science, *the stock of knowledge already acquired, or the views of a dominating School or personality, will to some extent block the recognition, or even the observation, of certain otherwise obvious facts that do not fit in with this knowledge or view, . . . old knowledge will often be rediscovered and presented under new labels.*" These characteristics aptly describe the progression of cyclone studies in America, as illustrated by the selective examples below.

---

1. The National Center for Atmospheric Research is sponsored by the National Science Foundation.

2.1 Espy and Ferrel: Thermodynamic and Dynamical Foundations

The early physical framework for cyclone studies was erected mainly by James Espy in the 1830s and by William Ferrel around 1860. Based on his own laboratory experiments and drawing upon the work of Dalton and others, Espy (K, 22–27) arrived (McDonald 1963) "at tolerably correct estimates of the dry and saturated adiabatic cooling rates over twenty years before Thomson's [Lord Kelvin's] theoretical treatment of these rates." Giving the first correct interpretation of the effect of the release of "latent caloric" in diminishing air density, Espy (1841) concluded that this was the key to the understanding of storms: "The result was an instantaneous transition from darkness to light." And indeed it was. As remarked by McDonald, it was earlier thought "that after condensation . . . air became *heavier*, an hypothesis that rendered cloud-growth rather mysterious," whereas "here was a new basis for understanding the way in which large-scale storms maintained themselves despite radial inflow."

Espy further realized that divergence aloft was essential to compensate the low-level inflow, so as to maintain the low pressure and to account for the movement of a storm. "The great expansion of the air in the cloud, will cause a rapid ascent and out-spreading above, which will cause the barom-

eter to fall under the cloud . . . and rise all round the storm. . . . But as there is known to be an upper current always, or almost always, moving in this latitude towards the north east or N.N.E., this current will cause the out-spreading of the air to be chiefly in that direction. . . . " (K, 26).

Ferrel's great contribution (1860; K, 37–41, 110–114) was the formulation of the equations for atmospheric motions on the rotating earth, and application of dynamical principles to describe circulations both on the hemispheric and cyclone scales. His introduction of the thermal wind relationship (in 1881) was fundamentally important since it provided a means of deducing the upper-level wind field from surface observations, and a formal basis for understanding the linkage between the thermal and dynamical processes that sustain circulation systems. His model of a warm-core cyclone (Fig. 1a) portrays a direct thermal circulation with cyclonically spiraling inflow in low levels and anticyclonically spiraling outflow at high levels.

In his initial essay, Ferrel (1856) espoused "Espy's theory of storms and rains" (condensation heating as the driving force), but rejected his contention "that there is only a rushing of the air . . . towards a center without any gyration." The violence and duration of a hurricane "depend on the quantity of vapor supplied by the currents flowing in below. Hence . . . hurricanes . . . do not abate their violence until they

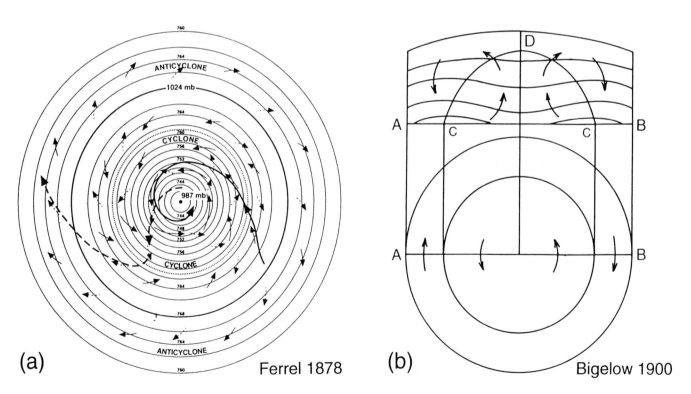

Fig. 1. (a) Ferrel's model of a warm-core cyclone in 1878 (K, 112; computer redrawn). Surface isobars 2 mm Hg interval; pressures at center and high-pressure ring shown in mb. Arrows, winds at surface (solid) and upper level (dashed), inflow and outflow streamlines added. (b) The model as illustrated by Bigelow (K, 40). At top, isobaric profiles and circulation in vertical section; cyclonic horizontal circulation is present inside *cDc*. Beneath, directions of geostrophic flow in inner and outer regions.

reach a high northern latitude where the atmosphere is cold and dry." Ferrel invoked "the principle of preservation of areas" (conservation of angular momentum) to account for the intense "gyratory motion" achieved by the inward-moving air, "which it still, in some measure, retains after ascending to the regions above, where the surrounding pressure does not prevail," causing the air to flow outward in upper levels. The inflowing air could not penetrate to the center, because "the gyrations become so rapid, that the centrifugal force nearly equals the centripetal [pressure force] . . . Hence, at a certain distance from the center, the hurricane has the greatest violence . . . " The circulation was viewed as a closed system (Fig. 1b), with the cyclone surrounded by a ring of anticyclonic circulation and thus isolated from the surroundings. While Ferrel's theoretical analysis related to a symmetrical cyclone, its physical principles would over time be modified to apply to extratropical cyclones. In the judgment of J. Bjerknes (1919), "Ferrel's convectional theory is confirmed in its essential part in as much as the ascending air in the cyclone is warm, only that this warm air does not form a central core, but comes from the side, at the ground covering a warm sector."

2.2 Loomis and Contemporaries on Storm Circulations

The forms taken by the wind fields in great storms were, from the beginning, a matter of principal concern from both scientific and practical standpoints, especially for the navigation and safety of sailing ships. Four contemporaneous conceptions are shown in Fig. 2. The first, by Redfield (1831; K, 16–17), is generalized from wind observations in storms that originated in the tropics and followed curved paths into the northeastern states. He proposed that the low central pressure was due to evacuation by centrifugal forces. Espy insisted that the airflow was centripetal (Fig. 2b) to fit his convective hypothesis. Tracy (1843), in a paper that anticipated this aspect of Ferrel's analysis (but carried the matter no further), deduced that the cyclonic circulation was caused by the rightward deviation (in the Northern Hemisphere) of inflowing airstreams owing to the earth's rotation (Fig. 2c).

The strikingly opposed views of Redfield and Espy led to a prolonged and heated controversy (K, 27–28, 42–43). Bôcher (1888), in a perceptive review of their arguments and fallacies, summed them up: "Espy looks at the forces; Redfield at the motions. Espy builds up theories on well-known physical principles and works over the facts until they fit the theories; Redfield looks at the facts first and then attempts to construct a theory for them . . . the discussion appears mainly to have served the purpose of stimulating each to strengthen every point of his theory, not to sift the true from the false." Bôcher concludes: "Meteorology is a young science. Indeed it is only within the last thirty years that it has become in any sense one of the exact sciences, that is, one susceptible of

mathematical treatment. It is Mr. Ferrel more than any one else, who has actually brought about the change which has occurred in this direction; but it is really due fully as much to the careful observations and sagacious theorizing which began 60 years ago."

Among early American investigators of the synoptic structure of cyclones, Elias Loomis (Miller 1931; K, 28–35, 123–125) was most prominent. Loomis (1846) attempted to conciliate the Espy-Redfield dispute by showing that both rotation and inflow are present, alluding (K, 34) "to the pressure gradient force as well as the deflecting force of the earth's rotation, whose action neither Espy nor Redfield had considered." His analyses of the wind field were fundamentally different from any of the earlier models, portraying for the first time (Fig. 2d) the structure of an extratropical cyclone. He demonstrated (see K, Fig. 8) the asymmetrical structure of the warm and cold air masses, and of cloud and precipitation, in relation to the isobars. Earlier, Loomis (1841) had analyzed the progression of a storm trough at 6-hour intervals from the Mississippi Valley over the West Atlantic, and deduced the circulation shown in Fig. 3. He concluded (K, 29–30) that "at least in this latitude, the most common cause of rain" is that: "When a hot and cold current, moving in opposite directions, meet, the colder, having the greater specific gravity, will displace the warmer, which is

(a) Redfield 1831

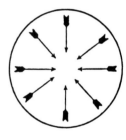
(b) Espy 1841

(c) Tracy 1843

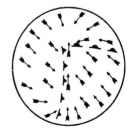
(d) Loomis1846

FIG. 2. Four conceptions of surface flow in storms, during the period 1831–1846 (see text). In (d), fronts have been added.

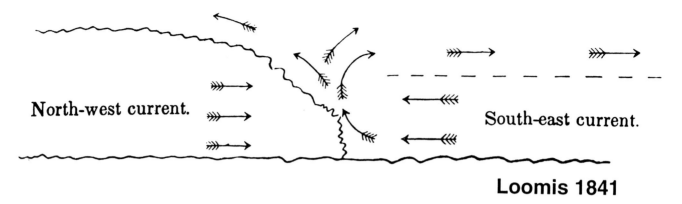

FIG. 3. Cold air displacing warm air "which is thus lifted from the surface of the earth, is cooled and a part of its vapor precipitated" (Loomis 1841; K, 30).

thus cooled and a part of its vapor precipitated." With snow behind and extensive rain ahead of the surface discontinuity, Loomis inferred that the warm air partly ascended the cold-air slope but "mainly turned back upon itself so that moisture as fast as precipitated fell through the lower current still blowing from the southeast." Loomis's demonstration of the value of synoptic charts for clarifying atmospheric phenomena was a stimulant for both research and forecasting services (K, 64–65).

Considering the prevalence of convective storms in some regions of America, we take note of the work of J. P. Finley (1884), who compiled an extensive climatology on tornadoes and proceeded to investigate the association of violent local storms with cyclones (Galway 1985). Finley found these to be concentrated in the "dangerous octant... to the south and east of the region of high contrasts [in temperature and dew point] of cool northerly and warm southerly winds...," being "more frequent when the major axis of the barometric troughs trend north and south." (His struggles to promote the issuance of tornado warnings to the public were squashed, and such warnings were not permitted until 1950.) Davis (1884), a supporter of Finley, composited his maps for a tornado outbreak to produce the remarkable analysis in Fig. 4, of the streamlines in relation to isobars and isotherms in a cyclone. The association with tornado locations (streaks) inspired the opinion that "with longer and more detailed study, the smaller storms may, a few years hence, be predicted with as much accuracy as the larger ones are now."

2.3 Bigelow: Counterflowing Airstreams, Upper Waves and Energy Conversion

Investigations of the 3-D structure of cyclones in America were initiated by Frank Bigelow in serial articles in 1902–1906 (K, 173–180). Compiling a massive collection of cloud-movement (nephoscope) data from the International Cloud Year 1896/97, supplemented by pilot balloon observations, and arranging these by sectors relative to cyclone and anticy-

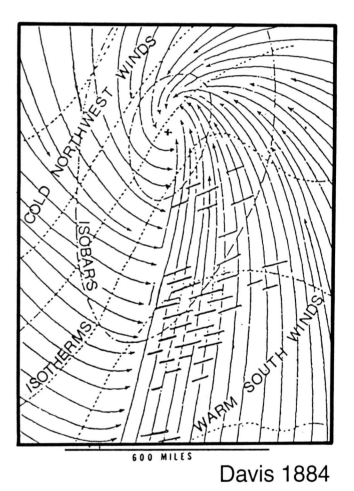

FIG. 4. Composite cyclone analysis by Davis (1884; taken from Galway 1985).

clone centers (for all cyclones east of the Rocky Mountains regardless of stage or intensity), Bigelow (1902) produced composites that showed that their perturbations persisted at levels up to 10 km. Removal of the "eastward drift," represented by the annual mean wind at each level, revealed

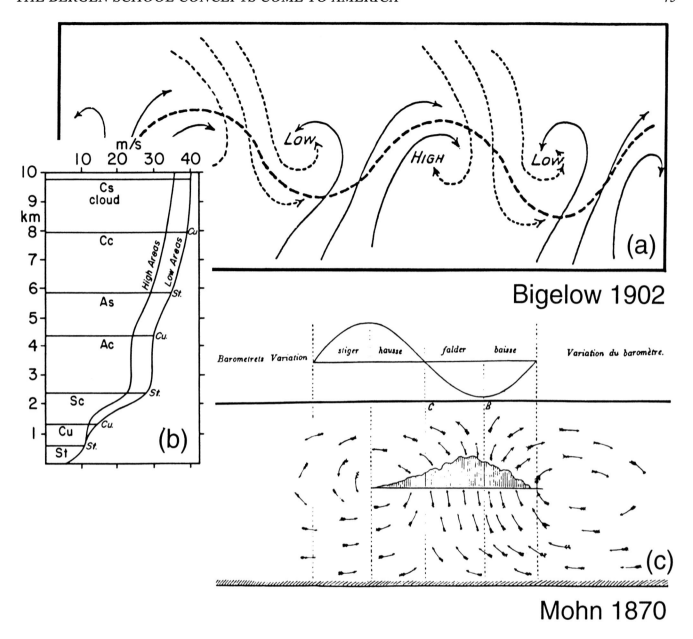

FIG. 5. (a) Bigelow's (1902) model of warm (solid arrows) and cold (dashed) airstreams in cyclones and anticyclones, in relation to waves at 3 km. (b) Mean zonal wind profiles in vicinity of highs and lows; assumed levels of cloud types indicated. (c) Vertical circulation in cyclone moving toward right (pressure change profile above, low center at C), according to Mohn (1870; K, 79).

that "separate streams from the north and from the south coalesce on the south side of the low area, . . . but the two streams have an origin outside the areas of high and low pressure. . . . The entire flow suggests . . . the conflict of two counterflowing, horizontal streams which tend to produce vertical rotation."

This view (Fig. 5a) was extended by Bigelow (1903) in a sweeping exposition on the links between the disturbances and the hemispheric general circulation. Alluding to the "two distinct lines of discussion," namely the Espy-Ferrel "thermodynamic theory" and the "hydrodynamic theory" in which

"the eastward drift simply curls up at places and forms eddies in the great current," he claimed that "no driving force sufficient to sustain a cyclone" had earlier been discovered. (Ferrel's "vortex with closed boundaries . . . does not . . . in the least satisfy the observations . . .") In late 1902, at Bigelow's instigation, the Weather Bureau had commenced daily analyses (constructed indirectly from surface data) of isobars at the 3500- and 10 000-foot levels. These further confirmed the scheme in Fig. 5a, establishing "the existence in the cyclone of the interaction of three practically independent currents of air," namely "the great overflowing eastward drift" in the upper waves, and the "underflowing" cold and warm airstreams near the ground.

Based on these features, Bigelow (1903) proposed that the source of "energy for circulation phenomena of the atmosphere . . . is to be attributed to the *thermal action due to the overflow of layers of cold air upon masses of warm air.* Abnormal stratification of air currents, where the relative cold is above the warm, necessarily involves an upward current, having an energy proportional to the difference of temperature." The warm and cold airstreams would both "curl strongly into the central vortex," with kinetic energy of horizontal motion being generated by cross-isobaric flow. Horizontal advection from lower and higher latitudes, as in Fig. 5a, thus furnishes a "powerful and persistent thermal source" for the process. In his fundamental theoretical treatise "Über die Energie der Stürme" (K, 186–194), published the following year, Margules acknowledged these opinions, quoting Bigelow: "The cyclone is not formed from the energy of the latent heat of condensation, however much this may strengthen its intensity; it is not an eddy in the eastward drift; but it is caused by the counterflow and overflow of currents of air of different temperatures." Although Bigelow's investigations became relatively well known in Europe, the "highly mathematical and often obscure character of his papers" isolated him from his Weather Bureau colleagues (K, 173).

Bigelow was presumably unaware of earlier works in Europe that were relevant to his own. Special among these was Mohn (1870; K, 76–83), whose empirical model (Fig. 5c) displays consequences of the entry of a warm and vapor-rich current into the forward part of a cyclone where latent heat release causes pressure falls, and penetration of a cold current into the rear. Bigelow did not consider the thermal asymmetry implied by the surface airstreams in his own Fig. 5a and by the release of latent heat in the warm air. His wind composites (Bigelow 1902) suggested that there was no significant shift of the vortex circulation with height. Accordingly, in illustrating "the penetration of a cyclone vortex into the upper strata," where this influence is expressed as a trough in the westerlies, Bigelow (1903, Fig. 28) regarded the disturbance axes as vertical (with, as he believed, a radial inflow component "inward toward the center at all levels of the cyclone."

Cloud-movement observations in Europe led to a markedly different picture, with a westward-tilted cyclone axis. W. Köppen (K, 151–152) had demonstrated in 1882 that this tilt is a hydrostatic consequence of thermal asymmetry, resulting from advection in the cyclone circulation. His construction [see Fig. 2 of Volkert's article in this volume] presaged important aspects of modern views; notably, together with M. Möller (K, 155–156), establishing the existence of upper-level wave divergence over an extratropical cyclone.

3. Introduction of the Polar Front

By the turn of the century, an impressive array of theoretical concepts and synoptic descriptions had been developed in the United States and Europe. These included the introduction of thermodynamic principles, with realization of the importance of latent heat release for atmospheric circulations, and of fluid dynamical principles that explained the rotation of cyclones and relations between the wind and pressure fields. Cold fronts had been described, along with the asymmetrical air-mass distribution in extratropical cyclones and the connection of surface disturbances with upper-level waves. The concept of cyclones forming in the shear zone between opposing warm and cold airstreams had been introduced, and the principle of conversion between potential and kinetic energy had been put forward, although the way this conversion leads to organized circulations in synoptic disturbances remained to be spelled out. Well into the twentieth century, professional academic training in meteorology was unavailable, and it seems unlikely that much of the above can have reached or influenced the typical forecaster, whose skills were acquired through on-the-job training and experience.

3.1   Existing Forecast Procedures

Acknowledging this situation, a board of senior forecasters was commissioned with the charge (Henry et al. 1916, preface): "Although the Weather Bureau has been successfully forecasting the weather for many years it is a rather notable fact that scarcely anything has been written to explain, more or less fully and in detail, the processes by which good forecasts can be made, and it is desirable to prepare . . . some sort of textbook or manual covering existing knowledge of this important subject." Accumulated rules that had been "communicated orally from the master to the pupil" would then be accessible to all. The resulting book, *Weather Forecasting in the United States*, gives a view of forecasting procedures just prior to the announcement of the polar front theory.

Forecasts were based entirely on surface weather maps, consisting of separate plots of current weather, isobars, isotherms, and isallobars (pressure changes in the previous 12 or 24 hours). Extrapolation of highs and lows was the basic method, guided by climatological storm tracks modified by isallobaric indications. A fundamental tool was the compilation of storm tracks and statistics on the seasonal frequency of occurrence and travel speeds of storms in various regions by Bowie and Weightman (1914), both of whom would later become early advocates of the Bergen school methods.

Types of storms (Fig. 6) were classified according to the regions in which they first appeared over the continent, recognizing that "many of them, especially in the winter season, have their origin in the Aleutian low and are offshoots therefrom." They generalized that "the variations in position

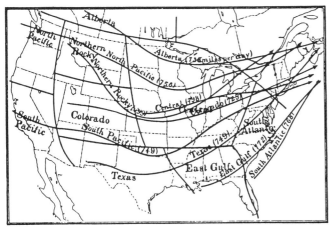

FIG. 6. Average paths of lows in January, with borders of regions in which each "type" is first observed (Bowie and Weightman 1914).

and magnitude of the elongated semipermanent area of low pressure that normally extends from southeastern Alaska westward over the Aleutian Islands to Kamchatka have a decided influence on the character of and courses followed by storms that cross the United States. If the Aleutian low is north of its normal position, lows will move along our northern border; whereas, if it is south of its normal position, lows will move far south of their normal tracks and stormy weather with great alternations in temperature will occur over the United States. . . . when the Aleutian low reaches its southernmost position, lows crossing the United States will make their appearance in the southern plateau region or over the Gulf of Mexico." Thus "abnormalities of pressure" in the "centers of action" (the Aleutian low and, for the region east of the Rocky Mountains, the Bermuda high) tended to control storm tracks for prolonged periods of time.

Although the *Weather Forecasting* book contains an introductory chapter on the general circulation in which cyclone hypotheses (driven eddy, convectional and counter current) are briefly described, its author (W.J. Humphreys) "is frank to say, however; that none of them conveys to him any workable conception of the origin, mechanism, or maintenance of the extratropical cyclone." No mention was made of U.S. contributions to the subject, including Bigelow's recent work. The reaction of Jerome Namias (1983), who at first attempted (in 1930) to learn about weather forecasting from the book, was that "there are hundreds of rules that I tried to memorize in order to learn to forecast. There were confusing and seemingly contradictory rules involving pressure changes, etc., and I soon had to give up."

The time was ripe for introduction of new organizing concepts. These came in the form of the polar front theory and air mass analysis. Their adoption, rather than being immediate, was hampered by the investment of experience among

forecasters and the reluctance of administrators to mandate a change to methods that were as yet unproven on the American scene.

3.2 The Initial Salvo from Bergen

The Bergen school proponents set the stage with a trilogy of papers in the *Monthly Weather Review*. Simultaneously with its publication in *Geofisiske Publikationer*, J. Bjerknes's article "On the structure of moving cyclones" appeared in the February 1919 issue of *MWR*. Along with it were two articles by V. Bjerknes (1919a,b): "Weather forecasting" and "Possible improvements in weather forecasting," subtitled "With special reference to the United States." The U.S. audience, accustomed to synoptic maps embodying comparatively amorphous isobars and isotherms exemplified by Fig. 7a, was introduced to analyses in terms of streamlines, with emphasis on the association of weather phenomena with discontinuities and convergences at the borders of cold and warm airstreams (Fig. 7b). All of the figures (including precipitation distribution) were schematic, intended to illustrate principles. The new concept was a radical departure from the methods of "isobaric geometry" used by forecasters.

In the brief third article, V. Bjerknes (1919b) made recommendations as to the improvement of forecasts: "Probably the most important step . . . will be the introduction in the daily weather service of good charts representing the lines of wind flow. . . . These charts will, however, have their full prognostic value only when the observations permit them to be drawn in such detail that the two fundamental lines of convergence of the cyclone, steering line, and squall line, can be accurately identified." This would require both a finer specification of wind direction (then reported by only eight compass points) and a denser network; "much could be accomplished by doubling or tripling the number of the stations. It will, however, be of great importance to have a network of stations along the Pacific Coast, with a closeness corresponding to that of the west coast of Norway [see Fig. 10 inset], in order to be able to catch the arrival of lines of convergence, and thus determine as early as possible the direction which the cyclones will take." A comparable density of reporting stations in the interior "would probably meet with difficulties from the point of view of the expense of the telegraphic service; and quite likely it will not be necessary . . . with the simpler topographic conditions in the United States, and the comprehensive view obtained from the great area of observations."

Bjerknes's suggestions were reiterated by Anne Louise Beck (1922), who had spent an American-Scandinavian Foundation fellowship year in the Bergen institute, and who commented further on the inadequacies of U.S. data for polar front analysis. This provided an opportunity for A. J. Henry, lead author of the 1916 forecasting book and now the influential editor of *MWR*, to add a "Discussion" (Henry 1922a)

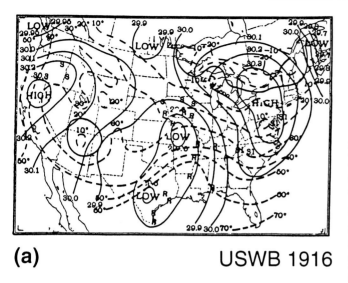

(a)   USWB 1916

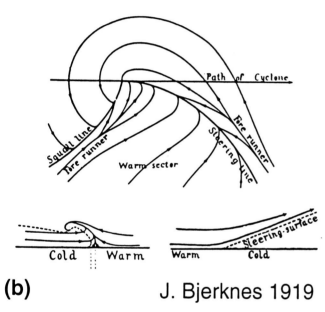

(b)   J. Bjerknes 1919

FIG. 7. (a) Synoptic map from the textbook of Henry et al. (1916). Isobars in inches Hg; isotherms, F°; S and R, snow and rain. (b) Bjerknes' (1919) streamline model, with sections across the warm and cold fronts. (The "Fore runner" divergence lines do not appear in later versions.)

cyclones are more clearly defined in the latitude of Norway (north 58° to 77°) than in the United States . . . [where they] arrive on its frontiers as fully developed systems. . . . The development of a primary cyclone . . . in the United States is a very rare occurrence . . . it seems highly improbable . . . that the occurrence of cyclones in families takes place in the latitudes of the United States with sufficient regularity to make the precept of definite value in long-range forecasting." Furthermore, "The area of precipitation in the advance of cyclones in the United States is rarely symmetrically distributed around the front of the cyclone, but is irregularly distributed according to geographic position of the cyclone and the season." Thus "the researches of the meteorologists of the Bergen Institute represent a decided step in advance, yet it can not be said that the theory they advance is complete and satisfying," calling for further study of U.S. situations.

3.3 Some Early Reactions

The Bergen school papers struck a spark with Clarence LeRoy Meisinger, a young Weather Bureau meteorologist. He quickly set out to apply the ideas on the U.S. scene, first in a study of sleet and glaze from the viewpoint of frontal ascent (Meisinger 1920a), then (Meisinger 1920b) in an analysis of a great cyclone that developed over Colorado and grew to major size and intensity. He concluded that "From the time the storm freed itself from the topographical hindrances of the Rocky Mountains the distribution of winds and precipitation during its eastward march conformed perfectly with the mechanical outline of Bjerknes." He observed that "Precipitation is closely related to [uplift at the 'steering line' and 'squall line']. . . . A third cause of rain is frequently present also, namely, the convection caused by the convergence of winds within the tongue of southerly air." Meisinger was apparently the first to use aerological (kite and pilot balloon) observations in support of frontal overrunning. He emphasized their potential in forecasting both precipitation and winds aloft for aviation: "One can not be content to study weather without a knowledge of what is going on in the third dimension. . . . These upper air data are indispensable and the establishment of more stations is certain to lead to an ever increasing return upon the investment."

Meisinger's dominant aim, pursued until his death in a scientific ballooning accident in 1924, was to devise methods to construct reliable upper-air pressure charts for aviation wind forecasting. To this end, he derived extensive statistics from kite soundings, and developed an operationally timely method for estimating upper-level pressures from surface data. In the opinion of Jerome Namias (1983), "Had Meisinger lived, he would have become one of the most prominent figures in American meteorology," In the event, however, "His contributions came at a time when the U.S. Weather Bureau exhibited scientific lethargy and rigid adherence to empiricism. . . . This promising line of research initiated by

in which he responded to Bjerknes's suggestions (ignoring the exception quoted above): "the number of telegraphic stations would be increased by about 4,500 . . . [from] slightly more than 200. . . . If the number . . . should be increased upward of twenty-fold it would be physically impossible to chart and generalize the data within a reasonable time after the observing hour, even if the present [5] district forecast centers . . . utilize reports from only such additional stations as would lie within their respective geographic districts."

Following his account of the Bjerknes and Solberg "life cycle" paper, Henry (1922b) remarked on the applicability of the scheme: "Undoubtedly the origin and development of

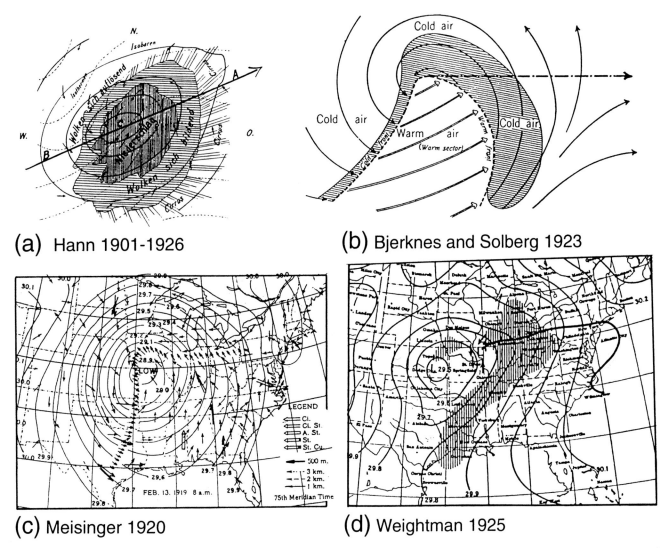

(a) Hann 1901-1926

(b) Bjerknes and Solberg 1923

(c) Meisinger 1920

(d) Weightman 1925

Fig. 8. At top, cyclone models (a) in Hann (1915, 522), not attributed but apparently from Mohn (1872); and (b) Bjerknes and Solberg (1923), showing frontal precipitation but omitting showers in warm sector and in cold air mass behind cold front. Beneath, analyses by (c) Meisinger (1920b), surface wind direction shown by thin arrows, cloud movements and pilot-balloon winds keyed at lower right; and (d) by Weightman (1925), showing precipitation during 4 h centered on map time.

Meisinger was abruptly terminated at the bureau after his death" (Lewis 1995).

A visit by V. Bjerknes and J. Bjerknes inspired senior forecaster R. H. Weightman (1925) to renew the subject of the polar front in the United States. Remarking that the existing 12-h map interval was inadequate, he presented a series of maps 4 h apart in which, curiously, the cold front was not shown. A prominent feature, alluded to as "cold-front rain," was clearly the warm-sector convection mentioned as a "third cause" by Meisinger. Panels from Meisinger's and Weightman's analyses are shown in Fig. 8, along with the Bergen model and an earlier model which, according to Bergeron (1959a), was "prevalent in Central Europe 1880–1930." This model appeared "during the 25 years 1901–1926 in four editions of the leading German textbook of Meteorol-

ogy by Julius von Hann. (These models, evidently, had as physical basis the assumption of a general frictional inflow and lifting of the air in the interior of any low pressure area.)"

The general lack of early enthusiasm for the polar front scheme in the United States was likely due to the same factors as in Europe. There "Skepticism and opposition greeted [the Bergen apostles] more often than not . . . part of the problem was that they were trying to market an incomplete product. . . . To adopt a substantially different form of analysis and thought, meteorologists needed to study the results carefully, which without detailed publications, they could not do . . . publications and lectures at conferences, moreover, tended to be dogmatic . . . experience showed that much of the craft involved could be learned only through direct contact with somebody who had mastered the methods. . . . The change amounted, after all, to a total shift in how cyclones were conceived" (Friedman 1989, 198–201).

Furthermore, among the Bergen meteorologists themselves, "there were many uncertainties as to the character and varieties of the fronts, the specifics of cyclonic development and evolution, and the reality of several secondary discoveries." While the Bjerkneses and Solberg tried to suppress details (such as secondary fronts) so as to preserve the unifying simplicity of their model, Tor Bergeron went about Europe analyzing such structures. In this way, as viewed by Schwerdfeger (1981), "he decisively helped his Norwegian colleagues to win worldwide recognition for the so-called polar front theory, not by theorizing a bit more, but rather by knowing how and where the observed facts of the real weather could fit into the idealized conceptual picture . . . the nice sketches in the papers of [J. Bjerknes and H. Solberg] were mostly based on surface observations and not accompanied by convincing case studies."

4. The Road to Adoption by Weather Services

Cleveland Abbe, one of the founders of the U.S. Weather Bureau and an indefatigable advocate of introducing mathematics and physics into meteorology, was the first true supporter of Vilhelm Bjerknes in the United States and from a practical point of view the most useful. Bjerknes gave a lecture on hydrodynamics at Columbia University in December 1905. Abbe was so impressed that he persuaded Bjerknes to give a second lecture in Washington, D.C., on his "vision for an exact physics of the atmosphere," and on that occasion introduced him to R. S. Woodward of the newly founded (1902) Carnegie Institution of Washington (Friedman 1989, 56–57). The result was a renewable Carnegie grant for Bjerknes, which enabled him to fund assistants for the rest of his active life and, as the first major undertaking, to carry out the kinematic-dynamic research that led up to the polar front concepts (Eliassen, this volume). The investment was well placed, leading, after a long train of events, to the official adoption of air-mass and polar front analysis by the U.S. Weather Bureau in 1938.

Fig. 9. Top: Carl-Gustaf Rossby in 1927, contemplating his rotating "dishpan" apparatus at the Weather Bureau; bottom: Lt. Francis Wilton Reichelderfer in 1923, as an aeronaut at the Gordon Bennett Balloon Cup Race in Brussels. (Photos courtesy of Patrick E. Hughes.)

4.1 The Rossby-Reichelderfer Alliance

The primary and in the long run most forceful advocates of the Bergen school methods were Carl-Gustaf Rossby and Francis W. Reichelderfer (Fig. 9). As characterized by C. C. Bates (1989), "They were a strange pair, although destined to become the 'yin and yang' of American meteorology. Both were ceaseless workers. After that, the similarity fades. Rossby could charm anyone when he wanted to get something started but after it was underway, he moved on, leaving many strings hanging in mid air. In contrast Reichelderfer was a modest, stubborn workhorse who deliv-

ered assignments on time and on budget with minimal fuss and muss."

Reichelderfer, educated as a chemist, volunteered for the U.S. Naval Flying Reserve Corps in 1917. He received three months training in meteorology at Harvard's Blue Hill Observatory and two years later secured his much-desired flight training. In 1921 (interview in Taba 1988, 90–91) he took part in a bombing exercise off the Virginia Capes for which he had forecast scattered thunderstorms. Instead they ran into a severe squall line—some planes had to land on the beach. "Now it happened that I had just received some of Bjerknes's papers on the structure of moving cyclones—up until then our concept of a depression was based on Abercromby's model [1885] which did not recognize fronts or wind-shift lines—and to me this was an exciting revelation and one which should permit us to be much more specific in our forecasting." From then on, Reichelderfer applied the polar front theory to his analysis of synoptic maps on a self-taught basis. (A decade later, the Navy sent him to Bergen, where he extended a two week visit, "to see just how Bergeron, Bjerknes, Petterssen, Solberg and the others did their analyses," to six months.) In 1923 he was put in charge of naval forecasting in Washington, D.C.

Rossby entered the scene in 1926 when he appeared at the Weather Bureau on an American-Scandinavian Foundation fellowship to do research and to "sell" the Bergen school philosophy to the United States. He had worked in Bergen in 1919, having been recruited from the University of Stockholm. There, as recalled by Bergeron (1959b), he arrived with no meteorological knowledge, but at the age of 20 "had an amazing persuasive and organizing faculty . . . practical map work was not his favorite job. . . . His forte appeared, instead, already then in being a constant source of penetrating general ideas." Having "managed to absorb the main aspect of V. Bjerknes' theoretical-hydrodynamic message already during this one year," he went on to the Lindenberg Aerological Observatory, then returned to Stockholm where he worked as a junior meteorologist and studied mathematical physics at the university. This "gave Rossby an indispensable fundament and tool for his later scientific work."

4.2 Rossby's Brief Encounter with the Weather Bureau

In the Weather Bureau, Rossby found a hotbed of resistance to new ideas, but also some valuable allies. The scene is described by Namias (1983): In the "map room" of the Washington forecast office "There was no central map, or 'map A' which the Bergen School developed, showing all weather elements on one chart. This was the status . . . eight years after [its] development. . . . In the corner of the room was [Reichelderfer] . . . analyzing weather maps according to Norwegian Methods . . . no one at that time (1926) in [the Weather Bureau office] practiced polar front analysis or, as a matter of fact, had even heard of it."

An exception was Weightman, with whom Rossby published quite a remarkable paper: "Application of the polar-front theory to a series of American weather maps." Their intention was to show "how and to what extent the Norwegian methods of analyzing synoptic maps could be applied to the study of American weather . . . , to ascertain what modifica-

tions . . . would be required and finally to determine whether any change in the present system of observations might be needed . . . [for application] to the daily forecast work in the United States."

To these ends, Rossby and Weightman (1926) chose a complex situation in which "at least four different air masses" were involved in a cyclone, shown in Fig. 10 at a stage when abundant moisture was being ingested from the Gulf of Mexico. At earlier times the fronts originated from the Pacific, Canada and the Gulf, illustrating the very different physiographic circumstances than over western Europe. Another important difference illustrated by their series was the widespread thunderstorm activity (outlined by heavy dashes) resulting from warm frontal lifting, in comparison with the typically uniform precipitation shields of Norwegian cyclones. To explain this "cyclonic convection," the authors called upon the generation of convective instability by differential advection. "A kite flight . . . , combined with a few pilot balloon observations and surface data from the Gulf region, can tell how much of a relative displacement of the different air layers is necessary to produce instability and free air convection."

Both the careful analysis of fronts and their interpretation as causes of precipitation, and the need to assess stability changes, called for augmentation and modification of weather reports. Remarking on the observational network, Rossby and Weightman noted that this was developed "during a period when forecasting was based on the experience accumulated by forecasters in their study of weather maps, rather than on an understanding of the dynamics of cyclones and anticyclones . . . [and] meteorological data in considerable detail were not needed." Their pleas to introduce more information into the observations, and to adopt the polar front theory which "enables us to explain phenomena which without a knowledge of the dynamics of the situation would hardly be understood," were not readily embraced and their paper was received with marginal interest.

It was, however, during this period that Rossby discovered Hurd Willett, who he thought had great potential, and arranged for him to visit the Bergen school in 1928 on leave from the Weather Bureau. Also this was when Rossby and Reichelderfer became good friends and colleagues. Reichelderfer visited the Bureau every day to pick up data so he could produce his own weather map and forecast, using the Bergen methods for the Navy flyers. He was elated to find somebody who had actually studied in the Bergen school and he was able to rescue Rossby from his remote desk in the library, thanks to the Daniel Guggenheim Fund for the Promotion of Aeronautics. Reichelderfer had become friends with Harry Guggenheim, a wartime Navy pilot, during ballooning expeditions.

This was the beginning of the "golden age of flying," and meteorology was hard pressed to provide the necessary forecasts to make flying safe. At the time, "The Weather

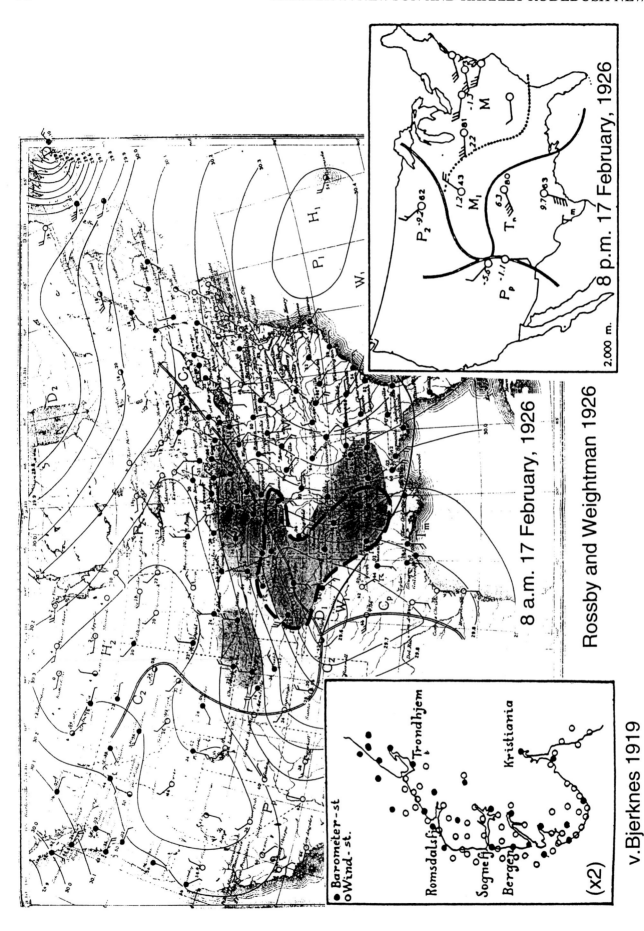

FIG. 10. Frontal and isobaric analysis by Rossby and Weightman (1926). Heavy dashes outline thunderstorm region. Lower right, pilot balloon and kite observations at 2~km (12 h earlier); letters, air mass types. Lower left, the density of observations (mostly wind stations) in the special Bergen institute network (V. Bjerknes 1919); double scale relative to U.S. chart.

Bureau had not yet got to grips with the main problems that faced aviators, for instance it was not the Bureau's practice to forecast fog!" (Reichelderfer, in Taba 1988, 91). In order to advance the art and science of meteorology for aviation, the Daniel Guggenheim Committee on Aeronautical Meteorology was formed with Rossby as chairman and representatives from the Weather Bureau, Army, Navy (Reichelderfer) and Department of Commerce (*Bulletin AMS* 1927, 122–124). One of the purposes was to promote more intimate exchange of views between pilots and forecasters—who had previously dealt primarily with agriculture and shipping. The second objective was to explore the present status of meteorological instruction in the United States.

This position enabled Rossby to stay at the Weather Bureau (because his grant was running out)—until he unfortunately responded to a request by Lindbergh for a forecast for a flight to Mexico City, which Lindbergh said was the best he had received from the Weather Bureau. Chief C. F. Marvin, already irritated by Rossby's exuberant schemes, was displeased that this had been done without his approval and Rossby was no longer welcome (Bates 1989). Reichelderfer then persuaded the Guggenheim Foundation to fund a model airway weather service for the flight between Los Angeles and Oakland, California, with Rossby in charge. Rossby arrived at the San Francisco office shortly after a letter from the Chief, warning Major Edward Bowie that Rossby was persona non grata. Bowie ignored the letter and welcomed Rossby to California (Byers 1959). During his very successful adventure in establishing the prototype weather reporting network, he was assisted by Horace Byers, a student at the University of California. This was the beginning of another long and productive relationship.

4.3   The Initiation of University Training

The next and most productive result of the coalition between Rossby and Reichelderfer was the establishment of the first U.S. department of meteorology at the Massachusetts Institute of Technology, originally funded by the Guggenheim Foundation. Selected to teach the second year of a two-year course for Navy aerologists (Bates 1989), Rossby initially was made instructor of this course but very soon it exploded into a true department of meteorology with Rossby as chairman. He persuaded Hurd Willett to leave the Weather Bureau and join him as his initial key staff member. The students of the Navy course (Reichelderfer 1928) had already studied physics, mathematics, and technical German in preparation for the main text, Exner's *Dynamische Meteorologie*, and beginning meteorology and oceanography. The curriculum at MIT began at the graduate level with Willett handling the synoptic studies and Rossby the dynamic studies. Students came from all sectors—Horace Byers was the first civilian to receive a Ph.D.

Boston, which had been prominent for meteorology with Harvard's Blue Hill Observatory and Department of Climatology, and the formation of the American Meteorological Society in 1919, now became the center for the transformation of meteorology in the United States from an art to a science, using and building on the methods developed at the Bergen school. When Bernhard Haurwitz arrived from Leipzig in 1932 he found, in addition to Rossby and Willett, the instrument specialist K.O. Lange, a pilot from Germany who flew daily ascents to aid Willett's study of air masses; three "lively" young men, Jerome Namias, Harry Wexler, and Athelstan Spilhaus; and Ethan Allen Murphy, who produced the daily weather map (Haurwitz 1985). Haurwitz himself had spent time in Bergen and received his Ph.D. from Professor Ludwig Weickmann, who was an early convert to the polar front. Therefore Haurwitz was a valuable proponent of the Bergen methods, although his primary interest was not synoptic meteorology. He also was in Toronto from 1935 to 1941 and contributed to the Canadian development of the theories of the Bergen school.

Rossby had both the desire and the ability to bring meteorologists from all over the world together and to create a truly international community—following the example of Vilhelm Bjerknes. He encouraged visits from J. Bjerknes, Holmboe, Haurwitz, and Sverdrup among others. This created a rich atmosphere for the students, most of whom went forth to become leaders and teachers of the Bergen school philosophy. A quotation from Rossby illustrates his philosophy during this period: "the principal task of any meteorological institution of education and research must be to bridge the gap between the mathematician and practical man, that is to make the weather man realize the value of a modest theoretical education and to induce the theoretical man to take an occasional glance at the weather map. The polar front theory, beyond a doubt, represents the most successful effort yet to bridge the gulf that separates the meteorological camps" (*Bulletin AMS* 1934, 266.)

Of note are two publications that reached larger numbers of meteorologists with the message of the Bergen School than any of the primary scientific papers. Namias (1940) published a collection of papers describing the air mass and polar front concepts in simple terms in a booklet called *Air Mass and Isentropic Analysis* with contributions by Haurwitz and Willett. Thousands of copies were sold by the AMS and were used in hundreds of weather offices around the world. Forecasters were hungry for a physical rationale on which to base weather forecasts. Sverre Petterssen, who had been an assistant at the Bergen school in 1924, succeeded Jacob Bjerknes as head of the forecasting service in Bergen in 1931. His interest was in forecasting and his doctor's thesis was on kinematics of the pressure field, conferring a quantitative aspect to applications of polar front analysis. In 1935 Reichelderfer invited him to the United States, where Pettersen created a syllabus on forecasting for the course Reichelderfer was running for naval aerologists. From this course eventually grew the book *Weather Analysis and Forecasting* (Petterssen 1940), which served to teach thousands of young U.S. students the theories of the Bergen school as they prepared for military forecasting service.

4.4   Revolution in the Weather Bureau

As mentioned earlier, the U.S. Weather Bureau was by nature resistant to new ideas. Forecasters were recruited from within, after a five- or six-year apprenticeship with a "seasoned" forecaster. With the exception of the department at MIT, meteorological education in universities including a strong background in mathematics and physics was nonexistent. The developing airline industry urgently required improved forecasts, weather being the major factor in safety and scheduled operations. The currently popular lighter-than-air dirigibles, viewed as important for both commerce and defense, were sensitive to updrafts in thunderstorms and squall lines, which were not adequately forecast without frontal and airmass analysis. The Navy airships *Shenandoah* and *Akron* had both had fatal crashes due to weather, leading to investigations (Bates 1989; Hughes 1995). The time had come to examine the hidebound and financially weakened Weather Bureau.

This evaluation was set in motion by President Roosevelt who established a special committee of the Science Advisory Board, which included Nobel Laureate Robert Millikan of Cal Tech, who had been head of the Army's World War I weather service and was now a firm advocate of the Norwegian methods, and Karl T. Compton, president of MIT, where he was Rossby's boss. The report of the committee (Millikan et al. 1933) was conciliatory in tone, reciting the range of valuable services of the Bureau, but compelling in its recommendations. These dealt "largely with the . . . introduction into all of the forecasting services . . . , of so-called air-mass analysis methods which merely supplement rather than replace the older methods. These new methods have so demonstrated their effectiveness [in Europe] . . . that there is the practical certainty that our whole forecasting service can be improved both as to accuracy and in reliability if the program . . . is followed." With interagency cooperation, these improvements could "easily be effected without prohibitive expenses, . . . if the whole meteorological service, including communications, is unified under the chief of the Weather Bureau."

The most important recommendations were "That provision be made at once for extending the so-called air-mass method over the United States; that operations be consolidated, and that current forecasters be educated in the methods." Observational data should be augmented to include information needed for the new analysis procedures, the international numerical code adopted, and surface maps increased from two to four per day. The maps should be synchronous with European maps; more frequent and de-

tailed ocean weather observations were desirable; and cooperation of other countries should be secured to "disclose the movement of major air-mass over all these areas." So as to "make possible . . . a daily upper air map of the whole country," aerological stations (at the time, airplane soundings) should be increased to 20–25, with the data "transmitted in a form which could be used without delay in the construction of the thermodynamical diagram . . . of air-mass vertical structure."

Adaptation of existing personnel and the infusion of new blood into the forecast system was a formidable challenge. This was not just a scientific problem, but one conditioned by increasing demands in the face of budget cuts due to the Great Depression. Chief Marvin had been obliged to dismiss a fifth of Weather Bureau employees (Bates 1989). The Board (Millikan et al. 1933) recommended that the transition be "made with caution" so as not to jeopardize existing services. An aim would be "decentralization of the general forecast work . . . with the ultimate assignment of a trained meteorologist to each of the principal airports." In a step-by-step process, one of the existing forecast centers would be started on frontal analysis methods, with forecasters "tested and apprenticed there before practicing elsewhere. About five years might be devoted to the extension . . . to perhaps seven [and eventually twelve] forecast centers." Part of the process would be assignment of experienced and academically qualified men "to an institution of recognized leadership in this field," including exchanges with "outstanding foreign meteorological institutes."

It is clear from the highly specific wording that the committee had been thoroughly briefed, and, as put by Bates (1989), "as far as Chief Marvin was concerned, it was a stacked deck." Marvin, who had been chief since 1913, promptly retired in 1934. Developer of a meteorograph employed for kite and later airplane soundings (Hughes 1970, 53–54), he had overseen the Bureau during a period of great expansion of observation services for aviation (Whitnah 1961, 170–171, 181–183). Forecasting procedures, however, had held to the course described in the *Weather Forecasting* book he had commissioned in 1916. He had responded to the committee, when queried about the Bureau's neglect of the Norwegian methods of weather analysis, "We didn't pursue it because the man [Willett] we sent to study in Norway left us" (Namias 1983).

Willis R. Gregg, the Bureau's aeronautical specialist, took over as chief and immediately set out to adopt the program recommended by the Science Advisory Board, which had sought his advice (*Bulletin AMS* 1934, 1; USWB 1934, 1–2). Under a new air-mass analysis section, Horace Byers joined the Bureau to teach incumbents the new methods, assisted by Harry Wexler and Charles Pierce. Gregg welcomed this effort but termed it a "sort of friendly competition between the practical and theoretical schools," and not much change was accomplished. When Gregg died of a heart attack in late 1938, the Air Transport Association demanded that the new chief be a "headknocker" from outside the Bureau (Bates 1989). Reichelderfer, then at sea as executive officer of the battleship *Utah*, was quickly appointed chief, which position he held until his retirement in 1963. His first actions were "to speed up and strengthen the changeover to air-mass analysis and forecasting," which at last became official in 1938; to institute "intensive, in-house training for Bureau personnel," and to initiate "four public forecasts a day, rather than just two" (Hughes 1983). Rather rapidly, the Weather Bureau was set on a course to join the world of science.

Reichelderfer persuaded Rossby to leave MIT and become assistant chief for research and education. Petterssen replaced Rossby as chairman at MIT. The threat of World War II was imminent, and the need to train meteorologists became imperative. This stimulated the development of new departments around the country. Graduate study was initiated at the California Institute of Technology, and Athelstan Spilhaus built up the department at New York University. Rossby and Byers began a program at the University of Chicago, J. Bjerknes and Sverdrup created a department at UCLA, and the United States finally reached a position of parity in the science of meteorology with European institutions. "Except for the department at Cal. Tech., . . . these new departments had the strong flavor of M.I.T. both in faculty and curricula" (Byers 1970). The polar front model, air-mass characteristics, and the rudiments of upper-air analysis, served as the main framework for training 7000 young meteorologists, among whom a relative few stayed on to serve the needs of postwar meteorology.

5. Expanded Views: Cyclones in Three Dimensions

Beginning in the mid-1940s there was a revitalized burst of research toward the understanding of cyclones and the general circulation, in which the Bergen school scientists continued to play a major role in U.S. meteorology. With J. Bjerknes and J. Holmboe at the University of California at Los Angeles, and C.-G. Rossby at the University of Chicago, Tor Bergeron and Erik Palmén paid extended visits to both places. Although the original polar-front concept was in principle three-dimensional, the recent enhancement of aerological observations enabled a new look at the hemispheric circulation and, in greater detail, the structures of disturbances over continental areas.

Dynamical principles relating to upper-level waves and to their connection with lower-tropospheric circulations had been formulated by Bjerknes, Rossby, and Sutcliffe, but these had been little exploited. The principles laid down by Jeffreys in 1926 concerning the role of eddies and mean circulations in the meridional transfers of momentum and heat had also never been tested. The field was ripe for harvesting and, with

many features of atmospheric circulations yet to be examined, students like ourselves were exposed to an atmosphere of ongoing discovery.

Foundations for aerological investigations had been laid by three interrelated contributions in 1937 and 1939. The monograph "Investigations of selected European cyclones by means of serial ascents" (Bjerknes and Palmén 1937) presented a comprehensive exposition of the methods, with the first 3-D analyses over the entire area of a cyclone. In a second paper in the same year, "Theorie der aussertropischen Zyklonenbildung," Bjerknes (1937; in 4 pages!) employed the gradient wind equation to derive the divergence in an upper-tropospheric wave pattern. He then related the divergence field to the structural features of a frontal cyclone, explaining its development, vertical motion field, and pressure changes associated with movement of the system.

The third foundation was Rossby's development of the theory of long waves in 1939, explaining in terms of conservation of absolute vorticity the existence of slow-moving or stationary long waves as the consequence of their retardation by advection of planetary vorticity. We shall illustrate further extensions of cyclone concepts to three dimensions by two papers, the first by Bjerknes and Holmboe at UCLA and the second by Palmén (Academy of Finland), resulting from his studies while affiliated with the University of Chicago as a visiting professor.

5.1 Bjerknes and Holmboe: Upper Waves and Cyclones

In the inaugural issue of the AMS *Journal of Meteorology*, Bjerknes and Holmboe (1944) greatly enlarged the theme of the Bjerknes (1937) paper. Their summary scheme for the properties of an extratropical cyclone is shown in Fig. 11. With gradient winds in an upper-level wave, there is no flow across the isobars nor acceleration along the flow, and hence horizontal divergence depends on differences of isobaric channel width at the trough and ridge (top left). The channel widths depend on the Coriolis parameter and centripetal accelerations in such a way that the magnitude of divergence is greatest for large wind speed and short wavelength (i.e., large and opposite curvatures at trough and ridge). At the level of nondivergence (Fig. 11a), wave speed $c = \bar{V} - V_c$ in which the "critical wind speed" $V_c = \beta L^2/4\pi^2$ (as in Rossby's formula).

The distribution of divergence, shown for air columns ahead of and behind the trough, then depends on the *relative wind* $(\bar{V} - c) \gtrless V_c$. "*The wave will travel with such a speed that the pressure tendencies arising from the displacement of the pressure pattern are in accordance with the field of horizontal divergence.*" With a westward tilt of the cyclone and trough axis as in Fig. 11b, the lower-tropospheric convergence necessary to sustain the cyclonic circulation, and its low pressure, are necessarily linked to the divergence down-

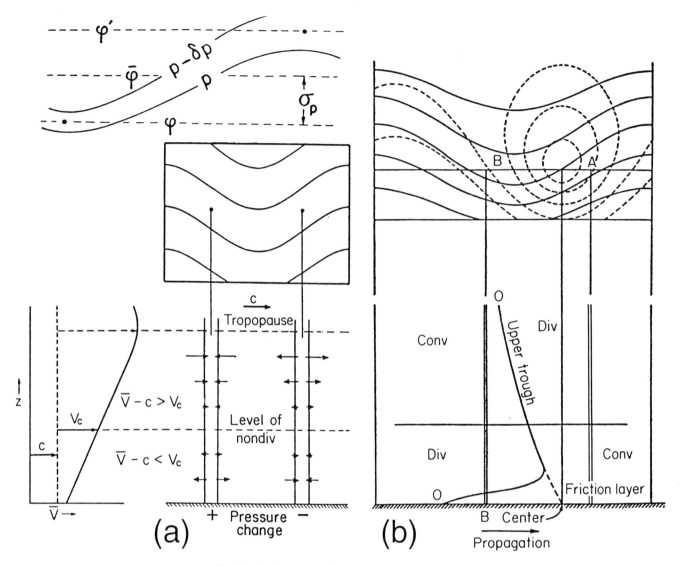

FIG. 11. Cyclone model by Bjerknes and Holmboe (1944).

stream from the upper-tropospheric trough. The intensity of divergence, and hence vertical motions, is greatest for large vertical shear (baroclinity) and short wavelength, and increases as the wave amplifies through thermal advection.

The 75th anniversary of the polar front is also the 50th anniversary of the Bjerknes-Holmboe paper, which can be seen as another turning point in the way of thinking about cyclones. Like the polar front cyclone concept, it had descriptive antecedents, at least in part. The new theory introduced a unifying concept, based on a single dynamical principle (the gradient wind), that explained in a straightforward way the features common to extratropical cyclones. This model was instrumental in a major change of the view of instability involved in cyclone formation: from the theory of a growing perturbation on a surface of discontinuity, to the baroclinity (vertical shear) of the troposphere as a whole.

5.2 Palmén: Three-Dimensional Fronts, Jet Streams and the General Circulation

The principles of three-dimensional frontal analysis were introduced in the United States by E. Palmén, who, early in his visits to Chicago, established the kinematic-thermal relationships among jet streams, frontal layers, and tropopauses. His ensuing investigations of 3-D circulations in this framework illuminated the role of extratropical disturbances in the meridional exchange of air masses and the energy processes that maintain the general circulation (Palmén 1951a,b).

Figure 12a illustrates contours of the frontal zone at two stages in the development of a major cyclone east of the Rocky Mountains. Questioning "whether it is, in principle, permissible to start from infinitely small perturbations in discussing the cyclone problem," Palmén (1951a) remarked

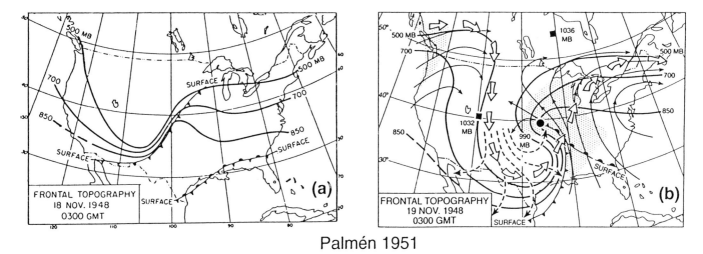

Palmén 1951

Fɪɢ. 12. Frontal contours of (a) developing and (b) mature upper wave and cyclone (Palmén 1951; some features added). Central pressures of sea-level systems; jet stream axis (short open arrows); 3-D streamlines at upper surface of warm frontal layer and (dashed) on isentropic surface in cold air, starting in middle troposphere.

that "it seems more likely that extratropical cyclones are induced by rather large migrating disturbances"; in this case, an upper-level trough approaching from the Pacific. In Fig. 12a there are two initially separate surface fronts: a cold front from the Pacific and a stationary front along the Gulf Coast (trailing from the preceding cyclone). This is ingested into the circulation to produce the structure in Fig. 12b, showing that (reflecting the differences in topographic circumstances in the United States compared to Europe) "the occluded frontal cyclone can be the result of processes other than the occlusion of wave-shaped frontal perturbations. . . . Essential for the whole development of a mature cyclone is the formation of the upper disturbance associated with the deformation of the upper front . . . with the formation of an upper cyclone or a very deep trough."

Following Fig. 12b the cold air at 500 mb was separated from its source, forming a cut-off cyclone with cold air subsiding at a lower latitude. *"The process of seclusion of the polar air at the 500-mb level thus corresponds to the occlusion process in lower layers.* In the upper atmosphere the warm air gains area, in the lower atmosphere the cold air gains area," with conversion of potential to kinetic energy. Palmén emphasized that this process could properly be understood only through an examination of 3-D air trajectories. Streamlines are shown in Fig. 12b at the upper surface of the warm front, along with (dashed arrows) characteristic streamlines on an isentropic surface in the polar air, that descend from the middle troposphere into the boundary layer.

These typify the "irreversible process" in which warm air ascends with release of latent heat and remains in upper levels, and cold air subsides as it streams into the tropics. Viewed relative to the polar-front zone around the hemisphere (arrows A and D in Fig. 13a), these transverse flows

in the disturbances correspond to a systematic direct solenoidal circulation, which generates kinetic energy and sustains the polar jet stream through a process connected with the disturbances, in distinction to generation by the mean circulation as in the subtropical jet stream (Fig. 13b). With this interpretation the general circulation scheme of Bjerknes and Solberg (1923), in which cyclone families and meridionally extensive anticyclones alternate around the hemisphere, was unified with the framework of global-scale jet streams.

Epilogue

The remarkable advances in observations of all kinds, together with theoretical-modeling investigations, have made it clear that there is great variety in the kinds of cyclones, the manners in which they form, and the processes that dominate from case to case. Concepts of cyclones have accordingly broadened and new features have been brought to light. Throughout the 75 years since its introduction, the polar front theory of cyclones has served as a keystone for modern meteorology; the newer concepts have served to enrich, rather than to supplant, the original concept.

The spirit of Jacob Bjerknes's contribution is aptly characterized by Petterssen (1974, 30–31): "Jack proposed a blueprint, or a model cyclone, and he went beyond the mere 'anatomy' of storms and described in broad outline their 'physiology'. . . . In long retrospect it may well be said that Jack's model was a remarkable contribution to progress, not so much because of the volume of firm knowledge that it conveyed as because it represented a daring and imaginative approach to the understanding of an important and exceedingly complex meteorological phenomenon."

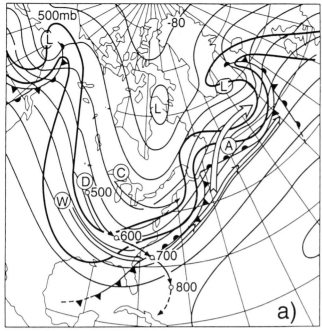

Palmén 1951

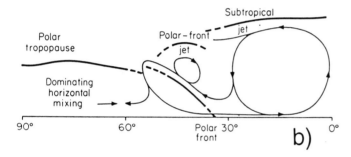

FIG. 13. (a) Schematic surface, 700- and 500-mb contours of frontal zone, and 500-mb geopotential contours, of a long wave with cyclone family. Broad arrow A (added), ascending air over warm front; D, descending trajectory in cold air, with successive pressure levels (mb). Air parcel remains in cold air as surface front advances. (Palmén 1951a.) (b) Quasi-meridional cells of the general circulation according to Palmén (1951b), in which airflow in middle latitudes is shown as if averaged relative to the meandering polar front zone, rather than the zonally averaged Ferrel cell.

Acknowledgments

We are indebted to Col. Robert C. Bundgaard for providing excerpts from Petterssen's autobiography; to Patrick E. Hughes for photos of Rossby and Reichelderfer; to Elizabeth Stephens for skillfully processing successive manuscripts; and to Suzanne Whitman for alterations to the figures.

References

Bates, C. C., 1989: The formative Rossby–Reichelderfer period in American meteorology, 1926–40. *Wea. Forecasting*, **4**, 593–603.

Beck, A. L., 1922: The earth's atmosphere as a circular vortex. *Mon. Wea. Rev.*, **50**, 393–401.

Bergeron, T., 1959a: Methods in scientific weather analysis and forecasting. An outline in the history of ideas and hints at a program. *The Atmosphere and the Sea in Motion*, B. Bolin, Ed., Rockefeller Institute Press, 440–474.

——, 1959b: The young Carl-Gustaf Rossby. *The Atmosphere and the Sea in Motion*, B. Bolin, Ed., Rockefeller Institute Press, 51–55.

Bigelow, F. H., 1902: Studies on the statics and kinematics of the atmosphere in the United States. III. The observed circulation of the atmosphere in the high and low areas. *Mon. Wea. Rev.*, **30**, 117–125.

——, 1903: IV. The mechanism of countercurrents of different temperatures in cyclones and anticyclones. *Mon. Wea. Rev.*, **31**, 72–84.

Bjerknes, J., 1919: On the structure of moving cyclones. *Mon. Wea. Rev.*, **47**, 95–99.

——, 1937: Theorie der aussertropischen Zyklonenbildung. *Meteor. Z.*, **54**, 462–466.

——, and H. Solberg, 1923: Life cycle of cyclones and the polar front theory of atmospheric circulation. *Geofys. Publ.*, **3**, 3–18.

——, and E. Palmén, 1937: Investigations of selected European cyclones by means of serial ascents. *Geofys. Publ.*, **12**, 5–62.

——, and J. Holmboe, 1944: On the theory of cyclones. *J. Meteor.*, **1**, 1–22.

Bjerknes, V., 1919a: Weather forecasting. *Mon. Wea. Rev.*, **47**, 90–95.

——, 1919b: Possible improvements in weather forecasting. *Mon. Wea. Rev.*, **47**, 99–100.

Bôcher, M., 1888: The meteorological labors of Dove, Redfield and Espy. *Amer. Meteor. J.*, **5**, 1–13. (Reprinted, *Bull. Amer. Meteor. Soc.*, **46**, 448–452.)

Bowie, E. H., and R. H. Weightman, 1914: Types of storms of the United States and their average movements. *Mon. Wea. Rev., Supplement No. 1*, 37 pp + 114 charts. (Abridged version, 12 pp, Inst. Meteor. Univ. Chicago, 1943.)

Byers, H. R., 1959: Carl-Gustaf Rossby, the organizer. *The Atmosphere and the Sea in Motion*, B. Bolin, Ed. Rockefeller Institute Press, 56–59.

——, 1970: Recollections of the war years. *Bull. Amer. Meteor. Soc.*, **51**, 214–217.

Davis, W. M., 1884: In "Notes and news." *Science*, **3**, 556.

Espy, J. P., 1841: *Philosophy of Storms*. Little and Brown, 552 pp.

Ferrel, W., 1856: An essay on the winds and the currents of the oceans. *Nashville J. Medicine and Surgery*, **11**, 287–301, 375–389.

——, 1860: The motions of fluids and solids relative to the earth's surface. *The Math. Monthly*, **2**, 72.

Finley, J. P., 1884: Character of six hundred tornadoes. Prof. Paper No. 7, U.S. Signal Service, 29 pp.

Friedman, R. M., 1989: *Appropriating the Weather: Vilhelm Bjerknes and the Construction of a Modern Meteorology*. Cornell University Press, Ithaca, N.Y., 251 pp.

Galway, J. G., 1985: J. P. Finley, the first severe storms forecaster. *Bull. Amer. Meteor. Soc.*, **66**, 1506–1510.

Hann, J. V., 1915: *Lehrbuch der Meteorologie*, 3d ed. Verlag Tauchnitz, 847 pp.

Haurwitz, B., 1985: Meteorology in the 20th century. A participant's view. *Bull. Amer. Meteor. Soc.*, **66**, 424–431.

Henry, A. J., 1922a: Discussion (of Beck, 1922). *Mon. Wea. Rev.*, **50**, 401.

——, 1922b: J. Bjerknes and H. Solberg on the life cycle of cyclones and the polar front theory of atmospheric circulation. *Mon. Wea. Rev.*, **50**, 468–474.

——, E. H. Bowie, H. J. Cox, and H. C. Frankenfield, 1916: *Weather Forecasting in the United States*. U.S. Weather Bureau, U.S. Government Printing Office, 370 pp.

Hughes, P., 1970: *A Century of Weather Service*. Gordon and Breach, 212 pp.

——, 1983: Francis W. Reichelderfer, 1895–1983. *EOS, Trans. Amer. Geophys. Union*, May 31.

——, 1995: The new meteorology. *Weatherwise*, **48**, 26–36.

Kutzbach, G., 1979: *The Thermal Theory of Cyclones: A History of Meteorological Thought in the Nineteenth Century*. Amer. Meteor. Soc., 254 pp.

Lewis, J. M., 1995: LeRoy Meisinger, Part I: Biographical tribute with an assessment of his contributions to meteorology. *Bull. Amer. Meteor. Soc.*, **76**, 33–45.

Loomis, E., 1841: On the storm which was experienced throughout the United States about the 20th of December, 1836. *Trans. Amer. Philos. Soc.*, **7**, 125.

——, 1846: On two storms which were experienced throughout the United States in the month of February, 1842. *Trans. Am. Philos. Soc.*, **9**, 161–184.

McDonald, J. E., 1963: James Espy and the beginnings of cloud thermodynamics. *Bull. Amer. Meteor. Soc.*, **44**, 634–641.

Meisinger, C. L., 1920a: Precipitation of sleet and formation of glaze in Eastern United States, 1920. *Mon. Wea. Rev.*, **48**, 73–80.

——, 1920b: The great cyclone of mid-February, 1919. *Mon. Wea. Rev.*, **48**, 582–586.

Miller, E. R., 1931: The pioneer meteorological work of Elias Loomis at Western Reserve College, Hudson, Ohio, 1837–1844. *Mon. Wea. Rev.*, **59**, 194–195.

Millikan, R. A., I. Bowman, K. T. Compton, and C. D. Reed, 1933: Preliminary report of the special committee on the Weather Bureau of the Science Advisory Board. *Bull. Amer. Meteor. Soc.*, **14**, 273–281.

Mohn, H., 1870: *Det Norske Meteorologiske Instituts Storm-Atlas*. B.M. Bentzen, Christiania.

——, 1872: *Om vind og Vejr*. Mallings Bogtrykkeri, Christiania.

Namias, J., 1940: *Air Mass and Isentropic Analysis*. 5th ed. Amer. Meteor. Soc., 232 pp.

——, 1983: The history of polar front and air mass concepts in the United States—An eyewitness account. *Bull. Amer. Meteor. Soc.*, **64**, 734–755.

Palmén, E., 1951a: The aerology of extratropical cyclones. In: *Compendium of Meteorology*, T. F. Malone, Ed., Amer. Meteor. Soc., 599–620.

——, 1951b: The rôle of atmospheric disturbances in the general circulation. *Quart. J. Roy. Meteor. Soc.*, **77**, 337–354.

Petterssen, S., 1940: *Weather Analysis and Forecasting*. McGraw-Hill Book Co., 505 pp.

——, 1974: *Kuling fra Nord*. H. Aschehoug (W. Nygaard), Oslo, 312 pp. [English translation, Excerpts on the "Bergen School," 10 pp. typescript, kindly furnished by R. C. Bundgaard.]

Redfield, W. C., 1831: Remarks on the prevailing storms of the Atlantic coast of the North American States. *Amer. J. Sci.*, **20**, 17–51.

Reichelderfer, F. W., 1928: Postgraduate course in aerology and meteorology for naval officers. *Bull. Amer. Meteor. Soc.*, **9**, 149–151.

Rossby, C.-G., and R. H. Weightman, 1926: Application of the polar-front theory to a series of American weather maps. *Mon. Wea. Rev.*, **54**, 485–496 + maps.

Schwerdtfeger, W., 1981: Comments on Tor Bergeron's contributions to synoptic meteorology. *Pure Appl. Geophys.*, **119**, 501–509. [Also in *Weather and Weather Maps* (Bergeron memorial volume), G. H. Liljequist, Ed., Birkhäuser Verlag, Basel, 1981.]

Taba, H., 1988: *The 'Bulletin' Interviews*. World Meteorological Organization, 405 pp.

Tracy, C., 1843: On the rotary action of storms. *Amer. J. Sci. and Arts*, **45**, 65–72. (Reprinted in Abbe, C., 1910: *The Mechanics of the Earth's Atmosphere*, 3rd collection, Smithsonian Misc. Coll., 51, 16–22.)

U.S. Weather Bureau, 1934: *Report of the Chief of the Weather Bureau 1933–34*. U.S. Government Printing Office.

Weightman, R. H., 1925: Some observations on the cyclonic precipitation of February 22–23, 1925, in the central and eastern United States. *Mon. Wea. Rev.*, **53**, 379–384.

Whitnah, D. R., 1961: *A History of the United States Weather Bureau*. University of Illinois Press, 267 pp.

Richardson's Marvelous Forecast

PETER LYNCH

Irish Meteorological Service, Dublin, Ireland

1. Introduction

While Vilhelm Bjerknes and his team were developing their synoptic models in Bergen, a radically different approach to forecasting was being pursued by Lewis Fry Richardson. Richardson's starting point was the system of fundamental physical principles governing atmospheric motion. He assembled the set of mathematical equations that represent these principles and formulated an approximate algebraic method of calculating their solution. Starting from the state of the atmosphere at a given time—the initial conditions—the method could be used to work out its future evolution.

Using the most complete set of observations available to him, Richardson applied his numerical method and calculated the changes in the pressure and winds at two points in central Europe. The results were something of a calamity: Richardson calculated a change in surface pressure over a six-hour period of 145 hPa, a totally unrealistic value. As Sir Napier Shaw remarked, the wildest guess would not have been wider of the mark! Despite the "glaring errors" in his forecast, Richardson was bold enough to publish his method and results in his remarkable *Weather Prediction by Numerical Process* (Richardson 1922; hereafter LFR). This profound, and occasionally whimsical, book is a treasure-store of original and thought-provoking ideas and amply repays the effort required to read it.

The application of Richardson's forecasting method involved an enormous amount of numerical computation. Even the limited results he obtained cost him some two years of arduous calculation (Lynch 1993). This work was carried out in the Champagne district of France where Richardson served as an ambulance driver during the Great War (Ashford 1985). His dedication and tenacity in the dreadful conditions of the war are an inspiration to those of us who work in more genial conditions.

In this chapter the results obtained by Richardson are examined and the causes of the errors in his forecast are explained. It is shown how a realistic forecast can be obtained by modifying the initial data. The study is based on the original observations for 20 May 1910, originally compiled by Hugo Hergessel and analyzed by Vilhelm Bjerknes. These are used to extend the table of values published by Richardson, to cover most of Europe. A numerical model is then constructed, keeping as close as possible to the method of Richardson, except for omission of minor physical processes. When the model is run with the extended data, the results are virtually identical to those of Richardson. In particular, an initial pressure tendency of 145 hPa in 6 hours is obtained at the central point, in agreement with Richardson. The tendency values are unrealistic, being generally about two orders of magnitude too large.

The reasons for the spurious tendencies will be discussed. They are essentially due to an imbalance between the pressure and wind fields resulting in large-amplitude high-frequency gravity wave oscillations. The "cure" is to modify the analysis so as to restore balance; this process is called *initialization*. An initialization method based on a digital filter will be outlined, and its application to Richardson's problem described. The forecast tendency from the modified data yields reasonable results. In particular, the tendency at the central point is reduced to 3 hPa per 6 hours—a realistic value! The chapter concludes with some speculations about *what-might-have-been* had Richardson been able to initialize his data.

2. Observations and Initial Fields

2.1 Observational Data

The forecast made by Richardson was based on "one of the most complete sets of observations on record" (LFR, p. 181). During the first decades of the century observations of conditions at the earth's surface were made on a regular basis, and daily surface weather maps were issued by several centers. Upper-air observations were made only intermittently, typically for one or a few days each month, as agreed by the countries participating in the work of the International Commission for Scientific Aeronautics (ICSA). The data were compiled and published by the Meteorological Institute of Strasbourg, under the editorship of Hugo Hergessel, Director of the Institute and President of ICSA.

A detailed analysis of the aerological observations was undertaken by Vilhelm Bjerknes at the Geophysical Institute in Leipzig. He produced a publication series consisting of sets of charts of atmospheric conditions at ten standard pressure levels from 100 hPa to 1000 hPa. These charts provided Richardson with the data required for his arithmetical forecasting procedure. The initial time and date chosen by Richardson for his forecast was 0700 UTC 20 May 1910. For this time there were 12 soundings and 18 reports of upper level winds over western Europe.[1] The observations are tabulated in a synopsized form in Bjerknes (1914).

2.2　Preparation of the Initial Fields

Richardson chose to divide the atmosphere into five layers, centered approximately at pressures 900, 700, 500, 300, and 100 hPa. He divided each layer into boxes and assumed that the value of a variable in each box could be represented by its value at the central point; we refer to such points as gridpoints. They were separated by $\Delta\lambda = 3.0°$ in longitude and $\Delta\phi = 1.8°$ in latitude. Richardson tabulated his initial values for a selection of points over central Europe. The area is shown on a map on page 184 of LFR, and the values are given in his "Table of Initial Distribution" on page 185.

In Section 9/1 of LFR, Richardson describes the various steps he took in preparing his initial data. He prepared the mass and wind analyses independently (univariate analysis). The initial fields used in the present study were obtained from the same source, but we did not follow precisely the method of Richardson; the procedure adopted is outlined below. In order that the geostrophic relationship should not be allowed to dominate the choice of values, the pressure and velocity analyses were performed separately (and by two different people; see Acknowledgment).

The initial pressure fields were derived from Bjerknes's charts of geopotential height at 200, 400, 600, and 800 hPa (his charts 6, 8, 10, and 12). A transparent sheet marked with the gridpoints was superimposed on each chart and the height at each point read off. Each level p_k corresponds to a standard height z_k with temperature T_k. Conversion from height z to pressure p was made using the simple formula

$$p = p_k\left(1 + \frac{z - z_k}{H_k}\right)$$

---

where $H_k = \Re T_k/g$. Sea-level pressure values were extracted in the same way as heights, from Bjerknes's chart number 1. His values in mm Hg were converted to hectopascals by multiplication by 4/3. Then the surface pressure p_S was calculated from

$$p_S = p_{SEA}\left(1 - \frac{\gamma h}{T_0}\right)^{g/\gamma\Re}$$

where h is orographic height at the point in question, and standard values $T_0 = 288$ K and $\gamma = 0.0065$ K m^{-1} were used for the surface temperature and vertical lapse-rate.

The initial values of momenta for each of the five layers are required. These were derived from the wind velocities at the intermediate levels 100, 300, 500, 700, and 900 hPa. The observed wind speeds and directions for each level were plotted on charts upon which isotachs and isogons were then drawn by hand. The gridpoint values of speed and direction could now be read off. It was necessary to exercise a degree of imagination as the observational coverage was so limited. The wind values were converted to components u and v and the layer momenta U and V were defined by

$$U = \frac{\Delta p}{g}u \quad V = \frac{\Delta p}{g}v$$

where Δp is the pressure across the layer (obtained in the pressure analysis).

The pressure, temperature, and momentum values, at a selection of points in the center of the domain, resulting from the reanalysis, are given in Table 1. The corresponding values obtained and used by Richardson, extracted from his "Table of Initial Distribution" (LFR, p. 185), are reproduced in Table 2. The stratospheric temperatures and orographic heights are also indicated (top and bottom numbers in each white block). To facilitate comparison, the orography values used by Richardson were used (where available) in the re-analysis.

There is reasonable agreement between the pressure and stratospheric temperature values in the two tables. In general, pressure differences are within one or two hectopascals. There is a notable exception at the point (48.6°N, 5.0°E), where the old and new values differ by 10 hPa. We will see below that Richardson's value at this point is suspect.

Comparing the momenta in Tables 1 and 2, we see more significant discrepancies. Although the overall flow suggested by the momenta is similar in each case, point values are radically different from each other, with variations as large as the values themselves and occasional differences of sign. These dissimilarities arise partly from the different analysis procedures used, but mainly from the large margin of error involved in the interpolation from the very few observations to the grid-points. In repeating the forecast, we

TABLE 1. Initial distribution, reanalyzed values

	5° E	8° E	11° E	14° E	17° E
54.0° N			106 -228 120 -144 0 -81 -97 0 -221 81 0		
52.2° N		-62 -138 -133 -135 -155 150	212 206 410 609 799 987 200	-25 -79 -107 -156 -181 100	
50.4° N	-175 208 -292 263 -249 174 -118 99 -88 51 200	216 205 409 607 796 983 200	-105 -182 -268 -38 -201 -18 -199 73 -127 73 400	212 206 410 608 798 976 300	-126 -218 -167 -213 -155 -130 -214 0 -175 82 300
48.6° N	221 204 406 605 793 984 200	-159 -275 -216 -131 -60 400	214 206 410 608 796 961 400	-131 -205 -147 -129 -81 400	213 206 410 608 798 989 200
46.8° N		217 204 405 604 794 872 1200	-208 18 -289 0 -172 172 -45 64 -32 38 1800	213 205 407 606 796 842 1500	
45.0° N			210 203 404 603 795 995 100		

TABLE 2. Initial distribution, Richardson's values

	5°E	8°E	11°E	14°E	17°E
54.0°N			-65 8 127 -104 81 -25 -81 0 -198 84 0		
52.2°N		-70 -62 -114 -91 -160· 150	214 205 409 609 798 988 200	-160 40 -60 -60 -219 100	
50.4°N	-30 -110 -245 300 -223 158 -91 87 -18 15 200	212 205 408 607 795 983 200	-56 -18 -146 -62 -95 29 -52 58 -110 55 400	214 205 409 609 798 976 300	-100 -32 0 -260 -55 -135 -25 48 -190 160 300
48.6°N	214 203 405 604 793 974 200	27 -328 -136 -33 48 400	212 205 409 608 796 963 400	0 -166 -95 -19 -65 400	214 204 408 607 798 988 200
46.8°N		214 204 406 605 795 875 1200	-50 80 -280 41 -175 150 -105 80 -155 40 1800	214 204 408 607 797 846 1500	
45.0°N			213 203 403 603 796 997 100		

have simply replaced the reanalyzed values of all fields by Richardson's original values at the few gridpoints where the latter are available. The values in Table 2 are thus the initial values for both Richardson's forecast and the forecasts described below.

3. The Fundamental Equations

The behavior of the atmosphere is governed by the fundamental principles of conservation of mass, energy and momentum. These principles may be expressed in terms of differential equations. The idea of solving the equations to calculate future weather was propounded in a famous address by Bjerknes (1904). The first attempt to put this idea into practice was that of Richardson.

Richardson was careful not to make any unnecessary approximations, and he took account of several physical processes, which had the most marginal effect on his forecast. He included in his equations many terms that are negligible; with the benefit of hindsight, we can omit most of these. We shall ignore all the effects of moisture and thermal forcing, and consider the adiabatic evolution of a dry atmosphere. One fundamental approximation was made by Richardson: the atmosphere is in a state of hydrostatic balance. This was an essential step, necessitated by the lack of observations of vertical velocity, and it enabled Richardson to derive his elegant diagnostic equation for this quantity.

We shall set out the basic equations as commonly used today, and then convert them to the form used by Richardson. Some of Richardson's notation is archaic and the modern equivalents will be used (a full table of his notation is found in Chap. 12 of LFR). The independent variables are latitude ϕ, longitude λ, height z, and time t. Distances eastward and northward are denoted x and y. The dependent variables are the eastward, northward, and upward components of velocity (u, v, w), pressure p, temperature T, and density ρ.

3.1. The Primitive Equations

The primitive equations may be found in standard texts on dynamic meteorology. The equations of motion are:

$$\frac{du}{dt} - \left(f + \frac{u \tan \phi}{a} \right) v + \frac{1}{\rho} \frac{\partial p}{\partial x} = 0$$

$$\frac{dv}{dt} + \left(f + \frac{u \tan \phi}{a} \right) u + \frac{1}{\rho} \frac{\partial p}{\partial y} = 0$$

$$\frac{\partial p}{\partial z} + g\rho = 0$$

The Earth's radius is a, its angular velocity is Ω, and $f = 2\Omega \sin\phi$ is the Coriolis parameter.

The continuity equation, expressing conservation of mass, is

$$\frac{\partial \rho}{\partial t} + \frac{\partial \rho u}{\partial x} + \frac{\partial \rho v}{\partial x} - \frac{\rho v \tan \phi}{a} + \frac{\partial \rho w}{\partial z} = 0$$

(a small term $2\rho w/a$ has been dropped). In combination with this equation, the horizontal equations of motion may be written in flux form:

$$\frac{\partial \rho u}{\partial t} + \frac{\partial \rho u^2}{\partial x} + \frac{\partial \rho u v}{\partial y} + \frac{\partial \rho u w}{\partial z} - \left(f + \frac{2u \tan \phi}{a} \right) \rho v + \frac{\partial p}{\partial x} = 0$$

$$\frac{\partial \rho v}{\partial t} + \frac{\partial \rho u v}{\partial x} + \frac{\partial \rho v^2}{\partial y} + \frac{\partial \rho v w}{\partial z}$$
$$+ f\rho u + \frac{\rho(u^2 - v^2)\tan\phi}{a} + \frac{\partial p}{\partial y} = 0.$$

The atmosphere is assumed to be a perfect gas:

$$p = \Re\rho T,$$

where $\Re$ is the gas constant for dry air. Using this "equation of state," the thermodynamic equation may be written

$$\frac{1}{\gamma p} \left(\frac{dp}{dt} \right) - \frac{1}{\rho} \left(\frac{d\rho}{dt} \right) = 0$$

where $\gamma = c_p/c_v$ is the ratio of specific heats.

3.2 "Finding the Vertical Velocity"

The vertical component of velocity in the atmosphere is typically two or three orders of magnitude smaller than the horizontal components. It is difficult to measure w and in general no observations of this variable are available. In particular, Richardson had no such observations for 0700 UTC 20 May 1910. Moreover, even if he had had such observations, he recognized the practical impossibility of computing the tendency $\partial w/\partial t$ which would have to be calculated as a tiny residual term in the vertical dynamical equation.

Richardson acknowledged the influence of Vilhelm Bjerknes's publications *Statics* and *Kinematics* (Bjerknes and Sandstrom 1910; Bjerknes et al. 1911) on his work. In his Preface (LFR, p. viii; Dover Edition, p. xii) Richardson states that his choice of "conventional strata," his use of specific momentum rather than velocity, his method of calculating vertical motion at ground level, and his adoption of the hydrostatic approximation are all in accordance with Bjerknes's ideas.

The hydrostatic equation results from neglecting the vertical acceleration, and other small terms, in the vertical dynamical equation. But this precludes the possibility of calculating the acceleration $\partial w/\partial t$ directly. It was a stroke of genius for Richardson not only to realize the need to evaluate w diagnostically from the other fields but also to construct a magnificent mathematical equation to achieve this.

To construct Richardson's w-equation we eliminate the time dependency between the continuity equation and the thermodynamic equation using the hydrostatic equation. Recall that the thermodynamic equation can be written in the form

$$\frac{1}{\gamma p}\left(\frac{\partial p}{\partial t} + \mathbf{V} \bullet \nabla p + w \frac{\partial p}{\partial z}\right) - \frac{1}{\rho}\frac{d\rho}{dt} = 0,$$

and that one of the various forms of the continuity equation is

$$\frac{1}{\rho}\frac{d\rho}{dt} + \left(\nabla \bullet \mathbf{V} + \frac{\partial w}{\partial z}\right) = 0.$$

We can eliminate the density between these and use the vertically integrated hydrostatic equation to get

$$\frac{1}{\gamma p}\left(-\int_z^\infty g\nabla \bullet \rho\mathbf{V}dz + \mathbf{V} \bullet \nabla p\right) + \left(\nabla \bullet \mathbf{V} + \frac{\partial w}{\partial z}\right) = 0.$$

Expanding the integrand and using the hydrostatic equation again we get

$$-\frac{1}{\gamma p}\int_z^\infty \left(g\rho\nabla \bullet \mathbf{V} + \frac{\partial \mathbf{V}}{\partial z} \bullet \nabla p\right)dz + \left(\nabla \bullet \mathbf{V} + \frac{\partial w}{\partial z}\right) = 0.$$

Since the upper limit of the integral is infinite, it is convenient to use pressure as the independent variable; this is done by using the hydrostatic equation once more, yielding the result:

$$\frac{\partial w}{\partial z} = -\nabla \bullet \mathbf{V} + \frac{1}{\gamma p}\int_0^p \left(\nabla \bullet \mathbf{V} - \frac{\partial \mathbf{V}}{\partial p} \bullet \nabla p\right)dp. \quad (1)$$

This corresponds to Eq. (9) on page 124 of LFR, save that we have omitted the effects of moisture and diabatic forcing, which were included by Richardson.

The solution of Eq. (1) for w is straightforward. The gradient $\partial w/\partial z$ is calculated for each layer, working downwards from the stratosphere since the integral vanishes at $p = 0$. Then w may be calculated at the interface of each layer, working upwards, once it is known at the earth's surface. Richardson followed Bjerknes in taking the surface value

$$w_S = (\mathbf{v} \bullet \nabla h)_S.$$

This is equivalent to the kinematic condition that the ground is impervious to the wind. However, Richardson does not state how he evaluates $\mathbf{v}_S$; in repeating his forecast we have assumed a simple relationship $\mathbf{v}_S = k\mathbf{v}_5$ where $\mathbf{v}_5$ is the lowest layer velocity and $k = 0.2$.

The vertical velocity equation was a major contribution by Richardson to dynamic meteorology. In recognizing its essential rôle in his forecast scheme he observed (LFR, p. 178) that "it might be called the keystone of the whole system, as so many other equations remain incomplete until the vertical velocity has been inserted."

3.3. Temperature in the Stratosphere

Richardson devoted a full chapter of 24 pages to the stratosphere. We shall not discuss the bulk of this, but we must consider the means by which the temperature of the uppermost layer is forecast. For, in the scheme adopted by Richardson, the vertical integral of pressure through the stratospheric layer depends on the temperature so that prediction of the latter is essential to ensure a "lattice-reproducing" scheme—that is, an algorithm which, starting with a set of variables at one instant, produces the corresponding set at a later instant.

Richardson calculated the change in stratospheric temperature using two different equations, his elaborate equation (8) on page 147 of LFR and a much simpler equation corresponding to (14) on page 143. The resulting temperature tendencies, given in his Computing Form P_{XIV} on page 201, were 9.1×10^{-4} K s^{-1} for the elaborate equation and 9.2×10^{-4} K s^{-1} for the simpler. In view of this close agreement, we shall confine attention to the simpler alternative

$$\frac{\partial T}{\partial t} = T\frac{\partial w}{\partial z}. \quad (2)$$

This equation is sufficient for predicting the stratospheric temperature as long as the assumptions of geostrophy and vertical isothermy are acceptable. We shall use this simple prognostic equation in the sequel.

4. "The Arrangement of Points and Instants"

Richardson chose to divide the atmosphere into five strata of approximately equal mass, separated by horizontal surfaces at 2.0, 4.2, 7.2, and 11.8 km, corresponding to the mean heights (over Europe) of the 800, 600, 400, and 200 hPa surfaces. He discusses this choice in LFR, Section 3/2. It is desirable to have a surface near the tropopause, one stratum for the planetary boundary layer and at least two more for the troposphere above the boundary layer. Taking layers of equal

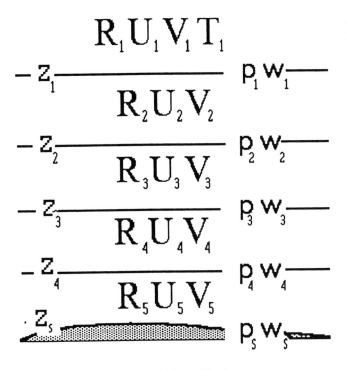

FIG. 1. Vertical stratification.

mass simplifies the treatment of processes such as radiation, and the particular choice of surfaces at approximately 2, 4, 6, and 8 decibars greatly facilitates the extraction of initial data from the charts and tables of Bjerknes. The strata are depicted in Fig. 1. Each horizontal layer is divided up into rectangular boxes or grid cells. Richardson selected boxes with sides of length $\Delta\lambda = 3°$ in the east-west direction and 200 km (or $\Delta\phi = 1.8°$) in the north-south direction.

The numerical integration of the equations is carried out by a step-by-step procedure—an algorithm—which produces later values from earlier ones. Richardson took pains to devise a numerical scheme such that where a particular variable was given at an initial time, the corresponding value at a later time at the same point could be computed. His scheme is best illustrated for the linear shallow water equations:

$$\frac{\partial U}{\partial t} - fV + \frac{\partial P}{\partial x} = 0$$

$$\frac{\partial V}{\partial t} + fU + \frac{\partial P}{\partial y} = 0$$

$$\frac{\partial P}{\partial t} + gH\nabla \cdot \mathbf{U} = 0.$$

The tendency of each component of momentum depends on the other momentum component and on the gradient of pressure. Thus, U and V should be specified at the same

points, and these points should be intermeshed with those where P is given:

U, V	P	U, V
P	U, V	P
U, V	P	U, V

This arrangement is also ideal for the continuity equation: the tendency of pressure P depends on the divergence of momentum which is (primarily) comprised of horizontal derivatives of momentum, so P should be evaluated at points intermediate between those where momentum is given. This staggered arrangement of points is known today as an E-grid. Platzman's proposal to call it a Richardson grid has much to commend it (Platzman 1967). When the values of a variable are specified on the discrete grid, spatial derivatives may be calculated approximately by means of finite differences. For example, the derivatives of P are, to second order accuracy,

$$\left(\frac{\partial P}{\partial x}\right)_{ij} = \left(\frac{P_{i+1,j} - P_{i-1,j}}{2a\cos\phi_j\Delta\lambda}\right), \quad \left(\frac{\partial P}{\partial y}\right)_{ij} = \left(\frac{P_{i,j+1} - P_{i,j-1}}{2a\Delta\phi}\right),$$

where $P_{i,j}$ is the value of P at the point $(i\Delta\lambda, j\Delta\phi)$.

The geographical coverage used in repeating Richardson's forecast is shown in Fig. 2. P-points are indicated by solid

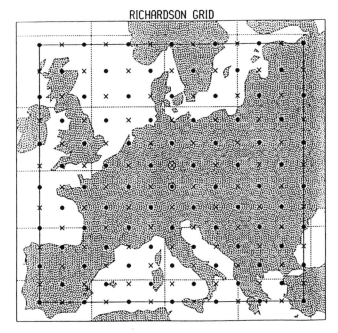

FIG. 2. Horizontal grid and geographical coverage.

circles and M-points by crosses. The region was chosen to best fulfill conflicting requirements: that it be as large as possible; that data coverage over the area be adequate, and that the points used by Richardson be located centrally in the region. The absence of observations precluded the extension of the region beyond that shown. The P-point and M-point for which Richardson calculated his tendencies are encircled.

The method of calculating the dependent quantities at each new time-level will now be described. It is performed by means of the familiar leapfrog scheme, called by Richardson the step-over method (LFR, p. 150). The prognostic variables are R, the mass per unit area, and U and V, the components of momentum per unit area, for each stratum; and the temperature T_1 of the stratosphere. Let Q denote a typical dependent variable; it is governed by an equation of the form

$$\frac{\partial Q}{\partial t} = F$$

where F, the tendency of Q, is a function of Q and the other dependent variables. Let us assume that all the dependent variables are known at time $t = n\Delta t$ so that $F^n = F(n\Delta t)$ can be computed, and that the value of Q at the previous time level $t = (n-1)\Delta t$ has been retained. Then the forecast value Q^{n+1} may be computed from the old value Q^{n-1} and the tendency F^n:

$$Q^{n+1} = Q^{n-1} + 2\Delta t F^n.$$

The first step forward cannot be made with the leapfrog scheme, since the variables are known only at $t = 0$. A simple noncentered step

$$Q^1 = Q^0 + \Delta t\, F^0$$

provides values of the variables at $t = \Delta t$; from then on, the leapfrog scheme can be used.

The calculations of Richardson were confined to the evaluation of the initial tendencies (at 0700 UTC 20 May 1910). He multiplied these by a time interval $2\Delta t = 6$ h to represent the change over the six-hour period centered at 0700 UTC. In modern terminology, the time-step is specified as the interval between adjacent evaluations of the variables; thus, the time-step used by Richardson was three hours, not six hours as so often stated. A three-hour step was also chosen by him in describing his fantastic forecast factory (LFR, p. 219).

5. The Equations for the Strata

As we have seen, Richardson divided the atmosphere into five "conventional strata" separated by horizontal surfaces at fixed heights 2.0, 4.2, 7.2, and 11.8 km. These heights will be denoted respectively by z_4, z_3, z_2, *and* z_1, all constants. The variable height of the earth's surface will be written $z_5 = h(\lambda, \phi)$. Variables at these five levels will be denoted by corresponding indices 1–5. Where convenient, values at the surface of the earth may be indicated by subscript S.

The equations of motion will be integrated with respect to height across each stratum, to obtain expressions applying to the stratum as a whole. Quantities derived by integrating in this way will be denoted by capitals:

$$R = \int \rho\, dz \quad P = \int p\, dz \quad U = \int \rho u\, dz \quad V = \int \rho v\, dz.$$

The stratum is specified by the index corresponding to the *lower* level; thus, for example,

$$R_3 = \int\limits_{z_3}^{z_2} \rho\, dz.$$

In differentiating mean values for the lowest layer, allowance must be made for the variation of the height h of the earth's surface. For the other layers, the limits are independent of x and y.

The continuity equation will now be integrated in the vertical. Taking, for example, the stratum between z_3 and z_2, and using the definitions of R, U, and V, we get

$$\frac{\partial R_3}{\partial t} + \frac{\partial U_3}{\partial x} + \frac{\partial V_3}{\partial y} - \frac{V_3 \tan\phi}{a} + [\rho w]_2 - [\rho w]_3 = 0. \quad (3)$$

The equations for the other upper layers are of similar form. For the lowest layer the slope of the bottom boundary must be considered, and the term $-[\rho w]_S$ is cancelled by a term $[\rho\mathbf{v} \cdot \nabla h]_S$.

The vertical integration of the horizontal equations of motion is performed in the same manner. To express the result in terms of the variables R, U, and V it is necessary to make an approximation in the horizontal flux terms (see LFR, p. 34). With this approximation the equations for the stratum (z_3, z_2) are

$$\frac{\partial U}{\partial t} + \frac{\partial}{\partial x}\left(\frac{U^2}{R}\right) + \frac{\partial}{\partial y}\left(\frac{UV}{R}\right) + [\rho uw]_2 - [\rho uw]_3$$
$$-\left(f + \frac{2U\tan\phi}{aR}\right)V + \frac{\partial P}{\partial x} = 0 \quad (4)$$

$$\frac{\partial V}{\partial t} + \frac{\partial}{\partial x}\left(\frac{UV}{R}\right) + \frac{\partial}{\partial y}\left(\frac{V^2}{R}\right) + [\rho vw]_2 - [\rho vw]_3$$
$$+fU + \frac{(U^2 - V^2)\tan\phi}{aR} + \frac{\partial P}{\partial y} = 0 \quad (5)$$

The equations for the other upper layers are of similar form. For the lowest layer the slope of the bottom boundary must be considered.

6. "Review of Operations in Sequence"

The title of this section is identical to that of LFR, Chap. 8, in which Richardson traces, step by step, the sequence of calculations necessary to carry his forecast forward in time. Let us assume that all the dependent variables are known at time $t = n\Delta t$. The advancement to the next time level, $t = (n + 1)\Delta t$, requires both prognostic and diagnostic components (these terms, borrowed from medicine, were introduced by Vilhelm Bjerknes). The prognostic variables are (at P-points) the mass per unit area R in each stratum and the stratospheric temperature T_1, and (at M-points) the components U and V of momentum in each stratum. Once these quantities are known for a particular moment, all the auxiliary fields (temperature, divergence, vertical velocity, etc.) for that moment can be calculated from diagnostic relationships.

The time-stepping calculations are done in a large loop, which is repeated as often as required to reach the forecast span. The sequence of calculations will now be given. For each step, the number of the relevant Computing Form in LFR is indicated [in brackets]. First we consider the P-points.

1. [P_I] The layer integral of pressure is calculated:

$$P = \Delta z \bullet \Delta p / \Delta \log p$$

Here Δ represents the difference in value across the layer. For the top layer $P = \Re T_1 / g$. The density integral is also calculated from

$$R = \frac{\Delta p}{g}.$$

2. [P_I] Mean values for each stratum are calculated for various quantities, e.g.,

$$\bar{p} = \frac{P}{\Delta z} \qquad \bar{\rho} = \frac{R}{\Delta z} \qquad \bar{T} = \frac{\bar{p}}{\Re \bar{\rho}}$$

3. [P_{XIII}] The divergence of momentum $\nabla \bullet \mathbf{U}$ is computed for each level.

4. [P_{XIII}] The values of $\nabla \bullet \mathbf{U}$ in the column above each P-point are summed up and the total multiplied by $-g$ to give the surface pressure tendency

$$\frac{\partial p_S}{\partial t} = -g \sum_{\text{all}}^{\text{strata}} \nabla \bullet \mathbf{U}$$

5. [P_{XV}] The divergence of velocity $\delta = \nabla \bullet \mathbf{V}$ is calculated using the following approximations for mean velocity in each layer:

$$u = U / R \qquad v = V / R.$$

6. [P_{XVI}] The vertical velocity gradient $\partial w / \partial z$ in each layer is now calculated using (1). The vertical velocity at the surface is determined from

$$w_S = \mathbf{v}_S \bullet \nabla h$$

where we assume $\mathbf{v}_S = k\mathbf{U}_5 / R_5$ with $k = 0.2$. Then it is a straightforward matter to calculate w at each interface, working upward from the bottom.

7. [P_{XIV}] The tendency of the stratospheric temperature T_1 is calculated next, using Eq. (2).

8. [P_{XVI}] The temperature at each interface is calculated by linear interpolation. Then the density there is computed using the gas law, after which the momentum ρw at each interface can be obtained.

9. [P_{XIII}] The tendency of the density integral R is now obtained using the continuity equation (3).

10. [P_{XIII}] The final calculation at P-points is the tendency of pressure at each interface, obtained from

$$\left(\frac{\partial p}{\partial t}\right)_K = g \sum_{k=1}^{K} \left(\frac{\partial R_k}{\partial t}\right).$$

The surface pressure tendency, already computed in step 4, is confirmed here.

This completes the calculations required at the P-points. We now list the operations at the M-points [LFR, Computing Forms M_{III} and M_{IV}]

11. The pressure gradient is evaluated by calculating the spatial derivatives of the integrated pressure P. For the lowest stratum there is an extra term due to orography. The x-component is given by

$$\frac{1}{2\Delta x}\left\{\delta P + \frac{\delta h \bullet \delta p_S}{\delta \log p_S}\right\}$$

where δ here represents the difference across a distance $2\Delta x$. The y-component is analogous.

12. The Coriolis terms and those involving $\tan \phi$ are evaluated. All the necessary quantities are available at the relevant points.

13. The horizontal flux terms are calculated. It is necessary to approximate the derivatives by differences over a distance $4\Delta x$ or $4\Delta y$.

14. The vertical flux terms are calculated. The momentum flux above the uppermost layer is assumed to vanish.

15. The tendencies of momenta, $\partial U/\partial t$ and $\partial V/\partial t$ may now be calculated, as all the other terms in Eqs. (4) and (5) are known.

The tendencies of all prognostic variables are now known, and it is possible to update all the fields to the time $(n + 1)/\Delta t$. When this is done, the entire sequence of operations may be repeated in another time-step.

7. "An Example Worked on Computing Forms"

This section title refers to the elaborate set of 23 forms drawn up by Richardson for the arrangement of his calculations. They are included in LFR, pages 188–210, filled in with the values relevant to the two points to which his forecast applied. Richardson arranged, at his own expense, to have sets of blank forms printed so that they could be used by anyone wishing to carry out similar forecasts. I do not know if they were ever put to their intended use.

A computer program has been written to repeat and extend Richardson's forecast. The same initial values were used, so that the calculated initial changes could be compared directly with the values in LFR. It will be seen that the computer model produces results consistent with those obtained manually by Richardson. In particular, the "glaring error" in the surface pressure tendency is reproduced almost exactly by the model.

7.1 The Initial Tendencies

Richardson's computations were confined to the calculation of the initial tendencies at a single pair of points. These calculations amount to evaluating the right-hand sides of equations of the form

$$\frac{\partial Q}{\partial t} = F.$$

The leapfrog method of integration in time amounts to approximating this equation by

$$\Delta Q = [Q(t + \Delta t) - Q(t - \Delta t)] = F(t) \times 2\Delta t .$$

It is important to note that the tendency F is independent of Δt, so that the change ΔQ is directly proportional to the time-step. The time-step between successive calculations is $\Delta t = 3$ h and the changes given in LFR are over a six-hour period centered at the initial time 0700 UTC 20 May 1910.

On page 211 of LFR, Richardson presents his results for the changes in the prognostic variables at the two points. We first consider the changes of the pressures at each of the four

interfaces and at the earth's surface for the central P-point. The values obtained by Richardson are given in Table 3. The corresponding changes produced by the numerical model are also given. The units of pressure change are hPa per 6 h. It is evident that the changes computed by the model are in close agreement with Richardson's calculations.

The changes of momentum calculated by Richardson and those computed with the model are given in Table 4. The agreement is not as close as for the pressure changes, but there is broad agreement between the two sets of forecast changes. The remaining prognostic variable is the temperature of the uppermost layer. The change tabulated by Richardson using Eq. (2) was $\Delta T_1 = 19.9°$. The computer model used the same equation and gave a forecast change of $\Delta T_1 = 19.6°$, in close agreement with Richardson.[2]

7.2 The Source of the Problem

Richardson ascribed the unrealistic value of pressure tendency to errors in the observed winds which resulted in spuriously large values of calculated divergence. This is true as far as it goes. However, the problem is deeper: even if the

TABLE 3. Six-hour changes in pressure (units: hPa per 6 h)

Level	Richardson	Model
1	48.3	48.5
2	77.0	76.7
3	103.2	102.1
4	126.5	124.5
5	145.1	145.4

TABLE 4. Six-hour changes in momentum components (units: 10^3 kg m^{-1} s^{-1} per 6 h)

| Layer | Eastward | | Northward | |
	Richardson	Model	Richardson	Model
I	−73.0	−71.8	−33.7	−39.7
II	−19.6	−19.9	+23.8	+29.0
III	−8.9	−10.3	−13.8	−15.9
IV	−15.3	−13.7	−4.3	−4.1
V	−17.9	−22.5	+6.3	+7.1

[2]Was Richardsons $\Delta T_1 \approx 80°$/day the first-ever forecast of a stratospheric sudden warming?!!!

winds were modified to remove divergence completely at the initial time, large tendencies would soon be observed.

A subtle state of balance exists in the atmosphere between the pressure and wind fields, ensuring that the high-frequency gravity waves have much smaller amplitude than the rotational part of the flow. Minor errors in observational data can result in a disruption of the balance, and cause large gravity wave oscillations in the model solution. They are avoided by modifying the data to restore harmony between the fields. We will describe a simple method of achieving balance, apply it to Richardson's data, and show that it yields realistic results.

8. Digital Filter Initialization

To obtain reasonable values for the tendencies, we must reduce the high frequency components implicit in the initial data to realistic amplitudes. This process is called initialization. There are several ways to achieve it, one of the simplest being to use a digital filter. Such a filter was used by Lynch (1992) to initialize Richardson's barotropic data (see LFR, Chap. 2). We will apply the same technique below to the full baroclinic case.

Consider a function of time $f(t)$ with low- and high-frequency components. To filter out the high frequencies, we may proceed as follows:

1. calculate the Fourier transform $F(\omega)$
2. set coefficients of high frequencies to zero
3. calculate the inverse transform

Step 2 may be performed by multiplying $F(\omega)$ by an appropriate weighting function $H(\omega)$. Typically, $H(\omega)$ is a step function, equal to one for $|\omega| \leq \omega_c$ and zero for $|\omega| > \omega_c$, with ω_c the cutoff frequency. The three steps are equivalent to a convolution:

$$f^*(t) = h * f(t) = \int_{-\infty}^{+\infty} h(t-\tau)f(\tau)d\tau \,,$$

where $h(t) = \sin(\omega_c t)/\pi t$ is the inverse Fourier transform of $H(\omega)$. To evaluate this integral approximately at $t = 0$, we calculate $f(t)$ at a finite set of times $\{-N\Delta t, ..., -\Delta t, 0, \Delta t, ..., N\Delta t\}$ and compute the sum

$$f^*(0) = \sum_{n=-N}^{N} f_n h_{-n} \qquad (6)$$

As is well known, truncation of a Fourier series may result in Gibbs oscillations. These may be greatly reduced by means of an appropriate window. The response is improved if h_n is modified by the Lanczos window

$$w_n = \sin[n\pi/N+1]/(n\pi/N+1).$$

The method outlined above was used to calculate filtered fields of height and wind at the initial time. The numerical model was integrated six hours forward and six hours backward from the initial time, providing a sequence of values centered on $t = 0$ for each variable at each gridpoint. The cutoff was set at $\tau_c = 2\pi/\omega_c = 6$ h and $\Delta t = 300$ s, so that $N = 72$. Filtered fields $f^*(t)$ could then be calculated using Eq. (6).

9. "Smoothing the Initial Data"

The idea of filtering in time goes right back to Richardson, who proposed several methods of smoothing the data, one way being to take the average value of observations made at successive times (LFR, Chap. 10). The digital filtering method is similar, but the time series are generated by the model, and the filter is designed for optimal selectivity.

Figure 3 shows the sea level pressure based on an extension of Richardson's values. The curious low near Strasbourg appears to be due to an error made by Richardson in converting sea level to surface pressure. This is confirmed by an examination of the values in Table 2. The surface pressure at the point (48.6°N, 5.0°E) is seen to be suspiciously low. The pressure analysis after filtering is shown in Fig. 4. The changes induced by the initialization are seen to be small. However, notice the absence of the erroneous low near Strasbourg.

Platzman (1967) examined Richardson's results and discussed two problems contributing to the large pressure tendency: horizontal divergence values are too large, due to lack of cancellation between the terms; and there is a lack of compensation between convergence and divergence in the vertical. Table 5 shows the six-hour changes in pressure thickness for each level, and the contributions from the horizontal and vertical parts of divergence. The values are rather large, and there is little cancellation between them. Table 6 shows the corresponding figures for the filtered initial data. The values are all reduced, generally by about a factor of two. However, the changes are such that the compensation in the vertical between horizontal divergence at different levels is now much more complete. The result is that the surface pressure change is dramatically reduced in size, from 145 to 3 hPa—a realistic value. (The vertical convergence integrates to zero, making no contribution to surface pressure tendency.) Clearly, the compensation in the vertical is vital in achieving balance.

Recently, a more sophisticated filter has been applied to Richardson's data. Lynch (1997) used an optimal filter based on the Dolph window, with a three-hour span. The initial pressure tendency was, in this case, further reduced to a value of –0.9 hPa per 6 hours. Richardson reported observations showing that the barometer was almost steady in the region of the central point. Thus, the value produced with the Dolph filter is the more realistic result.

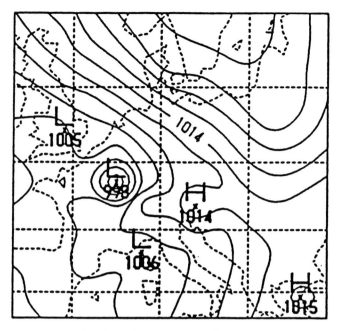

FIG. 3. Sea level pressure: original data

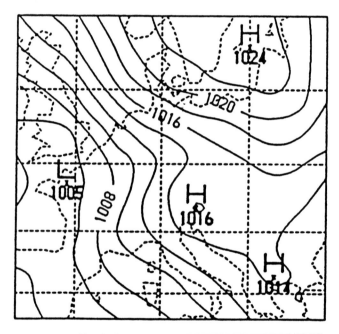

FIG. 4. Sea level pressure: filtered data

10. Concluding Remarks

The numerical model was used to extend the forecast to 24 hours. It was found that spatial smoothing was required to maintain stability. Moreover, a time-step consistent with the Courant-Friedrichs-Lewy stability criterion was required. Thus, lack of initialization is not the only shortcoming of the method devised by Richardson. The results of the extended forecast will be reported elsewhere.

TABLE 5. Six-hour changes in pressure thickness (Richardson's values: no initialization)

Layer	$(\partial \Delta p / \partial t)\Delta t$	Horizontal convergence	Vertical convergence
I	48.5	65.9	−17.4
II	28.4	−23.7	52.1
III	25.3	47.6	−22.3
IV	22.3	7.5	14.8
V	20.8	48.0	−27.2
Sum	145.4	145.4	0.0

TABLE 6. Six-hour changes in pressure thickness (after initialization by digital filter)

Layer	$(\partial \Delta p / \partial t)\Delta t$	Horizontal convergence	Vertical convergence
I	−1.0	12.4	−13.4
II	−2.4	−33.3	30.8
III	−0.9	11.0	−11.9
IV	−0.5	−10.4	9.8
V	1.7	17.0	−15.4
Sum	−3.2	−3.2	0.0

But let us suppose that Richardson had applied some filter, however crude, to his initial data. His results might well have been realistic, and his method would surely have been given the attention it certainly deserved. There can be little doubt that the failure of his trial forecast persuaded most meteorologists to ignore his work, so that his wonderful book gathered dust for many years. A more encouraging demonstration might have led his colleagues to consider his ideas more carefully and to investigate the potential usefulness of numerical forecasting in greater depth.[3] However, his fantastic forecast factory would hardly have come into being: even making no allowance for the short time-step required for stability, his figure of 64,000 "computers" required to keep pace with the weather was a serious underestimate (Lynch 1993).

Richardson claimed that his prediction was "a fairly correct deduction from a somewhat unnatural initial distribution." The model results presented above confirm that he was fully justified in making this claim, and that what he presented in his book was indeed a marvelous forecast.

---

[3]His venture in printing sets of blank forms might also have been more profitable.

Acknowledgment

Adrian Simmons kindly provided a review of this chapter. My thanks to Elías Hólm, who analyzed the winds without the luxury of seeing the heights.

References

Ashford, O. M., 1985: *Prophet—or Professor: The Life and Work of Lewis Fry Richardson.* Adam Hilger Ltd., 303 pp.

Bjerknes, V., 1904: Das Problem der Wettervorhersage, betrachtet vom Standpunkte der Mechanik und der Physik. *Meteor. Z.,* **21**, 1–7.

——, 1914: *Veröffentlichungen des Geophysikalischen Instituts der Universität Leipzig.* Erste Serie: *Synoptische Darstellungen atmosphärischer Zustände,* Jahrgang 1910, Heft 3.

——, and J. W. Sandström, 1910: Dynamic meteorology and hydrography. Part I: Statics. Carnegie Institution, Washington, Publication 88 (Part I), 146 pp. and tables.

——, Th. Hesselberg and O. Devik, 1911: Dynamic meteorology and hydrography. Part II: Kinematics. Carnegie Institution, Washington, Publication 88 (Part II), 175 pp. and Atlas of 60 plates.

Lynch, P., 1992: Richardson's barotropic forecast: A reappraisal. *Bull. Amer. Meteor. Soc.,* **73**, 35–47.

——, 1993: Richardson's forecast factory: The $64,000 question. *Meteor. Mag.,* **122**, 69–70.

——, 1997: The Dolph–Chebyshev window: A simple optimal filter. *Mon. Wea. Rev.,* **125**, 655–660.

Platzman, G. W., 1967: A retrospective view of Richardson's book on weather prediction. *Bull. Amer. Meteor. Soc.,* **48**, 514–550.

Richardson, L. F., 1922: *Weather Prediction by Numerical Process.* Cambridge Univ. Press, 236 pp.

Personal Recollections about the Bergen School

ERIK PALMÉN

Preface

In 1979 I succeeded in convincing Erik Palmén (1898–1985) that he should give an informal lecture about his participation in the activities of the "Bergen school" and "Chicago school" of meteorology. The lecture was organized in our department on February 21, 1979, and recorded. This report contains Palmén's personal recollections about the Bergen school. The second part of the lecture dealing with the Chicago school is not included here. The substance of this latter part is well reflected in the interview, published in 1981 by H. Taba (*WMO Bulletin*, Vol. 30, No. 2, pp. 92–100). Some additional comments by Erik Palmén on the Bergen school and the Chicago school have appeared in his article 'In my opinion . . .' from year 1985 (*Geophysica*, 21, 5-18), which also contains a list of Palmén's publications.

In his lecture, Erik Palmén spoke Finnish. Transcribing the tape recording first in Finnish and translating it then to English was done in 1993 by Mrs. Susanne Autio under my supervision. I have added some subtitles and, in certain places, also some explanatory words [in square brackets]. The English of the translation was checked by Mel Shapiro and Chester Newton.

Eero Holopainen

Introduction

When Eero Holopainen asked me to tell about my collaboration with the so-called Bergen school and the Chicago school, I was at first doubtful. I felt I'm nowadays just an old 'fossil' and all younger persons here are waiting to hear a lecture introducing something new. An old person just remembers what happened in the past. I want to emphasize that when I talk about the collaboration with these schools, I'll possibly talk too much about my own works. This does not mean that I would consider myself to be so special a person in this collaboration.

My first contact to the so-called Bergen school was in 1923. It was that year that one of the well-known members of the Bergen school, Tor Bergeron, visited Finland and gave a lecture on the new cyclone theory and the so-called polar front theory. Naturally I was already familiar with some publications concerning these theories.

I want to point out that at the time I was studying in the University there was no possibility of taking a degree in meteorology; in fact I have never officially studied meteorology. I felt I still was an amateur in this field of science. I also had just become an assistant in the Institute of Marine Research, so formally meteorology was not my field.

I became very interested in this new cyclone theory, and a while after Bergeron's visit I prepared a small article concerning the theory for the publication series of the Finnish Scientific Society. Later, in connection with some other works, I got more acquainted with this theory and in the year 1926 I defended my doctoral thesis, the name of which was "Über die Bewegung der aussertropischen Zyklonen." I did not have personal contact with the Bergen school until 1928. Before that, however, I had been corresponding with some of the members of the school.

But what actually was the Bergen school and how did it develop? I find it quite remarkable that three generations have worked in about the same field. Carl A. Bjerknes was the professor of mathematics in the University of Oslo (that time called Christiania). He became known for his hydrodynamic experiments, the purpose of which was to find an analogy to the effects of gravity. At that time it was generally assumed that there was so-called "ether" in space and that gravity and for example, light interact with it. He made many kinds of experiments with liquids, particularly with water. In the beginning of this century however, the ether concept was totally rejected.

His son, Vilhelm Bjerknes, began his career as an assistant to his father. Then he moved to study in France and later

Translated and edited form a lecture given by Erik Palmén at the Department of Meteorology, University of Helsinki, on February 21, 1979.

in Germany, where he became the first assistant to Heinrich Hertz. As an assistant he published 10–15 articles that dealt with the Hertz waves. Hertz died at a very early age, and a while afterward Vilhelm Bjerknes became the professor of Stockholm University (Norway and Sweden formed one country). He started to examine hydrodynamic problems and, a couple of years later, he published a very important article named "Über einen hydrodynamischen Fundamentalsatz und seine Anwendung besonders auf die Mechanik der Atmosphäre und des Weltmeeres" and, a few years later, "Zirkulation relativ zur Erde." In these publications he placed emphasis on the circulation that builds up in the atmosphere, as also in the ocean, when the air density distribution makes the situation baroclinic. He used the term "solenoid," and particularly he wanted to emphasize that the vertical circulation is caused by the solenoid field and the horizontal circulation is due to the Coriolis force. Among his students in Stockholm there were Sandstrom, who was a well-known meteorologist, and Ekman, whose name is familiar to meteorologists even today.

From Leipzig to Bergen

In 1907 Vilhelm Bjerknes moved to Oslo to be the professor of the university there, and in 1912 he was offered a professorship in geophysics in Leipzig University. Before that, in the year 1910, he published with his collaborators a notable work called "Dynamic Meteorology and Hydrography." The book dealt with the ways of developing meteorology and oceanography using mathematical principles. When he was invited to Leipzig University he gave a lecture. On that occasion he said that he would consider his task as accomplished if in one year he could correctly calculate the change of weather during one day. "It may take a year to drill a tunnel through a mountain, but later others may make a passage with an express train."

The principle was that the thermodynamic and dynamic primitive equations can be used in weather forecasting. Of course it was quite daring to say that after one year it would be possible to correctly calculate the development of weather during one day. Something like this tends to happen in meteorology: one promises a bit too much. Naturally this was an impossible task in those days, as there were no computers, but even today when it's possible to calculate this in a few minutes or hours, the correct calculation is often not within the bounds of possibility. This has always happened in meteorology: people tend to be too optimistic. (In fact I don't really believe that he expected that to be possible!)

A geophysical institute had been founded in Leipzig and V. Bjerknes was the director of it. As assistants from Norway he first had T. Hesselberg, who later became the director of the Norwegian Meteorological Institute, and H. Sverdrup,

later a well-known oceanographer and finally the director of the La Jolla Institute [Scripps Institution of Oceanography]. Many kinds of experiments were made in Leipzig, in order to better understand movements in the atmosphere, but after a couple of years, World War I broke out and the circumstances became difficult: all German pupils had to join the military forces. Hesselberg and Sverdrup had to leave Leipzig. Jacob Bjerknes, the son of Vilhelm Bjerknes, and H. Solberg, a young talented mathematician, took their places as assistants. When times got even harder the whole Norwegian group decided to move back to Norway, and a university chair was created for Vilhelm Bjerknes. The intention was to improve the weather service, because during the war there was a scarcity of food. This is how the young group finally ended up in Bergen's Lyseum, later Bergen University, where they had a small weather office, *Vervarslinga på Vestlandet*. Tor Bergeron and later also Carl-Gustaf Rossby from Sweden took part in this group as assistants.

V. Bjerknes: "Better Not To Read Too Much"

A few years later the polar front theory was developed, and I have often wondered why it was invented in the coastal part of Norway. If one takes a look at the weather chart, it can be seen that in the temperature field the front is more distinct on the continent than over the ocean. This, I think, showed how this group had actually somehow got rid of the old standpoint. One reason for this must be that they didn't have too much information about earlier meteorology. I remember how V. Bjerknes once said that it's better not to read too much, better not to be too dependent on what has been written already. In 1919 Jacob Bjerknes developed the front model, in which a cyclone consisted of the warm and cold front and there was a wave-like perturbation. When I later analyzed this first cyclone I discovered (and so did many others!) that it actually had already occluded. This shows that the new theory was not discovered from observations, but invented in brains; one then tried to find cases that show that the theory is correct.

The first important publications of this group were: J. Bjerknes (1919): On the structure of moving cyclones (*Geofys. Publikasjoner.*, **1**, No. 1, 1–8); J. Bjerknes and H. Solberg (1922): Life cycle of the cyclones and the polar front theory of atmospheric circulation (*Geofys. Publikasjoner*, **3**, No. 1, 1–18; the polar front theory was mentioned here for the first time!); Bergeron and Swoboda (1924): Wellen und Wirbel in einer quasistabonaren Grenzfläche über Europa (Veröffenti. Geophys. Inst. Univ. Leipzig **[2] 3**, No. 2, 63–172); Bergeron (1928): Über die dreidimensional verknupfende Wetteranalyse (*Geofys. Publikasjoner*, **5**, No. 6, 1–111. These publications dealt with the cyclone theory and how cyclones affect the general circulation, emphasizing, among other things, the energy problem.

As I already mentioned, this far I only had been corresponding with this group. But after publishing my doctoral thesis in 1926 on similar kinds of problems I decided to travel abroad. In the circumstances as they then were this was, of course, not so easy; no student awards were granted as nowadays. In 1928 I decided to make a journey to Sweden, Norway, Germany, and Austria to meet the leading meteorologists. I spent a few days in Stockholm, then I went to Oslo. My intention was to stay a couple of weeks in Norway. During my stay in Oslo, V. Bjerknes came and told me that if I go to Bergen I could stay with his son, because he was a bachelor and had a big apartment. He promised to call his son and tell about the situation. Partly for this reason Jacob and I became good friends and this friendship lasted until he died.

In Bergen I stayed about one month. I still remember how extraordinary this group was, consisting of meteorologists such as Jacob Bjerknes, Tor Bergeron, H. Hansen, and H. Svedrup, among others, and all the time there were foreign visitors. (Later also Sverre Pettersen came there to be the director of *Vervarslinga på Vestlandet*; V. Bjerknes had already moved back to Oslo to be the professor of theoretical physics). Usually the intellectual atmosphere was on a very high level and I first got an inferiority complex, because I came from Finland where in the Department of Meteorology there was usually Oscar Johansson and sometimes he had a temporary assistant, and that was all.

We talked a lot about all kinds of problems and especially I remember talking with Bergeron. He was just working on "Über die dreidimensional verknupfende Wetteranalyse', when he told me how all rain that forms in the atmosphere first requires ice crystals, only drizzle is possible to form without them. I told him how strange it is that I didn't come to think of it before, because I had often observed that, for example, on a warm summer day the rain is forming in the cumulus clouds only when their tops reach a level high enough, but in autumn when it's relatively cold the rain is already formed from cumulus clouds much lower. He also claimed that the upper part of the cloud had to reach the level, where the temperature is near $-10°C$ for the ice crystals automatically to form. As we notice, many can make observations about a phenomenon, but only few come to think about the reasons why things happen! (This was a very important part in Bergeron's book, because later all trials to produce artificial rain were based on it.)

After this first visit to Bergen, I continued to Berlin, Leipzig, Prague, Wien, and finally to Innsbruck to meet the leading meteorologists. In Berlin I met von Ficker, and especially I remember meeting with Alfred Wegener. He had just published an important textbook *Thermodynamik der Atmosphäre*, which nobody knows nowadays but which in those days really was one of the most excellent textbooks. I wanted to talk with Wegener about problems related to this subject, but he refused. He had a few years earlier published a well-known book about how continents move, and he complained that geologists were opposing it almost by common consent. It is this theory that has made Wegener's name well-known to future generations. He died in Greenland a couple of years after this.

In Leipzig I met L. Weickman and B. Haurwitz, in Wien F.M. Exner, whose textbook *Dynamische Meteorologie* was compulsory for students of meteorology at the university. In Innsbruck I met A. Defant. This journey took for me three months. And as I already mentioned, journeys abroad were very rare, so as I came back to Finland I was surrounded by journalists. That is not happening nowadays!

J. Bjerknes and the Vorticity Equation

My cooperation with Bergen school started especially in 1930, when I was visiting Bergen. Jack Bjerknes had made a series of observations of the atmosphere, by using Jaumotte's small meteorographs, and a series of pilot balloons had been released in Uccle, Belgium. He presented his observation series in 1930, but the paper was published only a couple of years later. The name of the publication was "Exploration de quelques perturbations atmosphériques à l'aide de sondages rapprochés dans le temps" *(Geofys. Publikasjoner. 9,* No. 9, 1–52). The perturbation was moving eastward and passed over Uccle [Palmén showed here Fig. 28 of the above-mentioned article, i.e., the synoptic map for 29 March 1928]. Bjerknes also presented a schematic vertical cross-section [Fig. 21 of the above-mentioned article] about the situation, where on the left side there is a warm front and on the right side a cold front, and between these fronts there is a cold polar-air mass. On the left side $\partial w/\partial z$ is negative and on the right side it is positive in the upper troposphere. It follows [from the continuity equation] that when $\partial w/\partial z$ is negative, then the divergence $\partial u/\partial x + \partial v/\partial y$ is positive. Using this he derived the formula, which in fact is nothing but the vorticity equation

$$\frac{d}{dt}\left(\frac{\partial v}{\partial x} - \frac{\partial u}{\partial y}\right) = -2\Omega_z\left(\frac{\partial u}{\partial x} + \frac{\partial v}{\partial y}\right)$$

which says that the vorticity change is directly proportional to (minus) the Coriolis parameter multiplied with divergence. I think this was in fact the first time that the vorticity concept appeared in literature. Of course, it was nothing new; it was only the consequence of the V. Bjerknes' circulation theorem. As we all know, the horizontal circulation is in fact the integral of vorticity. When V. Bjerknes talked about a certain horizontal curve and circulation changes, in differential form it, of course, included the vorticity equation. Usually it is thought that the vorticity concept was born in the investigations of Rossby, but in fact it happened already a few years earlier.

What interested me in these studies was the waves that appear to be connected with the perturbations in the atmosphere. It was assumed (which was a mistake) that these were only due to some kind of "orographic" effect connected with the cold-air mass. This was, of course, an "orographic" effect moving with the system. This was quite a primitive concept, but anyway something new.

"We Almost Found [in 1935] the Jet Stream"

We talked a lot about these problems and came to a conclusion that we should make observations with sondes. A few years later, in 1933, when I was a couple of months in Bergen, we made that kind of observations. We sent up balloons, not very many, but we didn't gain any remarkable results. I had to publish these because, as I think, Bjerknes didn't consider the results good enough. Anyway, we decided to make an experiment covering a larger area of Europe, and to send letters to the meteorological services in Europe asking them to make balloon observations during a couple of days after receiving a telegram. We did get a lot of compliant answers, and finally in the winter 1935 we thought the time to be appropriate for carrying out the experiment.

Our intention was to release up a large group of Jaumotte's sondes in Norway. I also made an agreement with my friend Rossi from Finland that he would release sondes from Lauttakylä where he lived. A large group of sondes was launched also in Uccle. As radiosondes had not yet been invented, you had to get back the sonde after the experiment, which was more difficult in Northern Europe than in Central Europe. Eventually, the experiment turned out to be successful. Problems appeared in planning the cooperation: if you want to catch a cyclone moving, for example, eastward and developing to a strong cyclone through the occlusion process, and you want to send a telegram to all the stations within the area, you are dealing with a difficult forecasting problem.

Our cyclone had a distinct warm and cold front, on the Atlantic Ocean there was about to be born a new wave-like perturbation [here Palmén showed Figs. 2 and 6 of J. Bjerknes and E. Palmén (1937): *Geofys. Publikasjoner*, **12**, 2, 1–62, i.e., the synoptic maps for 0700 GMT of the 15th and 17th of February 1935]. It was hard to decide to whom to send telegrams. The center of the low was 965 mbars, in fact it was the deepest cyclone during that winter period. The forecast appeared to be a good one: the depression took the route we predicted. A cross section through the system all the way from Madrid to the northern part of Sweden was prepared [Fig. 28 in the above mentioned article]. The front and the tropopause were analyzed, and it was especially emphasized that the tropopause is not a continuous surface. The topogra-

phy of various pressure levels was analyzed [Fig. 36 of the above-mentioned article]. The wave-like upper flow that builds in connection with the perturbation could be seen. I want to emphasize that we came to a conclusion that at the 300 mb level the geostrophic wind [in the wave trough] was 130 m s^{-1}. Because of this large calculated velocity we became doubtful of the pressure field analysis. Anyway, we almost found the "jet stream"; that is, of course, if we had been brave enough.

"I Should Have Been More Brave"

There was an event, when I should have been more brave. We were discussing the wave-like perturbations in the upper atmosphere, when (Jack) Bjerknes said to me that he had an idea. If you have sinusoidal isobars in the upper troposphere, at a given latitude the flow between equally spaced isobars is constant, but the flow is weaker in the north [due to the effect of the Coriolis parameter]. But because the centrifugal force is pointing to opposite directions in the trough and in the ridge, it can often be that more air is flowing through the ridge than through the trough. [Palmén showed here Fig. 6.1 from Palmén & Newton (1969): *Atmospheric Circulation Systems*. Academic Press]. Thus on the eastern side of the trough there is divergence and on the western side there is convergence, and the wave is moving with a speed depending on the difference between the Coriolis and centrifugal forces.

My answer to this was that it's a very nice thought, but I think it's too simple an assumption, so if you are going to publish something on it, I would rather not to be a coauthor. I thought that the flow was too ageostrophic and, for that reason, the assumption was too simple. He published it anyway [J. Bjerknes (1937): *Meteorologische Zeitschrift*, **54**, No. 12, 462–466], the result of which was that when maybe the most important paper by Rossby appeared [Rossby et al. (1939): *J. Marine Res.*, **2**, 38–55], on the first pages he wrote: 'In the attempt to understand the dynamics of the upper level trough over the United States, the author found great help in a remarkable paper by J. Bjerknes, which offers a simple explanation for the displacement of perturbations superimposed upon the zonal pressure distribution which normally prevails in the upper part of the troposphere." He [Rossby] also states that more air is transported through the ridge section than through the trough section depending on the speed of the wind and the wave length [Palmén showed here Fig. 11 from the above-mentioned article by Rossby et al.]. In fact Rossby's assumption was somehow more primitive than the one by J. Bjerknes, because in Bjerknes's assumption it was accepted the flow can vary in the meridional direction, but Rossby assumed the situation to be barotropic and the basic zonal velocity to be constant. This shows the course of development: Rossby did directly use the concept of vorticity, while in Bjerknes's scheme it was not mentioned, even though he had already used it.

I have always felt my action very foolish when I refused to cooperate with Bjerknes on this issue, but things like that happen.

So this is how the Rossby theory about the Rossby waves was born. Rossby et al. presented the following formula for the zonal speed of the wave-like perturbations

$$c = U - \frac{\beta L^2}{4\pi^2}$$

where c is the zonal phase speed of the perturbation, U the speed of the zonal flow, β the derivative of the coriolis parameter with respect to latitude, and L is the wavelength. Thus short waves move to east and waves long enough move to west. This was the theory of Rossby waves.

My collaboration with J. Bjerknes still lasted for about three synoptic studies. Finally World War II interrupted our joint work. It came to an end also for the reason that just before the war the radiosonde technique was developed and it became possible to operationally get synoptic information about the conditions in the free atmosphere; earlier it naturally took us a long time to get the equipment back and the results analyzed. Collaboration was impossible during the war and I had to temporarily leave research because I had many other things to carry out, especially since I had become director of the Institute of Marine Research.

Sutcliffe and His Development Theory

BRIAN J. HOSKINS

Department of Meteorology, University of Reading, Reading, United Kingdom

1. Introduction

Reggie Sutcliffe joined the U.K. Meteorological Office as a professional assistant in 1927 with an undergraduate degree in mathematics and a Ph.D. in statistics. The Meteorological Office, at that time, was a place where research was not seen as part of the job. The new recruit was not very impressed with what he saw about him (Sutcliffe, 1981–83).

After one year in London, Sutcliffe was posted to a forecasting outstation on Malta. By a stroke of good fortune, Tor Bergeron was at that time in the middle of a six-month stay in Malta sponsored by the U.K. Meteorological Office to study Mediterranean weather in the light of the Norwegian ideas. Sutcliffe had the benefit of seeing one of the masters at work and interacting with him. This was particularly beneficial because the Norwegian ideas were slow to gain acceptance by the forecasters in the United Kingdom: fronts did not appear in the Daily Weather Report until 1933. However, Sutcliffe was certainly influenced by his contact with the Bergen school. The success of the Norwegian frontal cyclone model and the need for such "mental models" was a recurring theme in his work (see, e.g., Sutcliffe 1948a).

Just as for V. Bjerknes and the Bergen school, the underlying drive in Sutcliffe's work was weather forecasting. "I regard prediction as the ultimate test of science and forecasting as the greatest discipline in meteorology" (Sutcliffe 1948b). He loved the challenge of the forecast problem. "As in all synoptic work the method must be to mould the whole model around the incomplete scaffolding of observations using theory, experience, flair, scientific judgement or even aesthetic sense to provide something which hangs together" (Sutcliffe 1942). Like V. Bjerknes before him, Sutcliffe stressed the need for a firm scientific basis. "The need for some dynamical or thermodynamical appreciation of the forecasting problem cannot be over-stressed. Present methods of general forecasting place great reliance on . . . extrapolating without much understanding of the physical process" (Sutcliffe 1947).

The mathematical analysis that Sutcliffe performed in the 1938–1947 period was a rigorous derivation of basic quasigeostrophic theory. However, it was clear that he required as the product of his mathematics not just an elegant theory but one that would be of practical use to forecasters.

The mental model of middle latitude synoptic development he created was a great advance and has been of lasting value.

2. The Late 1930s

Sutcliffe returned to London in 1935 and was posted to a Royal Air Force station in 1937. He soon saw the need to explain the basic ideas and was given six months leave to write *Meteorology for Aviators* (Sutcliffe 1938b). This gave him an opportunity to think about the flow of the atmosphere in depth and its relevance for both surface- and aircraft-level weather forecasting. Sutcliffe, as he later recalled (Sutcliffe 1981–83), was dissatisfied with cyclogenesis arguments based on the hydrostatic equation. He noted (Sutcliffe 1939) that a depression, far from being caused by a light stratosphere, was actually associated with a high tropopause pressure. He also felt that "the Norwegian theory . . . told us nothing about the dynamics. They told us that warm air went up cold fronts and cold air undercut the warm air, but not why" (Sutcliffe 1981–83). He realized that the basic question was one in hydrodynamics: How did the circulation develop? He also considered that the concentration was all on cyclones. It was known that there was subsidence in anticyclones, but there was no idea of the three-dimensional field of motion. "You wanted the field theory—not the theory of depressions but the theory of the circulation of the atmosphere" (Sutcliffe, 1981–83).

In Part I of his 1910 book, V. Bjerknes had mapped the atmosphere on 100-mb pressure levels up to 300 mb and introduced the idea of thickness. The regular use of upper-air, pressure-level, contour, and thickness charts was introduced by the Deutsche Seewarte in 1934, and in the same year,

Scherhag (1934) originated what became known as his divergence theory. The proposition, which was a development of the ideas of Dines (1914 and 1925), was that divergent upper winds must produce in general a fall of pressure if they are not compensated by a strong convergence below. Sutcliffe appreciated the argument but pointed out (Sutcliffe 1938a) that the geostrophic wind is non-divergent and that there appeared to be some confusion in the applications by Scherhag and others between horizontal divergence and difluence in the so-called delta region of the upper flow.

Sutcliffe's own ideas emerged in his 1939 paper on cyclonic and anticyclonic development. In this paper he noted the importance of the study of Brunt and Douglas (1929) relating surface pressure change, nongeostrophic isalobaric wind, and lower tropospheric convergence and divergence. He also noted the implication from the work of Brunt (1939) that descent and lower tropospheric divergence in an anticyclonic region must be compensated by convergence elsewhere in the column.

Sutcliffe then made the vital deduction that cyclonic development must be associated with lower tropospheric convergence and upper tropospheric divergence with only a small residual between them. The opposite would hold for anticyclonic development. He was thus able to make the giant step from previous ideas with the proposition that the diagnosis of surface pressure development was best performed by investigating the difference between upper and lower tropospheric divergence. In p coordinates and modern notation, he proposed consideration of a development indicator:

$$\mathcal{D} = \nabla \bullet \mathbf{v}\big|_{P_U} - \nabla \bullet \mathbf{v}\big|_{P_L} \qquad (1)$$

$$= -\frac{\partial \omega}{\partial p}\bigg|_{P_U} + \frac{\partial \omega}{\partial p}\bigg|_{P_L}$$

$$\cong \Delta p \frac{\partial^2 \omega}{\partial p^2} \qquad (2)$$

using a two-layer approximation. In a simple situation $\mathcal{D}$ being positive is associated with ascent and low-level creation of cyclonic vorticity by stretching.

Sutcliffe then pursued the hydrodynamics of the problem by investigating the divergence of the ageostropic wind at each level using the horizontal momentum equations. He proposed the use of the facts that the divergence is associated purely with the nongeostrophic wind, and that the latter is proportional to the acceleration and directed at right angles to it:

$$\mathbf{v}_{ag} = \frac{1}{f}\mathbf{k} \times \frac{D\mathbf{v}}{Dt}. \qquad (3)$$

Thus
$$\mathcal{D} = \nabla \bullet \left(\mathbf{v}_{agU} - \mathbf{v}_{agL}\right), \qquad (4)$$

and
$$\mathbf{v}_{agU} - \mathbf{v}_{agL} = \frac{1}{f}\mathbf{k} \times \mathbf{A}, \qquad (5)$$

where
$$\mathbf{A} = \left(D\mathbf{v}/Dt\right)_U - \left(D\mathbf{v}/Dt\right)_L. \qquad (6)$$

He then took $\mathbf{v}_L$ to be the wind just above the friction layer (i.e., the low-level geostrophic wind), and set

$$\mathbf{v}_U = \mathbf{v}_L + \mathbf{v}_s.$$

Sutcliffe did not like the term "thermal wind," and thus called $\mathbf{v}_s$ the "shear," although he related it directly to the thermal gradient. It follows that Eq. (6) may be written

$$\mathbf{A} = \mathbf{v}_s \bullet \nabla \mathbf{v}_L + \frac{D}{Dt}\mathbf{v}_s, \qquad (7)$$

where the final material derivative is taken moving with the upper flow.

Sutcliffe (1939) then considered the implications of the two terms in Eq. (7) on the differential ageostrophic motion, Eq. (5), and thus on his development indicator $\mathcal{D}$, Eq. (4). He called the first term the shearing term. He deduced that for a thermal wind oriented across the low-level trough (Fig. 1) this vector would be poleward and "the corresponding . . . geostrophic departure therefore is perpendicular to the trough, opposing $\mathbf{v}_s$. This implies upper divergence ahead of the trough and upper convergence behind, that is surface convergence ahead and divergence behind." $\mathcal{D}$ is positive ahead and negative behind, implying movement of the system in the direction of the thermal wind.

Sutcliffe (1939) referred to the second term in Eq. (5) as the thermal development term. He then considered the cases of diabatic heating, and increasing thermal gradients by confluence and shear (his cases 4, 5 and 6, respectively).

Diabatic heating leads to anticyclonic $\mathbf{v}_s$, upper ageostrophic winds outwards, upper divergence, lower con-

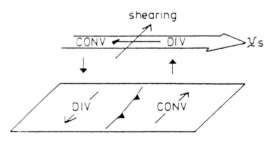

FIG. 1. The shearing term in Sutcliffe's theory. The shear $\mathbf{v}_s$ crossing the low-level pattern leading to a shearing term along the trough, ageostrophic motion toward the rear (solid arrowhead), convergence and divergence, and vertical motion.

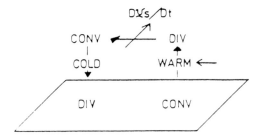

Fɪɢ. 2. Sutcliffe's Case 5, the thermal development term, warm air moving toward cold leads to an accelerating shear, ageostrophic motion toward the cold air (solid arrowhead), convergence and divergence, and vertical motion.

vergence and cyclonic development. This agreed with previous work.

Sutcliffe's verbal description of the result of confluence in a thermal pattern may be visualized as in Fig. 2. He described how, associated with the crowding of the isotherms, the upper level "geostrophic departure is directed from warm towards cold giving in the upper air convergence towards the colder side, divergence in the warmer side, and requiring as a necessary accompaniment, low-level divergence on the colder side and convergence on the warmer side."

The thermal shear discussion given by Sutcliffe identifies the winds from warm to cold with cyclonic development.

Sutcliffe finishes by noting the complexity of the problem and "the possible importance of convergence and divergence, vertical motion and thermodynamical processes in determining the thermal development."

In order to relate Sutcliffe's analysis to modern theory, we assume a simple vertical structure and write the shear term

$$\mathbf{v}_s \bullet \nabla \mathbf{v}_L = (\Delta p / f) B \partial \mathbf{v} / \partial s , \qquad (8)$$

where the baroclinicity $B = |\nabla \mathbf{B}|$, with

$$\mathbf{B} = \hat{R} \nabla \theta = \mathbf{k} \times f(\partial \mathbf{v} / \partial p),$$

s is along the thermal contours, and $\mathbf{v}$ is the wind at any level. In terms of the Q-vector studies of Hoskins et al. (1978) and Hoskins and Pedder (1980), the contribution to $(f^2 / \Delta p)$ times the differential ageostrophic velocity (Eq. 5) is seen to be

$$B\mathbf{k} \times \frac{\partial \mathbf{v}}{\partial s} = -\mathbf{Q} , \qquad (9)$$

this being the form of $\mathbf{Q}$ given by Hoskins (1982).

The thermal term and its contribution to $(f^2 / \Delta p)$ times the differential ageostrophic velocity is, using the simple vertical structure, quasigeostrophic theory and the thermodynamic energy equation,

$$\mathbf{k} \times f \frac{D}{Dt} \frac{\mathbf{v}_s}{\Delta p} = -\frac{D}{Dt} \mathbf{B} = -\frac{D}{Dt} \hat{R} \nabla \theta = -\mathbf{Q} - \sigma \nabla^2 \omega - \hat{R} \nabla \dot{\theta}$$
$$(10)$$

where $\dot{\theta}$ represents the diabatic heating and $\sigma^2 = -\hat{R} d\theta / dp$ is the static stability. By collecting together Eqs. (2), (4), (9), and (10), we have

$$f^2 \frac{\partial^2 \omega}{\partial p^2} = \left(f^2 / \Delta p \right) \mathcal{D} = -\nabla \bullet \mathbf{Q} - \left(\nabla \bullet \mathbf{Q} + \sigma^2 \nabla^2 \omega + \hat{R} \nabla^2 \dot{\theta} \right).$$
$$\text{shear} \qquad\qquad \text{thermal} \qquad (11)$$

Rearranging gives the omega equation

$$\sigma^2 \nabla^2 \omega + f^2 \frac{\partial^2 \omega}{\partial p^2} = -2\nabla \bullet \mathbf{Q} - \hat{R} \nabla^2 \dot{\theta} . \qquad (12)$$

Thus the following points are seen about Sutcliffe's analyses in this seminal paper:

- The shear term was a simple two-layer version of the Q-vector in the form later given in Hoskins (1982).
- The $\mathbf{Q}$ in the thermal term was qualitatively discussed in particular in both the idealized confluence and shear cases. However, no relation with the shear term was noted.
- The clear discussion of the frontal circulation in the confluence case came some 20 years before the detailed analyses of Sawyer (1956) and Eliassen (1959), which also included advection by the ageostrophic circulation.
- The quasigeostrophic response to diabatic heating was correctly analyzed.
- Apart from in his final comments, the adiabatic warming and cooling term, $\sigma^2 \nabla^2 \omega$, was neglected, largely because this was seen as part of the answer. In fact, the elliptic nature of the omega equation means that this usually gives only a quantitative, not qualitative, error in the vertical motion. Sumner (1950) later noted the damping of vertical motion by the static stability term.

3. Development Theory in the 1940s

With the outbreak of World War II in 1939, Sutcliffe was commissioned in the RAF. He played an important role in weather forecasting for the RAF and directly after the war he was involved in helping set up the new weather service in Germany. The challenge and the intensity of the work during this period severely curtailed any research, but the significant increase in upper air data during the war gave Sutcliffe the possibility of trying out his ideas.

For the purposes of the teaching he was performing, Sutcliffe recorded the geostrophic, hydrostatic and thermal wind equations in *p*-coordinates (Sutcliffe 1942).

Sutcliffe and Godart (1942)* gave the momentum and temperature equations in isobaric coordinates and discussed the relative importance of terms in the thickness/temperature equation. They split the horizontal advection into terms associated with the geostrophic and ageostrophic velocity, and again noted that with a simple state (or two-layer model) the surface geostrophic wind could be used for the former. The final recipe given for prediction of the thickness field was to first do a surface forecast "by any means available," then to do the geostrophic advection of the thickness field, and finally to consider where ageostrophic advection, vertical motion, or diabatic processes might be significant.

Sutcliffe was demobilized in 1946, and in his own words he "was doing nothing . . . waiting for the reorganisation" (Sutcliffe, 1981–83). He used this time to write ideas he had since his 1939 paper. In Sutcliffe (1947), he developed the same general argument as in the 1939 paper, but, following Rossby (1939, 1940) and others, he moved to consideration of vorticity and the vorticity equation rather than the momentum equation in diagnosing horizontal divergence. He used the *p*-coordinate equations of Sutcliffe and Godart (1942) and deduced the form of the equation for the "isobaric vorticity." After a careful consideration of the size of terms, he produced the consistent quasigeostrophic vorticity equation

$$\left(\frac{\partial}{\partial t} + \mathbf{v} \bullet \nabla\right)(f + \xi) = -f\nabla \bullet \mathbf{v}, \qquad (13)$$

where ageostrophic motion is neglected except in the right-hand side. Using Eq. (13) to substitute for the divergence, the development indicator $\mathcal{D}$ may be written

$$f\mathcal{D} = -\frac{\partial}{\partial t}\left(\xi_U - \xi_L\right) - \mathbf{v}_U \bullet \nabla\left(f + \xi_U\right) + \mathbf{v}_L \bullet \nabla\left(f + \xi_L\right). \qquad (14)$$

Sutcliffe then noted that

$$\xi_U - \xi_L = \frac{1}{f}\nabla^2 \Phi', \qquad (15)$$

where Φ', the "thickness," is the difference in geopotential between the upper and lower layers. Thus he saw that devel-

opment was associated with an imbalance "between the changes in the temperature field and the vertical difference in vorticity as determined by the different rates of advection at different levels."

Again the development associated with nonadiabatic processes was easily discussed. In Sutcliffe (1947), adiabatic warming and cooling were discussed explicitly. Their importance was recognized, but again they were omitted from the analysis. The concentration was then on the horizontal thermal (thickness) advection:

$$-\frac{\partial}{\partial t}\left(\xi_U - \xi_L\right) = \frac{1}{f}\nabla^2\left(\bar{u}\frac{\partial \Phi'}{\partial x} + \bar{v}\frac{\partial \Phi'}{\partial y}\right)$$
$$= \nabla^2(\bar{u}v' - \bar{v}u'),$$

where the bar and prime refer to layer averages and differences. Sutcliffe, by neglecting some deformation terms "connected with the processes of frontogenesis and frontolysis" that "require separate attention," reduced the thickness tendency contribution in Eq. (14) to

$$\left(\mathbf{v}_L \bullet \nabla\right)\xi_U - \left(\mathbf{v}_U \bullet \nabla\right)\xi_L.$$

Therefore, the development indicator, Eq. (14), was given by

$$f\mathcal{D} = -V'\frac{\partial}{\partial s}\left(f + \xi_U + \xi_L\right), \qquad (16)$$

or

$$f\mathcal{D} = -V'\frac{\partial}{\partial s}\left(f + 2\xi_L + \xi'\right), \qquad (17)$$

where $V' = |\mathbf{v}'_s|$ and $\xi' = \xi_U - \xi_L$, as in Eq. (15).

Sutcliffe went on to analyze each of the terms in Eq. (17). He had not found the first term, the β term, to be important. The second term implied cyclonic development ahead, anticyclonic to the rear of a surface cyclone, thereby leading to displacement in the direction of the shear, an effect that was "well known." He referred to this as the thermal steering effect.

The third term in Eq. (17), "the thermal vorticity effect," was determined entirely by the thickness chart. It predicted low-level "cyclonic (ascent) type development where the thermal vorticity decreases in the direction of the shear," in particular ahead of a cold thermal trough. He discussed how this analysis accorded with experience and helped the understanding of some forecasting problems. He also observed a link between the deductions from this term and the arguments used by Bjerknes (1937, 1940) relating cyclogenesis with the movement of upper air through upper cyclonic and anticyclonic patterns.

---

*It is noted in Sutcliffe (1947) that O. Godart (a Belgian who later became a professor in astronomy) also derived the mass conservation equation $\partial \upsilon/\partial \xi + \partial \varpi/\partial \psi + \partial w/\partial p = 0$.

This version of Sutcliffe's development theory has already been related to other versions of the omega equation by Hoskins et al. (1978). Following them, Eq. (17), omitting the β term but including the adiabatic warming term, can be written assuming a simple vertical structure, as

$$f^2 \frac{\partial^2 \omega}{\partial p^2} = \frac{f^2}{\Delta p} \mathcal{D} = 2B \frac{\partial}{\partial s} \xi - 8f\mathcal{D}_{ef}^2 \frac{\partial \alpha}{\partial z} - \sigma^2 \nabla^2 \omega .$$

Here $\mathcal{D}_{ef}$ is the magnitude of the deformation and α the angle of the dilatation axis. The deformation term, which was neglected by Sutcliffe, depends on the rotation of the dilatation axis with height. It is indeed important in frontal regimes, but the retained term usually gives a qualitatively correct version of the synoptic-scale field. As discussed above, the latter term again usually produces only a quantitative change in the field. Thus it remains true that Eq. (17) is qualitatively helpful and very easy to apply.

Sutcliffe was one of those who presented papers to a meeting of the Royal Meteorological Society on 18 February 1948. In the discussion (Sutcliffe, 1948b), E. T. Eady welcomed the paper as being on the "right lines," and commented on the importance of the thermal pattern from the point of view of their being "potentially available energy." That this energy is realizable would be shown in his solutions "(results not yet published)."

Sutcliffe and Fosdyke (1950) produced a summary of the 1947 theory and its application to 500 mb contour and thickness maps. Amongst the idealized patterns that are discussed is the self-development process involved in a classical warm-sector depression. The circulation of the depression creates a thermal trough to the rear and ridge ahead, which induces further cyclonic development in the region of the original depression, giving positive feedback and continued growth. They state that this "is perhaps more easily understood than any of the current and rather closely argued "theories," although it has much in common with the Bjerknes-Holmboe (1944) ideas."

4. Concluding Comments

The contributions of Sutcliffe in his 1939 paper and over the next decade can be summarized:

- He showed that forecasting middle-latitude cyclogenesis was a problem in hydrodynamics as well as thermodynamics and could be investigated as such.
- He emphasised the role of ascent with balancing upper-level divergence and lower-level convergence.
- He moved away from a consideration of pre-existing frontal discontinuities but showed the importance of thermal gradients for development.
- Like Charney (1947) and Eady (1949), he independently developed basic quasigeostrophic theory, but in Sutcliffe's case he did it on the back of a weather map.
- In 1939 he produced a qualitative discussion of the dynamics of cross-frontal circulations that is still seen as basically correct.
- He derived what were in effect approximations to both forms of the omega equation used today—the Q-vector form in 1939, and the vorticity and thermal advection form in 1947, producing also a very practical approximation to the latter form.
- He created a framework for practical consideration of three-dimensional flow and synoptic development in middle latitudes.

Sutcliffe encouraged the use of electronic computers as soon as they became available. Charney (Charney et al. 1950) appreciated that baroclinic effects should be included to improve his barotropic model predictions and realized that his suggestions were in accord with the reasoning of Sutcliffe. Bushby (1952) turned Sutcliffe's 1947 equation and the modified form with the static stability term included (Sumner 1950) into a vertical velocity equation. He then showed the first numerical solutions of the omega equation. Bushby (1952) found that the reduced vertical circulation with the static stability term included had maximum ascent ($\sim 4 \, \mathrm{cm\,s^{-1}}$) consistent with rainfall rates. Sawyer and Bushby (1951) noted that the thickness tendencies implied using the equation in Sutcliffe and Fosdyke (1950) were also realistic and went on to suggest a complete two-parameter model with the mean layer height being given by Charney's equivalent barotropic equation. Sawyer and Bushby (1953) showed the results from a two-parameter model.

However, Sutcliffe (1981–83) later said that computers came too soon. "Dynamicists would have altered weather forecasting quite radically, in quite a revolutionary way . . . because the ideas don't need a high speed computer." The relative roles of computation and basic understanding remains an interesting question.

Sutcliffe's theory was complementary to the Bergen concepts of fronts and air masses. It was a large step forward in the understanding of vertical circulation and synoptic development. In my opinion it is vital that such theoretical frameworks be used to guide the interpretation and use of numerical results and be used in the improvements of the models themselves.

References

Bjerknes, J., 1937: Theorie der aussertropischen Zyklonen bildung. *Meteor. Z.*, **54**, 462–466.
——, 1940: The deepening of extra-tropical cyclones. *Quart. J. Roy. Meteor. Soc.*, **66** (Suppl.), 57–59.

————, and J. Holmboe, 1944: On the theory of cyclones. *J. Meteor.*, **1**, 1–22.

Bjerknes, V., .1910: Dynamical meteorology and hydrography, Part 1.

Brunt, D., 1939: Physical and dynamical meteorology. 2nd ed., Cambridge University Press, 384 pp.

————, and C. K. M. Douglas, 1929: On the modification of the strophic balance. *Mem. Roy. Met. Soc.*, **3**(22).

Bushby, F., 1952: The evaluation of vertical velocity and thickness tendency from Sutcliffe's theory. *Quart. J. Roy. Meteor. Soc.*, **78**, 354–362.

Charney, J. G., 1947: The dynamics of long waves in a baroclinic westerly current. *J. Meteor.*, **4**, 135–163.

————, R. Fjörtoft, and J. von Neumann, 1950: Numerical integration of the barotropic vorticity equation. *Tellus*, **2**, 237–254.

Eady, E. T., 1949: Long waves and cyclone waves. *Tellus*, **1**, 33–52.

Eliassen, A., 1959: On the formation of fronts in the atmosphere. *The Atmosphere and the Sea in Motion*, Rockefeller Institute Press, 277–287.

Hoskins, B. J., 1982: The mathematical theory of frontogenesis. *Ann. Rev. Fluid Mech.*, **14**, 131–151.

————, I. Draghici, and H. C. Davies, 1978: A new look at the ω-equation. *Quart. J. Roy. Meteor. Soc.*, **104**, 31–38.

————, and M. A. Pedder, 1980: The diagnosis of middle latitude synoptic development. *Quart. J. Roy. Meteor. Soc.*, **106**, 707–719.

Rossby, C.-G., 1939: Relations between variations in the intensity of the zonal circulation of the atmosphere and the displacements of the semi-permanent centers of action. *J. Mar. Res.*, **2**, 38–55.

————, 1940: Planetary flow patterns in the atmosphere. *Quart. J. Roy. Meteor. Soc.*, **66** (Suppl.), 68–87.

Sawyer, J. S., 1956: The vertical circulation at meteorological fronts and its relation to frontogenesis. *Proc. Roy. Soc. Ser. A*, **234**, 346–362.

————, and F. Bushby, 1951: Note on the numerical integration of the equations of meteorological dynamics. *Tellus*, **3**, 201–203.

————, and F. Bushby, 1953: A baroclinic model atmosphere suitable for numerical integration. *J. Meteor.*, **10**, 54–59.

Scherhag, R., 1934: Zur theoric der Hoch-und Tiefdruckgebiete. Die Bedeutung der Divergenz in Druckfeldern. *Meteor. Z.*, **51**, 129–138.

Sumner, E. J., 1950: The significance of the vertical stability in synoptic developments. *Quart. J. Roy. Meteor .Soc.*, **76**, 384–392.

Sutcliffe, R. C., 1938a: A remark on divergent winds. *Meteor. Mag. Land.*, **73**, 44.

————, 1938b: *Meteorology for Aviators*. His Majesty's Stationery Office.

————, 1939: Cyclonic and anticyclonic development. *Quart. J. Roy. Meteor. Soc.*, **65**, 519–524.

————, 1942: Construction of upper isobaric contour charts. Meteor. Office, S.D.T.M. No. 50 (Internal Memorandum), Part 1.

————, 1947: A contribution to the problem of development. *Quart. J. Roy. Meteor. Soc.*, **73**, 370–383.

————, 1948a: The use of upper air thickness patterns in general forecasting. *Meteor. Mag. Land.*, **77**, 145–152.

————, 1948b: Report of discussion at a meeting of the Royal Meteorological Society. *Quart. J. Roy. Meteor. Soc.*, **74**, 178–187.

————, 1981–83: Three interviews with Professor R. C. Sutcliffe, FRS during 1981–1983; interviewer J. M. C. Burton. Owned by Royal Meteor. Soc., tapes and transcript in Meteor. Office Library.

————, and O. Godart, 1942: Construction of upper isobaric contour charts. Meteor. Office, S.D.T.M. No. 50 (Internal Memorandum), Part 2.

————, and A. G. Fosdyke, 1950: The theory and use of upper air thickness patterns in forecasting. *Quart. J. Roy. Meteor. Soc.*, **76**, 189–217.

HISTORICAL PHOTOS

To the young scientist, the publications of Vilhelm Bjerknes, Jacob Bjerknes, Tor Bergeron, Erik Palmén, Carl-Gustaf Rossby, July Charney, Eric Eady, Arnt Eliassen, Ernst Kleinschmidt, Edward Lorenz, and Joseph Smagorinsky provide a historical perspective of some of the fundamental contributions to the advancement of atmospheric science. To the not-so-young scientist, revisiting these manuscripts is often accompanied by the remembrance of faces and times past, and the camaraderie with colleagues from hone and abroad that spans a professional lifetime. Our scientific heritage encompasses not only intellectual accounts and practical applications but also humanistic qualities of the individuals who gave rise to the contributions—a family heritage recounted through the telling of stories, often inspired by photographs in an album, a gathering of friends. This photographic essay was prepared as a visualization of some of our colleagues who have contributed to the advancement of our science and to the enrichment of our lives and of humanity.

The photographs presented herein were generously provided by Mrs. Jacob (Hedvig) Bjerknes, Arnt Eliassen, John Green, Chester Newton, Hans Volkert, Robert Bundgaard, John Lewis, David Stuart, Akira Kasahara, the Norwegian Meteorological Institute, the University of Bergen, National Center for Atmospheric Research (NCAR), and the Royal Meteorological Society. We are especially indebted to NCAR Photographics for the extensive processing and printing of these photographs.

Melvyn Shapiro

Vilhelm Bjerknes

Jacob Bjerknes

Carl-Gustaf Rossby

Tor Bergeron

Erik Palmén

Reginald C. Sutcliffe

Eric Eady

Jule Charney, Tokyo NWP Symposium, 1960.

The International Aerological Commission, Bergen, Norway, 1921

1. Norinder
2. J. Bjerknes
3. Martin Knudsen
4. Sandstrøm
5. Hesselberg
6. Richardson
7. Mercanton

8. Chaumotte
9. de Quervain
10. Taylor
11. Wallén
12. Cruz Conde
13. van Bemmelen
14. V. Bjerknes

15. Sir Napier Shaw
16. van Everdingen
17. Gold
18. Schereschewski
19. Cave
20. Calvagen
21. Edlund

22. H. Köhler
23. Malmgren
24. Fujiwara
25. Sekiguchi
26. Gullstrøm

Hermann von Helmholtz

F.M. Exner

Allegaten 33. The home of V. Bjerknes and family (lower floor).
The work place for the Bergen meteorologists (upper floor).

Allegaten 33. Tor Bergeron (l.), Jacob Bjerknes (r.), and assistant at work, 1919

A young Jacob Bjerknes in the electrostatics laboratory.

Jacob

C.L. Godske, Stoermer, J. Bjerknes (l.r.), 1939.

Allegaten 33, Bergen, 1930.
From right: C.-G. Rossby, J. Bjerknes, H. Solberg, Mrs. Rossby, Hedvig Bjerknes, V. Bjerknes, C.L. Godske, Baiet?, Schweiz?

V. Bjerknes lecturing in Tromsø, Norway, 1930.

*Leipzig Reunion. Heinz Lettau, Paul Mildner,
Vilhelm Bjerknes, and Ludwig Weickmann ((l.)*

Standing: H. Solberg, H.U. Sverdrup, S. Pettersen (l.r.).
eated: T. Bergeron, V. Bjerknes, J. Bjerknes, C.L. Godske (l.r.), Bergen, 193⁴

Bergen Researchers (ca. 1920). Jacob Bjerknes, Tor Bergeron, unknown, Svein Rosseland, unknown (l.r.).

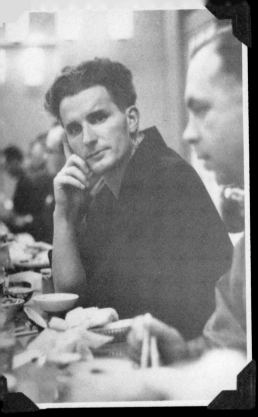

Ragnar Fjørtoft, Tokyo, 1960.

Norman Phillips,
Tokyo, 1960.

Jacob Bjerknes and Dave Fultz.
The 'Dishpan' Atmosphere, U. of Chicago (ca. 1953).

Erwin Biel with hat, Jule Charney with shoe, and Carl-Gustaf Rossby standing, Chicago waterfront, 1948.

Arnt Eliassen and Ed Lorenz, Norefjell, Norway, 1963.

Conference on Large-Scale Processes in the Atmosphere, MIT, 1952.

1. Loeser	8. White	15. Martin	22. Kuo
2. Keegan	9. Wahl	16. Hess	23. Solot
3. Thompson	10. Long	17. Benton	24. Shapiro
4. Lettau	11. Mintz	18. Wexler	25. J. Bjerknes
5. London	12. Widger	19. Charney	26. Lorenz
6. Dolezel	13. Fultz	20. Salmela	27. Starr
7. Cooley	14. Lowell	21. Willett	

Sir Napier Shaw and Vilhelm Bjerknes

Cyclone Climatology and its Relationship to Planetary Waves: A Review of Recent Observational Studies

EERO O. HOLOPAINEN

Department of Meteorology, University of Helsinki, Helsinki, Finland

1. Introduction

A main characteristic of atmospheric circulation is its large variability in time. This "large-scale turbulence," which is vividly illustrated in any sample of consecutive weather maps, has an important effect on the time-mean circulation. Extratropical cyclones are a fundamental part of this turbulence.

An arbitrary quantity can be formally represented as the sum of contributions from the zonally averaged time-mean flow, stationary (i.e., time-mean) eddies (SE) and transient eddies (TE). Extratropical cyclones are part of the TE's, which cover a wide range of frequencies. Because the spectrum of atmospheric eddies is essentially a continuous one, the definition of a "cyclone" is somewhat arbitrary. The flow field associated with such a scale is often obtained from the observed total field by applying some band-pass (BP) or high-pass (HP) temporal filter; an example of the former is found in Blackmon (1976), and of the latter in Hoskins et al. (1989). The study by Wallace et al. (1988) showed that many characteristics of the cyclone-scale eddies are relatively insensitive to the exact form of the temporal filter used, provided that disturbances with periods ranging from 2 to 6 days are retained.

The question concerning the role of cyclone-scale eddies in the general circulation is essentially the eddy–mean flow problem. It has at least two aspects: (1) the dependence of eddy activity and eddy properties on the mean flow, and (2) the feedback from eddies to the mean flow. This chapter is a general review of recent observational studies on the latter. In essence, it is an update of Holopainen (1990; hereinafter H).

2. Storm Tracks

2.1 Synoptic and Dynamical Storm Tracks

The path of the minimum pressure on a weather map is a synoptic definition of "storm track." In dynamically oriented studies a storm track ("waveguide") is currently inferred from the maximum of BP- or HP-filtered isobaric height variance or kinetic energy, or from lag-correlation maps (e.g., Wallace et al. 1988). If only unfiltered time series are available, the latitudinal position of the dynamical storm track can be identified by the maximum in the vertical velocity variance, for which maps are shown, for example, in Trenberth (1992) and, more approximately, in the meridional wind variance (see, e.g., Chang 1993).

Figure 1 is an illustration from Wallace et al. (1988) showing that synoptic and dynamical storm tracks are not the same: whereas the dynamical storm track is more or less zonal, the synoptic storm tracks or cyclone tracks typically curve poleward. The synoptic storm tracks were investigated by the original Bergen school scientists, and in his doctoral thesis from 1926, Prof. Erik Palmén also addressed this issue by studying the propagation direction of 245 cyclones. Due to shortage of data then, the climatology of the Southern Hemisphere synoptic storm tracks has naturally been revealed more recently (e.g., van Loon et al. 1972; Jones and Simmonds 1993).

The synoptic storm tracks are of primary importance for instantaneous weather conditions because they largely determine the wind and weather at the surface. The dynamical storm tracks, on the other hand, are more important in the discussion of the role of cyclone-scale eddies in the

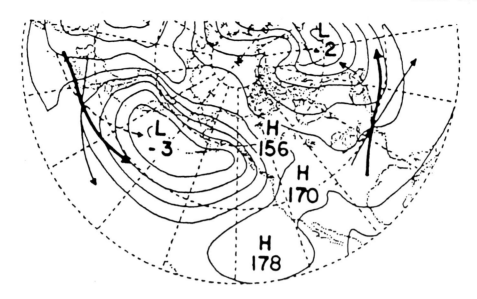

Fɪɢ. 1. Illustration of the difference between the synoptic and dynamical storm tracks for two grid points (40°N, 70°W and 35°N, 150°E), located on the axes of the two major storm tracks. The heavy arrows show the dynamic storm tracks inferred from lag-correlation study of the high-pass filtered 1000 hPa height anomalies. Synoptic tracks of cyclones and anticyclones, as inferred from unfiltered 1000 hPa height data, are shown by dashed and light solid arrows, respectively. Contours show the climatological mean wintertime 1000 hPa height field with contour interval 30 m (adapted from Wallace et al. 1988, Fig. 20).

atmospheric general circulation. This chapter addresses only dynamical storm tracks.

In recent years, several compilations of global atmospheric circulation statistics have appeared, including quantities that, in one way or another, also illustrate the long-term mean distribution of the dynamical storm tracks (e.g., Hoskins et al. 1989; Trenberth 1992). Figure 2 (from Hoskins et al. 1989) shows the ten-year wintertime mean temperature and the high-pass eddy heat flux at 700 mb. Because the cyclone-scale eddies effectively transport heat poleward across the storm tracks, the latitudinal position of the maximum meridional heat flux also indicates the location of the storm track. The well-known Pacific and Atlantic storm tracks of the Northern Hemisphere can be seen in Fig. 2. Note the strong baroclinicity in the upstream end of the storm tracks in the western parts of the oceans.

The storm tracks, as the circulation of the atmosphere in general, exhibit a large amount of low-frequency variability. Lau (1988) was a pioneer in exploring the dominant patterns of the month-to-month variability of the BP-filtered isobaric height variance. Over the Atlantic Ocean the first eigenvector (shown later in Fig. 8) appeared to depict the meridional migration of the eastern part of the Atlantic storm track from its time-mean position. The variability in the position of the storm tracks was found to be closely tied with the variability of the monthly mean flow in the middle troposphere. The latter can be thought as a kind of steering current for the cyclone-scale eddies, but also as being influenced by the presence of eddies.

Fraedrich et al. (1993) investigated the storm track changes associated with the dominant modes of the monthly mean surface pressure changes over Europe. This work confirmed the results by Lau (1988) concerning large meridional shifts of the storm tracks in the eastern part of the Atlantic. Such shifts occur, for example, in connection with the onset and breakdown of blocking highs in the eastern Atlantic or Europe. The symbiotic relation between planetary- and cyclone-scale waves has been illustrated in many other studies including Nakamura and Wallace (1990), Cai and Van den Dool (1991, 1992), Higgins and Schubert (1993), Sheng and Derome (1993), and Lin and Derome (1995).

2.2 Structure of the Cyclone-Scale Eddies in the Storm Tracks

The structure of transient cyclone-scale eddies has recently been documented in several studies by using one-point regression maps. For example, Chang (1993) used, as the reference time series, unfiltered time series of normalized 300-hPa meridional wind perturbations at a grid point in the most active region of the Pacific storm track (40°N, 180°E). He then derived regression statistics for perturbations in the horizontal wind, geopotential height, temperature, and the vertical velocity in other grid points. The statistical results, shown in Fig. 3, are in agreement with synoptic experience: the Z' and v' patterns tilt westward with height, warm air is rising and cold air sinking, there is a net meridional transfer of heat by the eddies, and so on. They also are broadly consistent with

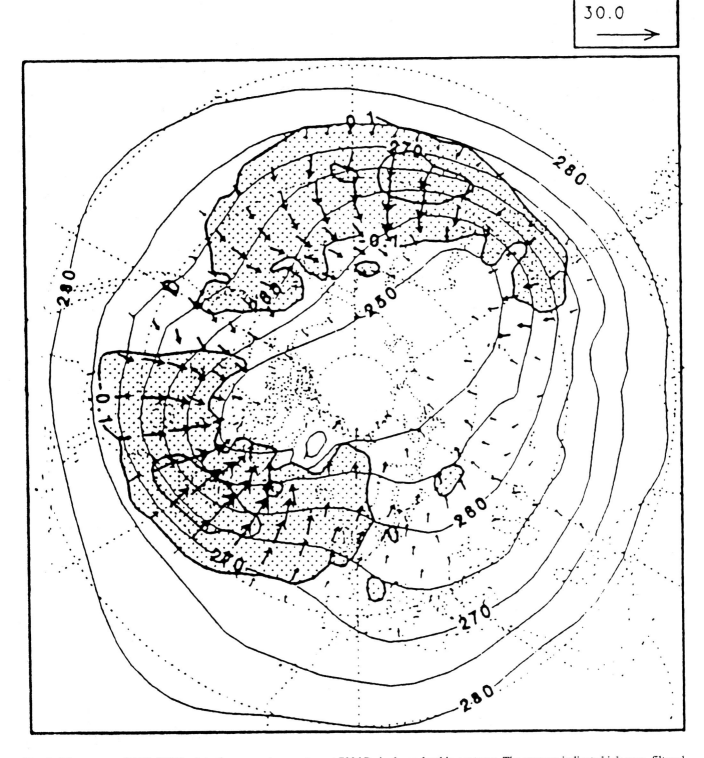

FIG. 2. The ten-year (1979–1989) wintertime mean temperature at 700 hPa is shown by thin contours. The arrrows indicate high-pass–filtered transient eddy temperature flux at 700 mb with the unit vector corresponding to 30 K m s^{-1}. The thick contours show the high-pass–filtered vertical temperature flux. The contour interval is 0.1 K Pa s^{-1}, and values less than -0.1 K Pa s^{-1} are shaded. Negative values indicate upward heat flux (Hoskins et al. 1989).

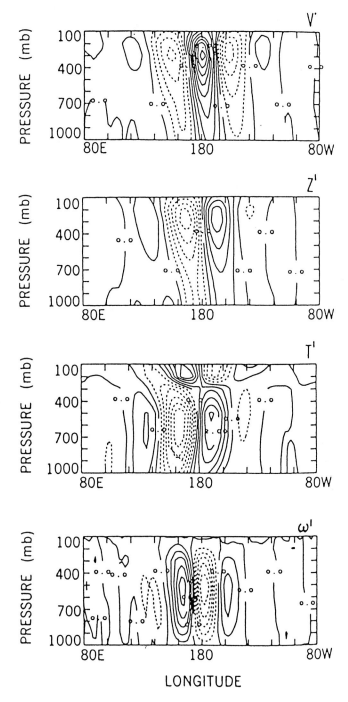

FIG. 3. Longitude–height plots of regression for v', Z', T', and ω' at 40°N over the Pacific sector based on time series of of v' at 300 hPa at 40°N, 180°W with zero time lag. Contour intervals are 2 m s^{-1}, 20 m, 0.5°C, and 0.02 Pa s^{-1}, respectively (Chang 1993).

those obtained in previous statistical studies (e.g., by Lim and Wallace 1991). If this kind of regression analysis is made with different time lags, the evolution from a more baroclinic structure at the western end of the storm track to a more barotropic structure at the eastern end can be seen (not shown here). An important dynamical issue concerns the effect of

the storm tracks on their environment. This interaction, caused mainly by the meridional transfers of heat and vorticity across the storm tracks, is discussed in Sections 3 and 4. Another issue concerns the dynamics of the storm track itself, involving, for example, factors that are important for the zonal length of the track. Some remarks on this issue are presented in Section 5.

3. Methods of Illustrating the Forcing Effect of Cyclone-Scale Eddies on the Mean Flow

There are several diagnostic methods of illustrating the forcing effect of cyclone-scale transient eddies on the larger-scale time-mean or slowly varying flow (see, e.g., Holopainen 1984 for a review). The net forcing consists of the sum of direct and indirect effects. In the conventional Eulerian framework, the direct forcing effects on temperature and momentum are given by the eddy flux convergence of these quantities, and the indirect eddy effect is created by the eddy-induced secondary circulations: the vertical motions induced by eddy fluxes affect temperature, and the horizontal divergent circulations that accompany these eddy-induced vertical motions affect the momentum balance. In general, the direct eddy effects are larger than the indirect eddy effects. As discussed in H, the direct and indirect eddy contributions depend on the framework used. The net effect, however, is independent of the framework.

3.1 Tendency Method

One of the methods used to illustrate the net forcing effect of the eddies on the time-mean flow is the "tendency method," introduced by Lau and Holopainen (1984). Here, the eddy effects on the time-mean geopotential are evaluated from the interior TE fluxes of potential vorticity and the TE heat fluxes at the lower and upper boundaries by inverting the associated three-dimensional Laplace operator. The net eddy forcing effect on mean temperature and mean geostrophic wind can be obtained from the corresponding geopotential forcing by using the hydrostatic and geostrophic relationships. If needed, the vertical motion can be obtained by solving the appropriate omega equation.

Figure 4 (adapted from Lau and Nath 1991) is a schematic illustration of the vertical motions induced by the meridional eddy fluxes of heat and vorticity across a storm track, and of the net zonal force that these fluxes create on the time-mean flow along the storm track. At the latitude of the storm track, the net force in the lower troposphere is directed eastward both in the case of poleward heat fluxes and vorticity fluxes. In the upper troposphere, however, the two contributions tend to oppose each other.

In several recent studies, the tendency method has been used in a simplified form by calculating only the forcing

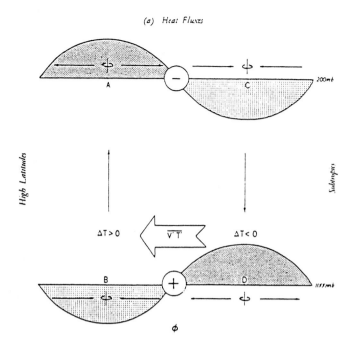

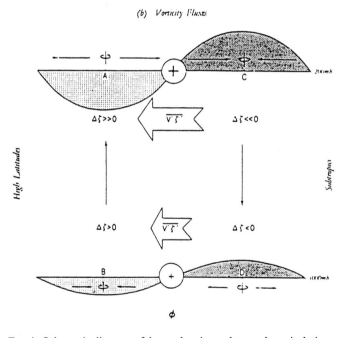

FIG. 4. Schematic diagram of the tendencies and secondary cirulations induced by (a) poleward eddy heat fluxes in the lower troposphere, and (b) poleward eddy vorticity fluxes with amplitudes increasing with height. The midpoint corresponds to the latitude of the dynamical storm track. Eastward (westward) zonal wind tendencies on the storm track due to cyclone-scale eddies are illustrated by plus (minus) signs within open circles (adapted from Lau and Nath 1991).

associated with cyclone-scale eddy vorticity fluxes (e.g., Lau 1988; Sheng and Derome 1993; Cuff and Cai 1995). In this case one only has to invert the two-dimensional (horizontal) Laplace operator to get the "barotropic" geopotential tendencies.

3.2 E vectors

Hoskins et al. (1983) introduced a quasi-vector **E** defined as

$$\mathbf{E} = (\overline{v'v'} - \overline{u'u'})\mathbf{i} - \overline{u'v'}\mathbf{j}.$$

E indicates the horizontal propagation direction of the eddy activity. A divergent (convergent) pattern of **E** vectors indicates a westerly (easterly) acceleration of the time-mean flow by the eddies. The **E** vector presentation is an extension in horizontal dimensions of the "Eliassen-Palm cross section" (Edmon et al. 1980), used to illustrate the propagation of eddy activity and the eddy-mean flow interaction in the mean meridional plane.

Compared with the tendency method, the **E** method suffers from the fact that it gives quantitative information of the direct eddy effect only (Holopainen 1984). This is enough if one studies only the barotropic interaction between the mean flow and eddies, because the indirect eddy effect is identically zero in this case. In the general baroclinic case, however, this is not true.

3.3 "Lorenz-energetics"

A classical method of studying the interaction between the mean flow and the eddies is the analysis of energetics. As seen in Fig. 2, the cyclone-scale eddies cause a smoothing effect on the mean temperature structure. This thermal smoothing effect of cyclone-scale waves on the time-mean flow has been known for a long time. Its implication for energetics is the damping of the available potential energy of the zonally averaged time-mean flow and of the SEs (e.g., Ulbrich and Speth 1991, Cuff and Cai 1995).

Regional studies of energetics are problematic, for example, because of ambiguities as to which terms should be considered as energy conversions and which as energy flux divergences. However, Chang (1993), briefly discussed in Section 5, may effect a revival of regional energetics studies.

4. Forcing Effect of Cyclone-Scale Eddies on the Time-Mean and Low-Frequency Components of Atmospheric Circulation

When we deal here with the geographical distribution of the forcing effect of the cyclone-scale TEs we mostly use the tendency method because it seems to provide the most illuminating way of quantitatively illustrating the net local forcing effect of the eddies.

4.1 Effect on the Climatological Time-Mean Flow

The wintertime long-term net climatological effect of the cyclone-scale eddies discussed, for example, in Lau and Holopainen (1984) and in H is an eastward-directed force in

TEND(TOT)   300 MB   DJF10 HIGH-PASS

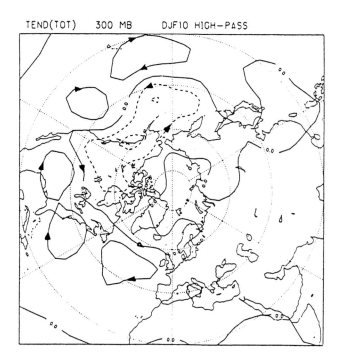

TEND(TOT)   1000 MB   DJF10 HIGH-PASS

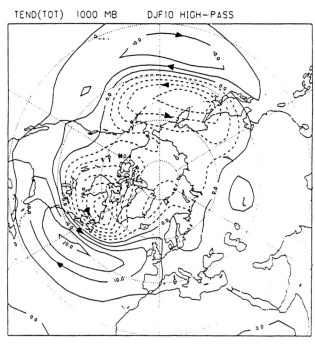

FIG. 5. Distribution of the geopotential tendencies associated with the cyclone-scale eddies in the northern extratropics in winter at 300 hPa (upper panel) and 1000 hPa (lower panel). The patterns are computed using the high-pass–filtered data from Hoskins et al. (1989). Unit: 10^{-4} m^2 s^{-3}. Contour interval 5×10^{-4} m^2 s^{-3}. Dashed lines indicate negative values. Arrowheads indicate the direction of the associated geostrophic wind tendencies.

the lower atmosphere. Figure 5 shows the wintertime geopotential height tendencies at 300 and 1000 hPa, calculated using the same algorithm as in H but using the high-pass filtered eddy statistics compiled by Hoskins et al. (1989). As expected on the basis of Fig. 4, a strong eddy-induced westerly force along the storm tracks off the eastern coasts of Eurasia and North America at about 40°N is seen at 1000 hPa. The effect of the cyclone-scale eddies obviously is of great importance for the maintenance of the long-term mean surface westerlies and the associated mean meridional pressure gradient in the midlatitudes.

As seen from the upper panel of Fig. 5, the time-mean force due to cyclone-scale eddies is much weaker in the upper troposphere than in the lower troposphere. Hence, the cyclone-scale eddies tend to enhance the barotropic component and to weaken the baroclinic component of the mean flow.

Figure 6 shows the forcing obtained for 300 hPa by using the two-dimensional (barotropic) version of the tendency method. This pattern should be compared with the fully three-dimensional results shown in the upper panel of Fig. 5. It is seen that the approximate tendency field has roughly the same pattern but an amplitude much larger than the total net forcing at the same level.

TEND(VORT)   300 MB   DJF10 HIGH-PASS 2-DIM

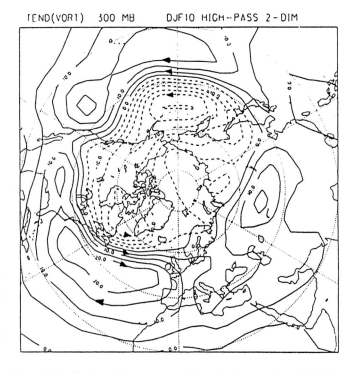

FIG. 6. Distribution of the geopotential tendencies associated with the cyclone-scale eddies at 300 hPa in the northern extratropics in winter, calculated with the two-dimensional ("barotropic") version of the tendency method. Unit and contours as in Fig. 5.

The time-mean eddy-induced westerly force due to cyclone-scale eddies on the storm tracks is so strong (of the order of 1 m s^{-1} per day) that it should be traceable to basic features of cyclone development, that is, to the occlusion process. Figure 7 (adapted from Godske et al. 1957) shows schematically an eastward-propagating extratropical cyclone in two stages of development. The coordinate system moves eastward along the dynamical storm track with the speed of the steering current. From panel a to panel b of Fig. 7, the cyclone center has moved (in accordance with Fig. 1) northward from the dynamical storm track, and the westerly geostrophic wind has increased in the map area. This increase is due to structural changes of cyclones in connection with the occlusion process. The statistical effect of such increases gives the climatological effect seen in the lower panel of Fig. 5. However, the increase of surface westerlies due to cyclone development in the area of the dynamical storm track can be best illustrated with numerical experiments, such as reported by Simmons and Hoskins (1979).

The long-term time-mean forcing seen in Fig. 5 can formally be split into the zonally averaged time-mean component and the stationary eddy (SE) component. The former tends to shift the zonally averaged time-mean jet stream poleward and to enhance (reduce) the barotropic (baroclinic) component of the zonally averaged time-mean flow in the middle latitudes. The effect of cyclone-scale eddies on the SE isobaric height variance was shown in H to be to maintain it, particularly in the lower troposphere.

4.2 Effect on the Low-Frequency Components of the Tropospheric Circulation

Most of the variance of the atmospheric circulation resides on time scales much longer than the lifetime of cyclone-scale eddies. The reasons for this low-frequency variability have not yet been fully revealed. In the extratropics, one of the potential causes is forcing by the cyclone-scale eddies.

When the storm track (and the associated time-mean steering current) changes, the concomitant forcing pattern also changes. Lau (1988) is a pioneer study on how the storm tracks and the associated mean-flow forcing relate to the low-frequency fluctuations of the general circulation. The principal modes of month-to-month variability of the wintertime storm tracks in the Northern Hemisphere were identified by an empirical orthogonal function (EOF) analysis of the bandpass-filtered geopotential height variance. The distributions of eddy-induced (barotropic) forcing indicated a near in-phase relationship between the synoptic-scale forcing and the quasi-stationary (monthly mean) flow pattern at 300 hPa. Lau and Nath (1991) expanded the results of Lau (1988) by calculating, for each winter month of the period 1966–1984, the full geopotential height tendencies due to cyclone-scale eddies. Throughout the troposphere, the patterns of the geopotential height tendencies appeared to exhibit a positive spatial correlation with the concurrent monthly averaged height anomaly. The characteristic time scale of this constructive eddy forcing in the storm track region ranged

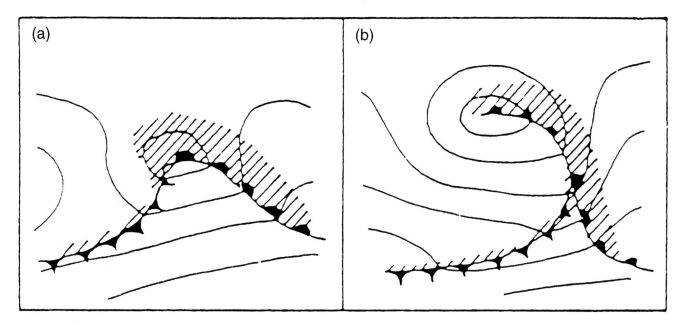

FIG. 7. Schematic surface map (isobars, fronts, and precipitation areas) for two development stages of an eastward-moving extratropical cyclone (adapted from Fig. 15.00.1 in Godske et al. 1957).

from several days at 1000 mb to 1–2 months near the tropopause.

The upper panel of Fig. 8 (from Lau 1988) illustrates the most important pattern (eigenvector A1) of the month-to-month variability of the bandpass-filtered 500 hPa height variance in the Atlantic. When the coefficient of this pattern is negative, the storm tracks in the eastern Atlantic are shifted northward from their climatological mean position. In this situation, which often occurs in connection with blocking in the eastern Atlantic, the cyclone-scale eddies take a northerly route around the blocking high. The lower panel of Fig. 8 shows the 1000 hPa geopotential tendency pattern associated with this mode. A negative coefficient on this pattern indicates enhanced westerly forcing in the lower troposphere along the storm track at about 60°N.

Fraedrich et al. (1993) studied the climate anomalies and simultaneous bandpass eddy activity anomalies in Europe. By using the **E** vector diagnostics, they studied the interaction between the mean flow and transient eddies at 500 mb north of 20°N. Their results agreed with those of Lau (1988) and Lau and Nath (1991). In particular, high-pressure systems over Europe were found to be supported by an anomalous transient eddy forcing, which enhances the zonal wind to the north of the high pressure and generates anticyclonic vorticity about 10 degrees upstream from its center.

Cai and Van den Dool (1991, 1992) studied the mutual dependence between the low-frequency flows and their attendant traveling storm tracks, using twice-daily tropospheric National Meteorological Center (NMC) data from ten winters. They found that this dependence resembles that between the climatological stationary waves and the climatological storm tracks. The barotropic feedback (due to eddy vorticity fluxes) of the traveling storm track tends to reinforce the low-frequency waves and to retard their propagation, whereas the baroclinic feedback tends to damp the low-frequency temperature field throughout the troposphere; this damping effect was recently examined in more detail by Lin and Derome (1995) using 700 hPa analyses for five winters.

Sheng and Derome (1993) examined the interaction between the low-frequency transients and synoptic-scale eddy forcing using 300 hPa data from 1981 to 1986, as analyzed at the European Centre for Medium-Range Weather Forecasts (ECMWF). They calculated the height tendency of the slow transients due to vorticity forcing by synoptic-scale eddies; that is, they used the approximate tendency method. They, too, found that the forcing by the synoptic-scale eddies plays an important role in the maintenance of the low-frequency eddies, particularly in the eastern part of the Pacific and Atlantic sectors, where the low-frequency variability is strong. The lag-correlation analysis between the forcing and the low-frequency height field indicated that the former leads the latter by a very small phase difference (about one day). A theoretical study by Branstator (1995) also nicely demonstates the two-way feedback between high- and low-frequency disturbances.

In the diagnostic studies mentioned above, the Northern Hemisphere conditions have been considered. In the Southern Hemisphere, essentially the same symbiotic relationship between the high- and low-frequency transients as has been found for the Northern Hemisphere (e.g., Cuff and Cai 1995) appears to exist.

The cyclone-scale eddies thus appear quite important in the maintenance of the low-frequency components of atmospheric circulation, and may sometimes even trigger their development. Nevertheless, these eddies may not be the most important mechanism for the initial growth of these components. The results of a composite study by Black and Dole (1993), for example, indicate that the primary mechanism for the growth of a large-scale, long-lived cyclone over the North Pacific can be the nonmodal instability of the time-mean flow.

4.3   Effect on Blocking

As discussed, for example, in H, there is a lot of evidence (e.g., that provided by Mullen 1987) that migratory cyclones and anticyclones are important in the maintenance of blocking. Much less is quantitatively known about their role during the onset and decay of persistent blocking patterns. Nakamura and Wallace (1990, 1993) have used a compositing technique to study the role of cyclone-scale eddies in the onset of blocking. According to Nakamura and Wallace (1990), the most striking common feature during the onset of blocking over the midlatitude eastern oceans is the enhancement of baroclinic wave activity along the upstream storm track. This enhancement tends to occur about five days before the blocking pattern is fully established.

One of the difficulties in studying the role of cyclone-scale eddies in the onset and decay of blocking is that the spectrum of atmospheric fluctuations is a continuous one. Therefore any partitioning of the total atmospheric flow in a given synoptic situation into "mean" and "eddy" components is always somewhat arbitrary. The conventional Reynods expressions for the eddy fluxes and their mean-flow effects are not valid unless the "mean" flow is clearly separated by a spectral gap from the "eddy" part of the flow. This is typically not the case in the onset and decay phase of blocking.

5.   Studies on the Storm Track Dynamics

Hoskins and Valdes (1990) used a linear stationary wave model to investigate storm track dynamics. They determined the forcing terms in the storm track region from data over several winters. As can be expected on the basis of Fig. 2, the direct effect of eddy heat transports was found to act against

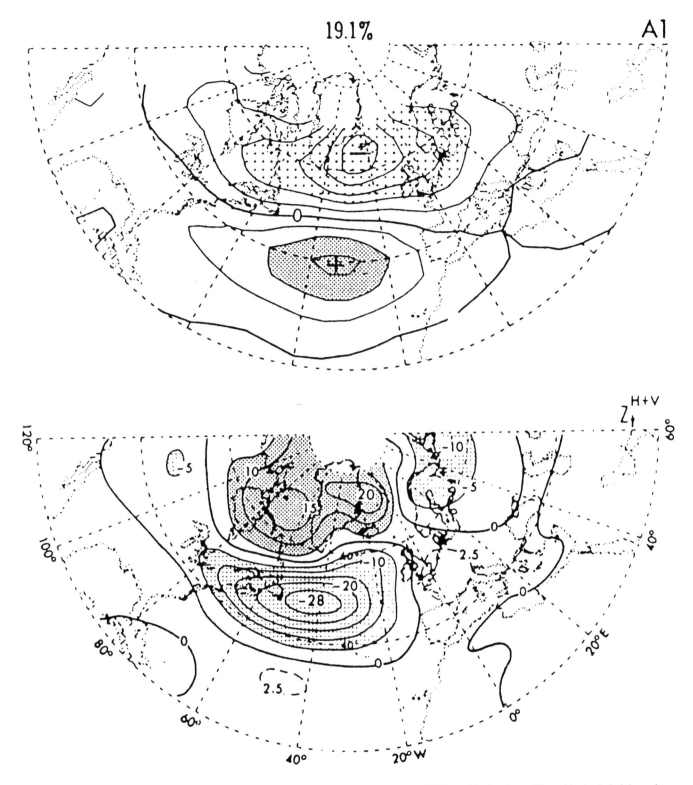

FIG. 8. The upper panel shows the most important mode (A1) of the month-to-month variability of the bandpass-filtered isobaric height variance at 500 hPa (Lau 1988). The lower panel shows the corresponding 1000 hPa height tendency associated with the bandpass eddy forcing (Lau and Nath 1991).

the existence of a strong baroclinic zone. They also found that the diabatic heating (an indirect eddy effect!) acted to maintain the baroclinicity in the storm track region.

In the studies by Chang and Orlanski (1993) and Chang (1993), the wave trains on the storm tracks were found to exhibit characteristics of downstream development, with successive perturbations developing toward the downstream side of existing perturbations (Fig. 9). An analysis of the eddy kinetic energy budget of the wave train indicated that downstream radiation of geopotential fluxes by existing perturbations triggers the development of new eddies downwind, with baroclinic conversion becoming important only during the later part of the life cycle of a downstream-developed wave. This study is perhaps the best observational validation so far of the group velocity concept, introduced in meteorology by the "Chicago school." Chang (1993) used unfiltered data and demonstrated that filtering of the time series was the likely reason the delicate signal of downstream development was not found in the earlier investigations (e.g., in Lim and Wallace 1991).

Branstator (1995) studied the organization of storm track anomalies by the background flow in a general circulation model. His storm track model may possibly serve as a useful means of parameterizing fluxes by cyclone-scale eddies in low-frequency models of the atmosphere.

6. Remarks

The structure and transfer properties of cyclone-scale waves and their interaction with the larger-scale flow are now relatively well known from observations. However, there are still many unanswered questions concerning the dynamics of the storm tracks. For example, whereas the traditional studies on the life cycles of baroclinic waves (e.g., Edmon et al. 1980; Hoskins et al. 1983) have emphasized the baroclinic growth of the eddies in the western part of the storm tracks and the barotropic decay at the eastern end of the tracks, the observational study by Chang (1993) shows that the dynamics of the storm tracks is likely to be more complicated. The convergence and divergence of energy fluxes may be as important a factor in the growth and decay of the eddies as the processes considered in earlier studies.

What determines the differences between the storm tracks in the Northern and Southern Hemispheres and the Atlantic and Pacific sectors of the Northern Hemisphere? What is the role played, for example, by ocean and land distribution, or sea-surface temperature distribution? How is the propagation of individual eddies and wave packets influenced by quasi-stationary waves? Answering these and related questions requires further modeling studies, and also field experiments, such as FASTEX (Fronts and Atlantic Storm Track Experiment).

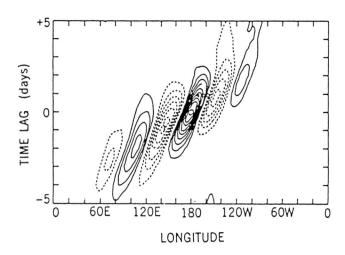

FIG. 9. Longitude–time lag correlation of 300 hPa meridional wind (unfiltered) averaged over 30°–60°N for seven winter seasons. The time series at the dateline is taken to be the reference for correlation (Chang 1993).

Acknowledgments

Jussi Kaurola and Jouni Räisänen are acknowledged for providing technical assistance in the preparation of the manuscript, and Mike Wallace for providing helpful comments.

References

Black, R. X., and R. M. Dole, 1993: The dynamics of large-scale cyclogenesis over the North Pacific Ocean. *J. Atmos. Sci.*, **50**, 421–442.

Blackmon, M. L., 1976: A climatological spectral study of the 500-mb geopotential height of the Northern Hemisphere. *J. Atmos. Sci.*, **33**, 1607–1623.

Branstator, G., 1995: Organization of storm track anomalies by recurring low-frequency circulation anomalies. *J. Atmos. Sci.*, **52**, 207–226.

Cai, M., and H. Van den Dool, 1991: Low-frequency waves and travelling storm tracks. Part I: Barotropic component. *J. Atmos. Sci.*, **48**, 1420–1436.

———, and ———, 1992: Low-frequency waves and travelling storm tracks. Part II: Three-dimensional structure. *J. Atmos. Sci.*, **49**, 2506–2524.

Chang, E. K., 1993: Downstream development of baroclinic waves as inferred from regression analysis. *J. Atmos. Sci.*, **50**, 2038–2053.

———, and I. Orlanski, 1993: On the dynamics of a storm track. *J. Atmos. Sci.*, **50**, 999–1015.

Cuff, T. J., and M. Cai, 1995: Interaction between the low- and high-frequency transients in the Southern Hemisphere winter circulation. *Tellus*, **47A**, 331–350.

Edmon, H. J., B. J. Hoskins, and M. E. McIntyre, 1980: Eliassen–Palm cross sections for the troposphere. *J. Atmos. Sci.*, **37**, 2600–2616.

Fraedrich, K., C. Bantzer, and U. Burkhard, 1993: Winter climate anomalies in Europe and their associated circulation at 500 hPa. *Climate Dyn.*, **8**, 161–175.

Godske, C. L., T. Bergeron, J. Bjerknes, and R. D. Bundgaard, 1957: *Dynamical Meteorology and Weather Forecasting.* American Meteorological Society and Carnegie Institution of Washington, 800 pp.

Higgins, R. W., and S. D. Schubert, 1993: Low-frequency synoptic-eddy activity in the Pacific storm track. *J. Atmos. Sci.*, **50**, 1672–1690.

Holopainen, E.O., 1984: Statistical local effect of synoptic-scale transient eddies on the time-mean flow in the Northern Hemisphere in winter. *J. Atmos. Sci.*, **41**, 2505–2515.

——, 1990: Role of cyclone-scale eddies in the general circulation of the atmosphere: a review of recent observational studies. *Extratropical Cyclones – Palmén Memorial Volume*, C. W. Newton and E. O. Holopainen, Eds., Amer. Meteor. Soc., 47–62.

Hoskins, B. J., and P. J. Valdes, 1990: On the existence of storm tracks. *J. Atmos. Sci.*, **47**, 1854–1864.

——, I. M. James, and G. H. White, 1983: The shape, propagation and mean-flow interaction of large-scale weather systems. *J. Atmos. Sci.*, **40**, 1595–1612.

——, H. H. Hsu, I. N. James, M. Masutani, P. D. Sardeshmukh, and G. H. White, 1989: Diagnostics of the global atmospheric circulation based on ECMWF analyses 1979–1989. WCRP-27, WMO/TD-No. 326.

Jones, D. A., and I. Simmonds, 1993: A climatology of Southern Hemisphere extratropical cyclones. *Climate Dyn.*, **9**, 131–145.

Lau, N.-C., 1988: Variability of the observed midlatitude storm tracks in relation to low-frequency changes in the circulation pattern. *J. Atmos. Sci.*, **45**, 2718–2743.

——, and E. O. Holopainen, 1984: Transient eddy forcing of the time-mean flow as identified by geopotential tendencies. *J. Atmos. Sci.*, **41**, 313–328.

——, and M. J. Nath, 1991: Variability of the barotropic and baroclinic forcing associated with monthly changes in the midlatitude storm tracks. *J. Atmos. Sci.*, **48**, 2589–2613.

Lim, G. H., and J. M. Wallace, 1991: Structure and evolution of baroclinic waves as inferred from regression analysis. *J. Atmos. Sci.*, **48**, 1718–1732.

Lin, H., and J. Derome, 1995: On the thermal interaction between the synoptic-scale eddies and the intraseasonal fluctuations in the atmosphere. *Atmos.-Ocean*, **33** (1), 81–107.

Mullen, S. L., 1987: Transient eddy forcing of blocking flows. *J. Atmos. Sci.*, **44**, 3–22.

Nakamura, H., and J. M. Wallace, 1990: Observed changes in baroclinic wave activity during the life cycles of low-frequency circulation anomalies. *J. Atmos. Sci.*, **47**, 1100–1116.

——, and ——, 1993: Synoptic behavior of baroclinic eddies during the blocking onset. *Mon. Wea. Rev.*, **121**, 1892–1903.

Sheng, J., and J. Derome, 1993: Dynamic forcing of the slow transients by synoptic-scale eddies: An observational study. *J. Atmos. Sci.*, **50**, 757–771.

Simmons, A. J., and B. J. Hoskins, 1979: The downstream and upstream development of unstable baroclinic waves. *J. Atmos. Sci.*, **36**, 1239–1254.

Trenberth, K. E., 1992: Global analyses from ECMWF and atlas of 1000 to 10 mb circulation statistics. NCAR/TN-373+STR, NCAR Tech. Note, 191 pp.

Ulbrich, U., and P. Speth, 1991: The global energy cycle of stationary and transient atmospheric waves: Results from ECMWF analyses. *Meteorol. Atmos. Phys.*, **45**, 125–138.

van Loon, H., J. J. Taljaard, T. Sasamori, J. London, D. V. Hoyt, K. Labitzke, and C. W. Newton, 1972: *Meteorology of the Southern Hemisphere. Meteorological Monographs*, No. 35, Amer. Meteor. Soc., 263 pp.

Wallace, J. M., G-H. Lim, and M. L. Blackmon, 1988: On the relationship between cyclone tracks, anticyclone tracks and baroclinic wave guides. *J. Atmos. Sci.*, **45**, 439–462.

Planetary Waves and Their Interaction with Smaller Scales

ISAAC M. HELD

Geophysical Fluid Dynamics Laboratory, NOAA, Princeton, New Jersey, USA

1. Introduction

Historically, dynamical meteorologists have grown comfortable with two classes of idealized dynamical systems. On the one hand, our understanding of baroclinic instability has evolved by studying the linear and nonlinear development of eddies on the scale of the deformation radius in simple larger-scale environments. On the other hand, there has been a continuing interest in planetary wave theory, aimed at understanding the climatological stationary waves, and a growing interest in the distinctive low-frequency variability observed on these larger scales. The realization that planetary scales have a potentially intricate dynamics of their own has encouraged the use of models in which instability on smaller scales is absent, so that this planetary dynamics can be isolated.

Two examples of papers that have heightened interest in the dynamics of planetary waves in isolation from deformation-scale eddies are Charney and DeVore (1979) on the possibility of multiple equilibria in nonlinear models of topographically forced waves, and Simmons et al. (1983) on the barotropic instability of the wintertime planetary wave field. These two papers also serve as examples of two rather distinct streams in research on planetary-scale dynamics—one focusing on the details of the linear dynamics around some mean flow, and the other taking a nonlinear systems perspective as its starting point.

We are much less comfortable with the complexity of models in which synoptic and planetary scales are both fully active. Our theoretical tools are typically less powerful in this more complex setting. For example, there have been attempts to generalize wave–mean flow interaction theory to the case of zonally asymmetric basic states, but the results are less useful than in the symmetric case. In particular, there typically is no simple relation between "non-conservative" effects on the eddies and their feedback onto the mean flow, as there is for waves on a zonally symmetric flow (Plumb 1986). One might hope that WKB theories dependent on scale separation between the waves and the mean flow would allow one to use some of the insights gained from the study of evolution on zonally symmetric basic states, and such theories are useful on occasion, but one rarely has much scale separation to work with.

The problem of the mutual interaction of planetary and deformation-scale eddies is being attacked from a variety of directions: linear instability and error growth are being studied in ever more complex flows; barotropic models are being used for studying wave-breaking scenarios in different larger-scale environments; the simplest two-layer quasigeostrophic model has a rich dynamics when planetary and synoptic waves are both present, of which we have had a few glimpses; and idealized GCMs are a powerful, as yet nearly untapped source of information about the interaction between storm tracks, low frequency variability, and stationary waves. While there is important ongoing work in all of these areas, much effort will be required to determine the relationships between different classes of models. A critical review at this time seems too daunting a task.

Instead, in the following I try to discuss some confusing aspects of planetary wave dynamics, most of which involve the interaction with smaller scales—starting with the perspective provided by the classical ideas of geostrophic turbulence.

2. Scale Separation and Geostrophic Turbulence

As one moves to larger scales in midlatitudes, the flow becomes more wave-like and, relatively speaking, linear. The simplest explanation for this tendency is based on the Rossby wave dispersion relation, and the comparison between typical eddy velocities u' on a given horizontal scale k^{-1} and the intrinsic Rossby wave phase speed, $c - U = -\beta/k^2$. Streamlines will overturn in the horizontal plane, in a frame of reference moving with a steady wave, if $u' > U - c$.

Therefore, one might hope to use the scale

$$k_\beta = (\beta/u')^{1/2} \qquad (1)$$

at which this ratio is unity, as the dividing line between planetary waves and strongly nonlinear eddies.

As described by Rhines (1975), k_β is also the scale at which a barotropic inverse energy cascade will be halted by the ß-effect, for it is at this scale that the nonlinear interactions weaken. The homogeneous turbulence picture of this cascade is familiar. Most important, the spectrum peaks at the scale at which the cascade stops, and not where it is being "stirred" by baroclinic activity near the deformation scale λ. The cascade stops when the energy-containing eddies become marginally linear; waves with $k < k_\beta$ are excited inefficiently and selectively.

It is of central importance to the Earth's atmosphere that the inverse energy cascade is, in fact, quite modest, and does not extend over a significant range of scales: the difference between k_β and λ^{-1} is subtle. If we want to understand why the cascade is weak, we have to understand the eddy energy level in the atmosphere. Recent work of V. Larichev and the author (Larichev and Held 1995; Held and Larichev 1996, HL hereafter) suggests that one should expect $u' \sim U_T(k_\beta\lambda)^{-1}$ if the cascade is stopped by the β-effect rather than the size of the domain, where U_T is the magnitude of the thermal wind through the troposphere, so that

$$k_\beta\lambda \sim (\beta\lambda^2)/U_T \qquad (2)$$

Since the right-hand side of this expression is of order unity in our atmosphere ($U_T = \beta\lambda^2$ is essentially the critical shear for instability in Phillips's two-layer model), this scaling is consistent with the fact that $k_\beta\lambda \sim 1$. Whether or not this scaling is completely accurate, it is clear that a very substantial cascade requires $u' \gg U_T$, which is not observed.

The estimate (2) of HL derives from a study of homogeneous turbulence, but similar results are found for baroclinically unstable jets. In particular, in the two-layer unstable jet model of Whitaker and Barcilon (1995) the wavenumber of the energy containing eddies is also found to be approximately proportional to β. Whether the dynamics underlying the choice of this scale is the same in homogeneous and jet-like flows is still unclear. In particular, Whitaker and Barcilon argue that the linear growth rates for unstable normal modes continue to play an important role in this scale-selection, even when the flow is strongly nonlinear, while these linear growth rates play no role in the discussion of HL.

The wavelength of a stationary external Rossby wave is $k_s = (\beta/U_b)^{1/2}$, where U_b is the mean zonal wind weighted in the vertical by the amplitude of the external mode geopotential (e.g., Held et al. 1985). But in the atmosphere U_b is comparable to U_T. If we take $u' \sim U_T \sim U_b$, then, $k_\beta \sim k_s$. From the perspective of this crude scaling, we might expect the inverse

cascade to carry energy from eastward propagating into quasi-stationary disturbances, perhaps, but not much further. However, this same scaling suggests that there is nothing fundamental that requires the energy-containing eddies to be quasi stationary. More generally, by accepting (2) we have

$$(\beta/k_\beta^2)/U_T \sim U_T/(\beta\lambda^2), \qquad (3)$$

implying that the energy-containing scales would propagate westward if the thermal wind were strong enough.

The cascade appears to be particularly weak in Southern summer. The eddy kinetic energy is often sharply peaked at zonal wavenumbers 6-7, and can be organized into coherently propagating wavepackets (Lee and Held 1993). Occasionally, energy makes its way to wavenumber 5, which can then dominate the circulation, evolving more slowly and propagating eastward at ~5 m/s (Salby 1982; Hamilton 1983). The flow appears to be more wave-like and "quasi-linear" when this occurs.

The larger energy levels in Northern winter imply a somewhat stronger inverse cascade; even in the absence of external sources of planetary waves, these energy levels should result in disturbances of larger scale that are quasi-stationary or even westward propagating. Of course, orography and quasi-stationary heating are major sources of planetary waves in Northern winter, primarily at scales near k_s within the troposphere. In addition, the weak low-level static stability in the eddy-generation regions of the oceanic storm tracks favors the growth of smaller-scale eddies, as does latent heat release. The result is a sense of scale separation, by a factor of two perhaps, between planetary and synoptic scales. But there is no clear energy gap between the scale of the quasi-stationary forced waves and the scales reached by the inverse energy cascade, and this is a major source of complexity.

It should also be emphasized that one can get a sense of scale separation between energy-containing "long waves" and weather-producing "short waves" even in an atmosphere with a single peak in its energy spectrum. The baroclinic eddies that produce the greatest vertical motion and divergence, and the most weather, will naturally tend to be of somewhat smaller scale than the eddies with the most energy, and standard quasi-geostrophic scaling tells us that this distinction can be substantial, since

$$w' \sim (H/L)R_o u' \sim k^2 u'$$

for a fixed strength of the horizontal flow and vertical scale (here R_o is the Rossby number). The spectrum of the vertical velocity differs by a factor of the *fourth* power of the wavenumber from that of the horizontal velocity.

The absence of clear scale separation suggests that we still have to contend with turbulent cascades when thinking about the interaction between these planetary waves and deformation-scale eddies, even when stationary forcing is providing a large direct source of planetary waves. In particular, the turbulence perspective reminds us not to be surprised to find deformation-scale eddies reinforcing the kinetic energy of structures with larger scales. Blocking anticyclones provide an important example. A tension exists in the literature on blocking between the role played by the large-scale flow itself, as it falls into a coherent blocked state with intrinsically small time tendency, and the role played by eddy driving. Given our dissipative and turbulent atmosphere, it seems a safe bet to assume that both are required for a long-lived pattern: a large-scale coherent structure *and* positive reinforcement from smaller scales. If the large-scale pattern has a strong tendency to evolve, there is little hope of creating a long-lived anomaly, whatever the interaction with small scales, while a coherent large-scale structure that is not actively maintained will quickly dissipate. In the idealized barotropic models in which eddies are successful in maintaining a blocked flow, the flow itself is generally close to a stationary state (Shutts 1983; Anderson 1995).

Jupiter's Great Red Spot (GRS) is a useful analogy. There is clearly an underlying coherent structure, but it is actively maintained by an inverse energy cascade as well. The manner in which the anticyclonic vorticity is resupplied is distinct in Jovian and terrestrial cases, however; the GRS swallows smaller anticyclonic vortices that are trapped in the same latitude belt; in terrestrial blocks the potential vorticity (PV) is instead resupplied by baroclinic waves that sweep in lower-latitude air as they approach the block.

The distinction between free and forced structures is of central importance to planetary wave–synoptic eddy interaction. If one forces some zonally asymmetric structure on the atmosphere, the storm tracks will be deformed, and the resulting PV fluxes may or may not feed back positively on the forced eddy. But from the perspective of the inverse energy cascade, we *expect* to see *free* large-scale structures generated and maintained by smaller eddies.

In addition to the inverse kinetic energy cascade, there is a less familiar direct cascade of available potential energy. In baroclinic geostrophic turbulence models with a substantial inverse cascade, there is a sharp distinction on large scales between the barotropic energy, which cascades upscale, and the baroclinic energy, which cascades downscale (Rhines 1977, Salmon 1980, Hoyer and Sadourny 1982). On scales much larger than the deformation radius, the baroclinic energy is essentially the available potential energy, or temperature variance. To the extent that the inverse cascade proceeds for a large range of scales, the flow becomes more and more barotropic, and the baroclinic potential vorticity (or the temperature at the surface and tropopause) is *passively* advected by this barotropic flow. The advection is passive because these fields are not inducing the bulk of the flow that is advecting them. Just as for any passive tracer, the temperature variance and available potential energy cascade to smaller

scales. This cascade changes its character at the deformation radius, where the flow is no longer predominately barotropic, and the advection is no longer passive.

Much of the atmospheric flow on planetary scales is *equivalent* barotropic and not fully barotropic. Yet some of this flavor carries over to the atmosphere. The planetary scales, whether excited externally or by an inverse energy cascade, deform the surface and tropopause temperature fields. The smaller scale, weather producing eddies see this deformed baroclinicity and, in their distinctive way, generate a cascade of temperature to even smaller scales as they simultaneously remove some of the baroclinic energy and transfer it to the barotropic component of the flow, re-energizing the inverse cascade. We speak of short waves as steered by and extracting available potential energy from the planetary scales, but perhaps it is useful to think of this interaction as part of a more continuous downscale cascade of *baroclinic* variance. On the smaller scales on which the surface and tropopause disengage, frontal instabilities can be thought of as another step in the cascade of surface temperature variance to small scales.

3. External Rossby Waves

Given that planetary-scale motions are somewhat Rossby wave-like, we are obligated to take a close look at the relevant dispersion relation. Most planetary waves of interest are stationary or quasi-stationary, and are strongly affected by horizontal and vertical shears in the mean zonal flow. At first glance, it is not self-evident which dispersion relation to focus on. To help decide, it is useful to think about the linear stationary response to a localized source (thermal or orographic) in the midlatitude westerlies.

A localized source excites disturbances of all scales, and the largest of these propagate through the tropospheric westerlies into the stratosphere. As one moves farther from the source, this vertically propagating component becomes less important than the part of the spectrum trapped in the *tropospheric planetary waveguide*. As the disturbance propagates downstream, the vertical structure within the troposphere quickly settles into the dominant vertical mode within this waveguide, the external Rossby wave. External Rossby wavetrains then move downstream as they are refracted by the spherical geometry and horizontal shears.

Explicit in this description is the claim that it is useful to think about modal structures in the vertical (the external Rossby wave, at least), but not in the horizontal. This is partly a question of time scale—the wave feels the vertical waveguide structure much sooner than it feels the global constraints on the horizontal structure. But more important, many disturbances never feel these global constraints at all; external mode wavetrains are often refracted into the tropics, where they break and can be efficiently absorbed. Although there may be cases in which quasi-stationary waves are effectively trapped in the large PV gradient along a strong jet, it seems to be a useful approximation, for quasi-stationary waves, to assume absorption due to wave-breaking in the tropics. Results consistent with this picture are obtained when idealized GCMs with a zonally symmetric climate are perturbed by a localized mountain, providing a relatively simple localized source of stationary waves. As described in Cook and Held (1992), the dominant ray paths change as the amplitude of the mountain increases (the "mountain" in this work has the scale of the Tibetan plateau), but in all cases the wavetrains propagate rather sharply into the tropics and do not emerge. More recent higher-resolution calculations confirm this picture.

Rapidly westward propagating waves, such as the "5-day wave," the gravest symmetric wavenumber 1 Rossby mode, hardly feel the horizontal and vertical shears and do not break in the tropics, and so are free to settle into global modal structures. These are clearly waves with "$k < k_\beta$". There exist several good summaries of the modal structure of the westward propagating variance (Madden 1979; Salby 1984; Venne 1989). As the westward phase speed of a disturbance decreases, the modal picture based on a zonally symmetric basic state becomes less useful; the "16-day wave," the second gravest symmetric wavenumber 1 mode, may be marginal in this sense, as argued by Branstator and Held (1995).

Are there any "internal" Rossby waves? There are vertically propagating waves, of course, but for realistic parameter settings there are no other neutral modes trapped in the tropospheric waveguide (Held et al. 1985). This has numerous consequences, the first of which is that quasi-stationary waves, removed from their source, should have a single vertical structure, that of the external mode. Therefore, we expect an equivalent barotropic structure with no nodes in the vertical, as long as the wave remains in the westerlies. The external mode's geopotential increases with height in the troposphere, and has maximum amplitude near the tropopause, resulting in cold lows and warm highs. A good rule of thumb exists as a consequence: warm lows are locally forced; warm highs are not.

One can understand this vertical structure by considering the dominant balance in the temperature equation for a wave on scales larger than the deformation radius:

$$(U-c)\frac{\partial}{\partial x}\frac{\partial \psi}{\partial z} - \frac{\partial \psi}{\partial x}\frac{\partial U}{\partial z} \sim 0, \qquad (4)$$

which implies that $\psi \sim U - c$. In the frame of reference of the wave, the zonal advection of the perturbation temperature $\partial \psi/\partial z$ is primarily balanced by the meridional advection of the mean temperature (with gradient $\partial U/\partial z$) by the perturbation meridional flow $\partial \psi/\partial x$: the high must be warm for the eastward advection of the warm air to balance the cooling in

the northerlies to the east of the high. The effects of vertical motion are relegated to a secondary role on these scales.

The external mode dispersion relation in vertical shear, and the accompanying modal structures, have not received as much attention as they deserve, given their central importance to planetary wave structure in the atmosphere. Due to the equivalent barotropic structure, we have been drawn toward thinking in terms of the simplest barotropic models, and have learned a great deal. However, for the scales of most interest, those that produce stationary or low frequency disturbances, external Rossby waves do not so much resemble simple barotropic Rossby waves as they do baroclinically unstable waves. The implications of this fact are, I think, most easily understood by thinking in terms of the pseudomomentum of the wave.

The pseudomomentum M is a quadratic measure of wave amplitude that is conserved in adiabatic, inviscid flow, and that tells one how the wave modifies the zonal mean flow when it propagates from one region to another or is dissipated. The flux of pseudomomentum in the meridional-vertical plane is also known as the Eliassen-Palm flux (Edmon et al. 1980). The pseudomomentum of a barotropic Rossby wave at a given latitude is simply

$$M = -\{\eta'^2\}\,\beta/2 \qquad (5)$$

where η' is the meridional particle displacement (the meridional displacement of a streamline in a steady wave). The brackets are an average over the phase of the wave. More generally, in the case of an unbounded atmosphere,

$$2M = \{\eta'^2\}\left(\frac{f^2}{N^2}\frac{\partial U}{\partial z}\right)\Big|_{z=0} - \int \rho\{\eta'^2\}\frac{\partial Q}{\partial y}\,dz \qquad (6)$$

where $\rho = e^{-z/H}$. In a flow without horizontal curvature, the potential vorticity gradient is $\partial Q/\partial y = \beta - f\rho^{-1}\partial(\rho S)/\partial z$, where S is the isentropic slope; therefore, if we idealize the tropopause as a discontinuity in isentropic slope, $\Delta_T S$, we can separate off its Eady-like contribution:

$$2M = \{\eta'^2\}\left(\frac{f^2}{N^2}\frac{\partial U}{\partial z}\right)\Big|_{z=0} - \{\eta'^2\}f\rho\Delta_T S|_{z=\mathrm{trop}}$$
$$- \int \rho\{\eta'^2\}\frac{\partial Q}{\partial y}\,dz \qquad (7)$$

The remaining integral is understood to extend over the smooth part of the PV gradient. In an unstable baroclinic mode, the pseudomomentum must sum to zero; otherwise it would increase exponentially, which is impossible since it is conserved. The positive contribution from the particle displacements at the surface exactly cancel the negative contribution from the interior (including the tropopause).

The pseudomomentum of a quasi-stationary external Rossby wave is not zero, but it is small. The negative contribution from the interior is dominant, so the mode's dynamics can be thought of as being somewhat more strongly controlled by upper level PV evolution than by the surface, but the compensation is close. Using mean shear and stability appropriate for midlatitude winter, one finds that a wave moving $5 \, \text{m s}^{-1}$ westward with respect to the surface wind has roughly 80% of its interior pseudomomentum compensated at the surface. By varying the phase speed of the wave, one finds that the phase speed must be westward and of magnitude comparable to the tropospheric thermal wind before the compensation drops below 50%.

When this compensation is strong, it is easy to excite the wave to large amplitude. To see this, define the operator M so that the pseudonorm $\{\eta, M\eta\}$ is the expression in (6) or (7). The external mode is orthogonal to all other modes using this weight (Held 1985). Therefore, an arbitrary initial condition $\eta_i = A \, \eta_e + \ldots$ can be projected onto the external mode by setting

$$A = \{\eta_e, M\eta_i\} / \{\eta_e, M\eta_e\} \qquad (8)$$

Suppose, for example, that η_i is confined to the upper troposphere. Then there will be no compensation between positive and negative values in the numerator, but there will be in the denominator. Therefore, the amplitude of the particle displacements in the external mode will be larger than those in the initial condition. The energy required for this growth is extracted from the mean available potential energy, just as in normal mode baroclinic instability. Other perspectives on this non-normal growth have been discussed by Farrell (1984).

In thinking about synoptic eddies, we are accustomed to focusing equally on upper tropospheric and surface disturbances. We should do the same when thinking about planetary waves. Focusing only on upper tropospheric PV, or thinking in terms of barotropic dynamics, overemphasizes the difficulty in exciting the external mode, and prevents one from appreciating that much planetary wave development must have a baroclinic character.

The fact that a wave as a whole has pseudomomentum of opposite sign to that of the surface disturbance may have other interesting implications. It is reasonable to assume that smaller-scale eddies try to damp the surface temperature signal in a planetary wave through their low-level PV (or heat) flux. Held et al. (1986) showed that the net effect of this low-level damping is to increase the amplitude of an external mode wavetrain. If the eddy damping is localized, the argument predicts larger amplitudes downstream of the damping. One can think (counter-?)intuitively of the damping as reducing the positive M near the surface, but the disturbance as a whole has negative M, so this actually increases its amplitude. One can also think of the damping as creating a vertical phase shift

that extracts more energy from the basic state. Quantitatively, this does not appear to be a very important effect for external mode wavetrains propagating from the tropics to higher latitudes, perhaps because their group velocity is large enough that they pass through the midlatitude storm track rather quickly. This catalytic effect of low level thermal damping may be of greater importance for localized blocks. To the author's knowledge, the potential importance of downgradient near-surface diffusive heat flux for maintaining blocks in baroclinic mean flows has not been examined.

Throughout the preceding discussion, the β-effect has played an important role. This is probably troubling to synopticians who see the β-effect as being at best of marginal importance on synoptic scales. Tropospheric PV-gradients are poorly defined and rarely resemble β. One expects baroclinic waves to mix PV near their midtropospheric steering levels, and they may be fairly successful at homogenizing it in some layers of the troposphere (e.g., Sun and Lindzen 1994). Therefore, it is important to ask if the external mode is sensitive to the tropospheric PV distribution.

Fortunately, it is not, as described by Swanson and Pierrehumbert (1995). If one starts with a profile in which the vertical shear and N^2 are independent of height, and then adjusts N^2 so as to counterbalance β and produce uniform PV, the changes in vertical structure of the mode are very small, as are changes in its wavelength. The compensation between the positive and negative contributions to the pseudomomentum of the wave is also nearly unchanged, even though the negative part is now concentrated at the tropopause.

In this homogenized atmosphere, the external mode feels both the δ-function PV gradients at the surface and the tropopause. Because of the elliptic relation between PV and the flow, the wave feels some average value of these gradients. For a flow with vertical shear only, the vertically averaged quasigeostrophic PV gradient equals β (if one integrates through the lower boundary, with the standard convention that the vertical shear vanishes beneath the surface), so one should not be totally surprised by the insensitivity to the redistribution of the gradient. The degree of insensitivity is impressive nevertheless. It is worth noting that the argument for the vertical structure of the mode based on (4) does not make any reference to the PV distribution or N^2.

4. Relevance of the Time Mean Flow

Given the strong zonal asymmetries in the climatology of the Northern Hemisphere, one is led to consider the dynamics in the neighborhood of, or even linearized about, a zonally asymmetric flow resembling the time-mean. How much can we learn about storm tracks, low-frequency variability, and the response to changes in boundary conditions within this linear framework? To what extent is this placing too much dynamical weight on the time mean flow?

To fix ideas, consider the extratropical response to the tropical heating anomaly associated with an El Niño event. From observational and modeling studies, the mean response in wintertime has the appearance of an equivalent barotropic wavetrain arcing across the North Pacific and North America (Horel and Wallace 1981). The simplest interpretation of this pattern (Hoskins and Karoly 1981), a linear stationary external Rossby wave propagating on the zonal mean winds, directly forced by tropical heating, is very appealing but has proven to be insufficient for several reasons.

This inadequacy is most clearly shown by experiments with a variety of models (from barotropic, Simmons et al. 1983; to GCMs, Geisler et al. 1985), which show that the extratropical response is not simply translated zonally when one shifts the longitude of the tropical heating. Rather, there is a preferred region in which the response is large, the PNA (Pacific–North American) region. This result has motivated the study of models linearized about asymmetric basic states. Although this may seem like a small modification to the theory, it raises the level of complexity substantially. In particular, nearly neutral normal modes that owe their existence to the zonal asymmetries of the basic state are created with which the stationary forcing can resonate. Despite the added complications, one is tempted to pursue this approach since a linear, or weakly nonlinear, model would allow one to go rather far in analyzing the details of the response.

But the planetary wave pattern itself undergoes dramatic fluctuations within a season, which are of more than enough amplitude to call into question the linearization about the time mean. One can raise the same concern about models linearized about zonally symmetric states, but variability in the zonal mean is less dramatic than in the planetary wave pattern. As a result, interest shifts to the structure of the probability distribution of the planetary wave pattern. If this distribution is more or less normally distributed about the mean, there might be some hope of justifying the study of dynamics in the neighborhood of the mean. In the other extreme, the picture in Fig. 1 of the response to El Niño is a possibility (e.g., Molteni et al. 1993). The principal response to the change in external conditions in this schematic is in the occupancy rate of two distinct "regimes." Under these circumstances, it is unlikely that linearizing about the time mean, which is not visited very often, will be of much value. There is no question that some idealized models will behave in this way when perturbed, but the jury is still out with regard to more realistic models and the atmosphere itself. A bimodal distribution is not needed to make this kind of picture relevant. A skewed distribution in which the mode differs significantly from the mean is sufficient, especially if the skewness changes when the system is perturbed.

Attempts at estimating the probability distribution of the large-scale flow continue to be controversial (Cheng and

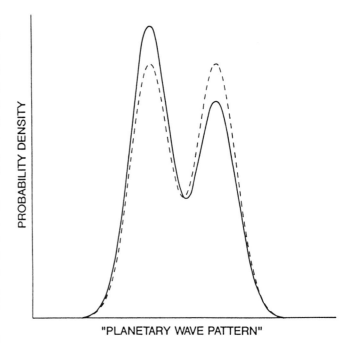

FIG. 1. A possible schematic of the probability distribution of the NH winter planetary wave pattern: unperturbed (dashed line), and as perturbed by El Niño (solid line).

Wallace 1993; Kimoto and Ghil 1993a,b; Hansen and Sutera 1986; Nitsche et al. 1994). Verifying model results against observations will be difficult if the details of the distributions are important, given the length of the time series needed to estimate these distributions.

In addition to the problems introduced by this potential for non-normal planetary wave statistics, several linear modeling studies of the extratropical response to tropical heating in GCMs and observations suggest that another ingredient must be taken into account: the response of the eddies in the Pacific storm track (Kok and Opsteegh 1985; Held et al. 1989; Hoerling and Ting 1994). Let us consider what such a linear diagnosis can and cannot tell us.

Suppose that our model is

$$\frac{\partial A_i}{\partial t} = N_{ijk}A_jA_k + Q_i,$$ (9)

so that

$$0 = N_{ijk}[A_j][A_k] + N_{ijk}[A_j'A_k'] + [Q_i],$$ (10)

where brackets represent the time mean and primes the deviations from this mean (transient eddies). Denoting the unperturbed climate by the subscript zero, and the perturbation in the climate by an asterisk, we have, in response to a perturbation in [Q],

$$0 = N_{ijk}([A_j]_0[A_k]_* + [A_k]_0[A_j]_*$$

$$+ [A_j]_*[A_k]_*) + N_{ijk}[A_j'A_k']_* + [Q_i]_* \qquad (11)$$

Suppose that the final two terms, the anomalous transient eddy fluxes and the anomalous forcing, are both thought of as given from models or observations. By disregarding the third term on the first line (stationary nonlinearity), one can solve the resulting linear problem for the perturbed climate. More ambitiously, one can iterate to a steady nonlinear solution so as to take this term into account. In the few partial diagnoses of El Niño responses that have been attempted along these lines, the direct response to tropical heating and the response to modified transient eddies are generally both important. The anomalies in the upper tropospheric momentum fluxes are the dominant part of the forcing by transients.

There is much that is confusing about this storm track response and its back effect on the mean flow. For example, by using an idealized GCM in which the unperturbed climate is zonally symmetric, Ting and Held (1990) found that the storm track eddies play only a minor role in shaping the response. Our suspicion is that the eddy momentum fluxes must be concentrated in the barotropic decay region of a localized storm track and this barotropic decay region must be strongly modified by the perturbation for these fluxes to play an important role. Recent work of Branstator (personal communication) appears to be consistent with this picture: in a GCM with realistic boundary conditions, he finds that forcing by anomalous transients is dominant for heating anomalies in the central Pacific, whereas the direct response to heating is dominant for anomalies in the western Pacific.

If a theory could be found for the storm track eddy fluxes as a function of the time-mean flow, then (11) would provide a deductive explanation for the perturbed climate, given $[Q]_*$. By linearizing this functional dependence, the eddy fluxes would become part of the linear operator, rather than the forcing. Branstator (1995) has examined a particularly simple algorithm for this purpose and tested it against GCM fluxes (although not in the context of an El Niño simulation). He linearizes the GCM dynamics about a given mean flow, then performs multiple integrations of this linear model for a specified length of time (which can be thought of as a fitting parameter), initializing with noisy initial conditions. The predicted eddy statistics are obtained by averaging over this ensemble. The results look promising.

The technique is closely related to that advocated by Farrell and Ioannou (1994), in which a model linearized about the time mean is driven stochastically. Extra damping is included so as to stabilize the linear model. The stochastic stirring and the damping are both thought of as mimicking the effects of nonlinearities, and the strength of the damping plays a role similar to that played by the length of the linear integration in Branstator's algorithm. As it is a direct attempt at simulating a statistically steady state, this approach may be more amenable to the elaboration of theories for the structure and parameter dependence of the required stirring and damping. But the hope would be that one could extract useful information from such a linear model without any detailed understanding of this nonlinearity.

But what if the actual dynamics of the planetary scale response is that pictured in Fig. 1? Presumably, the anomalous transients would show up as an important forcing term in a diagnosis based on (11), but the implication would be that one could not understand this change in transients in the context of a model linearized about the time mean, so that the stationary wave model would be of little use in predicting the response. For that matter, one could argue that the response of the mean flow is of no great interest under these circumstances. Whether one could ever predict or "understand" the changes in probability distribution of interest in this case without conducting the full model computation is unclear. However, to the extent that the program outlined in the previous paragraph is successful—if a robust algorithm can be developed for predicting transient eddy fluxes given the mean flow, which can then be used in the steady-state model (11) to predict the mean response—one could argue that the complexity of regime diagrams and probability distributions will be of secondary importance for understanding the perturbed climate. It will be interesting, and important, to see which of these perspectives emerges as most useful. Our understanding of the response of planetary waves and storm tracks to global warming, as well as to El Niño, may be strongly influenced by the outcome.

The physical meaning of a "mean flow" and of the distinction between high- and low-frequency transients may itself be questioned. (The following discussion was motivated by discussions with I. Orlanski and R. Panetta.) The lower panel in Fig. 2 shows a simple vortex street—a double row of isolated vortices, with the row of cyclones displaced northward from the row of anticyclones. The center panel shows the streamfunction associated with this vorticity distribution, assuming two-dimensional flow. In addition to the localized highs and lows, there is a westerly jet along the center line of the vortex street.

Suppose that this pattern is translating eastward. A time mean flow then exists whose main feature is the westerly jet along the center line. The transient eddies, the deviations from this mean, are displayed in the upper panel (by symmetry, the time mean is equal to the zonal mean). The transient highs and lows are now nearly symmetric about the center line. Furthermore, if this entire pattern undergoes a north-south oscillation on long time scales, there will be a low-frequency flow, an oscillating jet, whose movement will be exactly correlated with the north-south movement of the variance of the high-frequency transients.

FIG. 2. A storm track as vortex street. Upper panel: eddy streamfunction; middle panel: streamfunction; lower panel: vorticity.

As observed by Wallace et al. (1988), when one removes the time-mean flow before examining the storm track eddies in the atmosphere, one finds highs and lows that are symmetrically placed along the storm track axis, just as in the upper panel of Fig. 2. It is only when one adds in the time mean flow that one sees the synoptically familiar latitudinal displacement of the highs and lows. The transients so defined give the appearance of being more linear and wave-like. This appears to be a legitimate motivation for removing the mean, at least to the extent that linear models are useful in explaining these symmetric transient eddy statistics. But in the case of the vortex street, it is clearly more natural to view the entire pattern as a coherent structure. (While the flow as pictured is not an exact solution of the barotropic vorticity equation, with some effort it could be so modified; the vorticity patches would no longer be circular.)

Lau's (1988) analysis confirms impressions based on familiarity with daily weather that the movement of the

variance of the high frequency geopotential height (the storm tracks) is strongly correlated with the variations in the low frequency flow itself. Indeed, this correlation is so strong as to suggest that one is really looking at two parts of the same pattern. Ting and Lau (1993) and Branstator (1995) show that an impressively large fraction of the low frequency variability of the flow in a GCM can be thought of as "forced," in the sense of Eq. (11), by vorticity fluxes generated by the high frequencies. But a simpler picture suggested by the vortex street analogy is that this diagnosis is confirming the internal consistency of a coherent structure (speaking loosely) that includes time-varying and steady parts. Low-frequency variations in this structure produce coherent low-frequency variations in the flow and in the variance of the high-frequency eddies.

5. A Final Remark

We have to be alert to the possibility that our analysis tools produce distorted descriptions of the atmosphere that, in isolation, can lead to misinterpretation of the underlying dynamics. Given this potential for distortion, and the complexity of atmospheric flows, a variety of perspectives, tools, and models are clearly required. And we must not be shy about ranging widely into "un-atmospheric" fluid dynamical regimes to test our understanding. Even configurations as extreme as homogeneous geostrophic turbulence have much to tell us about planetary wave-synoptic eddy interactions.

References

Anderson, J., 1995: A simulation of atmospheric blocking with a forced barotropic model. *J. Atmos. Sci.*, **52**, 2593–2608.

Branstator, G., 1995: Organization of stormtrack anomalies by recurring low-frequency circulation anomalies. *J. Atmos. Sci.*, **52**, 207–226.

———, and I. M. Held, 1995: Westward propagating normal modes in the presence of stationary background waves. *J. Atmos. Sci.*, **52**, 247–262.

Charney, J. G., and J. G. DeVore, 1979: Multiple flow equilibria in the atmosphere and blocking. *J. Atmos. Sci.*, **36**, 1205–1216.

Cheng, X., and J. M. Wallace, 1993: Cluster analysis of the Northern Hemisphere wintertime 500-hPa height field: Spatial patterns. *J. Atmos. Sci.*, **50**, 2674–2696.

Cook, K. H., and I. M. Held, 1992: The stationary response to large-scale orography in a general circulation model and a linear model. *J. Atmos. Sci.*, **49**, 525–539.

Edmon, H. J., B. J. Hoskins, and M. E. McIntyre, 1980: Eliassen–Palm cross-sections for the troposphere. *J. Atmos. Sci.*, **37**, 2600–2616.

Farrell, B. 1984: Modal and non-modal baroclinic waves. *J. Atmos. Sci.*, **41**, 668–673.

———, and P. J. Ioannou, 1994: A theory for the statistical equilibrium energy spectrum and heat flux produced by transient baroclinic waves. *J. Atmos. Sci.*, **51**, 2685–2698.

Geisler, J. E., M. L. Blackmon, G. T. Bates, and S. Munoz, 1985:

Sensitivity of January climate response to the magnitude and position of equatorial Pacific sea surface temperature anomalies. *J. Atmos. Sci.*, **42**, 1037–1049.

Hamilton, K., 1983: Aspects of wave behavior in the mid- and upper troposphere of the Southern Hemisphere. *Atmos.-Ocean*, **21**, 40–54.

Hansen, A. R. and A. Sutera, 1986: On the probability distribution of planetary-scale atmospheric wave amplitude. *J. Atmos. Sci.*, **43**, 3250–3265.

Held, I. M., 1985: Pseudomomentum and the orthogonality of modes in shear flows. *J. Atmos. Sci.*, **42**, 2280–2288.

———, and V. D. Larichev, 1996: A scaling theory for horizontally homogeneous, baroclinically unstable flow on a beta-plane. *J. Atmos. Sci.*, **53**, 946–952.

———, S. W. Lyons, and S. Nigam, 1989: Transients and the extratropical response to El Nino. *J. Atmos. Sci.*, **46**, 163–174.

———, R. L. Panetta, and R. T. Pierrehumbert, 1985: Stationary external Rossby waves in vertical shear. *J. Atmos. Sci.*, **42**, 865–883.

———, R. T. Pierrehumbert, and R. L. Panetta, 1986: Dissipative destabilization of external Rossby waves. *J. Atmos. Sci.*, **43**, 388–396.

Hoerling, M. P., and M. Ting, 1994: Organization of extratropical transients during El Nino. *J. Climate*, **7**, 745–766.

Horel, J. D., and J. M. Wallace, 1981: Planetary-scale atmospheric phenomena associated with the Southern Oscillation. *Mon. Wea. Rev.*, **109**, 813–829.

Hoskins, B. J., and D. J. Karoly, 1981: The steady linear response of a spherical atmosphere to thermal and orographic forcing. *J. Atmos. Sci.*, **38**, 1179–1196.

Hoyer, J. M., and R. Sadourny, 1982: Closure modeling of fully developed baroclinic instability. *J. Atmos. Sci.*, **39**, 707–721.

Kimoto, M., and M. Ghil, 1993a: Multiple flow regimes in the Northern Hemisphere winter. Part I: Methodology and hemispheric regimes. *J. Atmos. Sci.*, **50**, 2625–2643.

———, and ———, 1993b: Multiple flow regimes in the Northern Hemisphere winter. Part II: Sectorial regimes and preferred transitions. *J. Atmos. Sci.*, **50**, 2645–2673.

Kok, C. J., and J. D. Opsteegh, 1985: On the possible causes of anomalies in seasonal mean circulation during the 1982/1983 El Nino event. *J. Atmos. Sci.*, **42**, 677–694.

Larichev, V. D., and I. M. Held, 1995: Eddy amplitudes and fluxes in a homogeneous model of fully developed baroclinic instability. *J. Phys. Oceanogr.*, **25**, 2285–2297.

Lau, N.-C., 1988: Variability of the observed midlatitude storm tracks in relation to low-frequency changes in the circulation pattern. *J. Atmos. Sci.*, **45**, 2718–2743.

Lee, S., and I. M. Held, 1993: Baroclinic wave packets in models and observations. *J. Atmos. Sci.*, **50**, 1413–1428.

Madden, R., 1979: Observations of large-scale traveling Rossby waves. *Rev. Geophys. Space Phys.*, **17**, 1935–1949.

Molteni, F., L. Ferranti, T. N. Palmer, and P. Viterbo, 1993: A dynamical interpretation of the global response to equatorial Pacific SST anomalies. *J. Climate*, **6**, 777–795.

Nitsche, G., J. M. Wallace, and C. Kooperberg, 1994: Is there evidence of multiple equilibria in planetary wave amplitude statistics? *J. Atmos. Sci.*, **51**, 314–322.

Plumb, R. A., 1986: Three-dimensional propagation of quasi-geostrophic eddies and its relationship with the eddy forcing of the time-mean flow. *J. Atmos. Sci.*, **43**, 1651–1678.

Rhines, P. B., 1975: Waves and turbulence on a ß-plane. *J. Fluid Mech.*, **184**, 289–302.

———, 1977: The dynamics of unsteady currents. *The Sea*, Vol. 6, F. L. Goldberg, I. N. McLane, J. J. O'Brien, and J. H. Steele, Eds., Wiley, 189–318.

Salby, M., 1982: A ubiquitous wavenumber-5 anomaly in the Southern Hemisphere during FGGE. *Mon. Wea. Rev.*, **110**, 1712–1720.

———, 1984: Survey of planetary-scale traveling waves: The state of theory and observations. *Rev. Geophys. Space Phys.*, **22**, 209–236.

Salmon, R. S., 1980: Baroclinic instability and geostrophic turbulence. *Geophys. Astrophys. Fluid Dyn.*, **15**, 167–211.

Simmons, A., J. M. Wallace, and G. Branstator, 1983: Barotropic wave propagation and instability, and atmospheric teleconnection patterns. *J. Atmos. Sci.*, **40**, 1363–1392.

Shutts, G. J., 1983: The propagation of eddies in diffluent jetstreams: eddy vorticity forcing of "blocking" flow fields. *Quart. J. Roy. Meteor. Soc.*, **109**, 737–761.

Sun, D.-Z., and R. S. Lindzen, 1994: A PV view of the zonal mean distribution of temperature and wind in the extratropical troposphere. *J. Atmos. Sci.*, **51**, 757–772.

Swanson, K., and R. T. Pierrehumbert, 1994: Potential vorticity homogenization and stationary waves. *J. Atmos. Sci.*, **52**, 990–994.

Ting, M., and I. M. Held, 1990: The stationary wave response to a tropical SST anomaly in an idealized GCM. *J. Atmos. Sci.*, **47**, 2546–2566.

———, and N.-C. Lau, 1993: A diagnostic and modeling study of the monthly mean wintertime anomalies appearing in a 100-year GCM experiment. *J. Atmos. Sci.*, **50**, 2845–2867.

Venne, D. E., 1989: Normal-mode Rossby waves observed in the wavenumber 1–5 geopotential fields of the stratosphere and troposphere. *J. Atmos. Sci.*, **46**, 1042–1056.

Wallace, J. M., G.-H. Lim, and M. L. Blackmon, 1988: Relationship between cyclone tracks, anticyclone tracks and baroclinic waveguides. *J. Atmos. Sci.*, **45**, 439–462.

Whitaker, J. S., and A. Barcilon, 1995: Low-frequency variability and wavenumber selection in models with zonally symmetric forcing. *J. Atmos. Sci.*, **52**, 491–503.

Advances in Cyclogenesis Theory: Toward a Generalized Theory of Baroclinic Development

BRIAN F. FARRELL

Department of Earth and Planetary Sciences, Harvard University, Cambridge, Massachusetts, USA

1. Introduction

Baroclinic cyclones are the primary weather-producing agents in midlatitudes, and the passing of lows and highs on the synoptic chart has long been recognized as heralding the change from storm to clearing skies. The association between a "falling glass" and deteriorating conditions gave impetus to the development of a variety of cyclogenesis theories, and the quest for a physical understanding of the cyclone has challenged some of the finest minds from the previous century to our own day. While all of the theories preceding our present understanding in terms of quasigeostrophic instability theory enjoyed an interval of ascendency, each predecessor ultimately suffered a more or less thorough eclipse; a historical observation that should at the least serve to forestall complacency.

In the "New World," atmospheric science enjoyed an early acceptance, and a surprising number of the pioneers in meteorology hailed from the brash provincialism and intellectual ferment of the post-colonial United States, including William Redfield, James Espy, Elias Loomis, and William Ferrel (Kutzbach 1979). Evidently, this interest was not parochial in character, as Vilhelm Bjerknes himself was quickly recognized across the Atlantic for his generalization of the circulation theorem and awarded a Carnegie Institution grant after his U.S. lecture tour in 1905, which supported some of his work in Bergen until 1941.

The first scientific theory of cyclogenesis is arguably that of Espy (1841) who maintained that the heating and resulting expansion of air when water vapor is condensed to form clouds must give rise to a warm core depression in a region of intense precipitation. Moreover, in Espy's view the wind field must converge upon the depression—a theory that contrasted with the centrifugal balanced winds circling the low pressure center envisioned in the contemporary theory of Redfield (1831). These opposing predictions were put to the test by Loomis (1841), who was an early advocate of the case study approach and claimed perceptively that "meteorology is to be promoted, not so much by taking the mean of long continued observations, as by studying the phenomena of particular storms." The results of his investigation of the Great Storm of 1836 did not support Espy's theory, and he was forced to conclude that "although a fall of the barometer is usually accompanied by an elevation of temperature, the reverse is sometimes the case."

In addition to the contradictory common observation that copious rainfall is not invariably associated with cyclogenesis, the prediction that winds blow into the center of the low was not supported by the case study undertaken by Loomis, although it must be admitted that the alternative of Redfield also failed full support. The debate between Espy and Redfield on the structure and origin of cyclones is instructive in that although evidence against the theory of Espy was immediately produced, the theory itself persisted even in the writings of the Bergen school (J. Bjerknes 1919). This persistence of Espy's thermal theory in spite of overwhelming observational evidence that midlatitude cyclones are primarily cold core provides additional confirmation that a logical and appealing theoretical construct can withstand a protracted siege of contradictory observation without apparent damage.

In the latter part of the nineteenth century, upper-level observations from mountain stations, kites, and balloons had a role in establishing the thermal structure of cyclones and bringing a weight of contradictory evidence to bear upon the predictions of the condensation theory of their formation; in the middle of the twentieth century, a parallel role occurred when extensive upper air observations collected after World War II began to show that the surfaces of discontinuity in temperature and wind envisioned by the frontal cyclone model did not extend into the upper atmosphere. By this time it was recognized from the work of Margules (1903) that the energy source for the cyclone was in fact potential energy associated with geostrophically balanced density contrasts arising in association with thermal contrasts. Drawing on the ideas of Helmholtz (1888), Margules (1906a) postulated an extended zero-order front making an angle α with the horizontal satisfying: .

$$\tan(\alpha) = \frac{f}{g} \frac{T_c v_w - T_w v_c}{T_w - T_c} \tag{1}$$

where $T_{c,w}$ and $v_{c,w}$ refer to the separate warm and cold layer temperature and front parallel velocity, f is the Coriolis parameter, and g is gravitational acceleration. This represented a lasting advance in that the baroclinic energy source for the cyclone had been identified, and Margules' advance constituted a revolutionary departure from the thermal theory of Espy. However, Margules (1906b) recognized that problems remained in that the mechanism by which this potential energy is realized to form the cyclone was as yet unclear. He perceptively indicated a possible resolution of this difficulty in terms of finite amplitude instability of the jet:

> Horizontal temperature gradients and masses with a great store of potential energy are everywhere but the storm is relatively rare. The masses, associated with moderate velocities, remain in a near-stationary state, which maintains itself. . . . Are there larger disturbances of the dynamic equilibrium, which grow on their own, feeding from the potential energy . . . (?)

The Bergen school sought a physical model of the process by which the release of potential energy identified by Margules takes place. Under the guidance of V. Bjerknes, a conceptual mechanism was advanced in which a process of transient frontal deformation produced structural evolution, ultimately taking the form of occlusion and lifting of the warm air from the surface. This frontal cyclone theory comprised an essentially transient finite amplitude description that made no necessary reference to linear perturbation theory. However, the well-known linear instability study of shear layers by Helmholtz (1886) strongly implied that geostrophically balanced frontal discontinuities should be perturbatively unstable. Conceptually useful as the Norwegian cyclone model may have been and continues to be, it ultimately failed to ground itself in a convincing perturbative baroclinic instability. In addition, events overtook the theory after World War II when upper air soundings made it increasingly clear that frontal discontinuities failed to penetrate into the midtroposphere but instead gave way to continuous jets. However, being superseded theoretically has not visibly diminished the influence of the polar front theory as a descriptive device, as a glance at any synoptic chart testifies.

While the thermal and polar front theories of cyclone formation were descriptive and phenomenological, the baroclinic instability theories of Eady (1949) and Charney (1947) adopted the program of Rayleigh (1880), according to which the explanation for the occurrence of waves at finite amplitude in a fluid was to be understood as the logical consequence of the existence of exponentially growing modes of the dynamic equations linearized about a representative background flow. With a flow chosen to model the recently discovered continuous baroclinic jets and with use of the recently formulated quasigeostrophic equations, Charney and Eady succeeded in finding unstable modes in highly simplified model problems. The normal mode paradigm envisioned modes growing exponentially from infinitesimal beginnings over a large number of e-foldings so that the mode of greatest growth emerged finally as a finite amplitude wave. This assumption of undisturbed growth was necessary to assure the asymptotic dominance of the most rapidly growing exponential normal mode and its emergence as the dominant structure at finite amplitude, which permitted the theory to make predictions concerning the structure of observed weather systems. In the words of Eady (1949): .

> An arbitrary disturbance (corresponding to some inhomogeneity of an actual system) may be regarded as analyzed into "components" of a certain simple type, some of which grow exponentially with time. In all the cases examined there exists one particular component which grows faster than any other. [and] by a process analogous to "natural selection" this component becomes dominant.

Despite this invocation of Darwin, baroclinic instability theory seems to have shared with its predecessor thermal theory an immunity to natural selection. Difficulties appeared immediately, and Eliassen (1956) was forced to conclude that:

> The present instability theories, while useful for explaining the broad features of the formation of growing disturbances, are of little use in synoptic analyses, except as background information.

This was, as Petterssen (1955) realized, due to the fact that:

> Cyclogenesis results not from the release of an infinitesimal perturbation by a dynamically unstable state, but rather from the release of some kind of instability by finite perturbations which can be identified with the wave-shaped motion patterns in the middle and upper troposphere.

In fact, the genesis and growth of midlatitude cyclones takes place primarily in association with preexisting disturbances that take the form of fronts or shallow centers at the surface and of short waves or jet streaks aloft. Forecasters have long recognized the importance of the interaction between these disturbances by pointing to correlates of this interaction, such as "positive vorticity advection over the low center" as associated with development. By contrast, the modal theory of cyclogenesis developed in the past 40 years deals exclusively with exponential growth of infinitesimal waves of normal mode form. As we have seen from Petterssen's remarks, the widespread acceptance of this paradigm by the theoretical community was never accompanied by equal enthusiasm among synoptic meteorologists. These reservations were accurately stated by Palmén (1951): .

> The disturbance caused by the supposed instability of the zonal current has been the subject of a large number of theoretical studies in the last twenty-five years. The instability of "infinitely small" disturbances superimposed upon the westerly current has been the usual starting point in these investigations. A study of these attempts to solve the cyclone problem, however, gives the impression that there still does not exist any theoretical solution fully applicable to the cyclone problem. Therefore the question arises whether it is, in principle, permissible to start from infinitely small perturbations in discussing the cyclone problem. If such an infinitely small perturbation could cause the development of strong cyclones, it would indicate that the atmosphere is extremely unstable. How such an instability associated with storing of useful potential energy could develop in an atmosphere where strong perturbations of all kinds are always present seems difficult to understand.
>
> If we consider this, it seems more likely that extratropical cyclones are induced by rather large migrating disturbances which were already in existence. These large disturbances must always be present. Since cyclones and anticyclones are probably cells for transforming potential energy into kinetic energy, the pre-existence of a large amount of useful potential energy is necessary for a strong cyclonic development. However, the pre-existing situation must correspond to some kind of potential instability that can be released only by the influence of finite perturbations.

Nor was this dissatisfaction with the modal theory of development confined to the meteorological community; in a discussion of results from modal stability analyses applied to the canonical engineering shear flow problems of Couette and pipe Poiseuille flow, which were well known to exhibit transition to turbulence when perturbed at sufficiently high Reynolds number, Sommerfeld (1946) voiced his doubts as follows:

> In every case the transcendental equation associated with the problem leads to the conclusion that the flow is stable. Thus we have a striking contradiction between theory and experiment. What conclusions should we draw? Should we suspect the method of small oscillations that has proved reliable in all other domains of mechanics including astronomy? Should we rather assume that for such a proof of instability finite instead of (infinitely) small

disturbances ought to be resorted to just because the object of investigation is the laminar motion? Or should we blame the Navier-Stokes equations as inadequate for our problem? This does not seem justifiable either, particularly since in the last analysis they form the foundation for all previous theoretical statements.

The failure of modal instability theory to account for observations of perturbation growth in the atmosphere and in laboratory flows was twofold: on the one hand, the postulated modes often failed to exist, especially in flows more realistic than those chosen to illustrate the theory, and even when instabilities could be found, modal growth rates were often far too slow to account for observed development timescales. However, an equally fundamental problem was that identified by Petterssen—observed development occurs in association with time-varying structures that do not resemble the fixed structure of normal modes.

Once the $t \rightarrow \infty$ asymptotic underpinnings of modal theory are abandoned, in light of observed temporal decorrelation time scales of synoptic disturbances on the order of six days, it is necessary to inquire if there are alternative transient structures with competing growth rates on the synoptically relevant time scales of a day to a week. It turns out that such disturbances can be found (Farrell, 1989a), and the most rapidly growing of these structures exhibit the characteristic development time and structural evolution (upper trough overtaking surface depression) that is the hallmark of midlatitude cyclogenesis. This transience of structure during baroclinic development and corresponding greater potential growth had been explained to generations of synoptic students in the insightful heuristic model of Sanders (1971).

Recognition that a large subspace of perturbations grows by exploiting both the barotropic and baroclinic energy of the meanflow, including the energy associated with downstream variation of jets in diffluence and confluence (Farrell 1989b), allows the great variety of transient development phenomena referred in synoptic parlance by colorful descriptions such as "digging troughs" and "phasing troughs" to be included under the aegis of stability theory. Freed of concentrating on the $t \rightarrow \infty$ asymptote, instability theory transcends the most telling objections of the synoptician: that the observed structures are not modal but highly variable both temporally and structurally.

There emerges from the study of transient development an instability theory affirming the revolutionary insight of Margules that the cyclone's energy source is primarily baroclinic, as well as that of Eady and Charney that quasigeostrophic perturbation theory is valid in flows with high shear such as midlatitude jets, but that incorporates these ideas in a synthetic extension of traditional modal stability theory. This generalization of stability theory explains individual examples of cyclogenesis and the variety of other transient development phenomena while also providing a theory for the growth of small forecast errors (Farrell 1990).

Moreover, the existence of a subspace of growing disturbances in baroclinic flow, as revealed by generalized stability theory, suggests a further generalization is possible. It can be shown that the net source of energy to the perturbation field attributable to nonlinear interactions among waves vanishes. From this it follows that the source of perturbation energy that arises from extraction of mean flow energy by the set of growing disturbances identified by generalized stability theory must be responsible for maintaining the fully developed turbulent state. Pursuing this line, it follows that generalized stability theory, valid for initial-value problems, can be extended by subjecting the flow to continuous perturbative forcing to provide a theory for the fully developed flow. The appropriate methodology is that of the stochastic dynamics of non-normal linear systems. The resulting theory for maintenance of variance in baroclinic jets is summarized below.

2. Generalized Stability Theory

Consider the general linear dynamical system that includes the governing equation for perturbation dynamics of the atmosphere: .

$$\frac{d}{dt}u = \mathbf{A}u, \tag{2}$$

in which $u(t)$ is the state vector of the system at time t, and $\mathbf{A}$ is the linear dynamical matrix. If the background state is steady, then $\mathbf{A}$ is not a function of time and the solution of Eq. (2) is explicit:

$$u(t) = e^{\mathbf{A}t}u(0). \tag{3}$$

In any case, there is a propagator matrix $\mathbf{R}_t$ that advances the system in time:

$$u(t) = \mathbf{R}_t u(0). \tag{4}$$

Only the existence of a propagator is required for the following development and stationarity is assumed for convenience only.

The central distinguishing attribute of $\mathbf{A}$ that determines its transient dynamics is its normality, that is, whether or not $\mathbf{A}\mathbf{A}^+ = \mathbf{A}^+\mathbf{A}$. If $\mathbf{A}$ commutes with its Hermitian transpose, here indicated by the plus sign superscript, then $\mathbf{A}$ has a complete set of orthogonal eigenvectors and perturbation growth rate is bounded by the member of the eigenspectrum of $\mathbf{A}$ with maximum real part, indicated by $\lambda_{\max}(\mathbf{A})$, explicitly: $\| e^{\mathbf{A}t} \| = e^{\lambda_{\max}(\mathbf{A})t}$. The stability of such a normal

dynamical system is determined for all time by its eigenspectrum. However, the dynamics of baroclinic and barotropic jet flows is governed by highly non-normal $\mathbf{A}$'s, so it is necessary to generalize ideas of perturbation growth by considering the growth, σ, of an arbitrary perturbation $u(0)$ over a time interval t as measured by the norm associated with the inner product represented by (u,v):

$$\sigma = \frac{(u(t), u(t))}{(u(0), u(0))}$$
$$= \frac{\left(e^{\mathbf{A}t}u(0), e^{\mathbf{A}t}u(0)\right)}{(u(0), u(0))} \tag{5}$$
$$= \frac{\left(e^{\mathbf{A}^+ t}e^{\mathbf{A}t}u(0), u(0)\right)}{(u(0), u(0))}.$$

where in the last step the definition of the adjoint has been used (Courant and Hilbert, 1962). It follows that a complete set of orthogonal perturbations $u(0)$ can be ordered in growth over time t by eigenanalysis of the matrix: .

$$e^{\mathbf{A}^+ t}e^{\mathbf{A}t}. \tag{6}$$

In particular, the greatest growth over time t is given by $\lambda_{\max}(e^{\mathbf{A}^+ t}e^{\mathbf{A}t})$.

Alternatively, a singular value decomposition of the propagator reveals the unitary initial (columns of U) and final (columns of $\mathbf{V}$) states as well as the growth (along the diagonal of Λ) realized by the complete set of disturbances:

$$e^{\mathbf{A}t} = \mathbf{V}\Lambda\mathbf{U}^+. \tag{7}$$

Two limits of this decomposition of the propagator are worth noting. In the limit $t \to \infty$ maximum growth occurs for the eigenfunction with associated eigenvalue of maximum real part as normal mode theory would suggest. To see this, consider the similarity transformation by the matrix $\mathbf{E}$, which has the eigenvectors of $\mathbf{A}$ arranged as columns in order of growth rate, and the diagonal matrix, Σ, of their associated eigenvalues:

$$e^{\mathbf{A}t} = \mathbf{E}e^{\Sigma t}\mathbf{E}^{-1}. \tag{8}$$

In the limit $t \to \infty$, the first column of $\mathbf{E}$ and the first row of $\mathbf{E}^{-1}$ exponentially dominate with amplification factor $e^{\mathrm{Re}(\Sigma_{11}t)}$:

$$\lim_{t\to\infty} e^{\mathbf{A}t}_{\alpha\beta} = \mathbf{E}_{\alpha 1}e^{\Sigma_{11}t}\mathbf{E}^{-1}_{1\beta}. \tag{9}$$

The initial condition of unit norm producing maximum growth over time t is the complex conjugate of $\mathbf{E}^{-1}{}_{1\beta}$, which is the conjugate of the biorthogonal of the leading eigenvector rather than the leading eigenvector itself:

$$\lim_{t \to \infty} e^{\mathbf{A}t}_{\alpha\beta}\left(\mathbf{E}^{-1}_{1\beta}\right)^* = \mathbf{E}_{\alpha 1}e^{\Sigma_{11}t}. \qquad (10)$$

Although intuition is vindicated in that in the limit $t \to \infty$, that eigenfunction dominates that has an associated eigenvalue with maximum real part; not so obvious is the fact that the optimal initial condition with which to excite that mode should be the conjugate of the biorthogonal of the dominant mode rather than the mode itself. In highly non-normal systems such as the atmosphere, a mode and its biorthogonal differ greatly and the perturbation that optimally excites a mode bears little resemblance to the mode itself (Farrell 1989a).

Given the observed time scale for cyclogenesis of 24 hr (Roebber 1984), the $t \to \infty$ asymptotic is not likely to provide an accurate description of the cyclogenesis process; of greater utility is analysis of the less familiar $t \to 0$ limit of Eq. (6), which controls the initial growth of perturbations. It is this limit that determines the maximum possible instantaneous growth and the structure that produces this maximum growth, in addition to supplying other information such as the rate of expansion of the error ellipse in the short-time forecast and those structures that contribute most to the short-time error growth.

The limit of (6) as $t \to 0$ is easily obtained by Taylor expansion:

$$\begin{aligned} e^{\mathbf{A}^+t}e^{\mathbf{A}t} &\approx (\mathbf{I}+\mathbf{A}^+t)(\mathbf{I}+\mathbf{A}t) \\ &= \mathbf{I}+(\mathbf{A}+\mathbf{A}^+)t+\mathrm{O}(t^2). \end{aligned} \qquad (11)$$

A tight upper bound on instantaneous growth rate and identification of the structure producing this limiting growth can be determined by eigenanalysis of the matrix $(\mathbf{A}+\mathbf{A}^+)$. Eigenanalysis of $(\mathbf{A}+\mathbf{A}^+)$ typically reveals that high growth rates over short times can be realized in baroclinic flows even though all normal modes of $\mathbf{A}$ are damped. Because these rapidly growing structures dominate development over the short time scales typical of cyclogenesis (12–48 hrs), the eigenfunctions of $(\mathbf{A}+\mathbf{A}^+)$ more accurately model the rapid development stage of cyclogenesis than do the eigenfunctions of $\mathbf{A}$.

The most relevant time scales for development in the atmosphere lie between the asymptotic limits $t \to 0$ and $t \to \infty$, and for these intermediate time scales the initial and final structures are found most easily from the SVD analysis of (6). Given that both asymptotic limits are subsumed, it is appropriate to refer to this analysis as the generalized stability analysis of the system (2).

3. Generalized Stability Theory Applied to a Baroclinic Jet

As an example of generalized stability analysis, consider stochastic excitation of perturbations on a zonal baroclinic flow. Harmonic scaled streamfunctions of the form $\phi(z,t) = \varepsilon^{-1/2}e^{z/2}\psi(z,t)$ with zonal wavenumber k and meridional wavenumber l in a β-plane channel obey a linearized quasigeostrophic potential vorticity equation with the form of Eq. (2):

$$\frac{d\phi}{dt} = \mathbf{B}\phi, \qquad (12)$$

where

$$\mathbf{B} = \frac{e^{z/2}}{\sqrt{\varepsilon}}\left(\Delta^{-1}(-ikU-R)\Delta - ik\Delta^{-1}Q_y\right)\frac{\sqrt{\varepsilon}}{e^{z/2}}, \qquad (13)$$

with

$$\Delta = \frac{d^2}{dz^2} - \left(\frac{\alpha^2}{\varepsilon}+S^2-\frac{dS}{dz}\right). \qquad (14)$$

Here U is the mean zonal wind, ε is the square ratio of the Coriolis parameter to the Brunt–Väisälä frequency, α is the total wavenumber, and R is a linear potential vorticity damping coefficient. In the above equations, time has been nondimensionalized by the reciprocal of the Coriolis parameter (f_0), vertical distance by the scale height (H), and horizontal distance by $H(\varepsilon_0)^{-1/2}$. The mean potential vorticity gradient is given by

$$Q_y = \frac{\beta}{\varepsilon}+2S\frac{dU}{dz}-\frac{d^2U}{dz^2}. \qquad (15)$$

and the stability parameter is defined as .

$$S = -\frac{1}{2}\left(\frac{1}{\varepsilon}\frac{d\varepsilon}{dz}-1\right). \qquad (16)$$

The operator Δ^{-1} is rendered unique by incorporation of the usual boundary conditions at the ground resulting from the vertical velocity produced by Ekman pumping associated with a coefficient of vertical diffusion ν. Values of parameters are chosen to be appropriate for the midlatitude atmosphere: $f_0 = 10^{-4}$ s^{-1}, $N = 10^{-2}$ s^{-1}, $H = 10$ km, and $\beta = 1.6 \times 10^{-11}$ m^{-1} s^{-1}. These values result in horizontal wavenumber $k = 1$ corresponding to 6300 km. The boundary condition at the top of the atmosphere is taken to be a vanishing vertical velocity at 4 scale heights. A realistic stratification and zonal wind distribution are also included (Fig. 1; see Farrell and Ioannou (1993) for details).

The continuous dynamical system is reduced to a finite dynamical system using standard finite differencing of Eq. (12).

The energy density is chosen as a perturbation measure:

$$E^t = \phi^+ \mathbf{M} \phi \qquad (17)$$

in which the energy metric is given for a grid of width δ by:

$$\mathbf{M} = -\frac{\delta}{8}(\mathbf{E}\mathbf{T}^2\mathbf{E} - \alpha^2\mathbf{P}), \qquad (18)$$

where T is the discretized d/dz operator, $\mathbf{P}_{ij} = \rho_i\,\delta_{ij}$, $\mathbf{E}_{ij} = (\rho_i\varepsilon_i)^{1/2}\delta_{ij}$, and ρ_i is the mean density at the ith grid.

The generalized stability analysis of this system can be compactly displayed by plotting the log of the maximum energy increase obtained by any initial perturbation over a given time interval as a function of that time interval (Fig. 2). The slope of this curve near $t = 0$ gives the maximum instantaneous energy growth rate, and the asymptotic slope at large times gives the rate at which the energy of the least damped eigenmode decays. The large increase in energy at intermediate times illustrates the potential for transient amplification realized in observed cyclogenesis occurring on these time scales.

4. Extension of Generalized Stability Theory to Constitute a Theory for the Statistical Equilibrium

The very great increase in energy obtained by a substantial subspace of perturbations suggests a further generalization of stability theory. The conventional goal of stability theory has

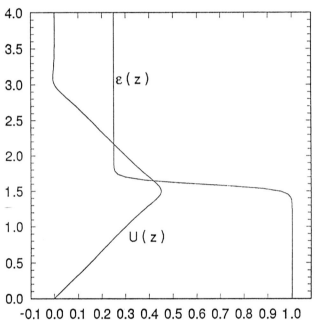

FIG. 1. The distribution with height of the zonal wind and static stability chosen for the examples. The nondimensional shear is $s = 0.3$, corresponding to a dimensional maximum zonal velocity of 45 m s^{-1}; the tropospheric value $\varepsilon = 1$ corresponds to a Brunt-Väisälä frequency $N = 10^{-2}$ s^{-1}.

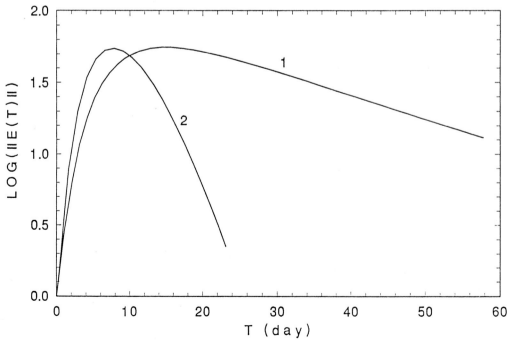

FIG. 2 Generalized stability analysis of the baroclinic jet shown in Fig. 1. The maximum energy obtained by any vertical structure is plotted as a function of the time, t, over which the growth is optimized. Asymptotic growth in the limits $t \to 0$ and $t \to \infty$ is clearly seen, as is the large growth at intermediate times. Ekman damping corresponding to $v = 20$ m^2 s^{-1} and Rayleigh damping with $R = 1/6$ day^{-1} has been used. (1) $k = 0.821$, $l = 1$; (2) $k = 4$, $l = 2$.

been to account for the existence of wavelike perturbations to the background flow field. However, in strongly sheared flows such as the midlatitude jet, the linearized equations embody the entire energetics by which the perturbation field is maintained, there being no net integrated contribution to perturbation energy from the nonlinear interaction of perturbations among themselves. Rather, the interaction among perturbations scatters energy, augments dissipation, and replenishes the subspace of growing perturbations as these are depleted over their transient growth cycles. Advantage may be taken of this pivotal role of the linear dynamical operator in maintaining the turbulent state to formulate a theory for the maintenance of turbulence together with its structures and transporting heat and momentum making use of the statistical dynamics of the non-normal dynamical operator governing perturbation dynamics. This is accomplished by stochastically forcing the dynamical system (2) (Farrell and Ioannou 1993, 1994a, 1994b, 1995; DelSole and Farrell 1995, 1996). This stochastic forcing is to be understood as a parameterization of the scattering of wave energy, which, together with an augmentation of dissipation to account for the disruption of waves by their mutual interaction, constitutes a parameterization of the effects of the nonlinear terms in the complete dynamics. To the extent that the statistically maintained variances and fluxes in the system are determined primarily by the transfer function of the strongly sheared jet, the exact distribution in space and time of the stochastic forcing is of secondary importance.

In order to solve for the stochastic dynamics of our system, it is convenient to transform (2) into generalized velocity variables $u = \mathbf{M}^{1/2}\phi$. In these generalized velocity variables, the stochastically forced perturbation potential vorticity equation takes the form:

$$\frac{du_i}{dt} = \mathbf{A}_{ij}u_j + \mathbf{F}_{ij}\xi_j\,, \qquad (19)$$

where $\mathbf{A} = \mathbf{M}^{1/2}\mathbf{B}\mathbf{M}^{-1/2}$ and ξ is the random forcing assumed to be a δ-correlated Gaussian white-noise process with zero mean and with unit variance:

$$\left\langle \xi_i(t)\xi_j^*(t')\right\rangle = \delta_{ij}\delta(t-t')\,, \qquad (20)$$

where $< \cdot >$ denotes the ensemble average. Note that the stochastic forcing excites independently and with equal unit magnitude each spatial forcing distribution as specified by the columns $f^{(j)}$ of the matrix F_{ij}. We want to determine the evolution of the variance sustained by (19), which in physical variables is the ensemble averaged energy $<E^t> = <u_i^*(t)u_i(t)>$.

The random response, u, is linearly dependent on ξ and consequently is also Gaussian distributed. Therefore, the statistics of the response of the dynamical system are fully characterized by the first two moments. The first moment vanishes for large times if $\mathbf{A}$ is asymptotically stable. The expression for the second moment, the ensemble average energy density, can be reduced to:

$$\left\langle E^t\right\rangle = \left\langle u_i^*(t)u_i(t)\right\rangle = \mathrm{trace}(\mathbf{F}^+\mathbf{K}^t\mathbf{F})\,, \qquad (21)$$

where

$$\mathbf{K}^t = \int_0^t e^{\mathbf{A}^+(t-s)}e^{\mathbf{A}(t-s)}ds\,. \qquad (22)$$

The evolution equation for $\mathbf{K}^t$ with initial condition $\mathbf{K}^0 = 0$ can be derived by direct differentiation of $\mathbf{K}^t$. It is

$$\frac{d\mathbf{K}^t}{dt} = \mathbf{I} + \mathbf{A}^+\mathbf{K}^t + \mathbf{K}^t\mathbf{A}\,, \qquad (23)$$

in which $\mathbf{I}$ is the identity matrix. When the potential vorticity damping, R, is chosen so that $\mathbf{A}$ is asymptotically stable, the asymptotic value of $\mathbf{K}^\infty$ can be determined by solving the asymptotic form of Eq. (23), which is the Liapunov equation:

$$\mathbf{A}^+\mathbf{K}^\infty + \mathbf{K}^\infty\mathbf{A} = -\mathbf{I}\,. \qquad (24)$$

Note that with an orthonormal set of forcing functions such that $\mathbf{F}\mathbf{F}^+ = \mathbf{I}$, the expression for the energy density simplifies to $<E> = \mathrm{trace}(\mathbf{K}^\infty)$ and the variance is independent of the specific forcing distribution.

It is also useful to determine the ensemble average correlation matrix of the response $C_{ij}^t = \left\langle u_i(t)u_j^*(t)\right\rangle$. The correlation matrix can be equivalently written as:

$$\mathbf{C}^t = \int_0^t e^{\mathbf{A}(t-s)}\mathbf{F}\mathbf{F}^+e^{\mathbf{A}^+(t-s)}ds\,, \qquad (25)$$

which satisfies in steady state the Liapunov equation:

$$\mathbf{A}\mathbf{C}^\infty + \mathbf{C}^\infty\mathbf{A}^+ = -\mathbf{F}\mathbf{F}^+\,. \qquad (26)$$

Note that because the operator is non-normal (i.e., $\mathbf{A}^+\mathbf{A} \neq \mathbf{A}\mathbf{A}^+$), Eq. (26) differs from Eq. (24) even for the case of unitary forcing. In any case, the ensemble average energy for full rank unitary forcing is:

$$<E> = \mathrm{trace}(\mathbf{C}^\infty) = \mathrm{trace}(\mathbf{K}^\infty)\,. \qquad (27)$$

If $\mathbf{A}$ were normal, as is the case in the absence of basic state shear, then all of the matrices, $\mathbf{A}$, $\mathbf{C}^\infty$, and $\mathbf{K}^\infty$, commute, and Eq. (26) can be immediately solved to yield, for unitary forcing, the steady-state ensemble average energy:

$$\begin{aligned}\langle E\rangle &= \mathrm{trace}(\mathbf{C}^\infty)\\ &= \mathrm{trace}\big(-(\mathbf{A}+\mathbf{A}^+)^{-1}\big)\\ &= \sum_i \frac{1}{2\,\mathrm{Re}\big(-\lambda_i(\mathbf{A})\big)}\,,\end{aligned} \qquad (28)$$

where $\lambda_i(A)$ are the eigenvalues of **A**. This expression for the variance maintained by normal dynamical operators has been extensively investigated in the past (cf. Wang and Uhlenbeck 1945). In the case of a normal dynamical operator, the motion can be resolved into orthogonal normal modes with the total variance found as the sum of contributions from each individual mode. Moreover, the variance contributed by each mode is inversely proportional to its damping rate. In such systems, the forcing is the only energy source, and the maintained variance is accumulated from the forcing, resulting in high variance if damping is small.

For non-normal dynamical operators, the non-orthogonality of the modes is indicative of the possibility for the perturbations to extract energy from the background flow field despite the absence of exponential instability. The energy balance in such a system is between the stochastic driving together with the induced extraction of energy from the background flow, on the one hand, and the dissipation on the other. Tapping the energy of the mean flow can consequently lead to levels of variance orders of magnitude larger than would have been expected to result in a normal system if the estimation of variance were made from the rate of dissipation of each mode

5. Determination of the Statistical Equilibrium Forcing and Response Functions

We showed that the ensemble average energy for a full-rank unitary forcing distribution can be derived either from the trace of $\mathbf{C}^\infty$ or, equivalently, from the trace of $\mathbf{K}^\infty$. Both $\mathbf{K}^\infty$ and $\mathbf{C}^\infty$ are by construction positive definite Hermitian forms with positive real eigenvalues associated with mutually orthogonal eigenvectors. Each eigenvalue equals the variance accounted for by the pattern of its corresponding eigenvector, and the pattern that corresponds to the largest eigenvalue contributes most to the variance. The decomposition of the correlation matrix into its orthogonal components is often referred to as the EOF decomposition.

The EOF decomposition of $\mathbf{C}^\infty$ determines the structures that contribute most to the ensemble average variance of the statistically steady state. These are the primary response structures of the dynamical system. They are determined by solving the eigenvalue problem:

$$\mathbf{C}^\infty u^{(i)} = \lambda^{(i)} u^{(i)}, \tag{29}$$

in which the variance accounted by the structure $u^{(i)}$ is given by $\lambda^{(i)}$.

The EOF's can be interpreted dynamically using the stochastic theory developed here, thus providing a link between observed atmospheric statistics and dynamical theory.

The leading EOF can be determined as the eigenfunction corresponding to the maximum eigenvalue of the operator:

$$\mathbf{C}^\infty = \lim_{t \to \infty} \int_O^t e^{\mathbf{A}(t-s)} e^{\mathbf{A}^+(t-s)} ds. \tag{30}$$

Note that due to the non-normality of **A** the eigenfunctions of Eq. (30) do not coincide with the eigenfunctions of the dynamical operator **A**. Only for normal dynamical operators, for which **A** and $\mathbf{A}^+$ commute, do the eigenfunctions of Eq. (30) coincide with the modes of the system.

Eigenanalysis of $\mathbf{K}^\infty$, on the other hand, allows ordering of the forcing distributions according to their contribution to the variance of the statistically steady state. This follows from (21) and the observation that the eigenvalues of $\mathbf{K}^\infty$ are stationary values of the Rayleigh-Ritz quotient:

$$I[f] = \frac{f^+ \mathbf{K}^\infty f}{f^+ f}. \tag{31}$$

Consequently, the forcings, $f^{(i)}$, obtained from eigenanalysis of

$$\mathbf{K}^\infty = \lim_{t \to \infty} \int_O^t e^{\mathbf{A}^+(t-s)} e^{\mathbf{A}(t-s)} ds \tag{32}$$

can be ordered according to their relative contribution to the stochastically maintained variance. The optimal stochastic excitation consists of the eigenfunction corresponding to the largest eigenvalue of Eq. (32). Note once again that because of the non-normality of **A** the eigenfunctions of **A**, $\mathbf{K}^\infty$, and $\mathbf{C}^\infty$ do not coincide. Examples of the differences in these structures are given in Fig. 3.

According to the theory presented here, eigenanalysis of (29) is expected to yield the observed EOF's if **A** is taken to correspond to the linearized operator of an observed atmospheric state. The set of primary forcing functions derived from eigenanalysis of $\mathbf{K}^\infty$ can in turn be found by solving Liapunov equation (24). It is clear from this analysis that the attempt to associate the EOF's with the normal modes of the system can succeed only when the dynamical system is governed by a normal dynamical operator.

6. Determination of the Statistical Equilibrium Spectrum

The Fourier transform of Eq. (19) readily yields that

$$\langle E^\infty \rangle = \frac{1}{2\pi} \int_{-\infty}^\infty F(\omega) d\omega, \tag{33}$$

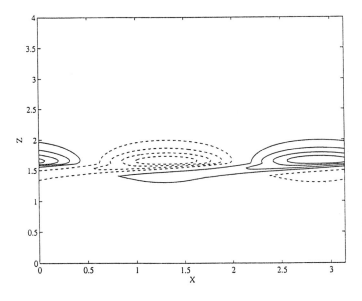

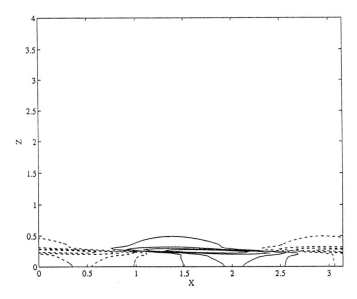

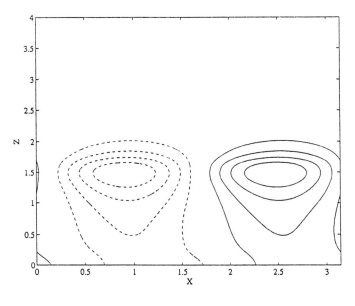

where the frequency response

$$F(\omega) = \text{trace}(\mathbf{R}^{+}(\omega)\mathbf{R}(\omega)) , \qquad (34)$$

follows from the resolvent

$$\mathbf{R}(\omega) = (i\omega\mathbf{I} - \mathbf{A})^{-1} . \qquad (35)$$

We may say that the power spectrum of the response to white noise forcing is given by the Frobenius norm of the resolvent $\mathbf{R}(\omega)$. On the other hand, $\|R(\omega)\|_2$ gives the maximum possible response at this frequency, and the optimal forcing structure to produce such a response can be obtained by singular value decomposition of the resolvent.

When $\mathbf{A}$ is normal, the magnitude of the response can be calculated by the distance of the eigenvalues from the forcing frequency ω. In fact, in that case we have exactly (Kato 1976):

$$\|\mathbf{R}(\omega)\| = \frac{1}{\text{dist}(i\omega, \Lambda(A))} , \qquad (36)$$

where $\Lambda(\mathbf{A})$ is the set of eigenvalues of $\mathbf{A}$. When $\mathbf{A}$ is non-normal, the maximum response may be seriously underestimated by the value predicted by the proximity to the eigenvalues of $\mathbf{A}$. Examples showing the difference between the response of normal and non-normal operators are discussed in Farrell and Ioannou (1993, 1994a, 1995, 1996).

7. Determination of the Ensemble Average Heat Flux

The ensemble average heat flux is defined as

$$H = c_p \int_0^\infty \rho\overline{v\theta}dz$$
$$= \frac{c_p T_g L_d}{gH} E_{in} \frac{h\,\text{trace}(\mathbf{H}^\infty)}{N_f} \qquad (37)$$

$$\mathbf{H}^\infty = \frac{k}{2}\text{Im}\left(\mathbf{ETM}^{-1/2}\mathbf{C}^\infty\mathbf{M}^{-1/2}\mathbf{E}^+\right), \qquad (38)$$

where h is the grid interval, N_f is trace($\mathbf{FF}^+$), C_p is the heat capacity, T_g is a representative temperature, H is the scale height, g is gravitational acceleration, $L_d = NH/f_0$ is the

FIG. 3 (left): (a) The first eigenfunction of $\mathbf{A}$; (b) the first eigenfunction of $\mathbf{K}^\infty$, corresponding to the first forcing function; (c) the first eigenfunction of $\mathbf{C}^\infty$, corresponding to the first response function. The wavenumbers are $k = 1$, $l = 1$. Ekman damping corresponding to $\nu = 10$ m^2 s^{-1} and Rayleigh damping with $R = 1/12$ day^{-1} have been used. The variable shown is the streamfunction f associated with the respective eigenfunctions.

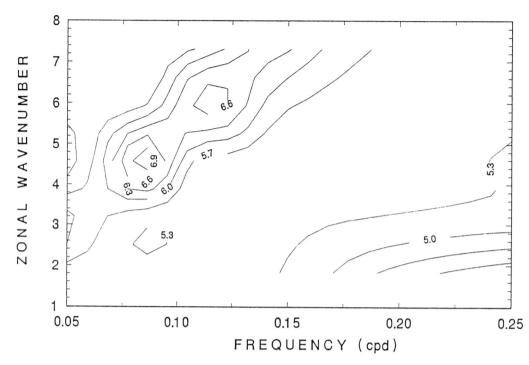

FIG. 4 The variance as a function of period and zonal wavenumber for the jet shown in Fig. 1. Ekman damping corresponding to $v = 20\,\mathrm{m}^2\,\mathrm{s}^{-1}$ and Rayleigh damping with $R = 1/8$ day^{-1} have been used.

Rossby radius of deformation, and $\mathbf{C}^\infty$ is the correlation matrix found by solving Eq. (26).

The correlation matrix of the generalized velocities determines the heat flux and, in particular, the diagonal elements of $\mathbf{H}^\infty$ give the height distribution of the heat flux.

8. The Statistical Equilibrium Energy Spectrum and Heat Flux in a Baroclinic Jet

Consider a baroclinic jet with stochastically forced waves of meridional wavenumber $l = 1$ (the meridional confinement implied by $l = 1$ is chosen to be consistent with the atmospheric state given by Salby (1982) (cf. Farrell and Ioannou 1993; Ioannou and Lindzen 1986)).

The full spectrum of the response as a function of period and zonal wavenumber is shown in Fig. 4. Comparison with observations (i.e., Hansen et al. 1989) reveals that the derived spectra are in remarkable agreement with the observed spectra.

Dimensional values of ensemble average energy and associated heat flux as a function of zonal wavenumber are shown in Fig. 5 for stochastic forcing of 1 W m^{-2} (for the dimensional forms of these quantities, refer to Farrell and Ioannou (1994a)). The maximum response is concentrated between zonal wavenumbers 5 and 6 (note also the reduction of the variance and the fluxes as dissipation increases). The fact that the atmosphere often exhibits enhanced variance at these scales suggests further that the atmosphere is stochastically maintained in a state of near neutrality. An example of the distribution with height of the ensemble average energy and heat flux is shown in Fig. 6.

9. Conclusion

We have described a generalization of stability theory that includes the transient development process recently studied using initial value problem approaches. This generalized stability theory emphasizes the central role of the nonnormality of the linear dynamical operator in determining stability. Generalized stability theory unites the $t \to 0$ asymptotic of the propagator that controls initial error growth and bounds the instantaneous growth rate of disturbances with the $t \to \infty$ asymptotic that is addressed by modal theory. These limits are connected by the singular value decomposition of the propagator, and it is the transient development at intermediate time scales of a day or two that is most relevant to cyclogenesis.

In addition, we have advanced a further generalization of stability theory building on the great potential for transient growth of perturbations in the highly non-normal system underlying dynamics of the midlatitude jet. In this theory, ensemble equilibrium statistics of the turbulent state are found with nonlinear terms in the dynamical equation parameterized as a combination of stochastic forcing to account for scattering of energy among waves and an augmentation of dissipation to account for disruption of wave coherence by wave/wave interaction. The spectral distribution of transient wave energy and the spatial distribution of variance and sensible heat flux obtained using this theory are found to be in remarkable agreement with observations.

The brief summary of generalized stability theory contained in this chapter necessarily leaves many issues open; some of these are addressed in a recent review (Farrell and Ioannou 1996). In addition, efforts to apply the stochastic

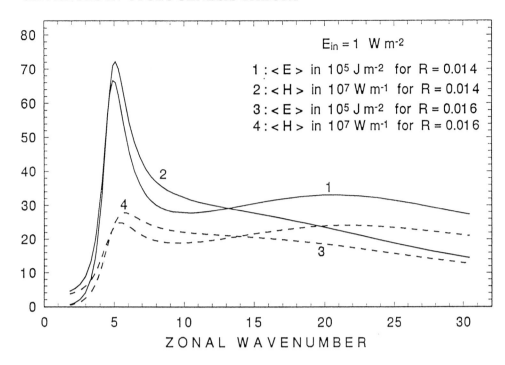

FIG. 5 The ensemble average energy and heat flux as a function of zonal wavenumber for $l = 1$. Ekman damping corresponding to n = 20 m^2 s^{-1} and Rayleigh damping with $R = 1/8$ day^{-1} (solid line) and $R = 1/7$ day^{-1} (dashed line) have been used.

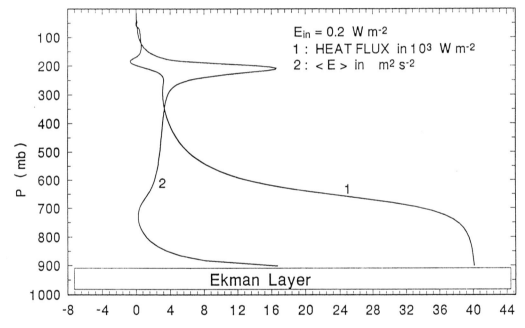

Fig. 6 The distribution with height of the ensemble average energy and heat flux for global zonal wavenumber 5 waves ($k = 0.821$ and $l = 1$). Ekman damping corresponding to n = 20 m^2 s^{-1} and Rayleigh damping with $R = 1/8$ day^{-1} have been used.

theory for equilibrium statistics to more realistic systems are ongoing (DelSole 1996).

Acknowledgment

The author wishes to acknowledge the esential contribution of Petros Ioannou to the theoretical results recounted in this paper. This work was supported in part by NSF ATM-92-16813 and NSF ATM-96-23539, and in part by the U.S. Department of Energy's (DOE) National Institute for Global Environmental Change (NIGEC) through the NIGEC Northeast Regional Center at Harvard University. Financial support does not constitute an endorsement by DOE of the views expressed in this chapter.

References

Bjerknes, J., 1919: On the structure of moving cyclones. *Geofys. Publ.*, **1**, 8.

Charney, J. G., 1947: The dynamics of long waves in a baroclinic westerly current. *J. Meteor.*, **4**, 135–163.

Courant, R., and D. Hilbert, 1962: *Methods of Mathematical Physics V. II.* John Wiley & Sons, New York.

DelSole, T. M., 1996: Can quasigeostrophic turbulence be modelled stochastically? *J. Atmos. Sci.*, **53**, 1617–1633.

——, and B. F. Farrell, 1995: A stochastically excited linear system as a model for quasigeostrophic turbulence: Analytic results for one- and two-layer fluids. *J. Atmos. Sci.*, **52**, 2531–2547.

——, and ——, 1996: The quasilinear equilibration of a thermally maintained, stochastically excited jet in a quasigeostrophic model. *J. Atmos. Sci.*, **53**, 1781–1797.

Eady, E.T., 1949: Long waves and cyclone waves. *Tellus*, **1**, 33–52.

Eliassen, A., 1956: Instability theories of cyclone formation. *Weather Analysis and Forecasting Vol. 1*, McGraw-Hill.

Espy, J. P., 1841: *The Philosophy of Storms*. Little and Brown.

Farrell, B. F., 1989a: Optimal excitation of baroclinic waves. *J. Atmos. Sci.*, **46**, 1193–1206.

——, 1989b: Transient development in confluent and diffluent flow. *J. Atmos. Sci.*, **46**, 3279–3288.

——, 1990: Small error dynamics and the predictability of atmospheric flows. *J. Atmos. Sci.*, **47**, 2409–2416.

——, and P. J. Ioannou, 1993: Stochastic dynamics of baroclinic waves. *J. Atmos. Sci.*, **50**, 4044–4057.

——, and ——, 1994a: A theory for the statistical equilibrium energy and heat flux produced by transient baroclinic waves. *J. Atmos. Sci.*, **51**, 2685–2698.

——, and ——, 1994b: Variance maintained by stochastic forcing of a non-normal dynamical system arising in fluid shear flow. *Phys. Rev. Lett.*, **72**, 1188–1191.

——, and ——, 1995: Stochastic dynamics of the midlatitude atmospheric jet. *J. Atmos. Sci.*, **52**, 1642–1656.

——, and ——, 1996: Generalized stability theor. Part I: autonomous operators. *J. Atmos. Sci.*, **53**, 2025–2040.

Fraedrich, K., and H. Bottger, 1978: A wavenumber-frequency analysis of the 500 mb geopotential at 50°N. *J. Atmos. Sci.*, **35**, 745–750.

Hansen, A. R., A. Sutera, and D. E. Venne, 1989: An examination of midlatitude power spectra: Evidence for standing variance and the signature of El Niño. *Tellus*, **41A**, 371–284.

Helmholtz, H., 1886: *Uber discontinuirliche Flussigkeitsbewegungen*. Berlin Akad. Monatsver.

——, 1888: *Ueber atmosphärische Bewegungen*. Sitzber. Akad. Berlin.

Ioannou, P. J., and R. S. Lindzen, 1986: Baroclinic instability in the presence of barotropic jets. *J. Atmos. Sci.*, **43**, 2999–3014.

Kato, S., 1976: *Perturbation Theory for Linear Operators*. Springer-Verlag.

Kutzbach, G., 1979: *The Thermal Theory of Cyclones*. Amer. Meteor. Soc., 254 pp.

Loomis, E., 1841: On the storm which was experienced throughout the United States about the 20th December, 1836. *Trans. Amer. Philos. Soc.*, **7**, 125.

Margules, M., 1903: *Uber die Energie der Sturme*. Jahrb. Zentralanst. Meteorol. Wien., 1–26.

——, 1906a: Ueber Temperaturschichtung in stationär bewegter und ruhender Luft. *Meteor. Z.*, **23**, 243–254.

——, 1906b: Zur Stormtheorie. *Meteor. Z.*, **23**, 482–483.

Palmén, E., 1951: The aerology of extratropical disturbances. *Compendium of Meteorology*, T. F. Malone, Ed., Amer. Meteor. Soc., 599–620.

Petterssen, S., 1955: A general survey of factors influencing development at sea level. *J. Meteor.*, **12**, 36–42.

Rayleigh, Lord, 1880: *Theory of Sound*. 2d ed. Macmillan.

Redfield, W. C., 1831: Remarks on the prevailing storms of the Atlantic coast of the North American States. *Amer. J. Sci.*, **20**, 29.

Roebber, P. J., 1984: Statistical analysis and updated climatology of explosive cyclones. *Mon. Wea. Rev.*, **112**, 1577–1589.

Sanders, F., 1971: Analytic solutions of the nonlinear omega and vorticity equations for a structurally simple model of disturbances in the baroclinic westerlies. *Mon. Wea. Rev.*, **99**, 393–407.

Salby, M. L., 1982: A ubiquitous wavenumber-5 anomaly in the Southern Hemisphere during FGGE. *Mon. Wea. Rev.*, **110**, 1712–1720.

Schafer, J., 1979: A space-time analysis of tropospheric planetary waves in the Northern Hemisphere geopotential heights. *J. Atmos. Sci.*, **36**, 1117–1123.

Sommerfeld, A., 1946: *Lectures in Theoretical Physics Vol. II*. Academic Press.

Wang, M. C., and G. E. Uhlenbeck, 1945: On the theory of the Brownian motion II. *Rev. Modern Phys.*, **17**, 323–342.

Numerical Simulations of Cyclone Life Cycles

ADRIAN SIMMONS

European Centre for Medium-Range Weather Forecasts, Shinfield Park, Reading, Berkshire, U.K. RG2 9AX

1. Introduction

Numerical simulation provides us today with both the basis for weather prediction and a powerful tool for gaining improved understanding of atmospheric motion. Its origin can be traced to a paper of 1904 in which Vilhelm Bjerknes set out the basic principles of the computation of atmospheric evolution. The conclusion of his article included the following words (taken from a translation of the original German by a later pioneer of numerical modeling, Yale Mintz): "It may be possible one day, perhaps, to utilize a method of this kind as the basis for a daily practical weather service. But however that may be, the fundamental scientific study of atmospheric processes sooner or later has to follow a method based on the laws of mechanics and physics. And thereby we will arrive, necessarily, at a method of the kind outlined here."

Bjerknes's visionary work "exercised a considerable influence" on the later pioneering practical studies of Richardson (1922). Quite apart from the basic principles of numerical prediction, Richardson's worked example used initial conditions derived in part from isobaric charts for 20 May 1910 that Bjerknes had prepared. Richardson was aware too of developments in the early days of the Bergen school, and recorded ideas on the numerical treatment needed to represent frontal discontinuities.

Richardson was uncompromising: his scheme was "complicated because the atmosphere is complicated." Indeed, it was not until some thirty years later that the more-or-less simultaneous development of the electronic computer and of rational simplifications of the governing equations made it possible for Charney et al. (1950) to complete the first successful numerical predictions, using a barotropic model. A few years later, Phillips (1956) used a two-layer quasi-geostrophic model to carry out the first numerical simulation of the general circulation. The papers reporting these ventures are noteworthy not only for the technical achievements they portray, but also for the insight they provide into the mechanisms at play during the life cycles of extratropical cyclones. Phillips discussed the process of baroclinic growth

of a wave and changes to the zonal-mean state, whereas Charney and his collaborators attributed their success to the essentially barotropic nature of the dynamics of mature waves. Platzman (1979) has given a fascinating account of that first barotropic forecast.

The subsequent years have seen enormous advances in our ability to build numerical models of the atmosphere and in the power of the computers on which these models are run. Detailed simulations of complete life cycles of extratropical cyclones can now be carried out for idealized situations, and they allow us to gain a good understanding of the processes involved. Moreover, the accuracy with which numerical weather forecasts often (although not always!) capture the evolution of actual systems over periods of several days provides a basis for further elucidation of mechanisms.

Simulations of the life cycles of both idealized and actual systems will be discussed in this contribution. Emphasis will be placed on the larger scales of motion, as important elements such as fronts, mesoscale substructures, and precipitation will be covered by other contributors. A review will be given of the original simulations of complete life cycles and of downstream and upstream development carried out some years ago by the author and Brian Hoskins at Reading University. Some results of related, more recent investigations will be discussed, and examples of similar behavior will be presented from comprehensive numerical predictions produced at the European Centre for Medium-Range Weather Forecasts (ECMWF). Further results from the ECMWF forecasting system will be used to illustrate some other aspects of the behavior and prediction of extratropical cyclones.

2. Numerical Simulations of a Basic Baroclinic Wave Life Cycle

2.1 Introduction and Summary

The earliest numerical simulations were carried out shortly after a major advance in the understanding of extratropical disturbances, the linear baroclinic instability theory of Charney (1947) and Eady (1949). Widespread development of this theory followed, with eventual extension to jet-like flows in spherical geometry with primitive-equation dynamics, using numerical methods (Gall 1976a; Simmons and Hoskins 1976). However, the structures of the linear solutions generally had upper-level amplitudes that, relative to surface amplitudes, were weak compared with what was seen in general circulation statistics, particularly for momentum fluxes. This motivated the use of numerical simulation to study the nonlinear evolution of baroclinic flows perturbed by small-amplitude disturbances of unstable normal-mode form. Simons (1972) and Gall (1976b) followed growth into the finite-amplitude regime, but with limited resolution. It was not until the studies by Simmons and Hoskins (1978, 1979, 1980), using a model with what was then relatively high resolution and weak diffusion, that baroclinic waves could be simulated over full life cycles of growth, maturity, dispersion, and decay.

The simplest form of behavior was found for a jet localized in middle latitudes and a perturbation comprising a regular train of waves extending around the midlatitude belt. Each wave grew by baroclinic instability, and developed a frontal structure at low levels in some agreement with the usual synoptic picture of an occluding cyclone. Wave growth slowed first at low levels. Upper-level growth continued for a while, and amplitudes thus became larger relative to low-level values than in the normal-mode solution, and thus more in accord with general circulation statistics. Barotropic processes became increasingly important during the occlusion of the disturbance, and eventually led to a decay of the wave at a rate similar to that of its earlier baroclinic growth. The decay was associated (taking a Northern Hemisphere view) with an increasing northeast to southwest tilt of the upper-level trough equatorward of the jet axis. The jet accelerated, and there was a transfer of kinetic energy from the eddy to the zonal-mean component of the flow. Figure 1 shows the conversions from zonal to eddy available potential energy and from eddy to zonal kinetic energy over a typical life cycle.

The basic life cycle experiment reported by Simmons and Hoskins (1978, 1980) was repeated by Thorncroft and Hoskins (1990) using substantially higher resolution. This provided a sharper picture of the frontal formation and a simulation of the development of a secondary frontal wave that was only hinted at in the original model run. The overall life cycle in this, and a second, simulation has been discussed in a second paper (Thorncroft et al. 1993). Here we comment on some particular aspects of the later stages of the upper-level growth and the subsequent barotropic decay.

2.2 Upper-Level Growth in the Nonlinear Regime

Edmon et al. (1980) diagnosed the basic life cycle experiment in terms of the Eliassen-Palm (EP) flux. In particular, they showed that following the nonlinear cessation of low-level

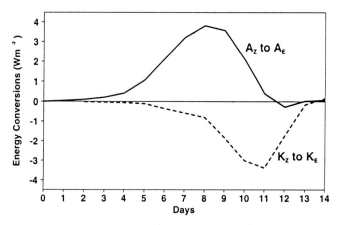

Fig. 1. Conversions from zonal to eddy available potential energy (solid) and kinetic energy (dashed) over a basic baroclinic-wave life cycle (after Simmons and Hoskins 1978).

growth, the region where the (predominantly vertical) EP flux is largest moves upward. They interpreted this as a sign of upward radiation from a saturating instability. This led Thorncroft et al. (1993) to refer to upward propagation of Rossby-wave activity into the strongest mean westerlies as taking over from baroclinically unstable growth as the dominant cause of the eddy kinetic energy increase. This view appears to differ from that suggested originally by Gall (1976b) and Simmons and Hoskins (1978), who noted that nonlinearity restricts low-level growth at a time when there has been little non-linear change in the upper troposphere. The wave then still possesses a structure that gives local growth at upper levels. This growth continues until the westward phase tilt with height, characteristic of baroclinic instability (but also upward Rossby-wave propagation), weakens sufficiently for the (increasingly active) barotropic decay process to dominate. An evolution of the upward component of the EP flux similar to that found by Edmon et al. (1980) is consistent with a loss of vertical tilt that occurs first at low levels and then spreads upward. Moreover, vertical Rossby-wave propagation may be an element of the upper-level growth of the normal-mode itself, just as the meridional structure of the mode at upper levels equatorward of the jet maximum can be regarded as due to horizontal Rossby-wave propagation from a midlatitude source (Simmons and Hoskins 1977). The extent to which a different growth process takes over at upper levels as the wave matures appears open to question.

It should also be recognized that the mature upper-level growth in the basic life-cycle experiment may be quite dependent on the initiation of the perturbation with normal-mode form. In particular, Petterssen and Smebye (1971) categorized the development of extratropical cyclones into two types. Type A resembles normal-mode development, and the phase lag between the developing upper-level trough and the surface cyclone remains essentially unchanged until the cyclone reaches peak intensity. Type B is associated with a pre-existing upper trough, and the distance of separation between this trough and the developing surface cyclone decreases rapidly during intensification, so that the trough axis tends to a vertical alignment as peak intensity is approached.

Furthermore, the normal-mode form of perturbation to a basic state is that which will dominate asymptotically for a large time provided linear theory still applies. Other forms of disturbance may grow more rapidly over a finite time interval, as discussed by Farrell (1982b). Adjoint techniques applied to numerical models now enable the determination of the small-amplitude perturbations that give the largest growth of specified measures of wave amplitude (which can be limited by domain or scale) over specified time intervals, for basic states that evolve in time (Buizza and Palmer 1994). The principal type of structure identified by Buizza and Palmer is localized in the horizontal with largest amplitude in the lower troposphere. An important factor in the energy growth is, in fact, the upward propagation of wave activity, this giving peak amplitudes in the upper troposphere at the end of the period over which growth is optimized. The predominant horizontal length scale increases as the perturbation develops. Combinations of such perturbations are currently used in practice at ECMWF to produce perturbed initial conditions in a novel ensemble forecasting procedure (Molteni et al. 1995).

2.3 Barotropic Decay

Simmons and Hoskins (1978) showed that the decay phase of the upper-tropospheric wave in the basic life cycle simulation could be reproduced with remarkable accuracy in an integration of the barotropic vorticity equation starting from the vorticity distribution of the mature wave at a model level close to 300 hPa. An exception was that the process occurred about twice as quickly in the barotropic integration. Diagnosis in terms of the classical omega equation showed that the forcing of vertical motion by thermal advection was unimportant at this stage. The slower response of the vorticity field in the full simulation was because of the vertical motion which adjusted the temperature field to maintain thermal wind balance in the presence of the strong vorticity advection. This slowness can be linked with the "long-wave correction" applied to improve early barotropic weather forecast models (e.g., Wiin-Nielsen 1959).

The barotropic behavior seen in the numerical simulation had in fact been identified in early studies of long waves in the upper westerlies. These studies combined synoptic description with interpretation in terms of the wave dynamics newly discovered by Rossby (1939). Simmons and Hoskins's analysis of the evolution of phase of the trough was in essential agreement with the work of Namias and Clapp (1944) on the mechanism of trough fracture, where the

northern part of a trough (in the Northern Hemisphere) moves faster than the southern part, in association with a sudden increase in zonal flow in the north. Cressman (1948) noted how the southern portion often formed a closed "cut-off" circulation. Early studies of the barotropic evolution of a large-amplitude disturbance superimposed on a zonal flow include those of Platzman (1952) and Kuo (1953); a more recent study is that of Held and Phillips (1987). Mention should also be made of studies of the instability of Rossby-wave motion (Lorenz 1972; Hoskins 1973; Gill 1974; and others).

Thorncroft and Hoskins (1990) presented the upper-level evolution in their higher-resolution simulation of the basic life cycle in terms of potential vorticity on the 310 K isentropic surface. The alternative view of potential temperature on the surface with potential vorticity equal to 2×10^{-6} $m^2 s^{-1} K\ kg^{-1}$ was given by Thorncroft et al. (1993). The combination of higher resolution and these choices of field provides a clear picture of the thinning of the trough as it decays in amplitude. Also, with higher resolution a weak cut-off forms, and the interaction between this and the surface front appears to be responsible for the secondary wave development noted earlier. Thorncroft et al. also discussed the mechanism of the trough thinning, and provided a comprehensive discussion of the mature stage of the life cycle in the light of recent work on the theory of wave/mean-flow interaction.

2.4 An Example from Operational Weather Prediction

To conclude this discussion of a basic cyclone life cycle, some aspects of an actual case are presented. When the writing of this chapter began, the most recent operational ECMWF forecast was found to have produced a development at short range that bore a striking similarity to the idealized life-cycle simulation. Moreover, the forecast turned out to be one of high accuracy. The upper panel of Fig. 2 shows the initial distribution of surface pressure over the North Atlantic. The disturbance in question already existed, and can be seen east of Newfoundland with a central low pressure of 1003 hPa. The forecast for three days later (shown in the central panel) was for the low to be located a little to the west of Norway, with a depth of 969 hPa, a value within 1 hPa of that subsequently analyzed (lower panel). Note also the successfully predicted development of a weak low-pressure feature in the subtropics some 20° of longitude west of the Canary Islands.

A comprehensive description of the development cannot be presented here, but we comment on two additional pictures that are relevant to the foregoing discussion. Figure 3 displays phase lines at 6-hourly intervals. The longitude of the trough position, determined from standard height maps, is plotted at the latitude of the center of the surface low. The low had a northward component to its track as it grew between the

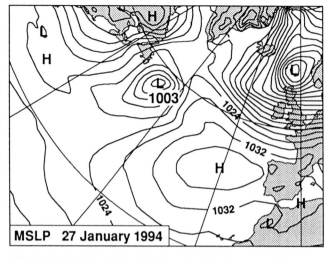

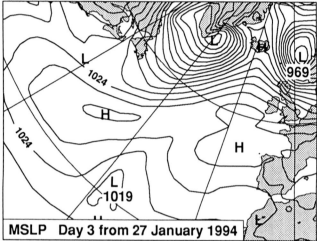

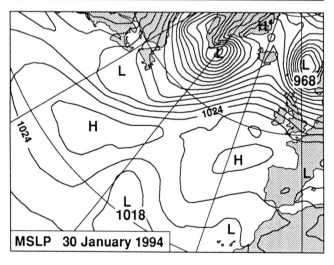

FIG. 2. Mean-sea-level pressure analysis (hPa) for 1200 UTC 27 January 1994 (upper), the three-day forecast starting from this time (middle), and the verifying analysis (lower).

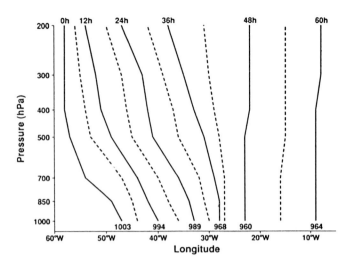

FIG. 3. Pressure/longitude section showing trough positions of Atlantic low at 6-hourly forecast intervals up to hour 60 for the forecast from 1200 UTC 27 January 1994. Also shown are the surface pressures at the center of the low, at 12-hourly intervals.

fall of 13 hPa, and the lowest pressure shown, 960 hPa at 48 hours, was reached first after 42 hours.

Three principal stages can be seen in Fig. 3. The first extends from about hour 6 until between hour 24 and hour 30. During this time there is very little change in the westward phase tilt as the low deepens. The disturbance moves with a rather uniform speed at all heights in a phase-locked manner similar to that of a normal-mode instability. The system did not, however, reach this stage from an infinitesimal perturbation. The preceding analyses show that the development was triggered by an incoming upper trough which had very little phase tilt with height. An indication of this can be seen in Fig. 3. The phase line for the initial time has much less tilt above 500 hPa than later during the growth phase.

The second stage lasts until between 42 and 48 hours, close to the time when the surface cyclone reaches peak intensity. Over this interval, movement slows near the surface and quickens aloft, culminating in vertical alignment of the system. The westward phase tilt is lost first near the surface. The tilt in the upper troposphere actually increases slightly between 24 and 36 hours as the wave moves faster at 500 hPa while losing low-level tilt. The upper tilt then disappears rapidly beyond 36 hours.

The third stage is again characterized by a phase-locked structure. Now, however, the system is equivalent barotropic, and at this northern latitude it moves rather uniformly eastward, and rather faster than before. The surface low fills slightly.

Figure 4 shows potential vorticity (PV) and winds on the 315 K isentropic surface after 24, 48, and 72 hours. The current operational ECMWF model (Ritchie et al. 1995) has a T213 spectral resolution in the horizontal and a 31-level

starting time of the forecast (hour 0) and hour 42, and thereafter it moved directly eastward. The minimum surface pressure is also shown in Fig. 3, at 12-hourly intervals. Deepening was largest between hour 30 and hour 36, with a

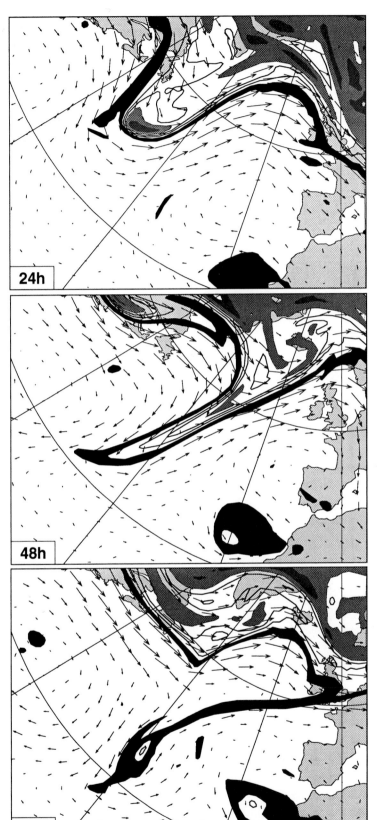

vertical resolution (T213L31 resolution), which enables the representation of finer-scale structure than in the idealized simulations discussed earlier. Figure 4 provides a particularly clear picture of the decay of the trough, with increasing phase tilt and reducing amplitude. The thinning and meridional extension associated with winds that blow parallel to the PV contours ahead of the trough, and across contours behind the trough, are evident. By 72 hours a narrow streamer connects the northern remnant of the trough with a small area of high PV well to the south and west, associated with which is the secondary surface low illustrated in Fig. 2. Appenzeller and Davies (1992) have linked such PV structures in high-resolution ECMWF analyses with corresponding METEOSAT water vapor images.

3. Other Types of Behavior of Mature Baroclinic Waves

Simmons and Hoskins (1980) examined the way the basic life-cycle simulation was changed when different barotropic components were added to the zonal-mean westerly flow. A significant sensitivity of the maximum level of eddy kinetic energy was found. In one case, weaker shear equatorward of the jet delayed the onset of the barotropic decay process. In another, modification of the cessation of growth at low levels was found to be a significant factor. James and Gray (1986) and James (1987) have examined this further.

In all but one of the cases reported by Simmons and Hoskins (1980) barotropic decay eventually occurred at a rate comparable with that of the earlier baroclinic growth, despite differences in maximum eddy amplitude. The anomalous case had larger cyclonic shear of the zonal-mean flow between 20°N and 50°N. Growth of the wave was followed by a much slower decay than in the basic case, as illustrated in Fig. 5.

A synoptic description of this case, from a rerun at higher resolution, has been given by Hoskins (1990). He showed how the mature wave at upper levels did not evolve into a thinning, trailing trough. Instead, it remained of broad zonal extent and was more meridionally

FIG. 4 (left). Potential vorticity (contoured) and wind (arrows) on the 315 K isentropic surface, for 24-, 48- and 72-hour forecasts from 1200 UTC 27 January 1994. The contour interval is 10^{-6} m^{2}s^{-1}K kg^{-1}, and the blackest shading denotes values between 1 and 2 × 10^{-6} m^{2}s^{-1}K kg^{-1}. Lighter shading denotes values between 5 and 6 × 10^{-6} m^{2}s^{-1}K kg^{-1}, and between 6 and 7 × 10^{-6} m^{2}s^{-1}K kg^{-1}.

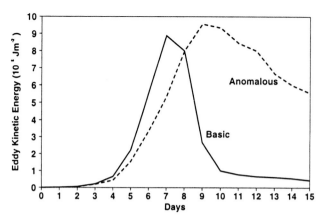

FIG. 5. The variation of eddy kinetic energy with time for "basic" and "anomalous" types of baroclinic waves (after Simmons and Hoskins 1980).

confined. The tongue of high PV air wrapped up cyclonically, and the feature cut off and persisted. The mechanisms involved have been examined in detail by Thorncroft et al. (1993). They viewed this case as involving a large-amplitude counterpart of nonlinear reflection from Rossby-wave critical layers. This was in contrast to the basic case, which could be regarded as involving nonlinear critical-layer absorption.

The Cleveland Storm of January 1978 provides an actual case of extreme cyclonic wrapping up. This storm was notable for a deepening of about 40 hPa over 24 hours, which, unusually, took place entirely over the continental United States and gave a record-breaking minimum surface pressure of 958 hPa. The development was associated with a "phasing" of two troughs, one moving eastward over the southern United States and the other southward from Canada. An ample supply of moisture was available from the Gulf of Mexico ahead of the southern trough (Wagner 1978; Bosart, personal communication).

A few years ago, data assimilation was carried out for this period at ECMWF using the then-operational T106L19 model and the extensive observational datasets gathered in preparation for the FGGE observing year. Diagnosis of the resulting analyses has been reported by Hakim et al. (1995, 1996). Forecasts for this case have recently been produced using T213L31 resolution. The forecast from 1200 UTC 25 January 1978 covered the period of rapid deepening in its first 24 hours, and provided an accurate simulation of the true development, with surface pressure minima of 962 hPa at 1200 UTC and 960 hPa at 1800 UTC on 26 January. The low thereafter slowly filled. The synoptic maps presented in Fig. 6 illustrate the rapid deepening and large spatial extent of the storm and its slow filling.

Of particular interest here is the upper-level evolu-

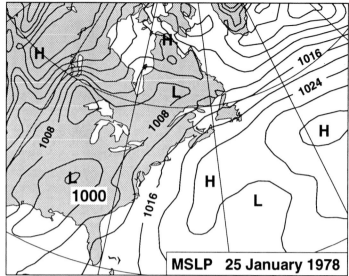

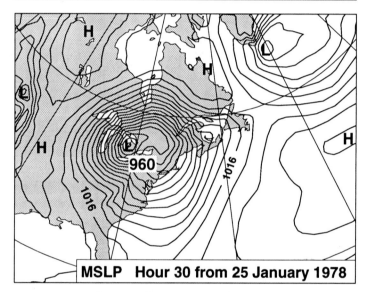

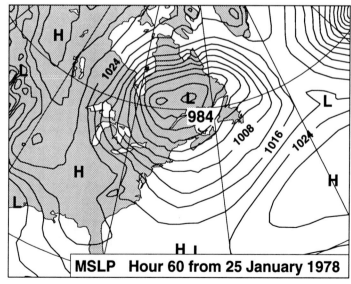

FIG. 6. Mean-sea-level pressure analysis (hPa) for 1200 UTC 25 January 1978 (upper), and 30-hour (middle) and 60-hour (lower) forecasts from this time.

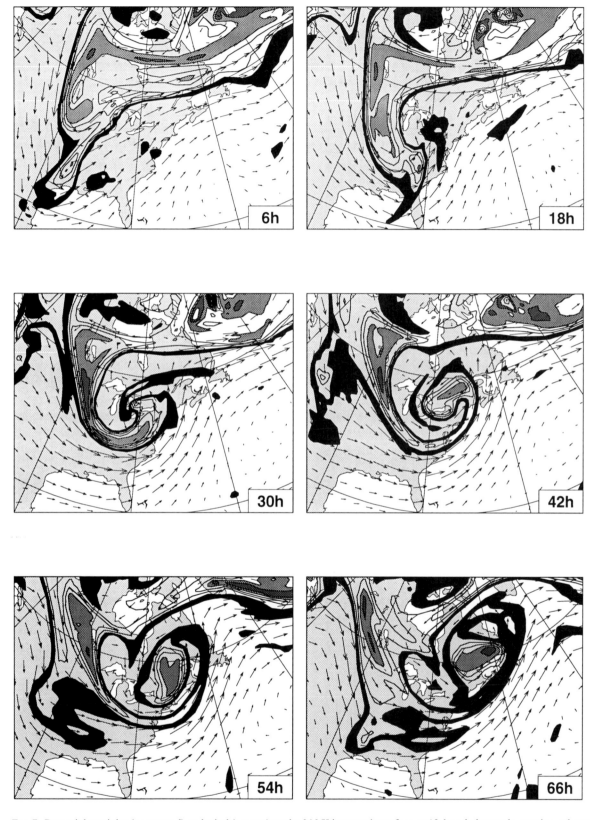

FIG. 7. Potential vorticity (contoured) and wind (arrows) on the 310 K isentropic surface, at 12-hourly intervals, starting at hour 6, from a forecast from 1200 UTC 25 January 1978. Contour intervals and shading are as in Fig. 4.

tion, and this is illustrated in Fig. 7 by maps of PV and winds on the 310 K isentropic surface. These are presented at regular 12-hourly intervals, starting 6 hours into the forecast.

At 6-hour range the two troughs involved in the development are identified by the PV maxima over Texas and west of Lake Superior. The base of the southern trough in fact is thinned and extended early on, rather as shown previously. However, the predominant feature of Fig. 7 is the elongation and wrapping up of the regions of high and low PV, which continues after the surface low has ceased to deepen. A gradual and rather complex merging takes place toward the end of the period shown and at later times. During this process there is a small-scale extrusion of high PV air that can be seen between Hudson Bay and the Great Lakes after hour 42. This extends southeastward and eventually connects with the border of the cut-off region. The 500 hPa height field presents a more bland picture of a cut-off low with strong gradients to the south and east. This reaches peak intensity at around hour 36, and decays slowly thereafter.

Unlike the idealized simulations discussed earlier, the forecast model contains a full set of parametrizations of physical processes. Latent heat release in particular has a significant influence on the development. Without it, a T106 forecast gives a surface pressure minimum some 20 hPa higher than from a corresponding forecast with the full parametrization. The basic pattern of cyclone development is, however, similar in the two forecasts, as in the cases of rapid Atlantic cyclogenesis studied by Reed et al. (1988).

In Fig. 7, the low PV of the air advected around the cut-off comes not from advection from the subtropics, but rather from lowering of PV due to strong latent heat release below the 310K surface in the vicinity of the Great Lakes. Earlier in the forecast, a patch of high PV forms ahead of the southern trough, and is then strung out, weakened, and at least partly drawn into the region of cyclonic wrap-up. Its origin too can be traced to latent heat release. The 310 K surface drops well into the lower troposphere ahead and to the south of the southern trough. It is thus located below the latent heating maximum associated with the incipient surface low. A similar behavior was noted by Uccellini (1990) for the Presidents' Day 1979 storm, some other aspects of which are discussed later in this article.

The operational forecast from 27 January 1994 discussed earlier provides several examples of mature development involving cut-off behavior falling between the extremes shown in Figs. 4 and 7. The Atlantic trough that developed after the one shown in Fig. 3 matured farther to the east, and developed less of a northeast to southwest tilt and more of a cut-off. The latter formed over the Iberian peninsula. Also, in the second half of the ten-day range, three lows were cut off in succession and moved to the south of a persistent blocking anticyclone over the northeastern Pacific. Successive maturing and elongating troughs had developed a northwest to

southeast tilt. Cutting off occurred as the base of each trough was advected by westerly flow in the subtropics, while the thinning shear-line to the north (the "umbilical cord" referred to by Palmén and Newton 1969) came under the influence of easterly flow ahead of it. The latter was associated with the anticyclonic circulation of the blocking high. The overall development in this region bore a striking similarity to the classical picture sketched by Berggren et al. (1949).

Palmén and Newton (1969, p. 277) identified five characteristic types of upper-level disturbances. The basic life cycle simulations discussed in Section 2 appear to fall into category (a), degeneration into a shear line, or (b), formation of a weak cut-off at the base of a trailing shear line. The other categories are various forms of more-intense and often longer-lived cut-off lows similar to those described in this section. In reality the cut-off occurs more often than might be assumed from the range of idealized experiments carried out by Simmons and Hoskins. This may be linked in part with a difference between the idealized jet flows and the climatological flow distribution. The relatively broad band of westerlies over the eastern Pacific, and the subtropical westerly jet located southeast of the Atlantic storm track, may favor cutoff formation in these regions. Orographic effects of the Rocky and southern European mountain ranges can also play a part. Inclusion of latent heat release in the idealized simulations may also promote cutting off (Jones 1992) by lowering PV to the east and north of the trough, as occurred in the Cleveland storm.

4. The Downstream and Upstream Development of Baroclinic Waves

The idealized simulations considered in the preceding sections represented growth and decay in the presence of a sequence of identical waves both upstream and downstream. Such simultaneous development is seldom observed in the atmosphere. Indeed, Bjerknes and Solberg (1922) described how surface cyclones typically occur in families, with new members forming behind (upstream of) existing members. Each new upstream cyclone tends, however, to develop somewhat to the east of where its predecessor developed. Conversely, major development in the upper troposphere is often followed downstream by either the development of a new system or the amplification of a pre-existing wave. Examples may be found in the forementioned works of Namias and Clapp (1944) and Cressman (1948), and illustrations of downstream intensification were discussed by Hovmöller (1949) in introducing the trough-ridge diagram that now bears his name. Two more recent studies are those by Chang (1993) and Lee and Held (1993), who discuss in particular how downstream development tends to occur within wave trains or packets of limited longitudinal extent.

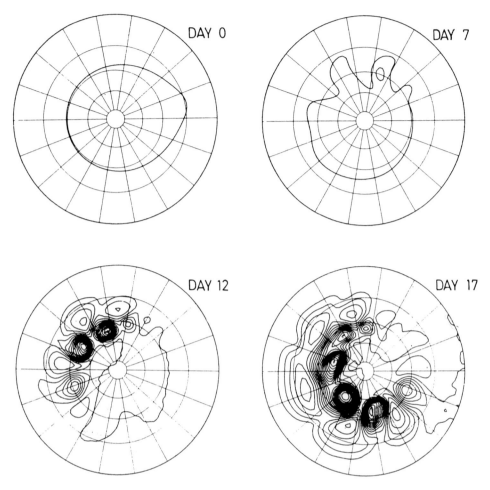

FIG. 8. Polar stereographic plot of surface pressure after 0, 7, 12, and 17 days illustrating downstream and upstream development of a sequence of baroclinic waves, triggered by an isolated initial perturbation. The contour interval is 4 hPa (Simmons and Hoskins 1979).

The early theoretical interpretation of downstream development was based on the idea of dispersion of Rossby waves (Rossby 1945; Yeh 1949), and interest in this type of development prompted further study of barotropic dispersion (Hoskins et al. 1977). The more general baroclinic problem was subsequently examined by Simmons and Hoskins (1979), partly by numerical simulation. One of the idealized midlatitude jet flows studied earlier was perturbed not by a regular train of identical waves, but rather by a local perturbation. This perturbation moved east, and grew rather as in the earlier life-cycle simulation. Ahead of it, new disturbances developed regularly in sequence. Behind it, shallower, smaller-scale disturbances also developed regularly. Each one of these formed a little to the east of the preceding one. Figure 8 shows the distribution of surface pressure at various stages of the simulation.

Each new downstream disturbance reached a larger amplitude than its predecessor, particularly in the upper troposphere. In contrast with the case of normal-mode initial

conditions, maximum amplitude was reached sooner at upper levels than at the surface. Downstream dispersion thus played an important role in the initiation of new disturbances and in the decay of old ones. A similar conclusion was reached by Orlanski and Katzfey (1991) in a case study of an actual cyclone life cycle.

Simmons and Hoskins(1979) also analyzed the spreading of an initially localized disturbance for quasigeostrophic f- and β-plane models. Farrell (1982a), Hoskins (1990) and Orlanski and Chang (1993) have examined this further. For the basic state whose stability was first studied by Eady (1949), the downstream limit of disturbance moves with the speed of the zonal-mean flow at the upper lid, and the upstream limit moves with the zonal-mean flow at the lower surface. Thus, for westerly surface flow each upstream disturbance develops to the east of the previous one. Hoskins (1990) discusses how development at the fringes could be regarded as occurring essentially through neutral Rossby-wave dispersion, the mean-state meridional temperature gradients at the upper and lower lids acting much as β effects of respectively positive and negative sign (for temperature decreasing poleward). Adding the actual planetary β effect modifies the picture somewhat. Downstream development becomes of longer wavelength, and spreading is somewhat slower. The upstream development occurs with a shorter wavelength. These changes bring the solution closer to the full numerical simulation. Further discussion of the possible relationship between upstream dispersion and secondary cyclogenesis can be found in the paper of Thorncroft and Hoskins (1990). The study by Lee and Held (1993) includes modeling of the baroclinic wave packet.

The downstream development process is of evident importance for numerical weather prediction. Miyakoda et al. (1971) recognized its relevance but met with limited success in its representation in their pioneering work on medium-range forecasting. A decade later, Hollingsworth et al. (1980) provided examples for which downstream development was accurately captured late in the medium range. Moreover, Hollingsworth et al. (1985) illustrated how the sensitivity of extratropical medium-range forecasts to differences in initial analyses tends to be large when the initial differences are located in the principal baroclinic zones. Differences subsequently propagate and amplify in the manner indicated by the idealized simulations of downstream development. Initial analysis errors over the central Pacific Ocean can spread to influence the forecast for Europe after around five days in winter, consistent with a propagation speed not much less than the speed of the maximum mean zonal flow.

5. Further Results from Numerical Weather Prediction

We conclude with three further examples drawn from model development work at ECMWF. The aim here is not to provide a general review of the topics being considered, but rather to illustrate sensitivity and thereby provide indications of the role that can be played by different processes in actual systems.

5.1 An Orographic Influence on Downstream Cyclogenesis

Jarraud et al. (1988) noted sensitivity of the prediction of cyclogenesis near the east Asian coast to the specification of orographic height in the forecast model. Here we discuss a particular case described by Simmons and Miller (1988) in which there was sensitivity also to the inclusion of a parametrization of unresolved gravity-wave drag. It was one of several test forecasts in which the largest medium-range impact of a change in orographic representation occurred over the Pacific.

In this case, the T106 operational forecast, which at the time used an "envelope" orography (Jarraud et al. 1988) but no gravity-wave drag, produced a spurious explosive deepening over the far-western Pacific, starting at about day 5. Deepening was even more intense in a forecast using grid-square mean orography. A much slower, smaller scale, and more realistic cyclogenesis occurred when the wave-drag parametrization was included (Fig. 9, lower panels).

Reference has already been made to the importance of the mobile upper-air trough in the process of cyclogenesis. Synoptic analysis in this case indicated that orographic influences on such a trough over central Asia earlier in the forecast range were primarily responsible for the different cyclogenesis several days later. In reality, and in the forecast with envelope orography and gravity-wave drag, a trough moving eastward to the north of the Tien Shan Range developed a northeast to southwest tilt, and weakened. This meridional tilt was much smaller in the forecast with mean orography and no gravity-wave drag, and the trough moved eastward as a pronounced feature (Fig. 9, upper and middle panels). Rapid surface cyclogenesis occurred when the trough later crossed the continental coastline. In this case, error growth was due more to advection of an erroneous feature into the area of interest than to the dispersive process discussed in the preceding section.

An influence of orography on the origin of upper troughs is suggested by synoptic analysis. Sanders (1988) tracked features in the 500 hPa height field during nine cold seasons, and showed that formation of new troughs occurred preferentially over and east of the Rocky Mountains and over the highlands of central Asia. Sanders also identified some "new" troughs as arising from the northern part of fracturing existing troughs of substantial meridional extent. Thus what we have regarded in earlier sections as the decay stage of a cyclone life cycle may leave a remnant that moves on in the westerlies to trigger new cyclonic life farther downstream.

5.2 The Presidents' Day Storm

A discussion of the evolution of the Presidents' Day storm of 19 February 1979 has been given by Uccellini (1990) as part of a more general review of processes contributing to the rapid development of extratropical cyclones. This case was studied originally at ECMWF as part of work with FGGE datasets, and was chosen by Bengtsson (1981) for inclusion in his contribution to the Memorial Volume for Tor Bergeron. Further studies have been made from time to time since then.

One component of the development of the storm was an upper trough that in the 48 hours preceding 1200 UTC 19 February 1979 moved from the Canadian Rockies to close to the eastern seaboard of the United States. Simmons and Miller (1988) presented maps of PV on the 310 K surface

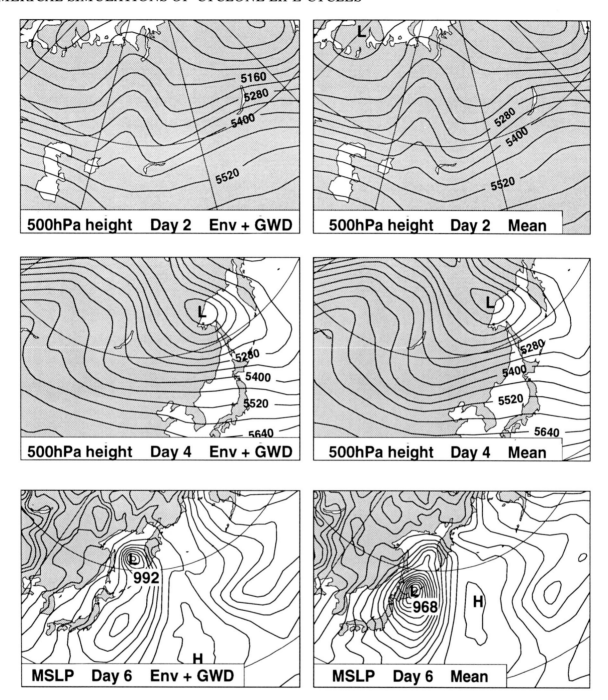

FIG. 9. Forecasts from 1200 UTC 15 January 1986 made with (left) and without (right) an "envelope" enhancement of orography and a parametrization of gravity-wave drag. Upper panel: 2-day forecasts of 500 hPa height; middle panel: 4-day forecasts of 500 hPa height; lower panel: 6-day forecasts of mean-sea-level pressure.

computed from ECMWF analyses. The maximum PV value associated with the trough was virtually unchanged during this period in reality, and concern was expressed at a loss of amplitude of the PV maximum in corresponding 48-hour forecasts. This situation now seems to have been remedied, as no appreciable loss of amplitude occurs in new forecasts with the current higher resolution (and lower diffusion) version of

the ECMWF model. This can be seen in the 24- and 48-hour forecasts included in Fig. 10, in which the initial maximum value of $7 \times 10^{-6} \, m^2 \, s^{-1} \, K \, kg^{-1}$ is clearly maintained as the perturbation moves southeastward from the Rockies to east of the Great Lakes.

Figure 10 also illustrates how the PV maximum decreases substantially in magnitude over the following 24 hours. In this

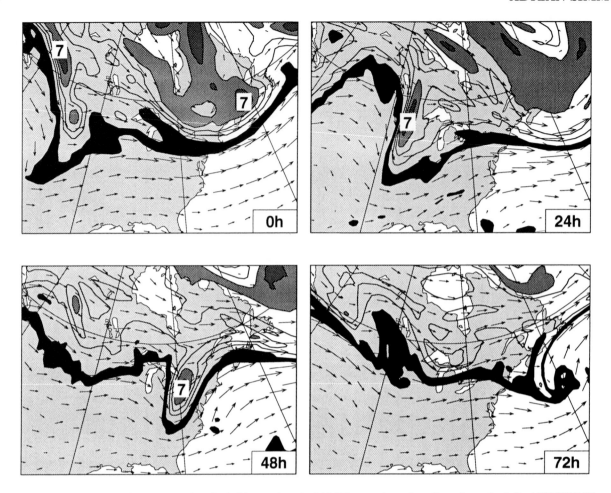

FIG. 10. Potential vorticity (contoured) and wind (arrows) on the 310 K isentropic surface, from the analysis for 1200 UTC 17 February 1979, and for 24-, 48-, and 72-hour forecasts starting from this time. Contour intervals and shading are as in Fig. 4.

case it appears that decay of the trough is due to latent heating in the vicinity of the surface low, rather than a predominantly dynamical process. The loss of amplitude is evident also in conventional 500 and 300 hPa height maps for the period.

Uccellini's (1990) review included numerical forecasts from 1200 UTC 18 February 1979, which showed the critical importance of boundary layer fluxes and latent heat release for the development of the surface cyclone. Simmons and Miller (1988) also showed that surface development was conspicuously absent when latent heating was suppressed in the ECMWF model. Study of other cases indicates that the role of the surface fluxes may be not as direct as that of latent heat release (Reed and Simmons 1991).

5.3   Sensitivity in the Prediction of Mature-Wave Behavior

Our final example is one that illustrates the pronounced sensitivity that can sometimes be found in medium-range forecasts of extratropical waves. Interest in forecast failures is often attracted to cases in which a significant cyclogenesis

was missed, but this example shows that there can also be major failure to represent accurately the mature stage of wave behavior.

Figure 11 shows the 500 hPa height analysis for 1200 UTC 16 November 1992 (upper left panel), and three day-5 forecasts valid at this time and date. The upper-right panel shows the operational T213L31 forecast and the lower-left panel a forecast carried out using T106L19 resolution. These two forecasts differ substantially in their representation of the trough, which in reality extends across the Greenwich Meridian. This was captured quite accurately in the T106L19 forecast, apart from a phase lag and failure to produce the weak cutoff over North Africa. Conversely, the higher resolution forecast produced an intense cutoff centered to the northwest of the Canary Islands, and the remnant of the northern part of the trough moved too fast to the southeast.

The trough in question was not directly associated with a strong surface development. Its growth appears to have involved downstream amplification of a weak pre-existing

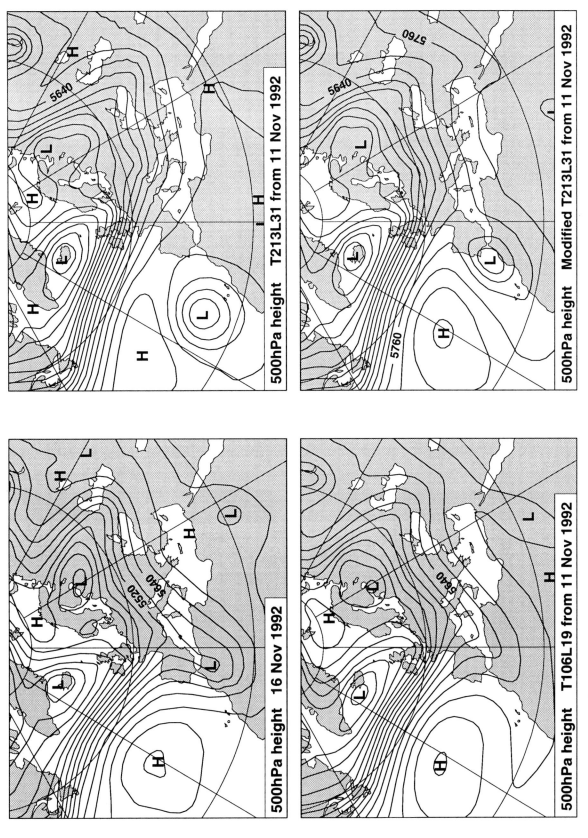

FIG. 11. 500 hPa height for 1200 UTC 16 November 1992, and 5-day forecasts verifying at this time from three different model versions.

wave as a system grew baroclinically upstream. This system in fact matured west of southern Greenland, and subsequent lee intensification produced the low that can be seen near Iceland in Fig. 11.

Attention was drawn to this case because the incorrect operational forecast of the maturing system over the eastern Atlantic was not matched by the operational forecast produced by Deutscher Wetterdienst using a variant of the T106L19 model previously operational at ECMWF. Major differences in operational forecasts can often be traced to differences in initial analyses, but model resolution appeared to be implicated here.

It appears that differences in radiative forcing played a crucial part in these forecasts. A T213L31 run with radiation suppressed gave no sign of formation of the erroneous cutoff. Studying the evolution of PV fields suggested it was radiative cooling in the upper troposphere early in the forecast range that was principally responsible. A forecast in which radiation was turned off above 500hPa for just the first day of the forecast is shown in the lower-right panel of Fig. 11. This change was evidently enough to improve substantially the higher resolution forecast.

It might be erroneous to conclude that there was a serious error in the cloud radiative forcing in the T213L31 forecast, as conceivably an analysis error could have been compensated by a parametrization deficiency of the T106L19 forecast. A sound, independently justified change to the forecasting system that improves this particular high-resolution forecast could well be found in the future. However, the existence of such a sensitive mature-wave behavior suggests that this behavior, which may be liberally categorized as Rossby-wave instability, can be as basic a factor as baroclinic instability in limiting predictability, a view expressed by Lorenz (1972).

Acknowledgments

Sakari Uppala carried out the data assimilation for the Cleveland storm, and Ernst Klinker helped investigate the case discussed in Section 5.4. Comments from Tony Hollingsworth, Tim Palmer and Anders Persson are gratefully acknowledged.

References

Appenzeller, C., and H. C. Davies, 1992: Structure of stratospheric intrusions into the troposphere. *Nature*, **358**, 570–572.

Bengtsson, L., 1981: The weather forecast. *PAGEOPH*, **119**, 515–537.

Berggren, R., B. Bolin, and C.-G. Rossby, 1949: An aerological study of zonal motion, its perturbation and breakdown. *Tellus*, **1**(2), 14–37.

Bjerknes, J., and H. Solberg, 1922: Life cycle of cyclones and the polar front theory of atmospheric circulation. *Geofys. Publ.*, **3**(1), 1–18.

Bjerknes, V., 1904: Das Problem der Wettervorhersage, betrachtet vom Standpunkte der Mechanik und Physik. *Meteor. Z.*, **21**, 1–7.

Buizza, R., and T. N. Palmer, 1995: The singular-vector structure of the atmospheric global circulation. *J. Atmos. Sci.*, **52**, 1434–1456.

Chang, E. K. M., 1993: Downstream development of baroclinic waves as inferred from regression analysis. *J. Atmos. Sci.*, **50**, 2038–2053.

Charney, J. G., 1947: The dynamics of long waves in a baroclinic westerly current. *J. Meteor.*, **4**, 135–162.

———, R. Fjørtoft, and J. von Neumann, 1950: Numerical integration of the barotropic vorticity equation. *Tellus*, **2**, 237–254.

Cressman, G. P., 1948: On the forecasting of long waves in the upper westerlies. *J. Meteor.*, **5**, 44–57.

Eady, E. T., 1949: Long waves and cyclone waves. *Tellus*, **1**(3), 33–52.

Edmon, H. J., B. J. Hoskins, and M. E. McIntyre, 1980: Eliassen–Palm cross-sections for the troposphere. *J. Atmos. Sci.*, **37**, 2600–2616.

Farrell, B. F., 1982a: Pulse asymptotics of the Charney baroclinic instability problem. *J. Atmos. Sci.*, **39**, 507–517.

———, 1982b: The growth of disturbances in a baroclinic flow. *J. Atmos. Sci.*, **39**, 1663–1686.

Gall, R., 1976a: A comparison of linear baroclinic instability theory with the eddy statistics of a general circulation model. *J. Atmos. Sci.*, **33**, 349–373.

———, 1976b: Structural changes of growing baroclinic waves. *J. Atmos. Sci.*, **33**, 374–390.

Gill, A. E., 1974: The stability of planetary waves on an infinite beta-plane. *Geophys. Fluid Dyn.*, **6**, 29–47.

Hakim, G. J., L. F. Bosart, and D. Keyser, 1995: The Ohio Valley wave-merger cyclogenesis event of 25–26 January 1978. Part I: Multiscale case study. *Mon. Wea. Rev.*, **123**, 2663–2692.

———, D. Keyser, and L. F. Bosart, 1996: The Ohio Valley wave-merger cyclogenesis event of 25–26 January 1978. Part II: Diagnosis using quasigeostrophic potential vorticity inversion. *Mon. Wea. Rev.*, **124**, 2176–2205.

Held, I. M., and P. J. Phillips, 1987: Linear and nonlinear barotropic decay on the sphere. *J. Atmos. Sci.*, **44**, 200–207.

Hollingsworth, A., K. Arpe, M. Tiedtke, M. Capaldo, and H. Savijärvi, 1980: The performance of a medium-range forecast model in winter—Impact of physical parameterizations. *Mon. Wea. Rev.*, **108**, 1736–1773.

———, A. C. Lorenc, M. S. Tracton, K. Arpe, G. Cats, S. Uppala, and P. Källberg, 1985: The response of numerical weather prediction systems to FGGE level IIb data. Part I: Analyses. *Quart. J. Roy. Meteor. Soc.*, **111**, 1–66.

Hoskins, B. J., 1973: Stability of the Rossby–Haurwitz wave. *Quart. J. Roy. Meteor. Soc.*, **99**, 723–745.

———, 1990: Dynamics of mid-latitude cyclones. *Extratropical Cyclones, The Erik Palmén Memorial Volume*, Amer. Meteor. Soc., 63–80.

———, A. J. Simmons, and D. G. Andrews, 1977: Energy dispersion in a barotropic atmosphere. *Quart. J. Roy. Meteor. Soc.*, **103**, 553–567.

Hovmöller, E., 1949: The trough-and-ridge diagram. *Tellus*, **1**(2), 62–66.

James, I. N., 1987: Suppression of baroclinic instability in horizontally sheared flows. *J. Atmos. Sci.*, **44**, 3710–3720.

———, and L. J. Gray, 1986: Concerning the effect of surface drag on the circulation of a baroclinic planetary atmosphere. *Quart. J. Roy. Meteor. Soc.*, **112**, 1231–1250.

Jarraud, M., A. J. Simmons, and M. Kanamitsu, 1988: Sensitivity of medium-range weather forecasts to the use of an envelope orography. *Quart. J. Roy. Meteor. Soc.*, **114**, 989–1025.

Jones, B., 1992: *Effects of Physical Processes in Baroclinic Waves.* Ph.D. thesis, University of Reading.

Kuo, H.-L., 1953: On the production of mean zonal currents in the atmosphere by large disturbances. *Tellus*, **5**, 475–493.

Lee, S., and I. M. Held, 1993: Baroclinic wave packets in models and observations. *J. Atmos. Sci.*, **50**, 1413–1428.

Lorenz, E. N., 1972: Barotropic instability of Rossby wave motion. *J. Atmos. Sci.*, **29**, 258–264.

Miyakoda, K., R.F. Strickle, C. J. Nappo, P. L. Baker, and G. D. Hembree, 1971: The effect of horizontal grid resolution in an atmospheric circulation model. *J. Atmos. Sci.*, **28**, 481–499.

Molteni, F., R. Buizza, T. N. Palmer, and T. Petroliagis, 1996: The ECMWF ensemble prediction system: Methodology and validation. *Quart. J. Roy. Meteor. Soc.*, **122**, 73–119.

Namias, J., and P. F. Clapp, 1944: Studies of the motion and development of long waves in the westerlies. *J. Meteor.*, **1**, 57–77.

Orlanski, I., and E. K. M. Chang, 1993: Ageostrophic geopotential fluxes in downstream and upstream development of baroclinic waves. *J. Atmos. Sci.*, **50**, 212–225.

———, and J. Katzfey, 1991: The life cycle of a cyclone wave in the Southern Hemisphere. Part I: Eddy kinetic energy budget. *J. Atmos. Sci.*, **48**, 1972–1998.

Palmén, E., and C. W. Newton, 1969: *Atmospheric Circulation Systems.* Academic Press, 603 pp.

Petterssen, S., and S. J. Smebye, 1971: On the development of extratropical cyclones. *Quart. J. Roy. Meteor. Soc.*, **97**, 457–482.

Phillips, N. A., 1956: The general circulation of the atmosphere: A numerical experiment. *Quart. J. Roy. Meteor. Soc.*, **82**, 123–164.

Platzman, G. W., 1952: The increase or decrease of mean-flow energy in large-scale horizontal flow in the atmosphere. *J. Meteor.*, **9**, 347–358.

———, 1979: The ENIAC computations of 1950—Gateway to numerical weather prediction. *Bull. Amer. Meteor. Soc.*, **60**, 302–312.

Reed, R. J., and A. J. Simmons, 1991: Numerical simulation of an explosively deepening cyclone over the North Atlantic that was unaffected by concurrent surface energy fluxes. *Wea. Forecasting*, **6**, 117–122.

———, ———, M. D. Albright, and P. Undén, 1988: The role of latent heat release in explosive cyclogenesis: Three examples based on ECMWF operational forecasts. *Wea. Forecasting*, **3**, 217–229.

Richardson, L.F., 1922: *Weather Prediction by Numerical Process.* Cambridge University Press, 236 pp.

Ritchie, H., C. Temperton, A. J. Simmons, M. Hortal, T. Davies, D. Dent, and M. Hamrud, 1995: Implementation of the semi-Lagrangian method in a high resolution version of the ECMWF forecast model. *Mon. Wea. Rev.*, **123**, 489–514.

Rossby, C.-G., 1939: Relation between variations in the intensity of the zonal circulation of the atmosphere and the displacements of the semi-permanent centres of action. *J. Mar. Res.*, **2**, 38–55.

———, 1945: On the propagation of frequencies and energy in certain types of oceanic and atmospheric waves. *J. Meteor.*, **2**, 187–204.

Sanders, F., 1988: Life history of mobile troughs in the upper westerlies. *Mon. Wea. Rev.*, **116**, 2629–2648.

Simmons, A. J., and B. J. Hoskins, 1976: Baroclinic instability on the sphere: Normal modes of the primitive and quasi-geostrophic equations. *J. Atmos. Sci.*, **33**, 1454–1477.

———, and ———, 1977: Baroclinic instability on the sphere: Solutions with a more realistic tropopause. *J. Atmos. Sci.*, **34**, 581–588.

———, and ———, 1978: The life cycles of some nonlinear baroclinic waves. *J. Atmos. Sci.*, **35**, 414–432.

———, and ———, 1979: The downstream and upstream development of unstable baroclinic waves. *J. Atmos. Sci.*, **36**, 1239–1254.

———, and ———, 1980: Barotropic influences on the growth and decay of non-linear baroclinic waves. *J. Atmos. Sci.*, **37**, 1679–1684.

———, and M. Miller, 1988: The prediction of extratropical weather systems—Some sensitivity studies. *Proceedings of 1987 ECMWF Seminar, II*, 271–315.

Simons, T. J., 1972: The nonlinear dynamics of cyclone waves. *J. Atmos. Sci.*, **29**, 38–52.

Thorncroft, C. D., and B. J. Hoskins, 1990: Frontal cyclogenesis. *J. Atmos. Sci.*, **47**, 2317–2336.

———, ———, and M. E. McIntyre, 1993: Two paradigms of baroclinic-wave life-cycle behaviour. *Quart. J. Roy. Meteor. Soc.*, **119**, 17–55.

Uccellini, L. W., 1990: Processes contributing to the rapid development of extratropical cyclones. *Extratropical Cyclones, The Erik Palmén Memorial Volume*, Amer. Meteor. Soc., 81–105.

Wagner, A. J., 1978: Weather and circulation of January 1978. *Mon. Wea. Rev.*, **106**, 579–585.

Wiin-Nielsen, A., 1959: On barotropic and baroclinic models, with special emphasis on ultra-long waves. *Mon. Wea. Rev.*, **87**, 171–183.

Yeh, T.-C., 1949: On energy dispersion in the atmosphere. *J. Meteor.*, **6**, 1–16.

A Planetary-Scale to Mesoscale Perspective of the Life Cycles of Extratropical Cyclones: The Bridge between Theory and Observations

MELVYN SHAPIRO

NOAA/Environmental Technology Laboratory, Boulder, Colorado, United States

HEINI WERNLI

Swiss Federal Institute of Technology, Zurich, Switzerland

JAIN-WEN BAO

CIRES, University of Colorado and NOAA/ETL, Boulder, Colorado, United States

JOHN METHVEN

University of Reading, Reading, United Kingdom

XIAOLEI ZOU

Florida State University, Tallahassee, Florida, United States

JAMES DOYLE AND TEDDY HOLT

Naval Research Laboratory, Monterey, California, United States

EVELYN DONALL-GRELL AND PAUL NEIMAN

NOAA/Environmental Technology Laboratory, Boulder, Colorado, United States

---

1. Introduction

The emergence of meteorology as a rational science began around the turn of the twentieth century when Max Margules, Hermann Helmholtz, Felix Exner, and Vilhelm Bjerknes formulated the theoretical basis for what was previously considered an empirical science with a qualitative application to weather forecasting and climatology. The concurrent synoptic studies of Sir Napier Shaw, Rudolph Lempfert, Johan Sandström, V. Bjerknes, and Heinrich von Ficker, among others, provided insight into the structure and evolution of weather systems, and an assessment of the represen-tativeness of the proposed theories. The synergy between dynamic and synoptic meteorology inspired new theories, observing strategies, conceptual models, and dramatic advances in weather forecasting. During the period 1913–1922, the Leipzig and Norwegian schools of meteorology made fundamental contributions to the advancement of the emerging science. With V. Bjerknes as their director and mentor, the research associates and students at the Geophysical Institutes in Leipzig, Germany, and Bergen, Norway, synthesized theory, observations, synoptic analysis and diagnosis in their quest for physical understanding and improved weather prediction. Their efforts gave rise to revolutionary paradigms for the theory, structure, and evolution of frontal cyclones, many of which remain widely applied in research and weather forecasting. A historical perspective of the science and the milieu of the period is reviewed in the works of Bergeron (1959), Kutzbach (1979), Friedman (1989), and in the historical chapters in this volume by Eliassen (1998), Friedman (1998), Newton and Newton (1998), and Volkert (1998).

In the years following the Leipzig and Bergen contributions, advances in theory, observing technology, and the advent of numerical weather prediction led to the development of a multitude of conceptual models. These models include baroclinic instability, upper-level fronts, tropopause folds, jet streaks, split flow, unbalanced jets, split cold fronts, cold fronts aloft, conveyor belts, baroclinic leaves, dry intrusions, frontal fractures, T-bone fronts, bent-back fronts, frontal seclusions, instant occlusions, inverted troughs, barrier jets, coastal fronts, frontal collapse, frontal gravity-current-like heads, and mesoscale vortices within synoptic-scale cyclones (see reviews by Keyser and Shapiro 1986; Browning 1990; Eliassen 1990; Reed 1990; Shapiro and Keyser 1990; Uccellini 1990). When confronted by such a variety of conceptual models, one might ask: Is the T-bone bent-back warm-front seclusion (Shapiro and Keyser 1990) similar to a back-bent warm frontal occlusion (Bjerknes and Solberg 1922; Bjerknes 1930; Bergeron 1934, 1937) or an instant occlusion (McGinnigle et al. 1988)? Do the results from numerical simulations of highly idealized cyclone life cycles (Simmons and Hoskins 1979; Thorncroft et al. 1993) describe the essence of complex planetary-scale to frontal-scale interactions occurring in nature? Are some conceptualizations simply alternative viewpoints of the same phenomenon or dynamical processes? Daniel Keyser posed this last question to Chester Newton, who replied, "Dan, now you understand why we call it 'research' instead of 'search.'"

This chapter focuses on building a conceptual bridge between theoretically idealized cyclone life cycles and those occurring in nature, with an emphasis on the influence of planetary-scale environmental flow on extratropical cyclone developments. Our methods are derived from established theories, applied both quantitatively and qualitatively to observations, and numerical simulations of actual weather events and their theoretical idealizations. We begin with a paradigm for progress in meteorology based on optimal methods of scientific inquiry, which forms the scientific framework for our search for further understanding of the life cycles of extratropical cyclones. This is followed by an overview of past and current research on idealized extratropical cyclone life cycles, the potential vorticity (PV) perspective of the tropopause and ensemble forecasts, upstream and downstream baroclinic development, and secondary cyclones.

2. A Paradigm for Progress in Meteorology

V. Bjerknes (1904, reprinted in this volume) proposed a rational approach for progress in meteorology, beginning with observations and their analysis, and culminating in weather prediction based on numerical integration of the differential equations of hydrodynamics and thermodynamics governing atmospheric motion. Bergeron (1959) considered the scientific method for progress as a systematic progression from observations, analysis, diagnosis, and physical understanding, to prediction, noting that to proceed from observations and their analysis directly to prediction, without supporting theory and diagnosis, would lead to incomplete physical understanding and limit advances in weather prediction. Hoskins (1983) described an "optimal" situation in which dynamical models of varying complexity interact with one another and with observations in the formulation of evolving conceptualizations of a broad spectrum of atmospheric phenomena and processes.

Figure 1 presents a graphic synthesis of the Bjerknes, Bergeron and Hoskins paradigms for scientific methods of inquiry, where *theory, observation,* and *diagnosis* are a synergy of elements, applied in pursuit of *physical understanding* and its expression through *conceptual models.*

Theory is based on the equations of atmospheric motion and the hierarchy of filtered approximations (e.g., quasi- and

FIG. 1. Physical understanding and conceptual representation through the union of theory, diagnosis, and observation.

semigeostrophy, and nonlinear balance) that reveal the dominant processes governing prescribed scales of atmospheric motion. Theories are realized through dynamical models of varying complexity using analytical, numerical, and laboratory methods of solution.

Observation constitutes the most diverse element of the paradigm. Meteorological observations are spatially and/or temporally inhomogeneous, of varying accuracy and precision, and with no specific observing system providing inclusive measurements of all dependent variables within the primitive form of the governing equations. Until recently, observations from operational networks and experimental field campaigns were the primary basis for improving conceptual models of the structure of atmospheric phenomena. However, the rapid advances in computer technology, when applied to numerical weather prediction, have led to the situation in which the spatial and temporal resolution of observations has been outpaced by that of the numerical models. As a consequence, fine-grid-resolution simulations initialized with operational, synoptic-scale data or idealized atmospheres are used to fill the subsynoptic-scale observational gaps. The high-resolution simulations provide dynamically consistent data sets that describe a broad spectrum of synoptic-scale though mesoscale phenomena. It should be recognized that without corroborating observations, a degree of uncertainty always remains as to whether numerical models are giving the correct solution.

Recent advances in meteorological observing technology are providing observations that challenge the present state of data assimilation and analysis. These observations include cloud and water-vapor drift winds from high-spatial- and temporal-resolution satellites; sea-surface winds from polar-orbiting microwave sensors; automated high spatial resolution, flight-level observations from commercial aircraft; winds and precipitation distributions from ground and airborne Doppler weather radars; winds and temperature from radar wind Profilers; and dropwindsondes deployed from aircraft using adaptive observing strategies. Field campaigns furnish calibrations and validations for these emerging technologies, and in addition provide the observations with which to evaluate the representativeness of simulated subsynoptic-scale phenomena, and are the basis for improving many aspects of numerical weather prediction (e.g., data assimilation methods and parameterizations of sub-grid-scale physical processes). Field campaigns quite often lead to unexpected discoveries of unexplored weather phenomena and their life cycles, which challenge the frontiers of theoretical understanding and weather prediction.

Diagnosis is the language of scientific communication, derived from the hierarchy of theories and performed on observations and on solutions from dynamical models. Diagnosis based on balanced dynamical theories (e.g., quasi- or semigeostrophy) uses analytical or numerical inversion methods (e.g., secondary circulations) to assess and to attribute flows to potential vorticity (PV) anomalies. Lagrangian diagnosis involves trajectory calculations to describe the transport of chemicals, aerosols, and water vapor, as well as the formation of clouds through the saturation of air parcels. Manipulations of the primitive equations of motion give Eulerian equations for assessing instantaneous Lagrangian tendencies of, for example, frontogenesis and potential vorticity change, as well as kinetic energy dispersion.

Conceptual models communicate the essence of scientific discoveries. They are expressions of *physical understanding*, idealized sketches of phenomena and their governing dynamical processes; they are the culmination of scientific methods of inquiry.

Examples of conceptual models that integrate elements of theory, observations, and diagnosis in their formulation and/or application include: the Norwegian model of the life cycle of extratropical cyclones (Bjerknes and Solberg 1922); baroclinic instability (Charney 1947; Eady 1949); the dynamics of upper-level frontogenesis (Reed and Sanders 1953; Newton 1954); the potential vorticity definition of the "dynamic" tropopause and its application to the study of the exchange of air and trace constituents between the stratosphere and troposphere (Reed 1955; Reed and Danielsen 1959; Danielsen 1964); the semigeostrophic balance perspective of transverse secondary circulations about fronts and jet streams (Sawyer 1956; Eliassen 1962; Shapiro 1981); balanced dynamics applied to the diagnosis of baroclinic life cycles through potential vorticity attribution (Kleinschmidt 1950a, 1950b; Bleck 1973, 1974; Hoskins et al.1985; Davis and Emanuel 1991); and Rossby-wave kinetic-energy dispersion and its relationship to downstream (upstream) baroclinic development of extratropical cyclone cycles (Simmons and Hoskins 1979; Orlanski and Chang 1993).

3. Planetary-Scale Influences on Extratropical Cyclone Life Cycles

One of the foremost challenges to the advancement of the physical understanding and the refinement of conceptual models of extratropical cyclones has been to formulate theoretical interpretations for the diversity of cyclone evolutions observed on weather charts, satellite cloud images, operational numerical weather predictions, and simulated with idealized atmospheres. Contemporary studies of cyclone development have addressed synoptic-scale and mesoscale processes, such as frontogenesis, secondary circulations, diabatic influences, quasi-Lagrangian air streams, and variations in surface physiography and friction, with a cursory consideration of the planetary-scale preconditioning of the synoptic-scale environment within which cyclones, fronts, jet streams, and the tropopause evolve. Past theoretical studies have shown that the evolution of cyclones is influenced by the mean flow within the environment of the developing cyclone. The interaction between planetary-scale mean flows and synoptic-scale disturbances is referred to as wave-mean interaction [reviewed in Held (1998, this volume)].

This section begins with a synopsis of theoretical studies of the influence of planetary-scale, environmental barotropic shear on the life cycles of idealized extratropical cyclones (Section 3.1). It is hypothesized that the variations in idealized and observed cyclone life cycles can also arise through the influence of different longitudinal phasing, meridional and vertical alignment of the subtropical, polar, and arctic jet streams and their associated PV anomalies in the environment of cyclone development (Section 3.2). This hypothesis is tested through simulations of observed cyclone developments (Section 5) and with idealized cyclone developments (Section 6) within varying meridional alignments of polar and subtropical jet streams.

3.1   *The Influence of Barotropic Shear on Idealized Cyclone Life Cycles*

The influence of planetary-scale flows on the life cycles of idealized extratropical cyclones was demonstrated through numerical simulations with adiabatic, inviscid, idealized atmospheres (e.g., Hoskins and West 1979; Simmons and Hoskins 1980; Hoskins 1990; Davies et al. 1991; Thorncroft et al. 1993; Wernli 1995; Methven 1996). These studies revealed that the evolutions of idealized cyclones and their associated fronts, tropopause potential vorticity, and Eliassen and Palm (1960) flux, are sensitive to extremely small environmental meridional barotropic wind shear, $\sim$10 m s^{-1} over 2000 km, superimposed upon the zonal basic state flow. The imposed environmental shear was both cyclonic and anticyclonic and independent of height, and hence referred to as barotropic shear. Those cyclones evolving without barotropic shear were referred to as Life Cycle 1 (LC1), and developed into "T-bone," bent-back warm-frontal cyclones as described by Shapiro and Keyser (1990) and Neiman and Shapiro (1993). Those cyclones that developed within cyclonic barotropic shear (Life Cycle 2; LC2) matured into classical "Norwegian" occluded cyclones as described by Bjerknes and Solberg (1922) and Bergeron (1928). The idealized cyclones forming under the influence of anticyclonic barotropic shear (Life Cycle 3; LC3) were open frontal-wave cyclones with well-defined cold fronts and weak warm fronts, which did not form into occluded cyclones. Thorncroft et al. (1993) introduced the LC1 and LC2 cyclone terminology.

Our first illustration of the influence of environmental barotropic shear on cyclone life cycles is taken from the semigeostrophic (SG) f-plane simulation of Wernli (1995), which extended the Davies et al. (1991) normal-mode experiments to the initial-value problem of upper-level induced

cyclogenesis. In addition to the nonsheared (LC1) and cyclonic-shear (LC2) life cycles, Wernli considered the influence of anticyclonic barotropic shear, hereafter referred to as the LC3 life cycle. The LC1 (nonsheared) cyclone (Fig. 2, top) developed T-bone frontal structure with a hammerhead-shaped cyclonic vorticity maximum that extended southward along the bent-back warm front. The LC1 warm and cold fronts have comparable baroclinicity, and the cold front has a north-south orientation. The weakening of the cold front at the T-bone, frontal triple point is referred to as frontal fracture (Shapiro and Keyser 1990). The structure of the Wernli (1995) LC2 cyclonic barotropic-shear cyclone (Fig. 2, middle) evolves as a classical Norwegian cyclone with a Bergeron warm-frontal occlusion and comma-shaped vorticity maximum that spirals inward into the warm secluded air at the cyclone center. The LC2 cyclone possesses a shorter cold front than the LC1 cyclone (Fig. 2, top) and no fracture of the cold front in the vicinity of the occlusion triple point. Note that the influence of cyclonic barotropic shear tends to rotate the cold front into a more northwest-to-southeast orientation than seen in the LC1 cyclone. The anticyclonic-shear LC3 cyclone (Fig. 2, bottom) contains an elongated cold front, relatively short and weak warm front, and prominent cold-frontal cyclonic vorticity filament. The influence of anticyclonic barotropic shear acts to orient the cold front into a northeast-to-southwest direction, and the axis of the warm sector slopes eastward to the north, without frontal occlusion. Wernli (1995) obtained similarly realistic variations in tropopause PV evolution in response to the differing barotropic shears, consistent with the results of Thorncroft et al. (1993), Methven (1996), and Davies (1998, this volume).

Our second illustration is taken from the study by Methven (1996), which presented primitive equation simulations on the sphere to illustrate the influence of environmental barotropic shear on idealized extratropical cyclone evolution. This study extended the work of Thorncroft and Hoskins (1990) and Thorncroft et al. (1993) using a model with high spectral resolution (T341) to address the influence of barotropic shear on the formation and deformation of PV and vorticity filaments in the vicinity of the tropopause and within surface fronts, including their cyclonic roll up and mesoscale instabilities. Methven (1996) simulated the nonshear (LC1) and cyclonic-shear (LC2) life cycles, but did not consider the effect of anticyclonic barotropic shear on cyclone evolution. Thorncroft et al. (1993) and Methven (1996) make the distinction between the categories LC1 and LC2 based upon anticylonic versus cyclonic Rossby wave breaking at upper levels, respectively, where the nonshear (LC1) life cycle on the sphere exhibits anticyclonic behavior at its late stages (days 6–10), developing backward-tilted, thinning tropopause troughs extending into the tropics, characteristic of anticyclonic Rossby-wave breaking. We suggest that the differences in the late stage of cyclone development between

no shear $(A = 0)$

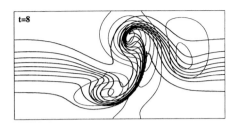
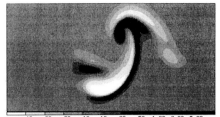

cyclonic shear $(A = +0.2)$

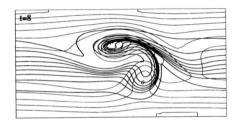
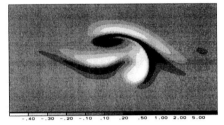

anticyclonic shear $(A = -0.2)$

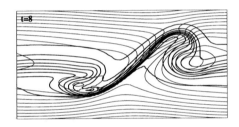
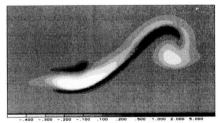

FIG. 2. Semigeostrophic f-plane simulations of three idealized cyclone life cycles at $t = 8$ (~day 4). Top panels: The nonshear cyclone (LC1). Middle panels: The cyclonic barotropic shear (+0.2 × 10^{-4} s^{-1}) cyclone (LC2). Lower panels: The anticyclonic barotropic-shear (–0.2 × 10^{-4} s^{-1}) cyclone (LC3). The panels on the left show the temperature and pressure fields with isopleth intervals of 2.1 K and 2.7 mb, respectively; the panels on the right show the distribution of relative vorticity in units of Coriolis parameter f, with cyclonic regions darker and anticyclonic regions brighter. The diagrams cover a domain of 14 640 × 7000 km (Wernli 1995).

the Wernli, and the Thorncroft and Methven simulations, are due, in part, to the effects of the earth's sphericity (Cartesian versus spherical geometry), and perhaps differences in initial conditions.

During its early phase of development (not shown), the Methven (1996) LC1 cyclone exhibited a well-defined surface cold front that perpendicularly intercepted the bent-back warm front at the T-bone frontal triple point. As the cyclone entered its mature stage (Figs. 3a–c), the bent-back warm front moved northward, while it extended westward into the northerly flow west of the center and then turned cyclonically southward into the cyclone center (Fig. 3a). At this stage of development (~0.5 days later than the Wernli example in Fig. 2, top panels), the cold and warm fronts converged toward occlusion well to the south of the old triple point (Fig. 3a). This occlusion resembles a frontal fracture between the much

stronger cold front and bent-back warm front. The accompanying relative vorticity field (Fig. 3b) has a positive vorticity filament along the cold front, which is enhanced within the occlusion and turns abruptly westward along the bent-back warm front. Note that there is an absence of cyclonic vorticity within the warm front south of the triple point because this segment of the warm front lies within anticyclonic shear to the east of the southerly low-level jet that flows northward at the leading edge of the cold front. It is not uncommon for warm frontogenesis to occur in regions of anticyclonic horizontal shearing thermal deformation (see, e.g., Hoskins and West 1979). The upper-level PV for LC1 (Fig. 3c) evolved into a large-amplitude wave with a weak PV gradient (on the 300-K surface) in the wave inflection between the positive anomaly of the trough and the negative anomaly of the downstream ridge.

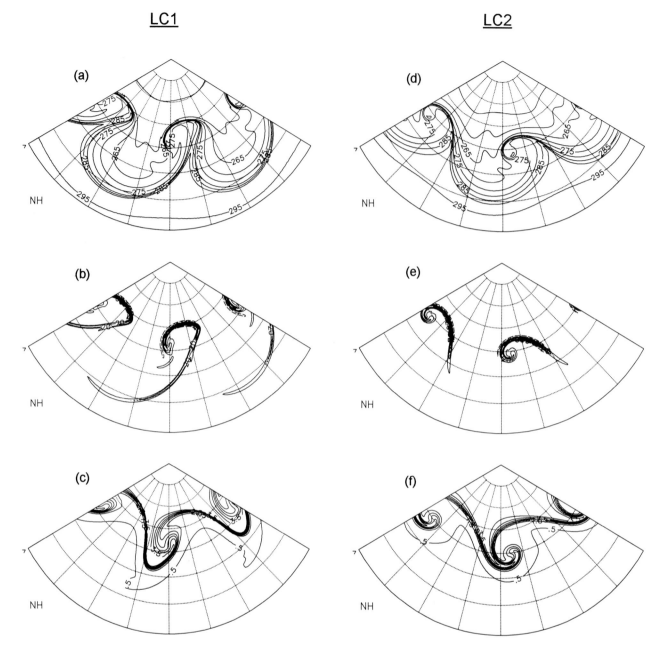

Fig. 3. Primitive-equation, spherical-domain simulations of two idealized cyclone life cycles at ~day 6. Left panels (a–c): The nonshear cyclone (LC1). Right panels (d–f): The cyclonic barotropic-shear (~0.2 × 10⁻⁴ s⁻¹) cyclone (LC2). Upper panels (a, d): Surface potential temperature at 5-K intervals. Middle panels (b, e): Surface relative vorticity at 10^{-4} s⁻¹ intervals. Lower panels (c, f): Potential vorticity on the 300-K isentropic surface at 0.5-PVU intervals (Methven 1996).

The structure of the Methven (1996) cyclonic-shear (LC2) cyclone (Figs. 3d–f) resembles the classical Norwegian frontal-occlusion cyclone. The surface thermal field (Fig. 3d) possesses a well-developed warm front that extends continuously from east of the cyclone warm sector to encircle the warm-air seclusion at the cyclone center. The Methven LC2 cold front is weaker than the warm front (unlike the semigeostrophic LC2 simulation in Fig. 2) and asymptotically merges with the warm front without the tendency for

frontal fracture or frontal T-bone orientation, as was clearly the case for LC1 (Fig. 3a). The companion surface vorticity field (Fig. 3e) contains a cyclonic vorticity filament that spirals into a serpentine coil at the center of the cyclone. The PV distribution on the 300-K surface for LC2 (Fig. 3f) contains a narrow extruded PV filament cyclonically coiled above, and displaced to the west of, the surface vorticity spiral. This spiral-shaped positive PV anomaly (Fig. 3f) corresponds to a deep, narrow PV tropopause trough, in contrast to the much broader tropopause trough of LC1 (Fig. 3c). Incidentally, the word "cyclone" is derived from ancient Greek for coiled serpent.

It should be noted that there is a difference in time scales between the development of some idealized cyclones and those occurring in nature. Those idealized cyclones evolving nonlinearly from normal-mode perturbations typically require up to 6 days to reach maturity, compared with the 1–2 day period for the rapidly developing cyclones in daily weather. It is likely that the comparatively longer time scale of the idealized cyclones results from the small amplitude of the initial perturbations to the zonal mean flow, and the absence of physical processes such as latent heating and friction. Nevertheless, the realism of the idealized cyclones cannot be denied.

3.2 A Conceptual Hypothesis: Jet stream Interactions and Cyclone Life Cycles

The degree to which the above variations in frontal structure within the idealized cyclones are representative of those occurring within nature was not resolved in the earlier theoretical studies. This is understandable considering that only recently have synoptic researchers begun to propose refinements or alternatives to the Bjerknes and Solberg (1922) Norwegian cyclone life cycle, exclusive of the effects of surface physiography (e.g., mountains; coastal zones). Is it possible to identify barotropic shears in daily weather maps or numerical weather forecasts? Thorncroft et al. (1993) attributed the differing idealized life cycles to the location of cyclone development relative to the shear vorticity of a zonal jet stream. If this jet were indeed only the zonal mean jet at ~30° N, then the cyclones of middle and high latitudes would be confined to the region of cyclonic shear north of the mean jet, negating the possibility of the nonshear, or anticyclonic barotropic shear life cycles at higher latitudes. The present task is to reconcile the theoretically idealized barotropic-shear cyclone evolutions with those occurring in nature.

Shapiro and Donall-Grell (1994) hypothesized that different phasing and vertical and/or meridional alignments of the arctic, polar, and subtropical jet streams and their respective PV anomalies were accompanied by differing environmental shears that modulate the life cycles of extratropical cyclones. Figure 4 illustrates this conceptual hypothesis for three meridional alignments of the polar and subtropical wind currents, and shows: (1) the LC3 frontal wave cyclone (left) within the anticyclonic shear to the south of the polar jet stream; (2) the LC1 (nonshear) T-bone, bent-back warm-frontal-seclusion cyclone (center) beneath the vertically aligned polar and subtropical jet streams and their associated tropopause PV folds, characteristic of the most extreme cyclonic events; and (3) the LC2 Norwegian cyclone (right) with its Bergeron warm-frontal occlusion evolving beneath

the region of cyclonic shear, and associated positive PV anomaly north of the subtropical jet stream. One can envisage similar interactions between a high-latitude arctic jet stream and a polar jet stream to its south. This perspective of frontal-cyclone life cycles can be perceived as a variant of wave-mean interaction, in which the low wave number and slowly evolving planetary waves of the arctic and subtropical jet stream characterize the mean flow with which the synoptic-scale more transient baroclinic waves of middle latitudes interact: a planetary-scale/synoptic-scale interaction with a mesoscale frontal overtone.

4. Jet Stream Interactions in the Vicinity of the Tropopause: A Potential Vorticity Perspective

The emphasis of this section is on representations of PV in the vicinity of the tropopause, its associated jet streams, and PV anomalies. Here we introduce the graphical representations of PV used to diagnose interactions between polar and subtropical jet streams in Sections 5 and 6. The Appendix to this chapter introduces the application of the PV perspective of baroclinic life cycles to the diagnosis of ensemble forecasts. Although a slight diversion from the basic theme of the chapter, the PV ensemble provides a dynamically based visualization of forecast uncertainty of the interactions between jet stream regimes that, in turn, impact on the predictability of cyclones.

4.1 Potential Vorticity in the Vicinity of the Tropopause

The Shapiro et al. (1987) adaptation of the Palmén and Newton (1969) idealization of the meridional structure of the Northern Hemispheric PV tropopause, primary jet streams, and associated fronts (Fig. 5) shows the vertical deformations of the tropopause associated with arctic, polar, and subtropical jet streams and their respective frontal zones. Vertical deformations of the tropopause will be referred to as tropopause "folds," or simply folds, when viewed in cross section, irrespective of whether PV becomes multivalued in the vertical.

Because of the broad range of potential temperature over which tropopause folds occur (~280 K for the arctic fold to ~340 K for the subtropical fold), PV representations on single isentropic or isobaric surfaces (e.g., 316 K or 300 mb, respectively) do not depict all tropopause folds and PV anomalies associated with the primary jet stream regimes. This is illustrated in Fig. 6, where at 0000 UTC 4 January 1989, the 340-K isentropic surface (Fig. 6a) intersects the typically small-amplitude waves and PV anomalies along the ~200-mb circumpolar subtropical jet stream, and the PV-gradient zone of the subtropical tropopause fold, whereas the 316-K isentropic surface (Fig. 6b) intersects the shorter

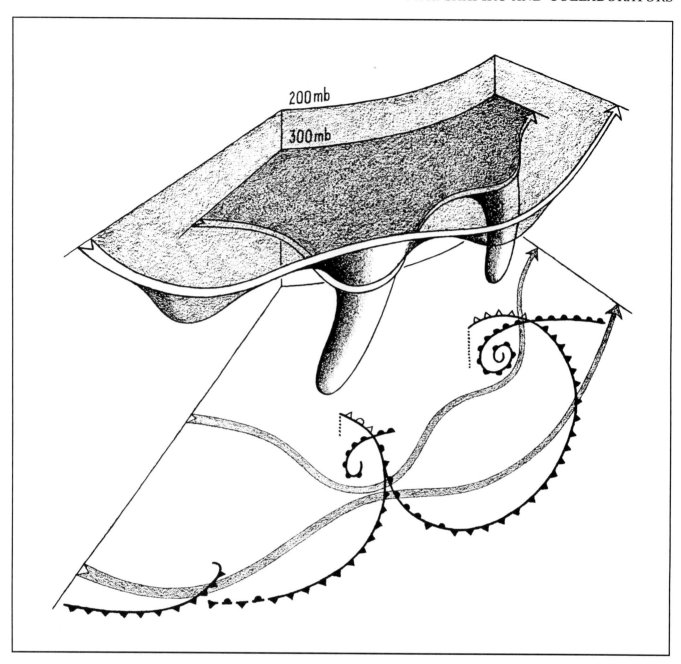

FIG. 4. A conceptual hypothesis for the influence of upper-level jet stream and PV alignments on frontal structure within extratropical cyclones. Upper plane (light shading): 200-mb planetary wave and subtropical jet stream (white ribbon) with associated PV anomalies suspended below. Middle plane (heavy shading): 300-mb synoptic wave and polar jet stream (white ribbon) with associated PV anomalies suspended below. Lower plane: Earth's surface with three characteristic cyclone frontal configurations; frontal symbols are conventional with fronts aloft entered as open symbols. Left cyclone: The anticyclonic shear cyclone (LC3) located south of the polar jet stream with its westward-trailing cold front and a weakly defined warm front. Middle cyclone: The nonshear cyclone (LC1) located beneath the vertically aligned polar and subtropical jet streams and associated coupled tropopause folds; the T-bone polar occlusion and bent-back warm-frontal seclusion cyclone. Right cyclone: The cyclonic-shear cyclone (LC2) situated north of the subtropical jet stream; the Norwegian frontal cyclone with its back-bent polar warm-frontal occlusion (Shapiro and Donall-Grell 1994).

wavelength, large-amplitude synoptic waves of the midlatitude ~300-mb polar jet stream and tropopause fold. Similar PV distributions are found on the ~280-K isentropic surface (not shown), which intercepts the ~400-mb arctic tropopause of the polar vortex and its associated arctic jet stream. The

pressure (altitude) of individual primary jet streams can vary by ± ~50 mb.

Hoskins and Berrisford (1988), Davis and Emanuel (1991), Hakim et al. (1995), Bosart et al. (1996), Morgan and Nielsen-Gammon (1998), among others, have utilized PV-

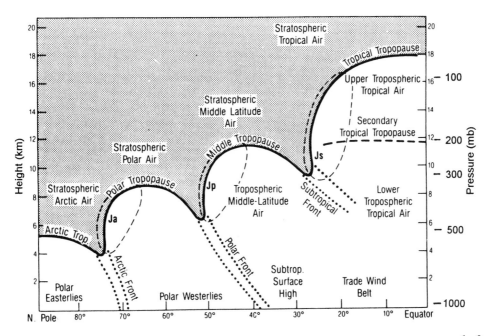

FIG. 5. The meridional structure of the tropopause. The dynamic 2-PVU potential-vorticity tropopause (1 PVU = 10^{-6} m^2 s^{-1} K kg^{-1}) is shown by the heavy solid line, and the stratosphere is stippled. The primary frontal zones are bounded by the heavy dotted lines and are labeled accordingly. The 40 m s^{-1} isotach (thin dashed line) encircles the cores of the three primary jet streams: arctic, Ja; polar, Jp; and subtropical, Js. The secondary (thermal) tropical tropopause is indicated by the heavy dashed line. Major tropospheric and stratospheric air masses, tropopause surfaces, and selected wind systems are labeled in the cross section. Individual cross sections may differ spatially and temporally from this idealized meridional model. With the exception of the addition of the arctic jet stream and the PV tropopause, all other primary wind systems, fronts, and air masses are as in the Palmén and Newton (1969, Fig. 47) cross section (Shapiro et al. 1987).

tropopause maps to chart upper-level PV anomalies and associated tropopause folds. Tropopause maps depict potential temperature, pressure, and wind velocity on the 2 ($\pm$0.5) PVU "dynamic" tropopause (1 PVU = 10^{-6} m^2 s^{-1} K kg^{-1}) and resolve vertical deformations of the tropopause, and jet stream meanders (upper-level waves) irrespective of their isentropic or isobaric levels. Cold and warm potential temperature anomalies that appear on tropopause maps are often referred to as tropopause troughs and ridges, respectively, analogous to positive and negative PV anomalies on θ surfaces. Unfortunately, the vertical inversion of PV surfaces within pronounced tropopause folds precludes the use of PV as a vertical coordinate for quantitative diagnosis through numerical inversion. Tropopause maps are used for qualitative diagnoses of baroclinic life cycles, where the atmosphere is approximated by a layer bounded by two material surfaces (i.e., the tropopause and the earth's surface), as in the quasigeostrophic Eady (1949) model, and for the depiction of interactions between arctic, polar, and subtropical jet streams.

4.2 Mean Potential Vorticity Maps

An alternative representation of PV distributions in the vicinity of the tropopause is obtained by preparing maps of mean (vertically averaged) potential vorticity (MPV) within the layer bounded by (1) an upper isentropic surface

($\theta \sim 340$ K) that passes through the core of the subtropical jet stream and its associated tropopause fold between $\sim$150 and 200 mb, situated below the tropical tropopause; and (2) a lower isentropic surface ($\theta \sim 280$ K) that passes just beneath the $\sim$500-mb arctic tropopause. Column-integrated PV normalized for the mass between isentropic layers is a representation of column-integrated PV substance (see, e.g., Haynes and McIntyre 1987) within the vertically bounded layer. Massacand et al. (1998) introduced pressure-weighted, mean-PV maps to study the dynamical relationship between stratospheric PV intrusions and flooding rain events in the Alps. From distributions of MPV and wind velocity, one can visualize vertically averaged PV advection, which is a leading component in baroclinic developments and θ-coordinate balanced diagnostics of secondary circulations (e.g., Davis et al. 1996). Because MPV integrates the PV anomalies of all upper-level jet streams and tropopause folds, it is useful for viewing interactions between the subtropical, polar, and arctic jet stream/frontal-zone systems, as is also the case for representations of thermal anomalies on 2-PVU tropopause maps.

An example of MPV and wind velocity at 0000 UTC 4 January 1989, for the isentropic layer 280–340 K (Figs. 6c and d, respectively) shows the signature for merging and diverging branches of subtropical and polar jet streams in the vicinity of 90°E, where these jet streams take respective

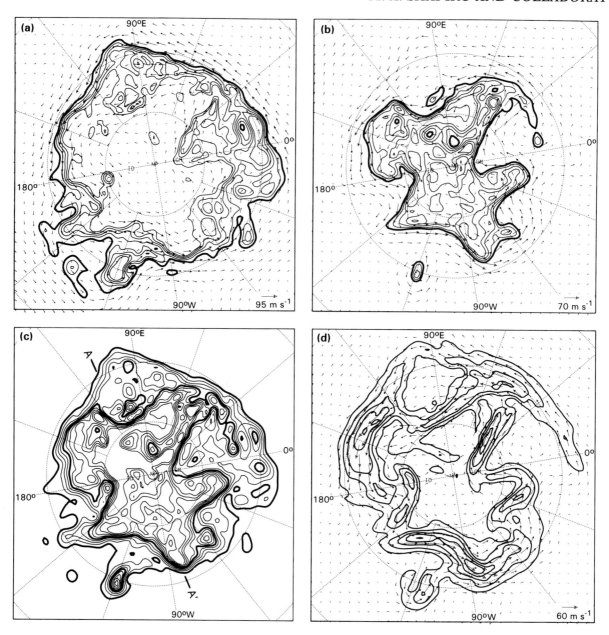

FIG. 6. Hemispheric perspective of potential vorticity (PV) and layer-mean (vertically averaged) PV (MPV) at 0000 UTC 4 January 1989. (a) PV on the 340-K isentropic surface (PVU, thin solid lines; 2-PVU isopleth, heavy solid line; 2-PVU contour interval); wind velocity vectors (arrows). (b) Same as (a) but for the 316-K isentropic surface. (c) MPV for the 280–340 K layer (outer heavy contour, 2 PVU; inner heavy contour, 10 PVU; *A* and *A'*, end points for Fig. 7 cross-section projection). (d) Layer-mean velocity for (c) (speed $\geq$ 30 m s^{-1}; 10 m s^{-1} contour interval, solid lines); velocity vectors (arrows).

southern and northern diversions around the Tibetan Plateau. Here, the negative (anticyclonic) MPV anomaly of a polar ridge is meridionally aligned with the positive (cyclonic) MPV anomaly of a subtropical trough. In contrast, the prominent positive anomaly at ~80°W, 35°N, with its concentrated southern gradient, marks the vertical alignment of polar and subtropical jet stream tropopause folds in a region where a polar trough is meridionally aligned with a subtropical ridge. The cross-section analysis across these features (Fig. 7) shows the multiple folding of the PV tropopause between

Asia and the North Pole and the notable single tropopause fold over the eastern United States in the region of vertical alignment of the polar and subtropical jet streams and their tropopause folds.

Figure 8 presents 48-h evolution of MPV preceding the onset of the 4–5 January 1989 North Atlantic cyclone development. This rapidly intensifying cyclone was observed in considerable detail during the Experiment on Rapidly Intensifying Cyclones over the Atlantic (ERICA (see Hadlock and Kreitzberg 1988) and identified as ERICA Intensive Observ-

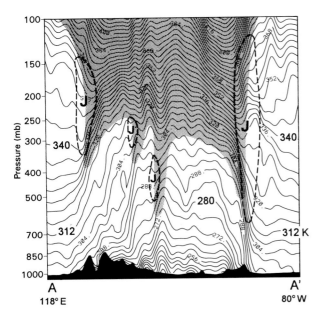

FIG. 7. Cross section between the points A and A' of Fig. 6c: potential temperature (K, solid lines); PV > 2-PVU (shaded); 40 m s^{-1} isotachs (dashed lines); jet stream cores (indicated by the letter J).

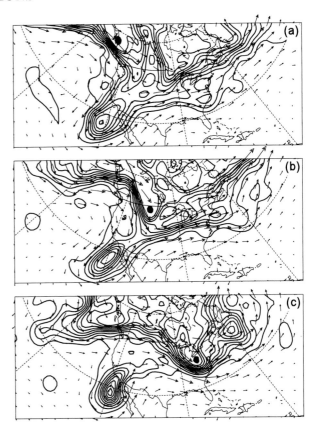

FIG. 8. Vertically averaged potential vorticity (MPV) for the 280–340 K layer: 2-PVU contour interval (solid lines); layer-mean velocity vectors (arrows); the center of a transient polar component of MPV (bold solid dot). (a) 0000 UTC 2 January 1989; (b) 0000 UTC 3 January 1989; (c) 0000 UTC 4 January 1989.

ing Period-4 (IOP-4). The IOP-4 development was investigated by Shapiro and Keyser (1990), Neiman and Shapiro (1993), Neiman et al. (1993), Chang et al. (1993, 1996), and Davis et al. (1996), among others. At 0000 UTC 2 January 1989 (Fig. 8a), a positive (cyclonic) MPV anomaly was centered at ~60°N at the west coast of British Columbia. The concentrated gradient at the southern edge of the anomaly marks the vertical alignment of polar and subtropical jet streams, their associated tropopause folds, and the meridional juxtaposition of a cyclonic anomaly of the northern polar trough (marked by the black dot over Vancouver Island.) with the anticyclonic anomaly of the subtropical ridge to the south. Positive and negative MPV or PV anomaly couplets can be viewed as the signature of a Bergeron (1928) confluence/diffluence deformation and its associated tropopause based jet stream wind maxima. Localized jet stream wind maxima within a Bergeron deformation were described by Riehl et al. (1952), later referred to as a "jet streaks" by Newton and Carson (1953), with subsequent applications by, for example, Uccellini and Johnson (1979), and reviewed in Uccellini (1990). By 0000 UTC 3 January 1989 (Fig. 8b), the positive MPV anomaly previously situated over the west coast (Fig. 8a) had propagated southeastward to the Montana-Canada border, where its polar component passed across and beneath the larger-scale cyclonic anomaly of the subtropical trough. The anticyclonic subtropical current across the southern United States, and its negative MPV anomaly to the south, remained unchanged from 24 h earlier. By 0000 UTC 4 January 1989 (Fig. 8c), the subtropical and polar tropopause folds were vertically aligned over the eastern

United States at ~40°N, 80°W (Fig. 7) with a substantial MPV gradient across the >70 m s^{-1} wind speed maximum in the northwesterly flow. Here the meridional juxtaposition of positive polar and negative subtropical MPV represent the PV anomaly signature of a Bergeron (1928) confluence/diffluence deformation, with a wind speed maximum (jet streak) oriented along the axis of dilatation. Bluestein (1993) noted that jet streaks could be considered as forming through PV-anomaly interactions among subtropical, polar, and arctic PV vorticity regimes.

It is recognized that the longevity of individual PV and MVP anomalies often exceeds that of the life cycle of individual jet streaks (e.g., Hakim et al. 1995). This is seen in Fig. 8, where a persistent, transient (mobile), positive MPV anomaly from the polar current participated in the development of two separate jet streaks within a 48-h period. This anomaly first phased with the quasi-stationary subtropical jet stream and the anticyclonic anomaly over the Gulf of Alaska (Fig. 8a) and propagated eastward to phase with the negative anomaly of the anticyclonic subtropical flow over the northern Gulf of Mexico (Fig. 8c).

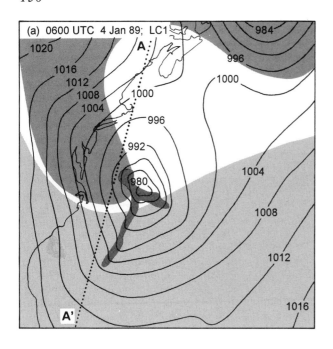

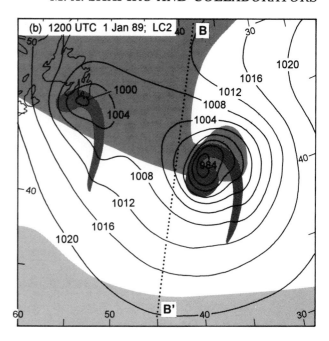

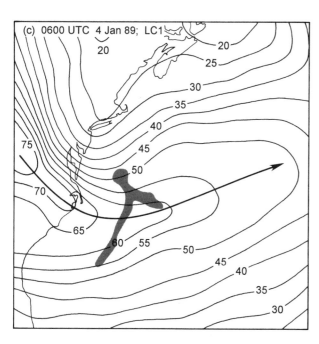

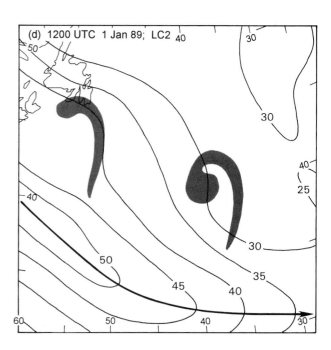

FIG. 9. Left panels (a,c): The 6-h simulation of the ERICA IOP-4 (LC1) cyclone at 0600 UTC 4 January 1989. (a) Sea-level pressure (mb, thin lines), 950-mb absolute vorticity (>2×10^{-4} s^{-1}, dark shading), and PV at $\theta = 316$ K (PV > 2 PVU, medium shading) and at $\theta = 340$ K (PV < 2 PVU, light shading). Boundaries between the white area and the medium and lightly shaded areas are the positions of the 2-PVU isopleth at the axes of the polar and subtropical jet streams and associated tropopause folds, respectively. Line *AA'* in (a) is the projection line for cross-section Fig.10a. (c) Wind speed (m s^{-1}, thin lines) and subtropical jet stream axis (long thin arrow) at $\theta = 340$ K; 950-mb absolute vorticity as in (a). Right panels (b,d): The 24-h simulation of the ERICA Pre IOP-4 (LC2) cyclone valid at 1200 UTC 1 January 1989. The isopleths and shading in (b) and (d) are same as in (a) and (c). Line *BB'* in (b) is the projection line for cross-section Fig. 10b.

5. Atmospheric Analogs to Idealized Cyclone Life Cycles: Simulations and Observations

Numerical simulations of three North Atlantic cyclones were performed to test the proposed conceptual hypothesis attributing the variations of cyclone evolution to differences in environmental shears due to differing meridional alignments of the polar, subtropical, and arctic jet streams and their associated PV anomalies. The first two cyclones developed off the northeast coast of North America within the western region of the North Atlantic storm track. The third cyclone began as an incipient frontal wave over the western Atlantic

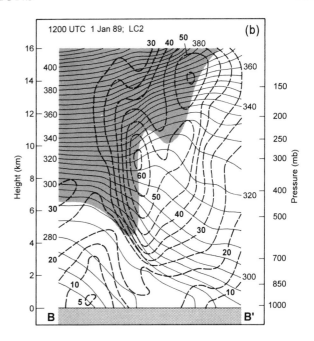

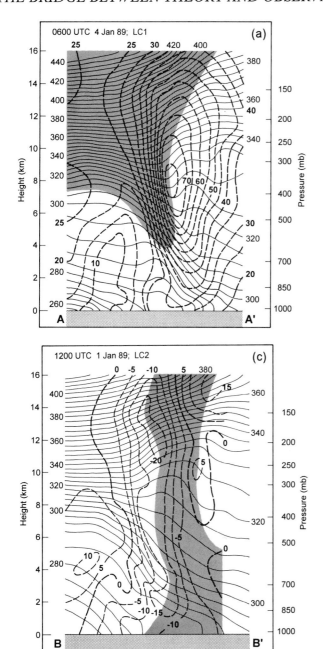

FIG. 10. (a) Cross-section analysis of potential temperature (K, thin lines) and wind speed (m s^{-1}, dashed lines) along the projection lines *AA'* of Fig. 9a; PV >2 PVU (shaded). (b) Same as (a), but for the projection line *BB'* of Fig. 9b. (c) The difference in cross-section-normal wind speed (Fig. 9b minus Fig. 9a), where the ~300-mb polar jet stream core is taken as a registration reference for the subtraction.

and developed into an end-of-storm-track primary cyclone north of Scotland. The simulated structure of these North Atlantic cyclones is compared with that of the idealized barotropic-shear cyclones in Section 3.1, and with observed cyclones.

5.1 Life Cycle 1: The Cyclone of 4–5 January 1989

The ERICA IOP-4 western North Atlantic cyclone of 4–5 January 1989 is presented as an analog to the nonshear idealized LC1 cyclone (Section 3.1). The surface development of IOP-4 commenced at ~0000 UTC 4 January 1989 over the warm waters of the Gulf Stream current to the east

of Cape Hatteras, North Carolina. The MPV and wind analyses for the 48-h period preceding the genesis of IOP-4 (Fig. 8) chronicles the longitudinal phasing of PV anomalies and vertical alignment of polar and subtropical jet streams that provided the single jet stream, nonshear (LC1) environment for this event. The 24-h simulation of IOP-4 was prepared for the period 0000 UTC 4 January to 0000 UTC 5 January 1989, with the Pennsylvania State University/National Center for Atmospheric Research (Penn State/NCAR) MM-5 prediction model, using a full-physics version of the model, and a 23-level, 25-km resolution, two-way interactive grid. The architecture of the MM-5 model is described in Grell et al. (1994).

Figures 9a and c present the 6-h simulation of the incipient phase of IOP-4. At 0600 UTC 4 January 1989, the polar and subtropical jet streams (marked by the position of the 2-PVU contour at $\theta = 316$ K and 340 K, respectively) are vertically aligned over the surface cyclone (Fig. 9a). At this time, the cyclone was a well-established circulation with a central MSL pressure of ~978 mb, including T-bone structure in its frontal vorticity (Fig. 9a). Note the larger amplitude and shorter wavelength of the polar jet stream PV wave compared to the lesser amplitude and longer wavelength of the subtropical jet stream. The concurrent 340-K wind speed analysis (Fig. 9c) shows the cyclone positioned directly beneath the subtropical jet.

The cross-section analysis of potential temperature, wind speed, and the 2-PVU tropopause (Fig. 10a) taken upstream from the developing cyclone along line *AA'* of Fig. 9a, shows the substantial vertical deformation of the PV tropopause with jet stream wind speeds in excess of 70 m s^{-1}. The single jet stream in the cross section is a manifestation of vertical alignment of the polar and subtropical wind currents above the region of surface cyclogenesis.

By 1800 UTC 4 January (18 h into the simulation) the IOP-4 cyclone was into its mature phase, ~1000 km east-southeast of Long Island, New York, with a central MSL pressure of 948 mb, with fully developed frontal T-bone structure in both potential temperature and vorticity (Figs. 11a and c, respectively). The largest vorticity >2 × 10^{-4} s^{-1} was within the bent-back warm front that encircled the cyclone center (Fig. 11c). The realism of the IOP-4 simulation (Figs. 9a and 11c) can be assessed by comparison with its verification (Figs. 12 and 13). Figure 12 shows the incipient formation of frontal T-bone and bent-back warm-frontal structure at 0600 UTC 4 January 1989. Figure 13a shows the subsequent fully-developed T-bone, bent-back warm front, and warm-air seclusion at the center of the mature cyclone at 1800 UTC 4 January 1989. The NOAA/Geostationary Operational Environmental Satellite (GOES) image (Fig. 13b) is concurrent with Fig. 13a. A comparison of the frontal structure of the mature IOP-4 cyclone (Figs. 11a and c) with the Methven (1996) LC1 idealized cyclone (Figs. 3a and b) shows significant similarities in temperature and vorticity structure between the observed and idealized cyclones.

5.2 *Life Cycle 2: The Cyclone of 31 December 1988 to 2 January 1989*

A proposed analog to the cyclonic barotropic-shear LC2 cyclone life cycle (Section 3.1) occurred over the western North Atlantic 4 days prior to the IOP-4 storm described in Section 5.1. This cyclone was investigated by Rotunno and Bao (1996), who referred to the storm as Pre IOP-4, because it preceded IOP-4 and was not designated an Intensive Observing Period for research observations. For the present study, Pre-IOP-4 was simulated for the 48-h period 1200 UTC 31 December 1988 to 1200 UTC 2 January 1989, with the MM-5 model configured as for IOP-4 (Section 5.1). Whereas the IOP-4 life cycle (Section 5.1) was one of rapid development, transcending inception to maturity in ~21 h, the Pre IOP-4 cyclone required ~48 h to reach its maximum intensity.

By 1200 UTC 1 January 1989, 24-h into simulation (Fig. 9b and d), the Pre IOP-4 surface cyclone and its upper-level PV structure were at their midpoint of development. The surface cyclone was situated ~1000 km southeast of Newfoundland with a central pressure of 978 mb, and had a well-defined comma-shaped cyclonic vorticity maximum with the comma head at the center of the cyclone (Fig. 9b). At this time, the cyclone was intensifying beneath a positive tropopause PV anomaly within the polar jet stream current, denoted by the 2-PVU contour at θ = 316 K (Fig. 9b). The secondary cyclone and 950-mb vorticity comma south of Newfoundland (Fig. 9b) was associated with a polar low vertically coupled to an arctic tropopause PV anomaly and was not a feature of interest in the present study. The axis of a subtropical jet stream, marked by the 2-PVU contour at θ = 340 K (Fig. 9b), flows within the broad subtropical trough positioned ~1000 km south of the cyclone center and its polar tropopause PV anomaly. The 340-K wind analysis (Fig. 9d) reveals the presence of cyclonic shear to the north of the subtropical jet stream axis above both the vorticity comma of the developing surface cyclone and the PV anomaly of the polar wave. The cross section of potential temperature, wind speed, and the 2-PVU tropopause (Fig. 10 b), taken upstream from the developing cyclone along the line *BB'* of Fig. 9b, shows the meridional separation of the northern 300-mb polar jet core and southern 150-mb subtropical jet core, and the double folding of the 2-PVU tropopause across the two distinct jet streams.

By 0130 UTC 2 January 1989, the Pre-IOP-4 cyclone was approaching maturity ~1000 km east of Newfoundland with a simulated minimum central pressure of 966 mb. At that time in its life cycle, its thermal structure (Fig.11b) possessed a pronounced bent-back warm front that encircled the secluded warm air at the cyclone center with a relatively weaker cold front extending outward from the center of the cyclone. The companion vorticity field (Fig. 11d) contains a cyclonic vorticity filament that extends from the cold front into the warm front and spirals inward into the cyclone center.

Unfortunately, there were no research observations to evaluate the representativeness of the simulated frontal evolution for this situation. As an alternative, we present analyses of a Gulf of Alaska cyclone that exhibited structural characteristics similar to those of the simulated Pre-IOP-4 cyclone. This cyclone was observed in detail with drop-windsondes from a NOAA P-3 research aircraft and reported in Shapiro and Keyser (1990). Figure 14a shows the NOAA polar satellite cloud image for this cyclone. The 850-mb temperature analysis at 0000 UTC 10 March 1987 (Fig. 14b) illustrates the inward spiral of the bent-back warm front that encircles the secluded warm air at the cyclone center. The vertical cross-section analysis through the warm seclusion (Fig. 14c) reveals the outward-sloping baroclinicity and the cyclonic wind shear of the encircling warm front and low-level jet. The similarities in structure among the simulations of the Pre IOP-4 cyclone (Figs. 11b and d), the idealized cyclonic barotropic-shear LC2 cyclone (Figs. 3d-f), and the observed Gulf of Alaska cyclone (Fig. 14) are notable.

For another perspective of the proposed analogy between observed and idealized barotropic-shear cyclone life cycles, let us consider the cross sections taken upstream of IOP-4 (Fig. 10a) and Pre-IOP-4 (Fig. 10b) as representative

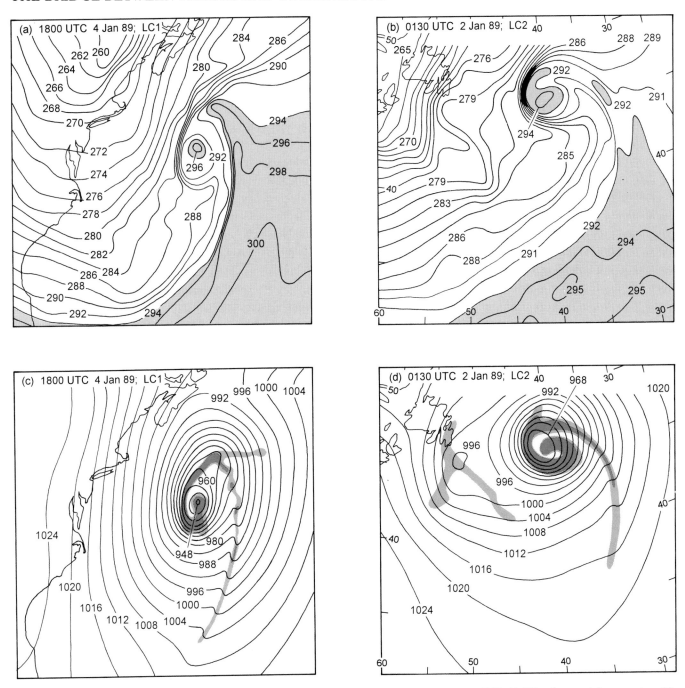

FIG. 11. Left panels (a, c): The 18-h simulation of the ERICA IOP-4 cyclone at 1800 UTC 4 January 1989. (a) 950-mb potential temperature (K, solid lines); potential temperature >294 K (shaded). (c) Mean sea-level pressure (mb, solid lines); 950-mb absolute vorticity >1 × 10⁻⁴ s⁻¹ (light shading) and >2 × 10⁻⁴ s⁻¹ (dark shading). Right panels (b, d): The 37.5-h simulation of the Pre-IOP- 4 cyclone at 0130 UTC 2 January 1989. (b) Same as (a), but with potential temperatures >292 K shaded. (d) Same as (c).

of the vertical structure of the environmental flow of non-sheared (LC1) and cyclonically sheared (LC2) cyclones, respectively. It is recognized that these cross sections are taken through the regions of maximum amplitude of positive upper-level PV anomalies of the polar and subtropical currents, and hence do not in the strictest sense constitute initial space- or time-mean basic states for the entire cyclone environment, as is the case for the idealized life cycles

discussed in Section 3.1. In an effort to identify a barotropic shear component for the Pre-IOP-4 cyclone, we have taken the difference in cross-section-normal wind speed between the two cyclones (Pre-IOP-4; (Fig. 10b) minus IOP-4 (Fig. 10a)), where the ~300-mb polar jet stream core is taken as the registration reference for the subtraction. Figure 10c shows that the difference between the Pre-IOP-4 and IOP-4 cross sections contains two vertical zones of horizontal shear,

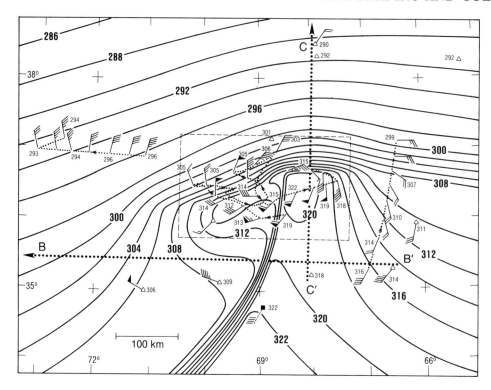

Fig. 12. The ERICA IOP-4 cyclone at 0600 UTC 4 January 1989: 350-m (AGL) equivalent potential temperature (K); wind vectors (flag = 25 m s⁻¹, full barb = 5 m s⁻¹, half barb = 2.5 m s⁻¹) (Neiman et al. 1993).

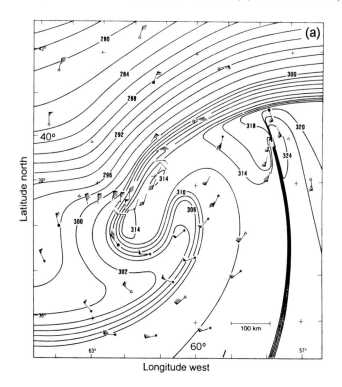

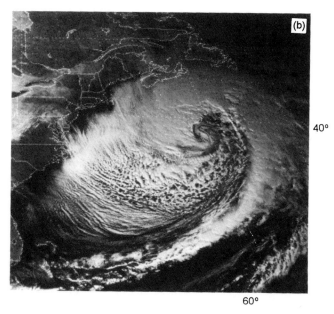

Fig. 13. The ERICA IOP-4 cyclones at 1800 UTC 4 January 1989: (a) 350-m AGL equivalent potential temperature (K); wind vectors same as in Fig. 12. (b) NOAA/GOES satellite 4-km-resolution visible imagery (Neiman and Shapiro 1993).

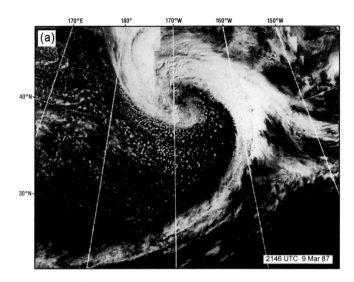

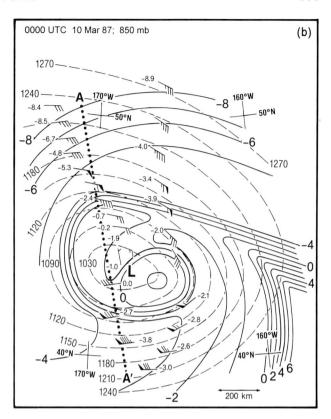

FIG. 14. The mature phase of a Gulf of Alaska cyclone. (a) 1-km-resolution visible polar satellite cloud image at 2146 UTC 9 March 1987 showing the cyclonic spiral cloud signature of the mature cyclone and warm-core seclusion with deep mesoconvective cloud development in advance of the cold front. (b) 850-mb temperature (°C, solid lines) and geopotential height (m, dashed lines) at ~0000 UTC 10 March 1987; wind vectors are as in Fig. 12; line *AA′* is the cross-section projection for (c). (c) Cross-section analysis at ~0000 UTC 10 March 1987 of potential temperature (K, solid lines); section-normal wind component (m s^{-1}, heavy dashed lines) along projection line *AA′* of (b); frontal and stable layer boundaries (thin dashed lines); wind vectors as in Fig. 12.

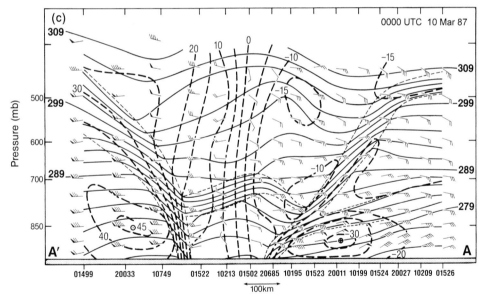

extending from the stratosphere to the lower troposphere, that are nearly uniform with height, and hence are approximately barotropic. The shaded area, ~400 km in width, denotes the region of near barotropic cyclonic shear, greater than 10^{-4} s^{-1}, in the region of the developing Pre-IOP-4 surface cyclone. The vertical structures of the observed IOP-4, and Pre-IOP-4 cyclones (Figs.10a and b, respectively) are used to define single and dual-jet stream basic states for the simulations of idealized cyclone-family life cycles presented in Section 6.2.

Piecewise PV inversion diagnostics could be used to quantify the influence of the relative positions of the primary jet streams and their PV distributions on the idealized and observed cyclone life cycles (e.g., Davis and Emanuel 1991; Davis et al. 1996). For example, the PV associated with the subtropical jet stream and its associated tropopause anomalies (situated above ~300 mb) can be inverted to assess its attribution to mid- and lower-tropospheric environmental shear on the life cycle of LC2 cyclones developing in higher latitudes within polar jet stream currents.

5.3   *Life Cycle 3: The FASTEX Frontal-Wave Cyclone of 15–17 February 1997*

A proposed analog to the anticyclonic barotropic-shear (LC3) idealized frontal-wave cyclone (Section 3.1) occurred over the North Atlantic ocean during 15–17 February 1997. This cyclone was the focus of intensive observations during the Fronts and Atlantic Storm-Track Experiment (FASTEX; see Joly et al. 1997) and was designated as Low 39a of IOP-16. Low 39a formed as the second of two secondary frontal wave cyclones following the 13–15 February primary baroclinic development (Low 38) of IOP-15. The mosaic of the NOAA/GOES-8 and METEOSAT infrared cloud imagery at 1200 UTC 16 February 1997 (Fig. 15) shows, (1) the occluded primary cyclone (Low 38) centered south of Iceland; (2) the first secondary-wave cyclone development (Low 39) at the southern tip of Greenland; and (3) the frontal wave (Low 39a) over the western North Atlantic east of Newfoundland.

This situation was simulated with the Penn State/NCAR MM-5 model for the 48-h period 1200 UTC 15 February to 1200 UTC 17 February 1997, configured with full physics, 23 vertical levels, and a horizontal resolution of 45 km. The present discussion shows only the 24-h simulation, verifying at 1200 UTC 16 February 1997 (Fig. 16). A detailed discussion of the full simulation appears in Section 7.1 (Fig. 24). The potential vorticity and wind velocity analyses on the 316-K isentropic surface situated near the level of the polar jet stream core are shown in Figs. 16a and b, respectively. The 950-mb vorticity and potential temperature analyses are shown in Figs. 16c and d, respectively. The cyclonic vorticity filament and associated frontal baroclinicity of the Low 39a frontal wave (Figs. 16c and d, respectively) are situated beneath the region of anticyclonic shear south of the jet

stream axis (Fig. 16b) and its sharp horizontal discontinuity of PV (Fig. 16a). A cross-section analysis (not shown) that transects the crest of the frontal wave, the upper-level jet stream, and its associated tropopause fold, contains the region of anticyclonic shear extending from the lower stratosphere down to 800 mb, above the surface position of the frontal wave. The 950-mb potential temperature analysis and wind vectors (Fig.16d) show the crest of the frontal wave southeast of Newfoundland with >40 m s^{-1} southwesterly winds within the wave warm sector. The MSL pressure and absolute vorticity fields (Fig. 16c) show the cyclonic vorticity filament of the frontal wave extending continuously from the warm front, east of the pressure minimum, to the crest of the thermal wave, and from there southwestward within the pressure trough of the trailing cold front. A comparison of the low-level vorticity and potential temperature fields for Low 39a (Figs. 16c and d, respectively) with those from the idealized anticyclonic barotropic-shear cyclone simulation (Fig. 2, lower panels), reveals the similarity between the observed and idealized LC3 cyclone simulations, respectively.

6.   The Downstream and Upstream Development of Idealized Cyclone Families in Single and Dual Jet Stream Environmental Basic Flows

This section opens with an overview of the earlier studies of downstream and upstream baroclinic development. This discussion is followed by numerical simulations of idealized cyclone families containing both downstream and upstream baroclinic developments. In these simulations, the basic state flows contain single and dual jet streams and cyclone developments that resemble LC1 and LC2 life cycles, respectively. Finally, the upstream baroclinic development of secondary frontal-wave cyclones in these simulations is diagnosed from the synoptic and PV-attribution perspectives.

6.1   *An Overview of Downstream and Upstream Baroclinic Development*

Numerical studies of idealized extratropical cyclones evolving from initially localized disturbances in the vicinity of the tropopause, superimposed on zonal westerly jet stream basic states have simulated sequential downstream and upstream baroclinic developments (see, e.g., Simmons and Hoskins 1979; Orlanski and Chang 1993). In these simulations, the downstream developments first appear in the vicinity of the tropopause followed by a downward development to the surface, whereas the upstream developments initially appear near the earth's surface as frontal waves, which develop upward to become PV waves in the vicinity of the tropopause. Downstream (upstream) developments propagate eastward (westward) in a frame of reference relative to an initial

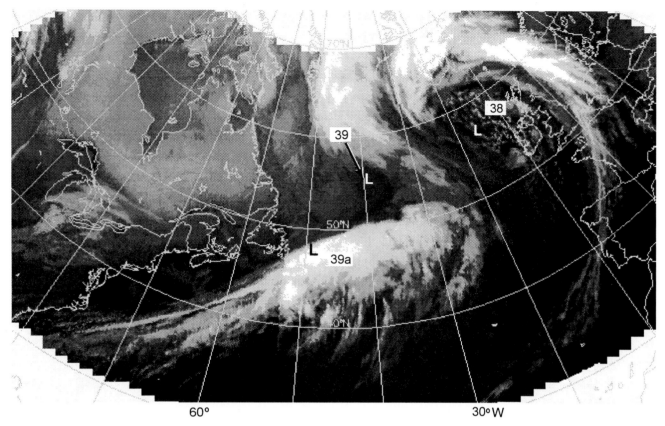

FIG. 15. Mosaic of the GOES-8 and METEOSAT infrared cloud imagery at 1200 UTC 16 February 1997. Positions of FASTEX cyclones are denoted by the letter L with their respective number designation.

disturbance in the westerlies. Under conditions of zero or weak surface winds, upstream developments cannot appear to the west of the initial position of the initial disturbance. However, as the initial disturbance propagates to the east, new developments appear upstream to its west in the same geographic region where the initial disturbance began its development. The dynamics of downstream and upstream development are equivalent, the specification being a manifestation of the reference frame. It has been suggested that differences in idealized cyclone evolution (i.e., downstream "top-down" versus upstream "bottom-up" development) are idealized analogs of Pettersen and Smebye's (1971) type-B versus type-A cyclone life cycle, respectively. Figure 17 llustrates the characteristic vertical structure of downstream and upstream development in sequential longitude/height diagrams of relative vorticity, where the downstream developments originate at the tropopause and the upstream developments first appear at the surface. Note that the upstream developments form sequentially at the same longitude as the initial disturbance.

Simmons and Hoskins (1979) remarked that

. . . the possibility of upstream development does not appear to have been given the same attention [as downstream development] by synopticians. The observed

growth of secondary depressions on trailing cold fronts and the formation of cyclone families may bear some relation to the ideas considered in this paper [Simmons and Hoskins 1979] but a clear identification is lacking. The tendency for upstream development to occur at one fixed location may enhance the role of local regions of increased baroclinicity in giving rise to preferred regions of cyclogenesis.

Simmons and Hoskins suggested that upstream baroclinic development should be considered a theoretical interpretation for the Norwegian conceptual paradigm of cyclone families.

Contemporary theoretical interpretations of downstream/upstream development are derived from the theory of atmospheric energy dispersion proposed by Rossby (1945) and Yeh (1949). This theory describes the response of a barotropic fluid to a localized source of vorticity. Figure 18 presents an example of kinetic energy dispersion from Orlanski and Chang (1993) that illustrates the eastward (downstream) and westward (upstream) expansion of a wave packet in a coordinate system relative to the initial perturbation, in which four waves emanate in both directions from the initial disturbance. Orlanski and Chang (1993), among others, noted that downstream (upstream) dispersion of kinetic energy is primarily

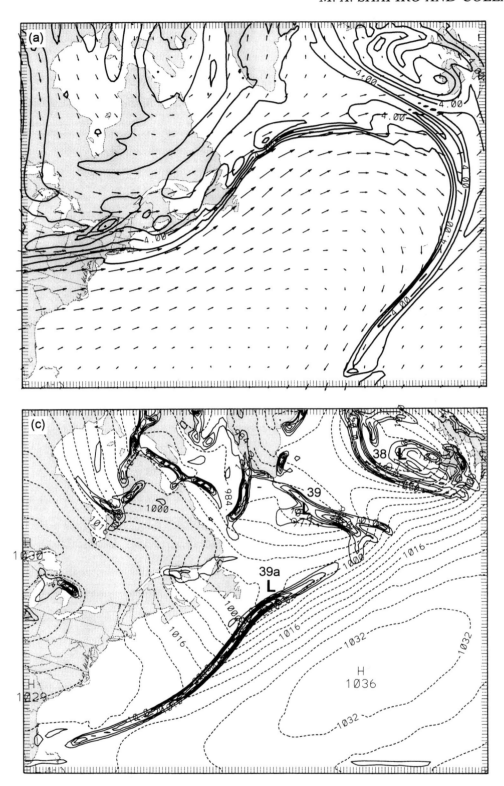

FIG. 16 (above and facing page). The 24-h simulation of a Life Cycle 3 (LC3) anticyclonic-shear frontal-wave cyclone, verifying at 1200 UTC 16 February 1997. (a) PV at $\theta = 316$ K (PVU, solid lines); wind vectors. (b) Wind speed at $\theta = 316$ K (m s^{-1}, solid lines), and wind vectors, as in (a). (c) MSL pressure (mb, dashed lines); 950-mb absolute vorticity (isopleth interval equals 0.5×10^{-4} s^{-1}, solid lines). Positions of FASTEX cyclones are denoted by the letter L with their respective number designation. (d) 950-mb potential temperature (K, solid lines); wind vectors.

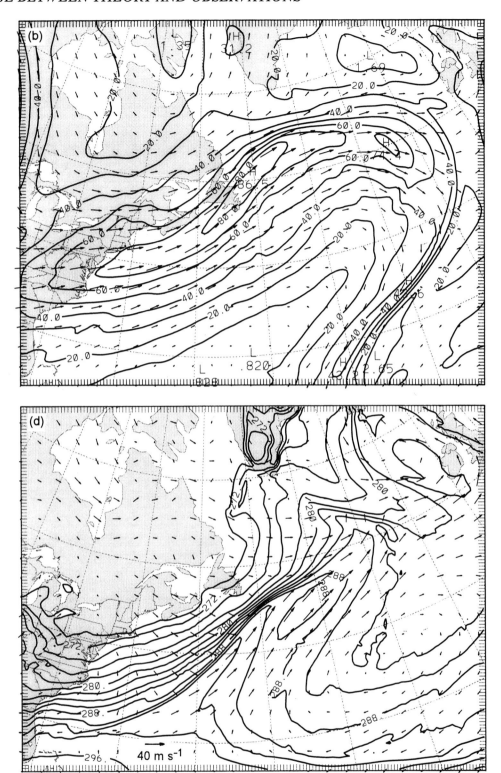

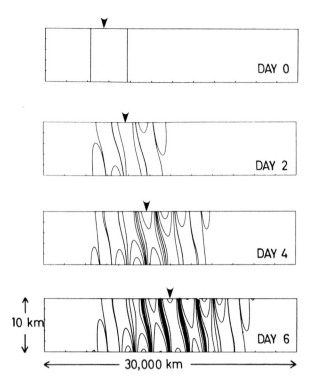

FIG. 17. Height (km) versus longitude (km) cross sections of relative vorticity on days 0, 2, 4, and 6 for the Eady basic state. Contours are drawn for values ±0.01, ±1, ±2.5, ±5, ±10, and 25 × 10⁻⁵ s⁻¹. The zero contour is not drawn to avoid the illustration of small-scale variability of negligible amplitude. Arrows mark the successive positions of the initial baroclinic development (Simmons and Hoskins 1979).

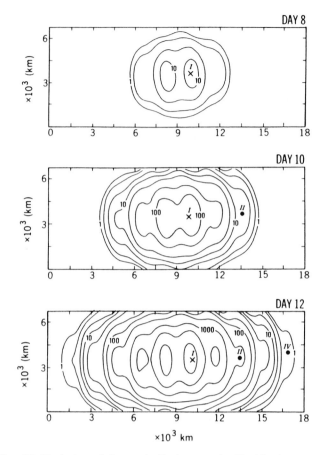

FIG. 18. Evolution of the vertically integrated eddy kinetic energy (m² s⁻¹) for the linear Eady flow simulation for days 8, 10, and 12. The crosses mark the positions of the initial wave (I), and the dots mark the positions of the second (II) and fourth (IV) downstream waves. Longitudinal positions of downstream and upstream wave fronts are relative to the initial wave (Orlanski and Chang 1993).

through the convergence (divergence) of the ageostrophic geopotential flux. From quasigeostrophic (QG) theory, the compensation between vorticity advection and divergence (stretching) maintains extratropical cyclones in the presence of the vertical shear, which would otherwise fracture the upper wave from the surface cyclone. Within this theoretical framework, Orlanski and Katzfey (1991) argued that, "the ageostrophic fluxes will be predominantly downstream in the upper troposphere where the wave lags the mean flow. . . . The ageostrophic fluxes in the lower atmosphere will be predominantly upstream, since the wave moves faster (in the relative sense) than the mean flow at that level." Simmons and Hoskins (1979), as well as others, have shown that the downstream fringe of the expanding wave packet travels with a group velocity near that of the upper-level jet stream, whereas the upstream fringe is nearly stationary when the surface mean flow is near zero. In these experiments, individual baroclinic disturbances (i.e., cyclones and anticyclones) propagate with a phase velocity of the mean flow at the steering level (~600 mb), as does the centroid of the expanding packet. When surface friction is included, the upstream developments are significantly damped at lower levels, with a lesser impact on the downstream developments (Orlanski and Chang 1993). It is perhaps because of such

frictional effects that the observational confirmation of upstream development has been so elusive, particularly over land.

From a historical perspective, Evjen (1936) was the first to document downstream baroclinic development. The phenomenon was later independently discovered by Namias and Clapp (1944), Rossby (1945), and Cressman (1948). The observational study by Miles (1959) addressed the upstream development of thermal troughs in the vicinity of Western Europe. Hovmöller (1949) linked the work of Evjen with later findings through the graphical representation that bears his name (the Hovmöller diagram). An overview of contemporary theoretical studies of this problem is provided by Chang (1993).

6.2 Idealized Cyclone Family Life Cycles within Single and Dual Jet Stream Flows

If the examples of observed cyclones evolving within single and dual jet stream regimes described in Section 5 are indeed the synoptic analogs of idealized cyclones evolving within differing environmental barotropic shears, it is then logical that we consider the simulation of idealized cyclones in which the initially perturbed basic state also contains dual jet streams in the vicinity of the tropopause. For these experiments, the basic zonal flow was derived from the meridional conceptualization of the primary jet streams, fronts and the tropopause (Fig. 5) and from observed cyclone events, such as discussed in Section 5. The simulations were performed with an adiabatic, frictionless, f-plane variant of the operational primitive-equation model of the German weather service (Majeweski 1991). The model domain was configured as a channel, 17 000 km long and 7000 km wide, and has 26 vertical levels, and a horizontal resolution of 75 km. Unlike previous numerical experiments of baroclinic life cycles, the present channel is noncyclic in its lateral (west-east) boundaries. The western inflow boundary remains unchanged from its initial basic-state parameters and the eastern outflow boundary is sponge absorbing. This computational configuration facilitates the simulation of upstream baroclinic developments uncontaminated by downstream developments entering into the upstream domain.

Two numerical simulations were performed, based on differing zonal basic state flows: (1) the single jet stream situation, analogous to meridionally merged and vertically aligned polar and subtropical jet streams, as described for the IOP-4 nonsheared LC1 cyclone life cycle (Section 5.1); and (2) the dual jet stream situation, analogous to the Pre-IOP-4 cyclonic-shear LC2 cyclone life cycle, with a meridional and vertical separation of the polar and subtropical jet streams and cyclogenesis within the baroclinicity of the polar jet stream/frontal zone system (Section 5.2). In both simulations, the initial baroclinic development was initiated by a finite-amplitude, three-dimensional perturbation superimposed upon the zonal basic state, with maximum perturbation amplitude ($\sim$9 m s^{-1}) in the vicinity of the polar tropopause ($\sim$300 mb) that weakens as it extends to the earth's surface. The longitudinal extent of the channel was sufficient for the development of cyclone "families."

In the following discussion, we refer to the cyclone that forms out of the initial perturbation as the *initial cyclone,* and the others as *downstream* or *upstream baroclinic developments* in reference to their longitudinal position relative to the initial cyclone, as in Simmons and Hoskins (1979) and Orlanski and Chang (1993). Upstream developments tend to be smaller in scale and of lesser intensity than downstream developments and are therefore referred to as *secondary cyclones* (discussed in Section 7).

6.2.1   The basic zonal flow

The vertical structure of the basic zonal flows for the idealized cyclone family experiments was analytically "sculpted" to mirror the vertical cross sections of wind and potential temperature of the observed LC1 and LC2 North Atlantic

cyclones shown in Figs. 10a and b, respectively. A triplet of circular PV anomalies was chosen as the initial perturbation condition, as in Schär and Wernli (1993) and added to the two-dimensional zonal flow near the 350-mb level: a central positive anomaly with an amplitude of $\sim$2 PVU and two flanking anomalies of opposite sign and half the amplitude. The flanking anomalies are shifted 2000 km upstream and downstream from the central anomaly, and they each have a horizontal scale of $\sim$1000 km. Balanced initial conditions suitable for the numerical integrations were derived from a QG PV inversion developed by Fehlmann (1997).

The idealized LC1 zonal basic state (Fig. 19a) contains a single jet stream extending well into the stratosphere with a pronounced vertical deformation (fold) of the PV tropopause across the jet stream core, a characteristic of deeply penetrating tropopause folds that accompany the vertical alignment of subtropical and polar jet streams, as in Fig.10a. The tropospheric baroclinicity of the LC1 basic state extends from below the upper jet down to the earth's surface. The longitudinal structure of the perturbed LC1 initial state across the center of the perturbation is shown in Fig. 19b.

The idealized dual-jet LC2 basic state (Fig. 19c) includes a southern subtropical jet stream at $\sim$250 mb and a northern polar jet at $\sim$350 mb. The LC2, PV tropopause is doubly folded across the two jet streams, as in Fig.10b, with each fold having less vertical deformation than the single jet stream PV fold in LC1 (Fig. 19a). The 305-K and 340-K isentropes intercept the PV anomalies of the polar and subtropical jet stream tropopause folds, respectively (Fig. 19c). The baroclinicity beneath the polar jet extends to the surface baroclinic zone within which the surface cyclones evolve. The baroclinicity beneath the subtropical jet stream does not extend to the surface. The longitudinal structure of the perturbed LC2 initial state across the center of the perturbation is shown in Fig. 19d.

6.2.2   Life Cycle 1: The single jet stream cyclone family

The top three rows of Fig. 20 present the 4.5-day (108-h) simulation of the LC1 cyclone family that developed from the initially perturbed single (polar) jet stream basic state (Figs. 19a and b). Figure 20 includes: (1) potential vorticity at $\theta = 305$ K and the superimposed 3-PVU isopleth at $\theta = 340$ K (row 1); (2) surface potential temperature and pressure (row 2); and (3) surface absolute vorticity and wind velocity vectors (row 3). The results are presented at 24-h intervals, for 4.5 days of simulation, where day 1.5 refers to the 36-h simulation, day 2.5 to 60 h, etc.

By day 1.5, an incipient cyclone has developed out of the initial perturbation to the basic state. A PV wave appears at the upper isentropic levels above the lower-tropospheric thermal wave and associated surface pressure depression of the incipient cyclone. Note the 90° westward phase tilt with

Life Cycle 1

Life Cycle 2

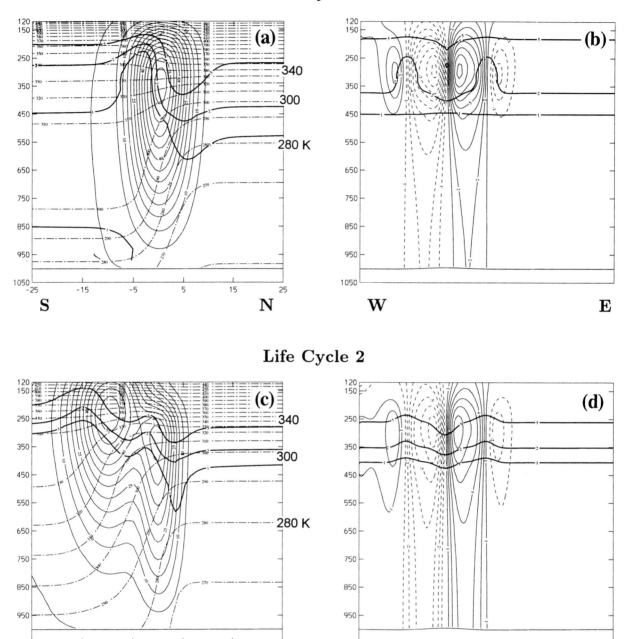

FIG. 19. The basic state and its finite-amplitude perturbation for the idealized single jet cyclone life cycle (LC1, a, b) and dual jet cyclone life cycle (LC2, c, d). Left panels (a, c): South-to-north oriented vertical cross sections at the western boundary of the channel for LC1 (upper) and LC2 (lower), with the zonal wind speed (isotach interval 5 m s^{-1}, solid lines) and potential temperature (isentropes interval equals 10 K, broken lines). Right panels (b, d): West-to-east vertical cross sections across the center of the initial perturbations for LC1 (upper) and LC2 (lower). Meridional wind velocity component, isopleth contour interval equals 2 m s^{-1} (>0.0, solid lines; <0.0, dashed lines). PV contours for 1, 2, and 5 PVU (heavy solid lines).

increasing height between the thermal wave below and the PV wave above, characteristic of amplifying baroclinic waves. The first downstream baroclinic development appears as a PV wave in the vicinity of the tropopause, with a lesser degree of development in temperature and vorticity structure below.

By day 2.5, the initial cyclone has developed a substantial surface circulation with further amplification of the PV wave aloft. There is a tendency for frontal fracture in the surface temperature field near the frontal T-bone triple point, and the associated surface vorticity has developed a T-bone configuration in the vicinity of the frontal triple point. Downstream and upstream baroclinic developments are in progress, with the downstream development being of greater amplitude near the tropopause and the upstream development having greater definition near the surface.

By day 3.5, the initial cyclone has entered into its occlusion phase, with a narrowing of the warm sector and the development of a bent-back warm frontal seclusion of warm air coincident with the circular vorticity maximum at the cyclone center. The downstream developing cyclone, east of the initial cyclone, has characteristics of the early stage of frontal T-bone structure, including frontal fracture of the cold-frontal baroclinicity at the apex of the warm sector, and coincident T-bone structure in vorticity. The amplification of the downstream upper-level PV wave continues, concurrent with the cyclone intensification at the surface. The developing upstream surface thermal wave continues to amplify, with an incipient secondary cyclone at its apex, and there appears the first evidence of PV wave amplification above the developing upstream secondary cyclone.

By day 4.5, the initial cyclone has reached the final phase of its occlusion. Its vertical structure is nearly equivalent barotropic, with the upper-level cyclonic (positive) PV anomaly centered directly above the near-surface vorticity maximum and warm core at the cyclone center. The downstream development has matured and possesses frontal T-bone structure in potential temperature and vorticity at the surface. Note, that at this time, the upper-level PV anomaly of the downstream development is in the process of cutting off from the narrowing PV cusp to its north. The upstream secondary cyclone has entered its T-bone frontal phase and is somewhat smaller in scale than the initial cyclone and the downstream development. The amplitude of the PV wave above the upstream cyclone is weak in comparison to the thermal, vorticity, and surface pressure features below. Note the formation of a new secondary frontal wave to the west (upstream) of the previous secondary development that, like its predecessor, has also developed without a precursor PV anomaly in the vicinity of the tropopause.

6.2.3 Life Cycle 2: The dual jet stream cyclone family

The lower three rows of Fig. 20 present the 4.5-day (108-h) simulation of the Life Cycle 2 cyclone family that evolved from the initially perturbed basic state (Figs. 19c and d) containing meridionally and vertically separated polar and subtropical jets streams. By day 1.5, the structure of the incipient LC2 cyclone appears quite similar to that of LC1, having a westward baroclinic tilt with height and a simultaneous amplification of the polar jet stream PV wave aloft and the thermal wave, vorticity, and surface pressure perturbations below. Note that the 3-PVU isopleth marking the axis of the subtropical jet stream shows the first indication of wave

amplification ~700 km south of the PV wave in the northern polar current.

By day 2.5, the initial cyclone has entered into the early phase of an occluding "Norwegian" cyclone, with the formation of a frontal thermal triple point in the vicinity of the cyclone center. Note that the LC2 cold front spirals inward to the cyclone center without loss of baroclinicity and "pinches" the warm sector to occlusion, whereas the LC1 cyclone (day 2.5) contains frontal fracture in potential temperature at the northern segment of the cold front. The low-level vorticity of the LC2 initial cyclone is characterized by a comma-shaped warm-frontal vorticity filament that spirals inward into the cyclone center. Note that there is little vorticity within the LC2 cold front, in contrast that associated with the T-bone cold front vorticity structure on the same day. The upper-level PV analysis (day 2.5) shows continued amplification of the polar jet stream PV wave above the developing initial cyclone and there is further amplification of the subtropical jet stream wave to the south. The development of upper-PV waves on the polar and subtropical currents is evident downstream from the initial baroclinic development. The upstream development west of the initial cyclone remains a frontal wave and continues to amplify with increasing baroclinicity and vorticity along its warm front, with evidence for upstream development in the vicinity of the tropopause. The first indication of a second upstream frontal wave also appears near the western boundary of the domain in the same location as the initial disturbance at the beginning of the simulation.

By day 3.5, the initial cyclone is well into occlusion, with a cyclonic roll up of warm-frontal vorticity at its circulation center and increased baroclinicity and vorticity within the cold front over that on day 2.5. The initial polar PV wave aloft has undergone a shortening of wavelength and a cyclonic roll up of positive (cyclonic) PV within its wave trough. The difference in wavelength between the downstream PV waves in the polar and subtropical currents has progressed to the point that the two wind regimes are about to phase into a single jet streak above the developing downstream cyclone. This longitudinal phasing and its resulting meridional alignment of a polar cyclonic-PV anomaly with a subtropical anticyclonic-PV anomaly has led to the formation of a localized single jet stream (jet streak) environment for the downstream development of an LC1 cyclone. This modification of the downstream basic state by the initial baroclinic development was unanticipated in the design of the LC2 experiment. Consistent with the single tropopause jet stream of the LC1 cyclone, the downstream cyclone has taken the form of the early stage of a T-bone frontal-fracture cyclone, quite similar in thermal structure to that of the downstream development in the preceding LC1 simulation at day 3.5 (Fig. 20). The upstream secondary surface cyclone has evolved into the early phase of occlusion, with concurrent amplification of a polar tropopause PV wave aloft. Note that the

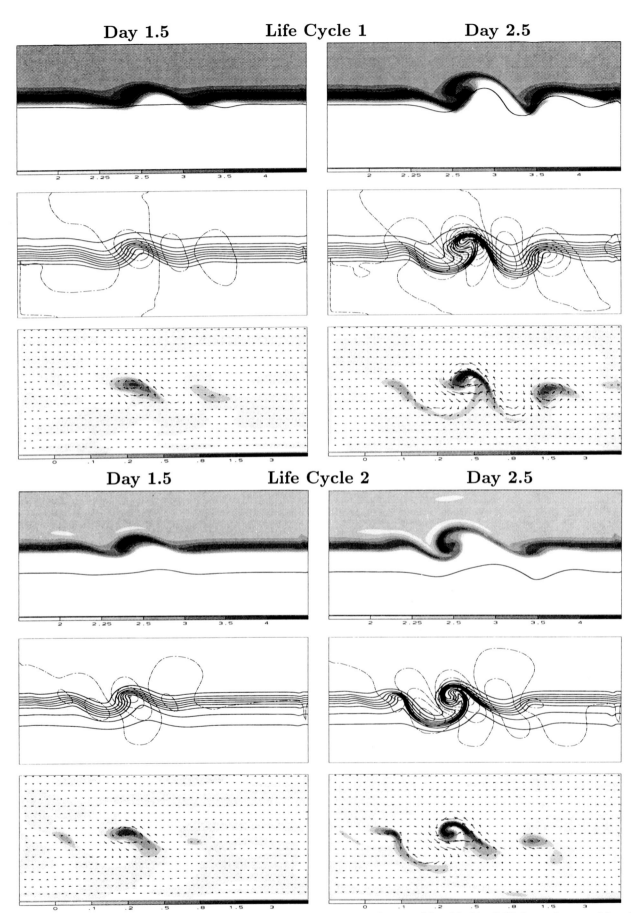

FIG. 20 (above and facing page). The 4.5-day simulation of the single jet (LC1, upper three panels) and dual jet (LC2, lower-three panels) cyclone family for days 1.5, 2.5, 3.5, and 4.5. Upper three panels (rows 1–3): (1) θ = 305 K PV (gray shading between (0–4 PVU) and the θ = 340 K, 3-PVU isopleth (solid line); (2) surface potential temperature (K, solid lines at 2-K intervals); (3) surface absolute vorticity (gray shading between 0 and 3×10^{-4} s^{-1}) and velocity. Lower three panels (rows 4–6): Same as the upper three panels, respectively.

Day 3.5   Life Cycle 1   Day 4.5

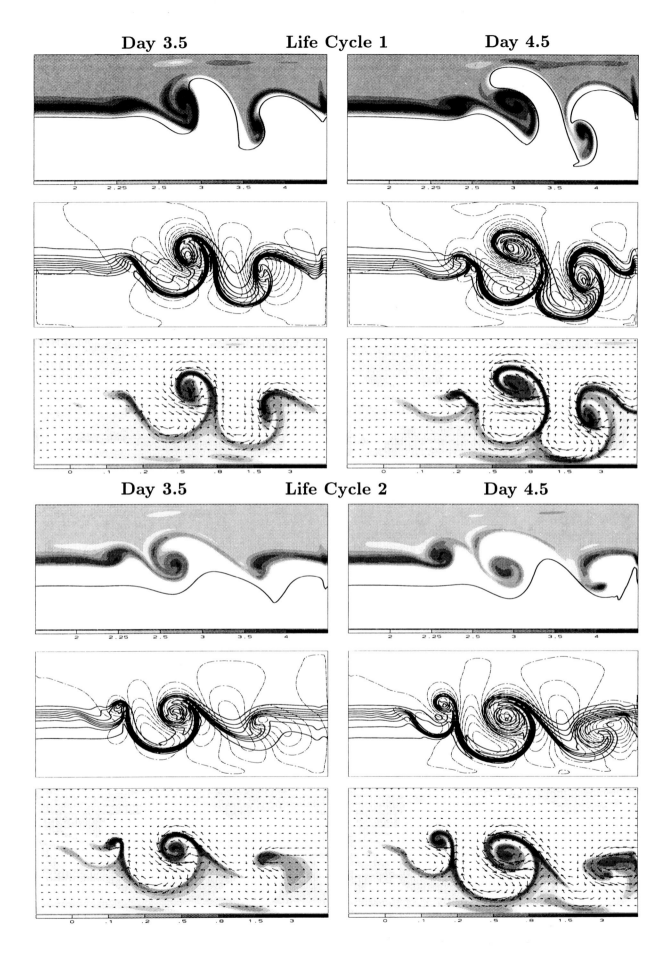

Day 3.5   Life Cycle 2   Day 4.5

upstream secondary cyclone is situated within the region of
cyclonic shear to the north of the subtropical jet axis and its
associated PV tropopause fold. The frontal wave upstream
from the secondary cyclone continues to amplify without
upper-level PV wave development in the vicinity of the polar
tropopause.

By day 4.5, the initial cyclone has fully occluded, with a
warm inner core that has expanded in radius from that on day
3.5. The vorticity within the warm and cold fronts of the
initial cyclone are merged to a single filament at the occlusion
triple point, which then spirals into the center of the cyclone.
Aloft, the initial polar PV wave has taken the form of a cusp,
with an extruded PV filament that connects to a circular
cyclonic PV anomaly almost detached from the stratospheric
polar PV reservoir to its north. The downstream surface
cyclone continues to evolve with the structural characteristics of an LC1 bent-back, warm-front, T-bone cyclone, including frontal fracture between cold and warm fronts in the
vicinity of the cyclone center. Aloft, the axes of the polar and
subtropical jet streams have become vertically aligned above
the downstream development, as evidenced by the merging
of their respective PV contours. By day 4.5, the absorbing
eastern outflow boundary has begun to effect the evolution of
the downstream cyclonic development. The secondary cyclone, west of the initial development, is fully occluded,
possessing a well-defined warm core and cyclonic vorticity
filament extending northward from the frontal triple point
that spirals into the center of the cyclone. The second frontal
wave, upstream from the secondary cyclone, continues to
amplify, and shows the first sign of a closed cyclonic circulation, without a companion PV wave on the polar tropopause.

The vertical structure of downstream/upstream development in the LC2 simulation is shown in the longitude versus
pressure diagrams of perturbation pressure (Fig. 21) associated with the simulation of the LC2 life cycles (Fig. 20, lower
three rows). The perturbation pressures (Fig. 21) are an
average across the middle ~3000 km of the channel to
compensate for the fact that the developments are not all at the
same latitude. The LC2 longitude/pressure diagrams exhibit
the same characteristics as those in Simmons and Hoskins
(1979) and Orlanski and Chang (1993), with downstream
developments originating at the tropopause and upstream
developments emanating upward from the surface.

6.3   The Synoptic and PV Perspectives of Upstream Baroclinic Development

The LC2 idealized cyclone-family simulation (Fig. 20, bottom three rows) can be used to illustrate the synoptic and PV-
viewpoints of upstream-developing frontal waves and secondary cyclones. On day 1.5 of the LC2 simulation, we draw
attention to the anticyclone centered west of the incipient
initial primary cyclone (day 1.5, row 5). Reference to the

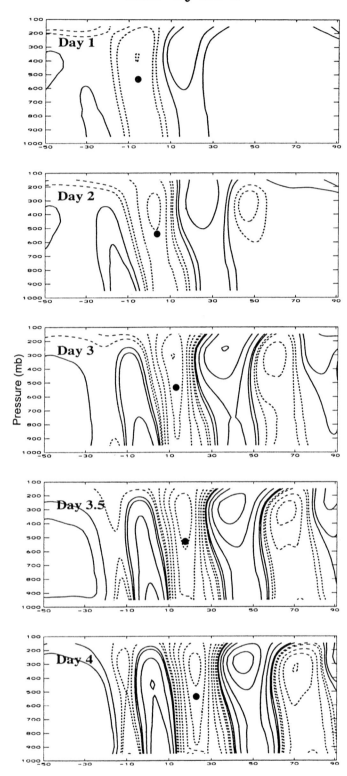

Life Cycle 2

FIG. 21. Pressure (mb) versus longitude (degrees) cross sections showing
the perturbation pressure on days 1, 2, 3, and 4 for the LC2 dual jet
stream cyclone family simulation shown in Fig. 20. Isobars of
perturbation <0.0, dashed lines, and >0.0, solid lines; isobar interval 10
mb. Pressure perturbations are deviations from the zonal mean averaged
across the inner 3000-km latitudinal belt, centered at the midlatitude of
the channel. Dots indicate the position of the initial perturbation.

longitude versus pressure diagrams of perturbation pressure (Fig. 21, days 1–3) shows this anticyclone as a surface-based, positive pressure perturbation, amplifying at –30° longitude, upstream (west) from the deepening low-pressure perturbation of the initial primary cyclone. The low-amplitude thermal wave west of the developing primary anticyclone (Fig. 20, days 1.5 and 2.5, row 5) is the incipient of a secondary cyclone developing on the end of the trailing cold front of the primary anticyclone. From a synoptic perspective, the upstream frontal wave formed in the region of localized warm advection, convergence, and frontogenesis associated with the flow around the anticyclone of the primary baroclinic development to its east. By day 4.5 (Fig. 20, row 5), the sequence is repeated as the thermal ridge of a second upstream frontal wave appears within the southerly flow and warm-air advection at the western flank of the anticyclone of the preceding secondary upstream baroclinic development. As noted in Subsections 6.2.2 and 6.2.3, the upstream developments in LC1 and LC2, respectively, were not initiated by a precursor finite-amplitude PV disturbance in the vicinity of the tropopause.

From a PV perspective, the development of thermal waves at the surface and/or the top of the planetary boundary layer is synonymous with the formation of boundary thermal (surrogate) PV anomalies (e.g., Mattocks and Bleck 1986; Hoskins et al. 1985). For example, the warm sectors (thermal ridges) of frontal waves and cyclones and adjacent cold-air outbreaks (thermal troughs) within the idealized primary and secondary baroclinic developments of LC2 (Fig. 20, day 2.5, row 5) represent boundary PV anomalies. When numerically inverted (e.g., Davis et al. 1996), these anomalies attribute cyclonic and anticyclonic circulations about their respective warm and cold anomalies. For the situation depicted in LC2 (Fig. 20, days 1.5 and 2.5, row 5), the negative (cold) anomaly west of the primary cyclone induces an anticyclonic circulation with southerly flow and localized warm advection at its western flank, which in turn initiates the amplification of the thermal ridge (warm sector) of the upstream frontal wave that subsequently develops into a mature upstream secondary cyclone on days 3.5 and 4.5. Similarly, the surface boundary PV (thermal) anomalies of the secondary cyclone (days 3.5 and 4.5) initiate the development of a second upstream frontal wave in the western portion of the domain. The above synoptic and PV interpretations also apply to the upstream baroclinic developments in LC1 (Fig. 20, rows 1–3). Thorncroft and Hoskins (1990) presented an equivalent PV perspective of upstream baroclinic development in their discussion of a two-dimensional (longitude/height) simulation of a neutral Eady wave model without an upper lid. Starting with an initially localized surface thermal disturbance, the simulation generated a sequence of lower-tropospheric edge waves upstream from the initial disturbance, without concurrent downstream developments. A conceptual schematic of this interpretation is shown in Fig. 22.

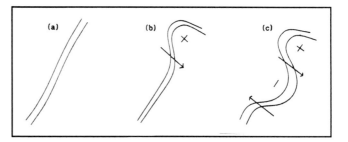

FIG. 22. Schematic of upstream development along a cold front. (a) Two isotherms in the initial cold frontal region. (b) The cold front with a warm anomaly toward the northeast that forms during the cyclonic wrap up of isotherms at low levels. (c) The cyclonic vorticity associated with this warm anomaly, which induces a disturbance upstream, that is, to the west of the initial warm anomaly (Thorncroft and Hoskins 1990).

Finally, downstream and upstream baroclinic development within barotropically sheared, and single and multiple jet stream idealized basic flows, can be diagnosed from the perspective of energy dispersion (e.g., Orlanski and Chang 1993), or the Eliassen and Palm (1960) energy flux perspective (e.g. Thorncroft et al.1993).

7. Secondary Cyclones

Extratropical cyclones are observed over a broad spectrum of spatial scales: synoptic-scale primary cyclones (~4000 km); intermediate-scale, secondary, frontal-wave cyclones (~1000 km); and the lesser mesoscale cyclones (~500–100 km). The fundamental difference between primary and secondary frontal-wave cyclones was described by Eliassen (1966):

> A natural conclusion from the various theories is that the large-scale cyclonic systems associated with the transient waves in the upper-westerlies are the primary perturbation systems, since their existence can be accounted for by baroclinic instability, even in the absence of fronts. These large cyclonic systems, in turn, produce the fronts, partly by setting up the deformation fields required in Bergeron's [1928] theory, and partly by the non-linear process demonstrated by Edelmann [1963]. Finally, the fronts will give birth to the smaller, frontal wave [secondary] cyclones.

The concept of frontal-wave cyclones forming within a confluent (stretching), frontogenetic Bergeron deformation was discussed in Godske et al. (1957, chapter 15, Sec.1). A contemporary perspective of secondary cyclone development along narrow, low-level, positive PV filaments within fronts was advanced by, for example, Joly and Thorpe (1990). These studies utilized the Charney and Stern (1962) theorem as the theoretical basis for the release of barotropic or baroclinic instabilities along frontal PV filaments. The influence of the environmental deformation on baroclinic development was noted, for example, by Farrell (1989) and

Thorncroft and Hoskins (1990). Subsequent investigations (e.g., Bishop and Thorpe 1994a and b) established the sensitivity of secondary frontal-wave development to along-front stretching deformation in the environmental flow, noting that wave growth was related to (1) the intensity of the frontal PV filament; (2) the scale of the disturbance compared to the scale of the filament; and (3) the magnitude of the environmental deformation. Bishop (1996a and b) used the technique of domain-independent vorticity and divergence attribution to separate the environmental stretching deformation rate from that of the smaller-scale frontal kinematics. The theoretical hypotheses of Dritschel et al. (1991), Bishop and Thorpe (1994a and b) and Renfrew (1995) were tested on actual situations over the North Atlantic in the study by Renfrew et al. (1997), which confirmed that, "sustained along-front stretching [deformation] is a sufficient condition for barotropic frontal stability; weak along-front stretching is a necessary condition for barotropic instability." The reader is referred to Renfrew et al. (1997) for an overview and examples of the influence of environmental flow on the development of secondary cyclones. The formation of secondary frontal cyclones was discussed from the perspective of upstream baroclinic development in Section 6.3.

We next present numerical simulations of three types of secondary cyclones: (1) secondary cyclone and frontal waves that form through upstream baroclinic development at the end of the trailing cold fronts of primary synoptic-scale cyclones in the absence of an initial localized disturbance in the vicinity of the tropopause; (2) secondary cyclones that form within the ascending branch of the transverse secondary circulation of a synoptic-scale, confluent, along-front stretching deformation field; and (3) secondary cyclones that evolve out of small (~100 km) initial disturbances.

7.1 North Atlantic Secondary Cyclone Developments during 14–16 February 1997

This section presents numerical simulations of two North Atlantic secondary cyclones that developed during the field phase of FASTEX. This discussion focuses on the role of (1) surface-based anticyclones, boundary surrogate PV anomalies in the initiation of upstream secondary cyclone development (as in Section 6.3), and (2) frontogenetic stretching deformation in the formation of frontal waves. We begin with a discussion of the satellite cloud structures associated with the 14–16 February 1997 North Atlantic primary cyclone development and subsequent secondary baroclinic developments. This satellite overview is followed by the presentation of a numerical simulation of the secondary baroclinic developments.

At 0000 UTC 14 February 1997 (Fig. 23a), the incipient of a primary cyclone development (Low 38 of FASTEX IOP-15) was centered southeast of Newfoundland with a central

pressure of ~998 mb. The composite mosaic of the NOAA GOES-8 and METEOSAT infrared cloud images shows the well developed cloud head of Low 38 east of Newfoundland, where the surface cyclone is centered near the cloud-free notch at the western edge of the cloud head. By 0000 UTC 15 February 1997 (Fig. 23b), Low 38 had undergone rapid intensification, with a minimum MSL pressure of ~975 mb, and by 0000 UTC 16 February 1997 (Fig. 23c) had entered its occlusion phase west of Ireland with a central pressure of ~ 978 mb. At this time, the bright (cold) cloud head east of Newfoundland is the cloud signature of the upstream baroclinic development identified as Low 39. This upstream development formed in the same location where the primary cyclone (Low 38) began its life cycle two days earlier (Fig. 23a). The satellite mosaic at 1200 UTC 16 February 1997 (presented in Section 5.3, Fig. 15) shows the spiral cloud system of the Low 39 upstream development off the southern tip of Greenland. A second secondary development (Low 39a of FASTEX IOP-16) appears as the baroclinic leaf cloud system southeast of Newfoundland.

A numerical simulation of the 14–16 February 1997 secondary developments was performed with the Penn State NCAR/MM-5 simulation model configured with full physics, a horizontal resolution of 40 km with 23 vertical levels, and initialized at 1200 UTC 15 February 1997. The initial conditions were prepared from the National Centers for Environmental Prediction (NCEP) global analysis, interpolated to the MM-5 higher-resolution grid. Selected results from the first 18 hr of this simulation (1200 UTC 15 February to 1800 UTC 16 February 1997, inclusive) are presented in Fig. 24, which includes: (1) potential vorticity and velocity vectors on the 316-K isentropic surface (Fig. 24, top rows); (2) sea-level pressure and 950-mb absolute vorticity (Fig. 24, middle rows); and (3) 950-mb potential temperature and velocity (Figs. 24, bottom rows). The final 6-h of the simulation, verifying at 1200 UTC 16 February 1997, was previously introduced in Fig. 16.

At 1200 UTC 15 February 1997 (Fig. 24, column 1), the mature primary cyclone (Low 38) was centered west of Ireland with an initialized central mean-sea-level pressure of ~973 mb, and its associated 1038-mb high positioned east of Cape Hatteras, North Carolina (Fig. 24b). This low–high pressure couplet represents the surface pressure signature of a fully developed synoptic-scale primary baroclinic development. We draw attention to the ~20 m s^{-1} southwesterly, low-level jet flow and substantial warm advection within the baroclinicity of thermal ridge at the western flank of the primary anticyclone southeast of Newfoundland (Fig. 24c). The initial amplification of the thermal ridge on the trailing cold front of Low 38 occurred during the 12-h period preceding the start of the simulation. The crest of this thermal ridge contains dynamical and microphysical characteristics considered as first-order components in the initiation of sur-

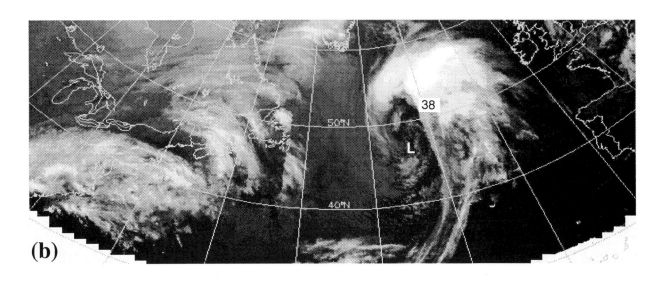

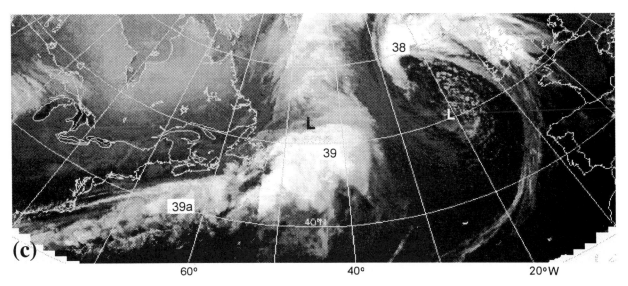

FIG. 23. Mosaic of the GOES-8 and METEOSAT infrared cloud imagery at 0000 UTC on (a) 14 February 1997, (b) 15 February 1997, and (c) 16 February 1997. Positions of FASTEX cyclones are denoted by the letter L with cyclone number as a subscript.

1200 UTC 15 Feb.   1800 UTC 15 Feb.

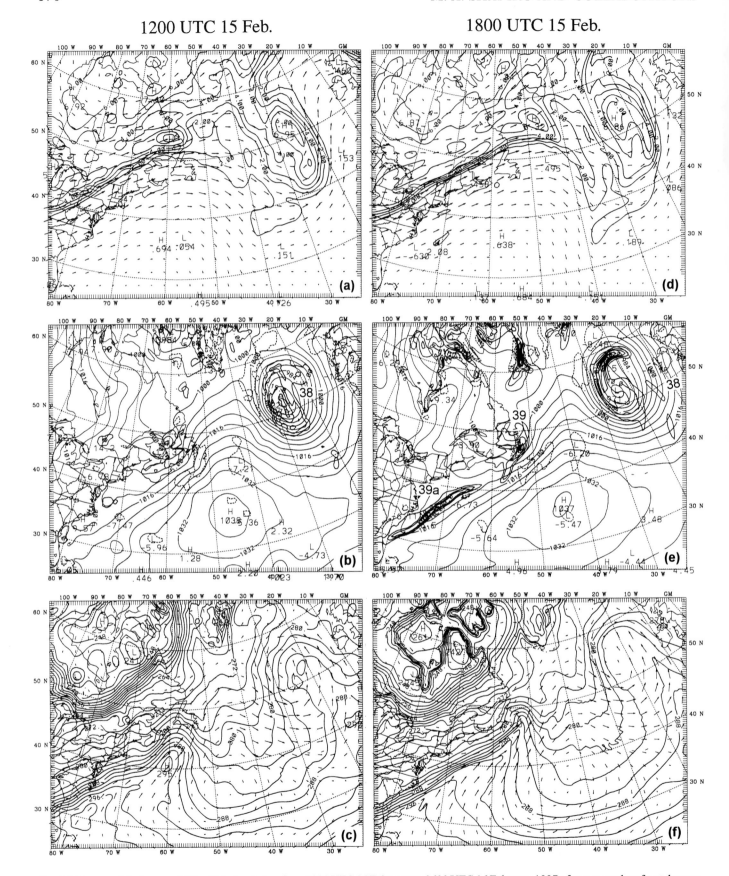

FIG. 24 (above and facing page). The 18-h simulation from 1200 UTC 15 February to 0600 UTC 16 February 1997 of two secondary frontal wave cyclones that occurred during FASTEX. Results are shown at 6-h intervals, with time increasing to the right with each successive column of panels. Upper panels: PV at $\theta = 316$ K (PVU, solid lines) and velocity vectors. Middle panels: MSL pressure (mb, dashed lines) and 950-mb absolute vorticity (isopleth interval of 0.5×10^{-4} s^{-1}, solid lines). Lower panels: 950-mb potential temperature (K, solid lines) and velocity vectors.

0000 UTC 16 Feb.   0600 UTC 16 Feb.

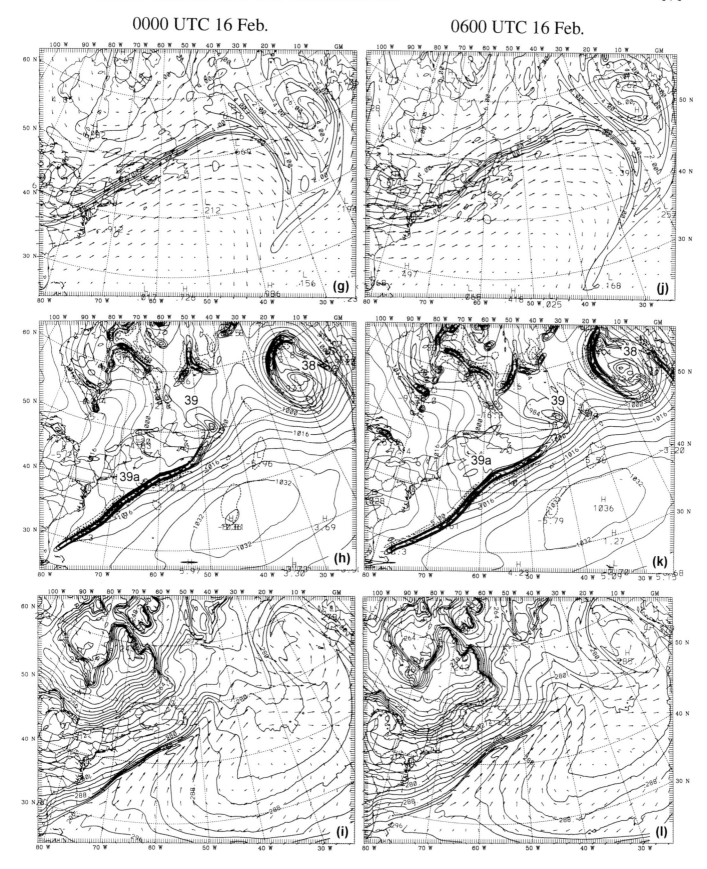

faced-based secondary frontal-wave cyclones: (1) localized warm advection that contributes to thermal wave amplification and positive boundary-PV development; (2) vorticity generation through vortex stretching (convergence) as inferred from the speed convergence in the absence of compensating directional diffluence south of Newfoundland; (3) latent heating through condensation in the region of maximum warm advection; and (4) upward sea-surface fluxes of latent and sensible heat into the wave warm sector, where relatively cooler and drier surface air circulated around the mid-Atlantic anticyclone and over the warmer sea-surface temperatures of the Gulf stream. From the PV viewpoint, the western- and eastern-Atlantic warm ridges and the mid-Atlantic cold trough (Fig. 24c) of the primary mid-Atlantic baroclinic development, constitute positive and negative boundary-PV anomalies, respectively, which from PV attribution, induce cyclonic and anticyclonic circulations about their respective thermal anomalies (see Section 6.1 and 6.3). From this perspective, the southwesterly flow in the crest of the upstream thermal ridge can be attributed to the negative (cold) boundary PV anomaly of the primary baroclinic development to the east. Note the similarity between the lower-tropospheric structure of the formative stage of the observed upstream secondary cyclone to that of the idealized LC1 and LC2 upstream developments (Fig. 20, days 1.5 and 2.5, rows 2 and 5, respectively). The PV analysis and wind vectors on the 316-K isentropic surface (Fig. 24a) show the large-amplitude, synoptic-scale (~4000 km) tropopause PV wave, with Low 38 centered beneath the positive PV anomaly of the upper-level wave trough. The lower-tropospheric thermal ridge of the incipient upstream baroclinic development (Fig. 24c) is positioned beneath the anticyclonic (negative) PV anomaly of the synoptic-scale ridge aloft and >500 km south of the upper-level jet stream axis, its associated PV gradient zone, and tropopause fold (Fig. 24a). At this time in the simulation, there was no signature of an intermediate-scale cyclonic PV anomaly in the vicinity of the western-Atlantic frontal wave; therefore, it is unlikely that a tropopause PV anomaly contributed to the initiation of this upstream baroclinic development.

By 1800 UTC 15 February 1997 (Fig.24, column 2) the upstream development had propagated northeastward over Newfoundland with a central pressure of ~995 mb (Fig. 24e) and had developed distinct warm and cold fronts (Fig. 25f). This cyclonic development was designated as Low 39 of FASTEX IOP-15. The 950-mb vorticity analysis (Fig. 24e) shows the comma-shaped cyclonic vorticity maximum of Low 39, with its head at the thermal wave crest and its tail trailing southwestward along the leading edge of the cold front. Of equal interest, is the formation of a frontal wave within the region of along-front stretching (confluent) frontogenetic deformation and cross-front shearing frontogenetic deformation off of the east coast of the United

States (Fig. 24f). This second secondary frontal wave is most evident as the cyclonic vorticity filament (Fig. 24e) at the leading edge of the contracting frontal zone. Reference to the 316-K PV analysis (Fig. 24d), shows that both secondary frontal waves formed on the anticyclonic shear side of the upper-level jet stream axis displaced ~1000 km from the PV gradient zone to their northwest. It is noteworthy that, whereas a single ~4000-km-wavelength, tropopause PV wave spans the North Atlantic (Fig. 24d), beneath it there are three distinct lower-tropospheric cyclonic vorticity features (Fig. 24e) associated with it: the mature eastern Atlantic primary cyclone (Low 38); and the two western-Atlantic secondary developments west of the primary cyclone (Lows 39 and 39a). Only the primary baroclinic development (Low 38) appears to be vertically coupled to a PV feature in the vicinity of the tropopause.

At 0000 UTC 16 February 1997 (Fig. 24, column 3), the Low 39 upstream secondary development was located ~500 km northeast of Newfoundland with a central pressure of ~992 mb, and its trailing cold-frontal vorticity filament had merged with the developing frontal wave off the U.S. east coast (Fig. 24h). The Low 39a development appears mostly in the form of localized frontogenesis (Fig. 24i) within the region of confluent and shearing deformation between the weak, <10 m s^{-1}, northwesterly flow and front-normal cold-advection off the North American east coast and the stronger (>20 m s^{-1}) southwesterly flow and along-front warm advection at the western flank of the primary anticyclone to the east. The recent studies by Renfrew et al. (1997) and Rivals et al (1998) have described the role of a localized frontogenetic geostrophic deformation and its coupled secondary circulation in the initiation of North Atlantic frontal waves along the trailing cold front of a preceding primary cyclone development.

At 0600 UTC 16 February 1997 (Fig. 24, column 4), the upstream development (Low 39) continued to intensify south of Greenland, with a central pressure down to ~982 mb (Fig. 24k). It was approximately at this time that the upstream-developing Low 39 passed beneath the axis of the upper-level jet stream and its associated sharp PV gradient, and when PV wave amplification was first noted aloft (Fig. 24j). The westernmost secondary frontal wave (Low 39a) was in the process of amplification in both vorticity (Fig. 24k), and potential temperature (Fig. 24l), with synchronous amplification of the tropopause PV wave aloft (Fig. 24j).

By 1200 UTC 16 February 1997 (see Fig. 16), the upstream secondary development (Low 39) had evolved into a mature secondary cyclone ~500 km southeast of Greenland with a central pressure of ~977 mb (Figs. 16c), ~21 mb lower than at 24 hrs earlier. The PV analysis and velocity vectors (Fig. 16a) show the further amplification of an intermediate-scale tropopause PV wave above Low 39, with positive (cyclonic) PV advection vertically coupled to the localized maximum of lower-tropospheric warm advection within its warm front (Fig. 16c), typical of the maturing phase of cyclone development. The Low 39a secondary development southeast of the Canadian Maritime Provinces had evolved into a primary frontal-wave cyclone (Low 39a), as evidenced by its synoptic-scale (~4000 km) lower-tropospheric frontal vorticity filament (Fig. 16c) and tropopause PV wave (Fig. 16a) aloft. Low 39a continued to propagate along the northeastward track, during the following 24 h, and passed beneath the upper-level jet stream axis and associated PV wall, as it entered its occlusion phase between Iceland and Scotland. This final phase of FASTEX Low 39a was observed in detail with dropwindsondes and Doppler radar by the NOAA/P-3 and UK Meteorological Service C-130 research aircraft on the following day.

7.2   Secondary Cyclone Development from a Small (~100 km) Initial Disturbance

The development of cyclones initiated by infinitesimally small disturbances within baroclinically unstable idealized jet flows has been the subject of theoretical studies by Eady (1949) and Charney (1947), and others (reviewed by Farrell 1998, this volume). However, when addressing cyclone development in nature, practical meteorologists perceive cyclones as evolving from synoptically identifiable finite-amplitude disturbances and their associated localized secondary circulations, rather than infinitesimal random disturbances. This difference in interpretation is addressed by Farrell (1998; this volume), who referred to the following quotations regarding the then-new theories of cyclone development:

> The present [baroclinic instability] theories, while useful for explaining the broad features of the formation of growing disturbances [cyclones], are of little use in synoptic analyses, except as background information (Eliassen 1956). Cyclogenesis results not just from the release of an infinitesimal perturbation by a dynamically unstable state, but rather from the release of some kind of instability by finite perturbations which can be identified with the wave-shaped motion patterns in the middle and upper troposphere (Petterssen 1955).

The identifiable synoptic precursors include upper-level jet streaks, tropopause PV anomalies, waves along pre-existing surface-based fronts, localized diabatic processes, and interactions between upper-level and low-level disturbances.

We next consider a North-Atlantic secondary cyclone development that originated from a very small (~100 km), low-level (~950 mb) vorticity disturbance. The situation is one of the historical cases studied in the preparation for the field phase of FASTEX (e.g., Renfrew et al. 1997). For the present study, the life cycle of this cyclone was simulated over the 48-h period 0000 UTC 20 January to 0000 UTC 22

January 1995 with the Naval Research Laboratory (NRL), Coupled Ocean Atmosphere Mesoscale Prediction System (COAMPS) described in Hodur (1997), configured with full physics, 30 vertical levels, and 60-km horizontal resolution. This disturbance developed along the diffuse trailing cold front of a preceding primary cyclone, without a significant near-surface frontal PV or vorticity filament as a source for the release of barotropic instability, and without the early influence of a finite amplitude PV disturbance in the vicinity of the tropopause. The results from the final 36-h of the simulation are shown in Fig. 25.

At 1200 UTC 20 January 1995 (12 h into the 48-h simulation) the MSL-pressure and 950-mb absolute vorticity analysis over the western Atlantic contained a Bergeron (1928) frontogenetic confluent (stretching), geostrophic deformation, and a barely discernable localized cyclonic vorticity disturbance indicated by the arrow in Fig. 25b. This initial disturbance first appeared 6 h earlier in the simulation off the southern tip of Greenland, within the cold, adiabatically stratified arctic boundary layer, and propagated southeastward to its 1200 UTC position on the west–east axis of dilatation of the stretching deformation (Fig. 25b). The disturbance could have originated as a piece of cyclonic vorticity "debris" that became disconnected from the West Greenland coastal cyclonic vorticity (PV) filament, which then migrated into the northerly flow of the low-level deformation. It has been suggested by Farrell (1989), and others, that residual debris left by previous decaying cyclones can serve as an initial finite (but small) amplitude disturbance that will evolve into a mature cyclone within a favorable unstable environment. Note that at this time, the vorticity disturbance is small, localized, and relatively weak ($\sim 1.5 \times 10^{-4}$ s^{-1}), and not tied to a lower-tropospheric frontal vorticity or PV filament, in agreement with the diagnosis of this situation by Renfrew et al. (1997). The companion ~250-mb PV and wind velocity analysis (Fig. 25a) shows a trans-Atlantic jet streak and its associated PV wall at the jet axis, situated ~600 km north of the low-level vorticity disturbance. Note that the disturbance (Fig.25b) is positioned beneath, and to the anticyclonic shear side of the upper jet (Fig. 25a). At this early stage, there is no discernible PV anomaly or jet stream entrance (exit) region aloft to dynamically couple to the developing vorticity disturbance below.

By 0000 UTC 21 January 1995, the low-level disturbance had propagated into a region of weaker stretching deformation, and was by then an identifiable MSL-pressure trough, with an increase in vorticity to 2.2×10^{-4} s^{-1} (Fig. 25e). The disturbance location relative to the upper flow remained south of the 250-mb PV wall, and beneath the right-front quadrant of the upper-level jet streak exit (Fig. 25d). A qualitative inference derived from the Sawyer (1956) and Eliassen (1962) secondary circulation equation suggests, that, at this time, the low-level disturbance was under the

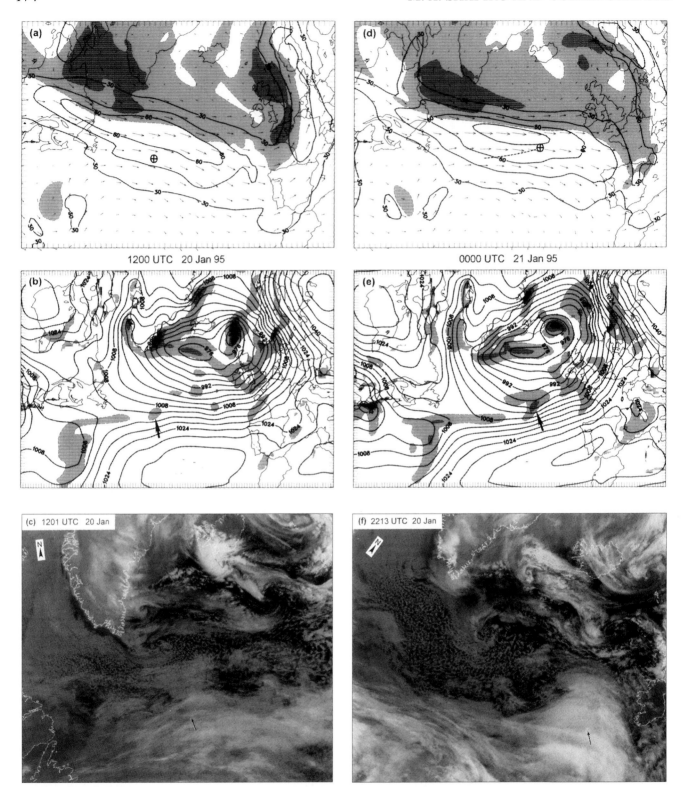

FIG. 25 (above and facing page). The 48-h simulation of a North-Atlantic secondary cyclone, presented at 12-h intervals beginning with the 12-h forecast at 1200 UTC 20 January and ending at 0000 UTC 22 January 1995. Upper panels: 250-mb PV (light shading, > 2 PVU; medium shading, > 3 PVU); wind speed (m s^{-1}, solid lines) and wind vectors;.crossed circles indicate the positions of the 950-mb vorticity maxima from lower panels; dashed line indicates the previous 12-hr track of the 950-mb vorticity maxima. Middle panels: MSL pressure (mb, solid lines at 4 mb intervals); 950-mb absolute vorticity (light shading, >1.5 × 10^{-4} s^{-1}; medium shading, >2 × 10^{-4} s^{-1}; and dark shading, >2.5 × 10^{-4} s^{-1}. Bottom panels: DMSP satellite infra-red cloud images depicting the observed development of the simulated secondary cyclone. Black arrows (Figs. 25a and b) indicate positions of the 850-mb vorticity maxima associated with the developing secondary cyclone.

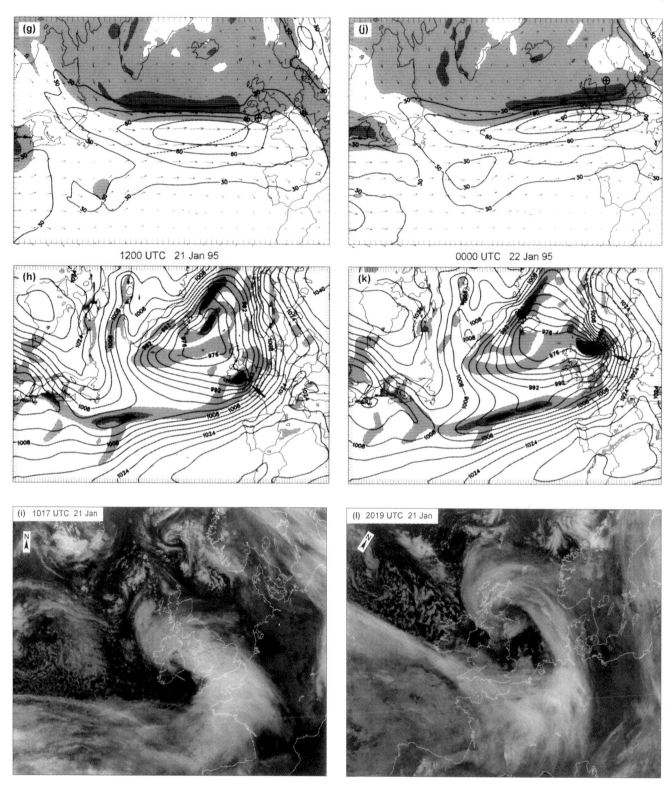

influence of upper-level convergence and mid-to-upper-tropospheric subsidence within the indirect secondary circulation of the jet-streak exit. It should be noted that, whereas 12 h earlier there were ~5 pieces of mesoscale vorticity debris over the central North Atlantic (Fig. 25b), only the disturbance under discussion was sustained. The sustained disturbance contained a localized area of enhanced condensation heating within the broader region of lower-tropospheric ascent along the deformation dilatation axis. We suggest that localized diabatic heating contributed to its sustenance.

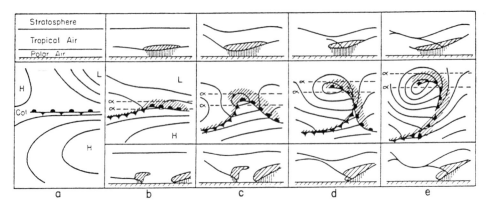

FIG. 26. Life cycle of cyclones from wave to vortex. Upper and lower panels: tropopause and frontal (air-mass) boundaries and clouds. Middle panels: surface pressure (solid lines), fronts. Cross-section projections for upper and lower panels (northern and southern dashed lines, respectively) (Godske et al 1957).

By 1200 UTC 21 January 1995, the disturbance was situated over Ireland (Fig. 25h), having propagated out of the lower-tropospheric confluent deformation zone, and expanded in size and intensity, with a minimum MSL pressure of ~982 mb located at the ~3.2×10^{-4} s^{-1} vorticity maximum. The disturbance had passed beneath the axis of the upper-level jet and associated PV wall at the tropopause fold (Fig. 25g), and was positioned beneath the region of cyclonic speed shear of the left-front (ascending) quadrant of the upper jet exit, downstream from the advancing positive PV anomaly of the upper-level jet. This vertical alignment (phase locking) between the upper-level PV anomaly and the low-level disturbance on the trailing baroclinicity of the parent cyclone cold front initiates the release of baroclinic instability with synchronous amplification of the surface cyclone and its coupled upper-level disturbance. It was near this time that this disturbance developed into a substantial secondary cyclone.

By 0000 UTC 22 January 1995 the cyclone had reached maturity over the North Sea between Norway and Scotland. Its central MSL pressure had decreased to ~964 mb, and the 950-mb vorticity had increased to ~4×10^{-4} s^{-1} (Fig. 25k). Note the amplification of the PV wave aloft and the presence of an anticyclonic PV anomaly and circulation over southern Norway and Sweden (Fig. 25j). The ~34-mb decrease in the cyclone central MSL pressure over 24-h categorizes this cyclone as "rapidly intensifying." The COAMPS 48-h simulation of the cyclone position and intensity at this time is in close agreement to that which was observed.

Limited corroboration of the above simulated secondary cyclone development is presented in the DMSP infra-red cloud images (Fig. 25, lower panels). The 1017 UTC 20 January 1995 DMSP image (Fig. 25c) shows the incipient disturbance as the "chevron" shaped middle-cloud system off the southern tip of Greenland. By ~2019 UTC 20 January (Fig. 25f), the disturbance had intensified (as in the simulation, Fig. 25e) and possessed a well-defined upper-level cloud shield. By 1201 UTC 21 January (Fig. 25i) one observes the cloud signature of the rapidly deepening phase of

secondary cyclone formation, with the surface cyclone centered over Ireland. The final satellite image for 2213 UTC 22 January (Fig. 25-l) shows the occlusion phase of the secondary development over northern Scotland, with the brightest (coldest) clouds situated at the occlusion triple point over the northern Mediterranean.

Historically, frontal-wave cyclone development, such as illustrated in Fig. 24 (Low 39a) and Fig. 25, was conceptualized by J. Bjerknes in his chapter on Climatology and Synoptic Meteorology prepared for Godske et al.(1957). This early conceptual model (Fig. 26) shows the frontogenetic stretching (confluent) deformation (Fig. 26a) as the region for the formation of the frontal wave in Fig.26b. Bjerknes referred to this initiation phase as "cyclogenesis at axis of dilatation east of a col in the westerlies." Note that the frontal wave forms in the lower troposphere without a concurrent upper-level disturbance in the vicinity of the tropopause (Fig. 26b). The leading edge of the surface front is positioned on the anticyclonic shear side of the polar front aloft (see Godske et al., Fig 15.04.3), as was the case for the confluent deformation frontal wave cyclone in Fig. 16. By the midpoint of the 24-hr life cycle (Fig. 26c), the disturbance has propagated out of the initial stretching deformation zone, and a closed cyclonic circulation appears at the crest of the frontal wave, with synchronous amplification of the tropopause wave. The mature phase of the cyclone (Fig. 26c) has the T-bone frontal alignment of the LC1 cyclone, and the decaying phase (Fig. 26d) is a classical Bergeron occluded cyclone. Note that the amplitude of the tropopause deformation is greater to the south of the occluded cyclone, consistent with the final position of the cyclone to the north of the upper jet.

In summary, the rapid development of this North Atlantic secondary cyclone represents another example of the interaction between an upper-level PV anomaly and a low-level disturbance within a baroclinically unstable environment. The baroclinic interaction phase of this development is similar to the idealized simulation by Thorncroft and Hoskins (1990). As also noted by Renfrew et al. (1997), "Initially there is not a strong enough low-level PV strip [filament] for barotropic instabilities to play an important role. However, the increase in amplitude does coincide with a decrease in environmental along-front stretching which would allow barotropic instabilities." The initial low-level disturbance originated out of debris ejected from the Labrador Sea west of Greenland and was most likely maintained by latent heat release. There, however, remains a key unresolved aspect of the above-described secondary cyclone life cycle. In spite of the realism of the numerical simulation, from the initial appearance of the small disturbance to its maturation as a substantial secondary cyclone, it was not established whether there was an observed initial low-level vorticity disturbance of the type appearing during the first 12 h of the simulation. As noted in the observation element of the Paradigm for Progress in Meteorology (Section 2), "without corroborating observations, a degree of uncertainty always remains as to whether numerical models are providing the correct solution." However, in spite these uncertainties in the details of the life cycle, the 48-h prediction of the mature phase of the cyclone (Fig. 25h) was quite accurate. The origin and significance of small initial disturbances that on occasion evolve into significant weather-producing systems remains a subject for further investigation.

8. Summary and Future Directions

8.1 Summary

The authors have aspired to construct a conceptual bridge between the theoretical perspectives of idealized extratropical cyclone life cycles and those occurring in nature, with an emphasis on the influence of planetary-scale environmental flow on primary and secondary cyclone development. Our approach was based was on the Bjerknes (1904), Bergeron (1959), and Hoskins (1983) pathways for progress in meteorology (Section 2). An overview of theoretical interpretations, derived from idealized numerical simulations, illustrated the sensitivity of cyclone frontal development and upper-level PV evolutions to variations in planetary-scale (~10 000 km) meridional barotropic shears. It was hypothesized that variations in cyclone life cycles in nature are attributable to different phasing, and meridional, and vertical alignments of the subtropical and polar jet stream (Section 3.3 and Fig. 5). These jet stream interactions were visualized and diagnosed within the dynamical framework of PV thinking and its graphical representations of the tropopause (Section 4). The numerical simulation of three North Atlantic cyclones, evolving within different meridional alignments of the subtropical and polar jet streams, were compared to the idealized environmental barotropic-shear cyclones. These simulations were verified by observations (Section 5) and were considered as analogs to the theoretical barotropic-shear idealizations. Vertical cross sections taken across the

North Atlantic cyclones were used to design perturbed basic states containing single and dual jet streams in the vicinity of the tropopause, from which idealized cyclone families were simulated (Section 6.2). The simulated cyclone families contained both downstream and upstream baroclinic developments, whose evolutions were discussed from the synoptic and theoretical (PV) viewpoints (Section 6.3). The results from the idealized and observed cyclone family simulations supported the hypothesis that differences in the environment of cyclones arising from phasing and meridional/vertical alignment of the subtropical and polar jet streams would lead to idealized cyclone life cycles similar to those observed in nature, and to those in the barotropic-shear numerical experiments. The final discussion addressed the development of secondary cyclones through numerical simulations of observed North Atlantic situations (Section 7). The first example (Fig. 24) developed through upstream baroclinic development at the end of the trailing cold front of the primary synoptic-scale cyclone in the absence of an initial localized disturbance in the vicinity of the tropopause. The synoptic and PV perspectives of upstream baroclinic development (Section 6.3) were used to diagnose the structural and dynamical similarities between this simulated and observed upstream development, and the idealized upstream frontal wave cyclones. The second example (Fig. 24) formed within the ascending branch of the transverse secondary circulation of a synoptic-scale, confluent, along-front stretching deformation field. The third example (Fig. 25) originated out of a small (~100 km) initial disturbance in a confluent stretching deformation zone, and underwent rapid development when it phased with a PV anomaly in the vicinity of the tropopause.

8.2 Remaining Issues and Future Considerations

We conclude this chapter by discussing some remaining issues for further investigation:

Environmental basic states and initial disturbances

There remain unresolved issues regarding the approach toward defining the environmental space-time basic states within which primary and secondary cyclones evolve. The time mean, spatially varying, Northern Hemispheric winter flow has been found to be linearly stable when realistic damping is imposed (Hall and Sardeshmukh 1998). If the cyclone environment is taken to be this time-mean state, which has rather small baroclinicity, then there are no growing modes with fixed structure, although in some regions (the storm tracks), transient growth and decay occur as finite perturbations amplify. In earlier idealized studies (e.g., Simmons and Hoskins 1979; Thorncroft et al. 1993), the basic states are very unstable to infinitesimal perturbations and are chosen because they resemble a snapshot of a section across a storm track, much in the spirit of the present study. In addition, one should also consider the effects of longitudinal variations in background stretching deformation along the storm tracks (see, e.g., Schultz et al. 1998). It remains a challenge to design idealized numerical experiments with spatially and temporally varying basic states of the complexity that occurs within the climatological storm tracks.

The idealized, longitudinally and temporally invariant, barotropic-shear and dual jet stream basic states are simplifications of the environments within which cyclones develop in nature. For example, cyclone development within the North Atlantic storm track often begins as a disturbance on the anticyclonic-shear side of an upper-level jet stream within the confluent deformation of the western entrance of an oceanic storm track. The surface disturbance then evolves into a frontal wave that becomes vertically aligned with the tropopause, PV wall at the axis of the upper jet as it enters its rapid intensification phase. During the final phase of its life cycle (occlusion), the surface cyclone resides on the cyclonic-shear side of the upper jet stream exit within the diffluent deformation of the eastern side of the Atlantic storm track. Within this life cycle scenario, a cyclone can exhibit the frontal characteristics of all three idealized barotropic-shear cyclones (LC3, LC1, and LC2). This is due, in part, to the interaction between the cyclone and its environment, as discussed below.

There is the issue of the sensitivity of cyclone development to the scale and amplitude of the initial disturbance, and to the meridional separation between the upper-level jets. Regarding the dual jet stream idealized cyclone experiment (Section 6), the large scale of the initial disturbance, compared to the relatively small separation between the two jet streams, led to a fully developed LC2 initial cyclone that did not differ substantially from the initial single jet stream LC1 cyclone. In contrast, the smaller LC1 and LC2 upstream secondary developments showed significant differences from one another. This is likely the result of the smaller meridional scale of the secondaries that permits their development without significant modification of the basic flow. Small amplitude disturbances will feel differences in the environment, such as the jet separation, to a greater degree, but are also more likely to propagate out localized baroclinically unstable regions before reaching maturity. Given an awareness of the environmental basic flows, operational forecasters may be able to anticipate the characteristics of subsequent cyclone life cycles.

The influence of cyclones on environmental background flows

There is the issue of how cyclones interact with background flows and with each other (both laterally and vertically). In the present study, there is no clear distinction between the spatial displacements of jet streams and the developing cyclones. The alignment of the jets clearly influences cyclone development, yet the cyclones themselves alter the strength and meridional alignment of the jets, which in turn impacts on lower tropospheric cyclogenesis. The location of upper-tropospheric jets will also be influenced by the planetary-scale waves that are influenced by zonal asymmetries in the earth's surface, as discussed in Held (1998). An illustration of the interaction between cyclone development and a dual-jet idealized LC2 basic state is shown in Fig. 20. Here baroclinic development in the polar-jet current leads to wave amplification in the subtropical current to the south, and subsequent vertical alignment of the polar and subtropical jets above the downstream cyclone development at day 4.5. In this numerical experiment, the interaction between the initial cyclone and the background flow led to the formation of a single jet environment for the development of an LC1 downstream cyclone, starting from an initial LC2 dual-jet basic state. The interaction between the primary cyclone and the jet location can influence downstream and upstream cyclogenesis.

Interactions between tropopause-based and surface-based disturbances

The interaction of upper-level and low-level disturbances, and resulting rapid development of extratropical cyclones remains a subject of considerable interest (e.g., the different interpretations of development of the Queen Elizabeth II (QE-II) cyclone presented in Gyakum (1983a, 1983b); Uccellini (1986); and Manobianco et al. (1992)). This debate centers on the relative importance of upper- versus lower-tropospheric disturbances, and latent heating in the initial development of rapidly deepening cyclones. Idealized cyclone simulations can be designed to address the interactions between transient upper-level disturbances in the westerlies (e.g., PV anomalies, jet streaks) and surface-based baroclinic disturbances (e.g., incipient frontal-waves, secondary cyclones, upstream baroclinic developments). For this purpose, simulations of idealized cyclone families could be performed within longitudinally-extended cyclic domains, permitting upper-level downstream disturbances to enter the western boundary and subsequently phase with low-level upstream disturbances. The resulting cyclone developments could be considered as idealized analogs of observed events such as the QE-II cyclone.

Upstream (downstream) baroclinic development

Upstream baroclinic development, although easily simulated with adiabatic, inviscid, idealized atmospheric numerical models, remains difficult to document in nature. This is due, in part, to the fact that surface friction acts to suppress the development of surface-based disturbances, more so over continents than over oceans. The relatively data-sparse oceans are the most likely regions for upstream baroclinic development, where there is greater uncertainty as to whether a weak undetected upper disturbance has contributed to their development. In addition, the likelihood that a given upstream development can complete a 2–4 day life cycle without being intercepted by a transient upper-disturbance in the westerlies, is not great. Present-day synoptic researchers and forecasters have not enthusiastically embraced the concept of Pettersen and Smebye (1971) Type A (bottom-up) cyclone development induced by upstream energy dispersion from a preceding baroclinic development to the east. Downstream development, although widely addressed in the contemporary literature from the theoretical and diagnostic perspectives, also remains an important issue for further consideration.

Planetary-scale influences on clouds, precipitation, and Lagrangian air streams

It remains for future research to develop the link among planetary-scale influences on cyclone life cycles, their fronts and associated cloud and precipitation systems, Lagrangian air streams, and secondary circulations. Although not shown, the differences in frontal structure between the North Atlantic LC1, LC2, and LC3 cyclones (Figs. 11a and c, 11b and d, and Fig. 16, respectively) were associated with significant differences in Lagrangian air streams, and cloud and precipitation distributions. Whereas the IOP-4 (single-jet, LC1) cyclone had strong ascent and precipitation along the bent-back warm front to the west of the cyclone center, and strong moist connection along its cold front, the warm-core seclusion of the Pre-ERICA IOP-4 (dual-jet, LC2) cyclone possessed subsidence to the west of the cyclone center, and forward (eastward) sloping clouds and precipitation ahead of its surface cold front. The FASTEX IOP-16 (anticyclonic environment, LC3) cyclone (Fig. 16) contained rearward sloping ascent along its cold front.

Observations and their analysis

Past field studies have required large expenditures and human efforts to gather voluminous data sets, but only static subjective analyses of a fraction of these data sets have been analyzed and reported. In order to extract the maximum information content from the experimental and operation observations, it will be necessary to utilize four-dimensional data assimilation (4DDA) methods to provide dynamically consistent gridded data sets with which to resolve the structure and life cycles of a broad spectrum of synoptic-scale and mesoscale weather systems critical to further advances in the understanding and prediction of extratropical cyclone life cycles.

Methods have been developed that utilize the predictive equations to identify regions sensitive to initial analysis errors. These methods have been successfully applied to the targeting of additional observations in remote, data-sparse regions (see, e.g., Bishop and Toth 1996; Langland and Rohaly, 1996; Snyder 1996; and Palmer et al. 1998) Targeted observing strategies provide an "intelligent" approach toward enhancing observations for the purpose of improved numerical weather prediction. These methods can also be used for the diagnostic assessment of the relative importance of specific atmospheric structures (e.g., tropopause-based PV disturbances, midtropospheric thermal anomalies, low-level fronts and associated PV/vorticity disturbances) and physical processes (e.g., latent and sensible heating) in the life cycles of extratropical weather systems.

Finally, fine-grid-resolution (mesoscale) numerical prediction models initialized with operational, synoptic-scale data or idealized atmospheres now provide extremely realistic simulations of cyclone life cycles, including their internal mesoscale structures and physical processes. It remains a topic of debate whether further improvements in intermediate range (24 to 72 h) numerical forecasts of the mesoscale substructures within extratropical cyclones will benefit to a greater degree from more accurate specifications of the planetary-scale and synoptic-scale flows than from higher-resolution localized observations on the mesoscale.

Acknowledgments

The authors express their appreciation to Chester Newton, Richard Rotunno, Greg Hakim, Tom Schlatter, and Anders Persson for their comments and helpful suggestions. We are grateful for the help of Rene Fehlmann for preparing the idealized basic state structures for the cyclone family experiments (Section 6.2). Special thanks to Christy Sweet and Sandra Rush for technical editing of the manuscript, to Jim Adams for preparation of the figures, and to David Serke and Mike Mariani (NOAA/National Geophysical Data Center) for preparing the DMSP satellite images for Fig. 25.

APPENDIX: A PV Representation of Uncertainties in Ensemble Weather Forecasts of Extratropical Cyclones

The errors associated with numerical weather forecasts are attributable to (1) the theoretical limit of atmospheric predictability; (2) sparse or inaccurate observations, data assimilation, and initialization procedures; and (3) model architecture and physical process parameterizations. From a predictability standpoint, the dispersion of these errors can lead to differences in the predicted life cycles. The differences between forecasts and their verification provide a measure of forecast skill for phenomena and processes such as cyclone position and intensity, 500-mb height error, and quantitative

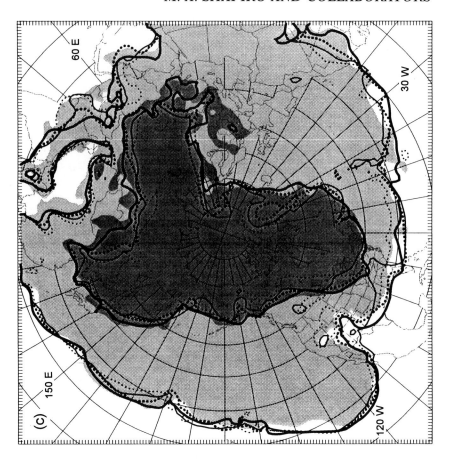

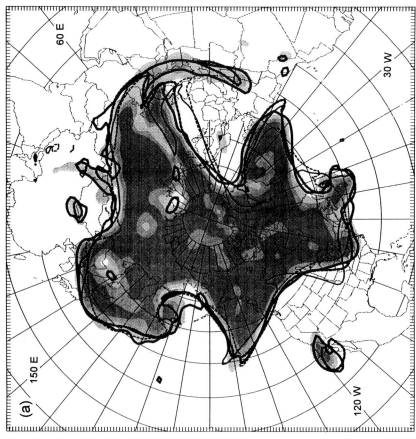

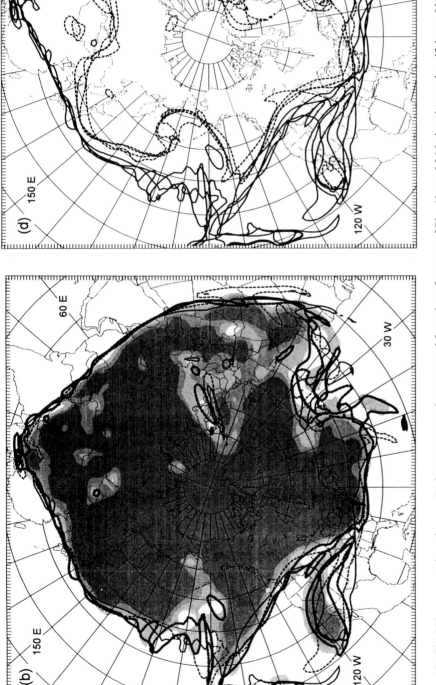

Fig. 27. The PV representation of ensemble forecast uncertainty prepared from three forecasts of 72-, 48-, and 24-h duration, and verifying at 0000 UTC 4 January 1989. (a) Polar jet stream ensemble 2-PVU members on the θ = 316 K surface for forecasts of 72 h (dashed line); 48 h (thin solid line), and 24 h (medium solid line); and the verification (heavy solid line). Ensemble mean PV is shown gray shaded at three contour levels of decreasing brightness for PV = 2, 4, and 6 PVU. (b) Subtropical jet stream ensemble 2-PVU members at θ = 340 K; members and shading are as in (a). (c) 950-mb potential temperature ensemble member strands for = 285 K (southern strands) and = 273 K (northern strands), approximating the ensemble forecast uncertainty in the leading edge of the polar and Arctic air masses, respectively, for forecasts of 72 h (small dotted lines), 48 hrs (heavy dotted lines), and 24 h (thin solid lines); and the verification (heavy solid lines). Light and dark shaded areas mark the <285 K and <273 K potential temperatures of the ensemble mean, respectively. (d) The composite of the polar (dashed lines) and subtropical (solid lines) ensembles shown in (a) and (b), respectively.

precipitation. An ensemble weather forecast provides a measure of forecast confidence by identifying ensembles of alternative flow regimes that are generated by small initial perturbations to the forecast for comparison with an unperturbed forecast with the identical forecast model. Toth and Kalnay (1993) have encouraged the construction of "spaghetti" diagrams to illustrate the degree of uncertainty of ensemble forecasts. These diagrams include selected 500-mb height contours from each perturbed member forecast, typically one contour from the midlatitudes and another from the subtropics. Height contours from the forecast members (viz., spaghetti strands) are charted for comparison with each other, the unperturbed control forecast, the ensemble mean, and/or the verification analysis. The degree to which the member strands are bundled versus spatially spread is a measure of confidence in the forecast when considered along with the ensemble variance. The more rapid the growth of forecast uncertainty, the faster the strands spread from the bundle and entangle as the forecast approaches its limit of usefulness. Although height contour ensembles clearly illustrate forecast error growth (see, e.g., Houtekamer and Derome 1995), they are less than optimal for identifying the dynamical characteristics of forecast uncertainty.

Considering that tropopause-based PV anomalies, lower-tropospheric thermal anomalies, and internal PV anomalies are leading components of the PV viewpoint of baroclinic life cycles, we have chosen to explore their utility for identifying the dynamical origins of ensemble forecast uncertainty. The PV perspective of baroclinic life cycles addresses the attribution of balanced flows to PV and its associated anomalies. PV anomalies arise from (1) horizontal wave-like displacements and/or vertical deformations of PV in the vicinity of the tropopause; (2) concentration, dilution, or redistribution of PV through spatial gradients in diabatic heating, turbulence and surface friction; and (3) localized surface and/or planetary boundary-layer thermodynamic anomalies referred to as lower-boundary (surface) surrogate PV anomalies (see reviews by Hoskins et al. (1985); Haynes and McIntyre (1987); and Bosart 1998, this volume).

Let us first consider two representations of ensemble forecast uncertainty in the vicinity of the tropopause, (1) a display of ensemble member 2-PVU contours taken from PV analyses on isentropic surfaces that intercept tropopause folds near the level of maximum wind of the arctic, polar, and subtropical jet streams; or equivalently, (2) isentropes taken from member 2-PVU tropopause maps. These ensemble isentropes depict forecast uncertainty for the polar, subtropical, and Arctic jet streams, their associated PV anomalies, and meridional interactions.

The second component of the PV ensemble illustrates forecast uncertainty for near-surface thermal anomalies, which act as surrogate PV anomalies. Here, we suggest the selection of two isopleths of potential temperature displayed on an isobaric surface (~950 mb): one isentrope at the leading edge of the polar air stream ($\theta \sim 285$ K) and a second at the leading edge of arctic cold-air flows ($\theta \sim 273$ K). The near-surface thermal ensemble visualizes forecast uncertainty of the polar and arctic fronts and their adjacent thermal anomalies. An alternative would be to consider equivalent potential temperature ensembles that include the potential contribution of lower-tropospheric moist diabatic processes to forecast uncertainty. The selection of member isentropes for the near-surface PV surrogate ensemble is not as straightforward as that for the tropopause, owing to seasonal, geographical, and diabatic influences on the thermal properties of near-surface polar and arctic flow regimes.

Finally, the degree of uncertainty of predicted internal (tropospheric) PV-altering processes attributable to spatial gradients in the phase change of water substance, internal turbulent mixing and surface boundary-layer processes can be represented by charting the ensemble of the 2-PVU member contours on a pressure surface within the mid to lower troposphere (~600–800 mb) on an internal θ surface (beneath the tropopause) between 280 and 300 K.

An illustration of the application of PV dynamics to the diagnosis of an ensemble forecast is presented in Fig. 27. The forecast ensemble was prepared from three Northern-Hemispheric forecasts of 72-, 48- and 24-h duration, between 0000 UTC 1 January 1988 and 0000 UTC 3 January 1989, inclusive, and their verification at 0000 UTC 4 January 1989 that coincides with the onset of the ERICA IOP-4 cyclogenesis (Section 5.3). This ensemble was initially prepared for the medium-range prediction study of this cyclone by Zou et al. (1998) using a 75-km-horizontal by 23-level resolution configuration of the Penn State/NCAR MM-5 multiscale prediction model. The ensemble includes the 2-PVU members on the 316-K and 340-K isentropic surfaces (Figs. 27a and b, respectively) and the near-surface (~950-mb) 273-K and 285-K isentropes marking the leading edge of the arctic and polar flows, respectively (Fig. 27c). The members are superimposed on their ensemble means. One can also overlay the member strands over their ensemble variance instead of the mean. Regions of large member spread and/or short-wavelength oscillations that coincide with weak gradients in the ensemble mean and/or small variance of the ensemble are of lesser significance than regions of diverging member strands adjacent to sharp gradients in ensemble mean and/or large variance. A composite of polar and subtropical PV ensembles (Fig. 27d) facilitates the simultaneous representation of ensemble spread of the forecast of subtropical and polar jet streams, and their meridional alignments.

The PV representation of ensemble forecasts can also be used to visualize vertical variations in ensemble uncertainty. For example, the 2-PVU members of the polar-jet flow (Fig. 27a) and the lower-tropospheric arctic and polar isentrope members (Fig. 27c) over the Eastern Atlantic Ocean show greater forecast consistency, that is, less spread, than the 2-PVU members of the subtropical-jet flow (Fig. 27b) in this region. Conversely, the ensemble spread over the eastern United States (Fig. 27a) exhibits greater uncertainty in the position and amplitude of the polar-jet 2-PVU members than the position of subtropical jet stream/tropopause members above and to its south (Fig. 27b). The forecast uncertainty in the lower-tropospheric thermal members (Fig.27c) lies mostly within the southern polar (285 K) isentrope, with lesser spread in the northern arctic (273 K) isentrope. This result is in agreement with the predictability study of this cyclone by Zou et al. (1998), showing uncertainty in the 4- to 5-day prediction of the ERICA IOP-4 cyclone due to initial analysis errors for both the polar tropopause PV anomaly over the Gulf of Alaska and the lower-tropospheric thermal anomaly over the southwestern United States.

References

Allart, M. A. F., H. Kelder, and L.C. Heijboer, 1993: On the relationship between ozone and potential vorticity. *Geophys. Res. Lett.*, **20**, 811–814.

Bergeron, T., 1928: Über die dreidimensional verknüpfende *Wetteranalyse. Geofys. Publ.*, **5** (6), 1–111.

——, 1934: Die dreidimensional verknüpfende Wetteranalyse. II. Teil: Dynamik und Thermodynamik der Fronten und Frontalstörungen. Übersetzung von W. I. Romanowskaja aus dem deutschen Manuskript, redigiert von S. P. Chromow. Ausgabe der Zentralverwaltung des Hydro-meteorologischen Einheitsdienstes, Moskau (Russ.).

——, 1937: On the physics of fronts. *Bull. Amer. Meteor. Soc.*, **18**, 265–275.

——, 1959: Methods in scientific weather analysis and forecasting. An outline in the history of ideas and hints at a program. *The Atmosphere and the Sea in Motion*, B. Bolin, Ed., Rockefeller Institute Press, 440–474.

Bishop, C. H., 1996a: Domain independent attribution. I: Reconstructing the the wind from estimates of vorticity and divergence using free space Green's functions. *J. Atmos. Sci.*, **53**, 241–252.

——, 1996b: Domain independent attribution. II: Its value in the verification of dynamical theories of frontal waves and frontogenesis. *J. Atmos. Sci.*, **53**, 253–262.

——, and Thorpe, A. J., 1994a: Frontal wave instability during moist deformation processes. Part I: Linear wave dynamics. *J. Atmos. Sci.*, **51**, 852–873.

—— and ——, 1994b: Frontal wave instability during moist deformation processes. Part II: The suppression of nonlinear wave development. *J. Atmos. Sci.*, **51**, 874–888.

——, and Z. Toth, 1996: Using ensembles to identify observations. likely to improve forecasts. Preprints, 11th Conference on Numerical Weather Prediction, Aug 19–23, Norfolk, Virginia, 72–74.

Bjerknes, J., 1930: Practical examples of polar-front analysis over the British Isles in 1925–1926. *Geophysical Memoirs* No. 50, Meteorological Office, 50 pp.

——, and H. Solberg, 1922: Life cycle of cyclones and the polar front theory of atmospheric circulation. *Geofys. Publ.*, **3**, 1–18.

Bjerknes, V., 1904: Das problem der wettervorhersage, betrachtet vom standpunkte der Mechanik und der Physik. *Meteor. Z.*, **21**, 1–7.

Bleck, R., 1973: Numerical forecasting experiments based on the conservation of potential vorticity on isentropic surfaces. *J. Appl. Meteor.*, **12**, 737–752.

——, 1974: Short-range prediction in isentropic coordinates with filtered and unfiltered numerical models. *Mon. Wea. Rev.*, **102**, 813–829.

Bluestein, H. 1993: *Synoptic-Dynamic Meteorology in Midlatitudes, Vol I: Observations and Theory of Weather Systems*, Oxford University Press, 431 pp.

Bosart, L. F., 1998: Observed cyclone life cycles, *The Life Cycles of Extratropical Cyclones*, M. A. Shapiro and S. Grönås, Eds., Amer. Meteor. Soc., 189–215.

——, G. Hakim, K.Tile, M. Bedrick, W. Bracken, M. Dickenson, and D. Schultz, 1996: Large-scale antecedent conditions associated with the 12–14 March cyclone ("Superstorm '93") over eastern North America. *Mon. Wea. Rev.*, **124**, 1865–1891.

Browning, K. A., 1990: Organization of clouds and precipitation in extratropical cyclones. *Extratropical Cyclones, The Erik Palmén Memorial Volume*, C. W. Newton and E. O. Holopainen, Eds. Amer. Meteor. Soc., 129–153.

Chang, M., 1993: Downstream development of baroclinic wave as inferred from regression analysis. *J. Atmos. Sci.*, **50**, 999–1015.

Chang, S. W., R. J. Alliss, S. Raman, and J.-J. Shi, 1993: SSM/I Observations of the ERICA IOP-4 marine cyclone: A comparison with in situ observations and model simulation. *Mon. Wea. Rev.*, **121**, 2452–2464.

——, T. R. Holt, and K. Sashegyi, 1996: A numerical study of the ERICA IOP 4 marine cyclone. *Mon. Wea. Rev.*, **124**, 27–46.

Charney, J., 1947: The dynamics of long waves in a baroclinic westerly current. *J. Meteor.*, **4**, 125–162.

——, and M. E. Stern, 1962: On the stability of internal barotropic jets in a rotating atmosphere. *J. Atmos. Sci.*, **19**, 159–172.

Cressman, G. P., 1948: On the forecasting of long waves in the upper westerlies. J. Meteor., 5, 44- 57.

Danielsen, E. F., 1964: Stratospheric-tropospheric exchange based on radioactivity, ozone, and potential vorticity. *J. Atmos. Sci.*, **25**, 502–518.

Davies, H. C., 1998: Theories of frontogenesis. *The Life Cycles of Extratropical Cyclones*, M. A. Shapiro and S. Grönås, Eds., Amer. Meteor. Soc., 139–162.

——, C. Schär, and H. Wernli, 1991: The palette of fronts and cyclones within a baroclinic wave-development. *J. Atmos. Sci.*, **48**, 1666–1689.

Davis, C. A., and K. A. Emanuel, 1991: Potential vorticity diagnostics of cyclogenesis. *Mon. Wea. Rev.*, **119**, 1929–1953.

——, E. Donall-Grell, and M. A. Shapiro, 1996: The balanced dynamical nature of a rapidly intensifying oceanic cyclone. *Mon. Wea. Rev.*, **124**, 3–26.

Donall-Grell, E., and M. A. Shapiro, 1994: Visualization of the life cycle of a rapidly intensifying marine cyclone. Proc., Intl. Symp. on the Life Cycles of Extratropical Cyclones, Bergen, Norway, University of Bergen, 202–207.

Dritschel, D. G., P. H. Davies, M. N. Juckes, and T. G. Shepherd, 1991: On the stability of a two-dimensional vorticity filament by adverse shear. *J. Fluid Mech.*, **230**, 647–665.

Eady, E. T., 1949: Long waves and cyclone waves. *Tellus*, **1**, 33–52.

Edelmann, W., 1963: On the behavior of disturbances in a baroclinic channel. Tech. Note No. 7, Contract AF61(052).

Eliassen, A., 1956: Instability theories of cyclone formation. *Weather Analysis and Forecasting, Vol. I*, S. Petterssen, Ed., McGraw-Hill, 428 pp.

——, 1962: On the vertical circulation in frontal zones. *Geofys. Publ.*, **24**, 147–160.

——, 1966: Motions of intermediate scale: Fronts and cyclones. *Advances in Earth Science*, E. D. Hurley, Ed., The MIT Press, 111–138.

——, 1990: Transverse circulations in frontal zones. *Extratropical Cyclones, The Erik Palmén Memorial Volume*, C. W. Newton and E. O. Holopainen, Eds., Amer. Meteor. Soc., 155–165.

——, 1998: Vilhelm Bjerknes's early studies of atmospheric motion. *The Life Cycles of Extratropical Cyclones*, M. A. Shapiro and S. Grönås, Eds., Amer. Meteor. Soc., 5–13.

——, and E. Palm, 1960: On the transfer of energy in stationary mountain waves. *Geophys. Publ.*, **22**, 1–23.

Evjen, S., 1936: Über die vertiefung von zyklonen. *Meteor. Z.*, **53**, 165–172.

Farrell, B. F., 1989: Transient development in confluent and diffluent flow. *J. Atmos. Sci.*, **46**, 3279–3288.

——, 1998: Advances in cyclogenesis theory: Toward a generalized theory of baroclinic development. *The Life Cycles of Extratropical Cyclones*, M. A. Shapiro and S. Grönås, Eds., Amer. Meteor. Soc., 111–122.

Fehlmann, R., 1997: Dynamics of seminal PV elements. Ph.D. Thesis, Swiss Federal Institute of Technology, Zurich, Switzerland.

Friedman, R. M., 1989: *Appropriating the Weather: Vilhelm Bjerknes and the Construction of a Modern Meteorology.* Cornell University Press, 251 pp.

——, 1998: Constituting the polar front, 1919–1920. *The Life Cycles of Extratropical Cyclones*, M. A. Shapiro and S. Grönås, Eds., Amer. Meteor. Soc., 29–40.

Gaza, R. S., and L. F. Bosart, 1990: Trough merger characteristics over North America. *Wea. Forecasting*, **5**, 314–331.

Godske, C. L., T. Bergeron, J. Bjerknes, and R. Bundgaard, 1957: *Dynamic Meteorology and Weather Forecasting*, Amer. Meteor. Soc., and Carnegie Inst. of Washington, 800 pp (see Chapter 15, p. 535)

Grell, G.A., J. Dudhia, and D. R. Stauffer, 1994: A description of the fifth-generation Penn State/NCAR Mesoscale Model (MM-5). NCAR/TN-398 + IA, National Center for Atmospheric Research, Boulder CO, 120 pp.

Gyakum, J. R.,1983a: On the evolution of the QE II storm. I: Synoptic aspects. *Mon. Wea. Rev.*, **111**, 1137–1155.

——,1983b: On the evolution of the QE II storm. II: Dynamic and thermodynamic structure. *Mon. Wea. Rev.*, **111**, 1156–1173.

Hadlock, R., and C. W. Kreitzberg, 1988: The experiment on rapidly intensifying cyclones over the Atlantic (ERICA) field study: Objectives and plans. *Bull. Amer. Meteor. Soc.*, **69**, 1309–1320.

Hakim, G. D., L.F. Bosart, and D. Keyser, 1995: The Ohio valley wave-merger cyclogenesis event of 25–26 January 1978. Part 1: Observations. *Mon. Wea. Rev.*, **123**, 2663–2692.

N. M. J. Hall and P. D. Sardeshmukh, 1998: Is the time-mean Northern Hemisphere flow baroclinically unstable? *J. Atmos. Sci.*, **55**, 41–56.

Haynes, P. H., and M. E. McIntyre, 1987: On the evolution of vorticity and potential vorticity in the presence of diabatic heating and frictional or other forces. *J. Atmos. Sci.*, **44**, 828–841.

Held, I. M., 1998: Planetary waves and their interaction with smaller scales. *The Life Cycles of Extratropical Cyclones.* M. A. Shapiro, and S. Grönås, Eds., Amer. Meteor. Soc., 101–109.

Hodur, R. M., 1997: The Naval Research Laboratory's Coupled Ocean/ Atmosphere Mesoscale Prediction System (COAMPS). *Mon. Wea. Rev.*, **125**, 1414–1430.

Hoskins, B. J., 1983: Dynamical processes in the atmosphere and the use of models. *Quart. J. Roy. Meteor. Soc.*, **109**, 1–21.

——, 1990: Theory of extratropical cyclones. *Extratropical Cyclones, The Erik Palmén Memorial Volume*, C. W. Newton and E. O. Holopainen, Eds., Amer. Meteor. Soc., 63–80.

——, and N. V. West, 1979: Baroclinic waves and frontogenesis. Part II: Uniform potential vorticity jet flows—cold and warm fronts. *J. Atmos. Sci.*, **36**, 1663–1680.

Hoskins, B. J., M. E. McIntyre, and A. W. Robertson, 1985: On the use and significance of isentropic potential vorticity maps. *Quart. J. Roy. Meteor. Soc.*, **111**, 877–946.

——, and P. Berrisford, 1988: A potential vorticity perspective on the storm of 15-16 October 1987. *Weather*, **43**, 122–129.

Houtekamer, P.L., and J. Derome, 1995: Methods for ensemble prediction. *Mon. Wea. Rev.*, **123**, 2181–2196.

Hovmöller, E., 1949: The trough and ridge diagram. *Tellus*, **1** (2), 62–66.

Joly, A., and A. J. Thorpe, 1990: Frontal instability generated by tropospheric potential vorticity anomalies. *Quart. J. Roy. Meteor. Soc.*, **116**, 525–560.

——, and the FASTEX Team, 1997: The Fronts and Atlantic Storm-Track Experiment (FASTEX): Scientific objectives and experimental design. *Bull. Amer. Met. Soc.*, **78**, 1917–1940.

Keyser, D., and M. A. Shapiro, 1986: A review of the structure and dynamics of upper-level frontal zones. *Mon. Wea. Rev.*, **114**, 452–499.

Kleinschmidt, E., 1950a: Über Aufbau und Entstehung von Zyklonen I (On the structure and formation of cyclones). *Meteor. Rundsch.*, **3**, Teil 1, 1–6.

——, 1950b: Über aufbau und entstehung von zyklonen II (On the structure and formation of cyclones). *Meteor. Rundsch.*, **3**, Teil 2, 54–61.

Kutzbach, G., 1979: *The Thermal Theory of Cyclones. A History of Meteorological Thought in the Nineteenth Century.* Amer. Meteor. Soc., 255 pp.

Langland, R. H., and G. D. Rohaly, 1996: Adjoint -based targeting of observations for FASTEX cyclones. Preprints, 7th Mesoscale Processes Conf., 9-13 Sept. 1996, Reading, U.K., Amer. and Roy. Meteor. Socs., pp. 369–371.

Majewski, D. 1991: The Europa-model of the Deutcher Wetterdienst. Numerical methods in atmospheric models, Vol. II., ECMWF, Reading UK, 147–191.

Manobianco, J., L. W. Uccellini, K. F. Brill, and Y.-W. Yuo, 1992: The impact of dynamic data assimilation on the numerical simulations of the QE II cyclone and an analysis of the jet streak influencing the precyclogenetic envrionment. *Mon. Wea. Rev.*, **120**, 1973–1976.

Massacand Alexia, C., H. Wernli, and H. C. Davies, 1998: Heavy precipitation on the Alpine south side: An upper-level precursor. *Geophys. Res. Lett.*, **25**, 1435–1438.

Mattocks, C. and R. Bleck, 1986: Jet streak dynamics and geostrophic adjustment processes during the initial stage of lee cyclogenesis. *Mon. Wea. Rev.*, **114**, 2033–2056.

McGinnigle, J. B., M. V. Young, and M. J. Bader, 1988: The development of instant occlusions in the North Atlantic. *Meteor. Mag.*, **117**, 325–341.

Methven, J., 1996: Tracer behaviour in baroclinic waves. Ph.D. Thesis, Reading University.

Miles, M. K., 1959: Factors leading to the meridional extension of thermal troughs and some forecasting criteria derived from them. *Meteor. Mag.*, **88**, 193–203.

Morgan, M. C., and J. W. Nielsen-Gammon, 1998: Using tropopause maps to diagnose midlatitude weather systems. *Mon. Wea. Rev.*, **126**, 241–265.

Namias, P., J., and Clapp, 1944: Studies of the motion and development of long waves in the westerlies. *J. Meteor.*, **1**, 57–77.

Neiman, P. J., and M. A. Shapiro, 1993: The life cycle of an extratropical marine cyclone. Part I: Frontal-cyclone evolution and thermodynamic air-sea interaction. *Mon. Wea. Rev.*, **121**, 2153–2176.

———, ———, and L. F. Fedor, 1993: The life cycle of an extratropical marine cyclone. Part II: Mesoscale structure and diagnostics. *Mon. Wea. Rev.*, **121**, 2177–2199.

Newton, C. W., 1954: Frontogenesis and frontolysis as a three-dimensional process. *J. Meteor.*, **11**, 449–461.

———, and J. E. Carson: 1953: Structure of wind field and variations of vorticity in a summer situation. *Tellus*, **5**, 321–339.

———, and H. R.-Newton, 1998: The Bergen school concepts come to America. *The Life Cycles of Extratropical Cyclones*, M. A. Shapiro and S. Grönås, Eds., Amer. Meteor. Soc., 44–60.

Orlanski, I., and E. K. M. Chang, 1993: Ageostrophic geopotential fluxes in downstream and upstream development of baroclinic waves. *J. Atmos. Sci.*, **50**, 212–225.

———, and J. Katzfey: 1991: The life cycle of a cyclone wave in the Southern Hemisphere. Part 1: Eddy energy budget. *J. Atmos. Sci.*, **48**, 1972–1998.

Palmén, E., and C. W. Newton, 1969: *Atmospheric Circulation Systems*. Academic Press, 603 pp.

Palmer, T. N., R. Gelaro, J. Barkmeijer, and R. Buizza, 1998: Singular vectors, metrics and adaptive observations. *J. Atmos. Sci.*, **55**, 633–653.

Petterssen, S., 1955: A general survey of factors affecting development at sea level. *J. Meteor.*, **12**, 36–42.

———, and S. J. Smebye, 1971: On the development of extratropical cyclones. *Quart. J. Roy. Met. Soc.*, **79**, 457–482.

Reed, R. J., 1955: A study of a characteristic type of upper-level frontogenesis. *J. Meteor.*, **12**, 542–552.

———, 1990: Advances in knowledge and understanding of extratropical cyclones during the past quarter century: An overview. *Extratropical Cyclones, The Erik Palmén Memorial Volume*, C. W. Newton and E. Holopainen, Eds., Amer. Meteor. Soc, 27–45.

———, and F. Sanders, 1953: An investigation of the development of a mid-tropospheric frontal zone and its associated vorticity field. *J. Meteor.*, **10**, 338–349.

———, and E. F. Danielsen, 1959: Fronts in the vicinity of the tropopause. *Arch. Meteor. Geophys. Bioklim.*, **A11**, 1–17.

Renfrew, I. A., 1995: The development of secondary frontal cyclones. Ph.D. Thesis, University of Reading.

———, A. J. Thorpe, and C. Bishop, 1997: The role of environmental flow in the development of secondary frontal cyclones. *Quart. J. Roy. Meteor. Soc.*, **123**, 1653–1675.

Riehl, H. and Collaborators, 1952: Forecasting in middle latitudes. *Meteor. Monogr.*, **1** (5), 80 p.

Rivals, H., J.-P. Cammas, and I. A. Renfrew, 1998: Secondary cyclogenesis: the initiation phase of a frontal wave observed over the eastern Atlantic. *Quart. J. Roy. Meteor. Soc.*, **124**, 243–267.

Rossby, C.-G., 1945: On the propagation of frequencies and energy in certain types of oceanic and atmospheric waves. *J. Meteor.*, **2**, 187–204.

Rotunno, R., and J.-W. Bao, 1996: A case study of cyclogenesis using a model hierarchy. *Mon. Wea. Rev.*, **124**, 1051–1066.

Sawyer, J. S., 1956: The vertical circulation at meteorological fronts and its relation to frontogenesis. *Proc. Roy. Soc. London*, **A234**, 346–362.

Schär, C., and H. Wernli, 1993: Structure and evolution of an isolated semi-geostrophic cyclone. *Quart. J. Roy. Meteor. Soc.*, **119**, 57–90.

Schultz, D. M., D. Keyser, and L. F. Bosart, 1998: The effect of large-scale flow on low-level frontal structure and evolution in midlatitude cyclones. *Mon. Wea. Rev.*, **126**, 1767–1791.

Shapiro, M. A., 1981: Frontogenesis and geostrophically forced secondary circulations in the vicinity of jet stream-frontal zone systems. *J. Atmos. Sci.*, **38**, 954–973.

———, T. Hampel, and A. J. Krueger, 1987: The arctic tropopause fold. *Mon. Wea. Rev.*, **115**, 444–454.

———, and D. Keyser, 1990: Fronts, jet streams, and the tropopause. *Extratropical Cyclones, The Erik Palmén Memorial Volume*, C. W. Newton and E. O. Holopainen, Eds., Amer. Meteor. Soc., 167–191.

———, and E. D-Grell, 1994: In search of synoptic/dynamic conceptualizations of the life cycles of fronts, jet stream and the tropopause. Proc., Int. Symp. on the Life Cycles of Extratropical Cyclones, University of Bergen, Bergen, Norway, 163–181.

Simmons, A. J., and B. J. Hoskins, 1979: The downstream and upstream development of unstable baroclinic waves. *J. Atmos. Sci.*, **36**, 1239–1254.

———, and ———, 1980: Barotropic influences on the growth and decay of nonlinear baroclinic waves. *J. Atmos. Sci.*, **37**, 1679–1684.

Snyder, C., 1996: Summary of an informal workshop on adaptive observations and FASTEX. *Bull. Amer. Meteor. Soc.*, **77**, 953–964.

Thorncroft, C. D., and B. J. Hoskins, 1990: Frontal cyclogenesis. *J. Atmos. Sci.*, **47**, 2317–2336.

———, ———, and M. E. McIntyre, 1993: Two paradigms of baroclinic-wave life cycle behavior. *Quart. J. Roy. Meteor. Soc.*, **119**, 17–55.

Toth, Z. and E. Kalnay, 1993: Ensemble forecasting at NMC: The generation of perturbations. *Bull. Amer. Meteor. Soc.*, **74**, 2317–2330.

Uccellini, L. W., 1986: The possible influence of upstream upper-level baroclinic processes on the development of the QE II storm. *Mon. Wea. Rev.*, **114**, 1019–1027.

———, 1990: Processes contributing to the rapid development of extratropical cyclones. *Extratropical Cyclones, The Erik Palmén Memorial Volume*, C. W. Newton and E. O. Holopainen, Eds., Amer. Meteor. Soc., 81–105.

———, and D. Johnson, 1979: The coupling between upper and lower tropospheric jet streaks and the implication for the development of convective storms. *Mon. Wea. Rev.*, **107**, 682–703.

Volkert, H., 1998: Components of the Norwegian cyclone model: Observations and theoretical ideas in Europe prior to 1920, *The Life Cycles of Extratropical Cyclones*, M. A. Shapiro and S. Grönås, Eds., Amer. Meteor. Soc., 15–28.

Wernli, J. H., 1995: Lagrangian perspective of extratropical cyclogenesis. Dissertation No. 11016, Swiss Federal Institute of Technology (ETH), Zurich.

Yeh, T.-C., 1949: On energy dispersion in the atmosphere. *J. Atmos. Sci.*, **6**, 1–16.

Zou, X, Y.-H. Kuo, and S. Low-Nam, 1998: Medium-range prediction of an extratropical oceanic cyclone: Impact of initial state. *Mon. Wea. Rev.* (in press).

Observed Cyclone Life Cycles

LANCE F. BOSART

Department of Earth and Atmospheric Sciences, The University at Albany/State University of New York, Albany, New York , USA

1. Introduction

The purpose of this chapter is twofold: (1) to provide an overview of observed cyclone and anticyclone behavior and life cycles, and (2) to offer a perspective on current and future research directions pertaining to the knowledge and understanding of extratropical cyclones and anticyclones. The reader is cautioned from the outset to appreciate the bias that inherently must accompany any such effort because the author's perspective on the field inevitably must differ from the reader's. Also, the author will always be open to the charge of "rewriting history" and/or misinterpreting or misunderstanding the work of others through errors of omission and commission. Although the author pleads *nolo contendere* to all such charges in advance, he will proceed accordingly under the assumption that readers of this paper will easily recognize his biases and mistakes. The review is necessarily restrictive. Important topics such as (1) numerical simulation studies of extratropical cyclogenesis processes, (2) life cycle studies of large-scale cyclogenesis processes and low-frequency phenomenon (including blocking), and (3) mesoscale convective processes and frontal life-cycle studies are mentioned only briefly as they are addressed in more detail elsewhere in this book. The interested reader is also urged to consult the excellent overview papers by Keyser and Uccellini (1987), Anthes (1990), Bengtsson (1990), Browning (1990), Eliassen (1990), Newton (1990), Reed (1990), Shapiro and Keyser (1990), and Uccellini (1990) for more detailed information on these and related subjects.

The paper begins with a brief description of the Norwegian Cyclone Model (NCM) as presented by Bjerknes (1919) and Bjerknes and Solberg (1922) in Section 2. Mention is made of how the invention of the radiosonde created new scientific opportunities to study the three-dimensional structure of cyclones and anticyclones and the middle-latitude westerly currents. This is followed in Section 3 by a synopsis of the results from classical and modern storm track analyses before a discussion of self-development ideas that antici-

pated quasigeostrophic theory in Section 4. The three-dimensional structure of cyclones and anticyclones is discussed in Section 5 and is followed in Section 6 with a review of air mass exchange concepts from an isentropic perspective. A spectrum of results from studies of observed explosive cyclogenesis events is presented in Section 7 and is followed by a description of planetary scale flow–orographic interactions in Section 8. A potential vorticity perspective on cyclone development is given in Section 9. Conceptualized models of idealized cyclones are compared with observations in Section 10. Trough fracture and trough merger and cyclone development are discussed in Section 11. Downstream development and cyclogenesis as an initial value problem is contained in Section 12 along with a discussion of scientific issues and opportunities.

2. The Norwegian Cyclone Model and Early Aerological Studies

Viewed from a historical perspective, the study of synoptic meteorology evidently was energized by the intentional and widespread installation of the telegraph in the middle of the nineteenth century. The deployment of this new technology made it possible to "see" the evolution and movement of often complex surface weather systems on continental and regional scales for the first time and provided the basis for the first comprehensive attacks on the structure of the cyclone. Given that the overwhelming majority of available observations were surface based, it is quite understandable that the early pioneers in observational meteorology approached the study of surface cyclones and anticyclones with special emphasis on documenting the evolution of temperature, wind, and pressure fields associated with the movement of the individual cyclone and anticyclone centers. In this mostly descriptive initial research, cyclones and anticyclones tended to be viewed as objects that were carried along with the air currents. The view of cyclones as objects was challenged in the seminal paper by Shaw and Lempfert

(1906), who argued that cyclone development was part of a process.

The cyclone-as-a-process viewpoint of Shaw and Lempfert (1906) was advanced, quantified, and polished in Bergen, Norway, where T. Bergeron, J. Bjerknes, and H. Solberg (see, e.g., Bjerknes 1919; Bjerknes and Solberg 1922; and Bergeron 1937, 1959) and colleagues assembled numerous synoptic observations of pressure, temperature and wind in cyclones approaching and crossing northwestern Europe and vicinity. They used the observations to deduce not only the existence of cold and warm fronts but to synthesize a complete cyclone life cycle, including the evolution of clouds and precipitation along the major frontal zones, by means of the idealized NCM. An excellent historical summary of developments during this period is contained in Namias (1983) and Reed (1990). One truly innovative aspect of the NCM was the incorporation of simplified air mass concepts ("air mass thinking") to explain the time-dependent behavior of warm and cold currents that became intertwined in the developing cyclonic circulation. An appreciation of the importance of viewing the cyclone as a fully three-dimensional entity arose from this innovation. Indeed, Shaw (1923) makes very clear that the Bergen school meteorologists were very successful in cementing the viewpoint that a cyclone is a process and not an object.

The early aerological studies in the 1920s and 1930s (see, e.g., the classical papers by Palmén 1931 and Bjerknes and Palmén 1937) that made use of temperature observations from kite- and/or balloon-borne instruments provided the first confirmatory evidence that the highly idealized conceptual models of cyclone life cycles of the Bergen School had a basis in fact. The invention and perfection of the radiosonde in the late 1930s and 1940s and its widespread operational deployment around the world in the late 1940s and early 1950s spearheaded a range of synoptic studies on cyclone behavior and cyclone life cycles (see, e.g., the papers by Palmén 1951a,b). Not surprisingly, the rapid progress in observational capability stimulated theoretical advances in dynamic meteorology. An important example of this process was the paper by Bjerknes (1937) in which he laid the theoretical groundwork (later updated in Bjerknes and Holmboe 1944) for understanding the generation of the important ascending and descending currents in cyclones on the basis of upper- and lower-level horizontal divergence patterns deduced from variations in the gradient wind along the flow direction. Undoubtedly these observational and theoretical advances played a very important role in stimulating Rossby's pioneering theoretical work on planetary wave dynamics (see, e.g., Rossby 1939, 1940) and the equally significant theoretical work by Charney (1947) and Eady (1949) ("normal mode thinking") to explain midlatitude cyclone evolution out of which came an appreciation of the crucial dynamical role played by the Earth's surface and the "quasi-rigid" tropopause in the process.

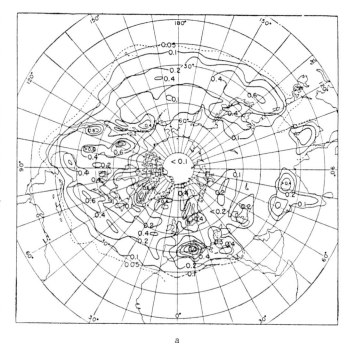

a

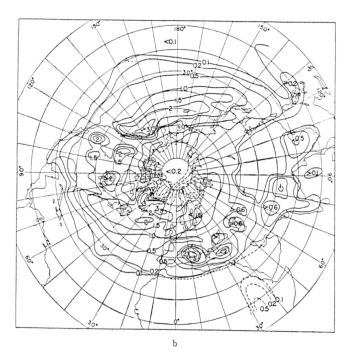

b

Fig. 1. (a) Percentage frequency of occurrence of cyclogenesis in squares of 100,000 km² in winter (1899 to 1939) (Petterssen 1956, Fig. 13.6.1). (b) Percentage frequency of cyclone centers in squares 100,000 km² in winter (1899 to 1939) (Petterssen 1956, Fig. 13.6.2).

3. Classical Versus Modern Storm Track Analysis

Work commenced at the University of Chicago in the late 1940s and 1950s to map the spatial distribution of extratropical cyclones and anticyclones around the Northern Hemisphere. The research was motivated by a desire to understand how topographic variations and land-sea thermal contrasts modulated the observed hemispheric distribution of cyclones and anticyclones. It was also anticipated that such a mapping would yield new insight on regional weather patterns. The database consisted of once-daily (typically 1200 UTC) manually prepared sea-level pressure analyses for the period 1899–1939. The resulting 40-year hemispheric dataset has been used by a number of investigators, most prominently Petterssen (1956) and Klein (1957, 1958), to construct a series of climatological analyses and atlases. Figure 1, taken from Petterssen (1956), shows his winter maps of cyclone frequency and frequency of cyclogenesis. The main cyclogenetic regions are located: (1) over the western Atlantic and Pacific ocean basins and extreme eastern portions of North America and Asia, (2) immediately downstream of major mountain barriers such as the Rockies and Alps, and (3) over smaller, relatively warm water bodies such as the northwest Gulf of Mexico and the west-central and eastern Mediterranean Sea (cyclones subject to orographic modification by the Alps and Turkish Plateau are found in this area as well). Similar patterns are seen in the summer with the appropriate poleward shift. The corresponding cyclone frequency maps show especially pronounced maxima downstream of the major mountain barriers, suggestive of the frequent occurrence of terrain-tied disturbances to the lee of these barriers. Comparison of maps of cyclone center frequency and frequency of occurrence of cyclogenesis shows that in general the regions of maximum cyclone frequency tend to lie downstream and somewhat poleward of the regions of maximum occurrence of cyclogenesis over the western Atlantic and Pacific ocean basins. This behavior is indicative of the downstream (downshear) development of mobile cyclones away from their genesis regions. Exceptions to this pattern are noted over the Mediterranean Sea (and selected other regions) in the vicinity of the Alps where the orographic trapping of cyclones, once formed, is more likely.

The corresponding anticyclone frequency and frequency of anticyclogenesis maps (not shown) clearly reveal the preference for anticyclone centers to be found over the subtropical oceans in all seasons and over higher latitude regions of the continents in winter. Given that anticyclonic vorticity is destroyed by sensible heating in the boundary layer, it is no accident that anticyclone centers tend to avoid warm inland water bodies and bays in winter with the reverse being true in summer. Subsequent investigations of Northern Hemisphere cyclone and anticyclone frequencies and the occurrence of cyclogenesis, and anticyclogenesis (see, e.g., Zishka and Smith 1980; Whittaker and Horn 1981, 1984; Hanson and Long 1985; and Chen et al. 1991, 1992) have confirmed and expanded upon the earlier findings of Petterssen and Klein. As an example, Fig. 2, taken from Whittaker and Horn (1981), shows a latitude-time plot of the frequency of North American cyclogenesis. Noteworthy is the pronounced cyclogenesis asymmetry in the transition seasons produced

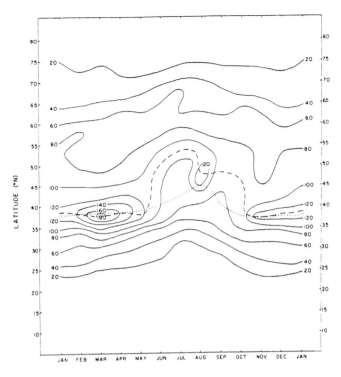

FIG. 2. Isopleths of the combined latitudinal and seasonal distribution of cyclogenesis for the North American sector. Dashed line indicates latitude of maximum activity; dotted line from Klein (1957) indicates maximum activity for Northern Hemisphere (Whittaker and Horn 1981, Fig. 5).

by the robust late February to early April maximum in the 35–40°N belt. This maximum is likely associated with the frequently observed cyclogenesis to the lee of the Rockies that is characteristic of the spring season when the westerlies are still strong at relatively low latitudes and the static stability is reduced due to the rapid increase in insolation, a case study example of which appears in Keshishian et al. (1994).

Similar studies in the Southern Hemisphere have been lacking because of the obvious data limitations, especially with regard to the observed behavior of fronts. Noteworthy among the published investigations in the Southern Hemisphere are the papers by Palmer (1942), Clarke (1954), Taljaard et al. (1961), Taljaard (1972), Gan and Rao (1991), Sinclair (1994, 1995, 1996) and the recent satellite-based climatologies of Carleton (1981, 1985) and Satyamurty et al. (1990). Although orographic cyclogenetic signatures are found downstream of the Andes over eastern coastal South America and the Southern Alps in New Zealand, overall cyclone frequencies tend to be more zonally homogeneous with less variation between summer and winter in comparison to the Northern Hemisphere.

The classical definition of storm tracks as used by, for example, Klein (1957, 1958) refers to the locus of paths followed by individual sea-level cyclones with no regard to the tracks of the flanking sea-level anticyclones or the mobile short wave troughs and ridges aloft. Blackmon (1976) proposed an alternative definition of the storm track based upon the location of variance maxima in the fields of geopotential height in the middle and upper troposphere arising from transient disturbances with periods less than about one week. The advantage of Blackmon's definition is that storm tracks are readily interpretable as variance maxima of both cyclonic and anticyclonic geopotential height perturbations and his results reveal that the prominent storm tracks in the Northern Hemisphere extend from the east coasts of Asia and North America eastward across the Atlantic and Pacific ocean basins, respectively. In general, the region of maximum sea level pressure variance lies equatorward (poleward) of the geopotential height variance maxima aloft in the storm track entrance (exit) regions, indicative that developing cyclones typically form preferentially equatorward of the principal polar jet stream and move to the poleward side of the jet stream as they mature and die. This behavior of cyclones is entirely consistent with expectations from quasigeostrophic theory in that deepening cyclones must be accompanied by ascent (driven largely by the upward increase of cyclonic vorticity advection; see, e.g., Bluestein 1992, 1993; and Holton 1992), low-level convergence and resultant cyclonic vorticity generation, and adiabatic cooling in the lower half of the troposphere over the cyclone center.

Typically, the sea-level pressure variance maxima over the oceans lies somewhat poleward of the classical storm track positions (compare Petterssen 1956 and Wallace et al. 1988), given that geopotential height falls and rises aloft are more prominent north of the jet stream axis and north of the surface cyclone and anticyclone positions. A similar relationship is also noted for the Southern Hemisphere storm tracks (see, e.g., Trenberth 1981, 1991; especially Fig. 1 in the latter paper; and Berbery and Vera 1996). Despite the widespread acceptance of Blackmon's (1976) modernized definition of storm track, it is important to remember that the classical storm track definition is still of considerable use to climatologists and weather forecasters alike because areas of clouds and precipitation ("weather") are most likely to be concentrated downshear of cyclones and upshear of anticyclones. As a summary to this discussion, Fig. 3 presents maps of the idealized baroclinic waveguides (storm tracks) at 1000 and 500 hPa as taken from Fig. 19 of Wallace et al. (1988). They define baroclinic waveguides on the basis of: (1) teleconnectivity (a measure of maximum negative variance correlation at one grid point with a distant grid point for a variable such as the geostrophic streamfunction), (2) the swath of observed geopotential streamfunction variance maxima, and (3) lag-correlation maps of high frequency streamfunction fluctuations. Noteworthy in Fig. 3 is that the 1000 hPa waveguides are considerably more amplified than the 500 hPa waveguides in the vicinity of the Rockies and the Tibetan Plateau, indicative of surface disturbances moving poleward (equatorward) relative to the 500 hPa flow west (east) of major mountain barriers.

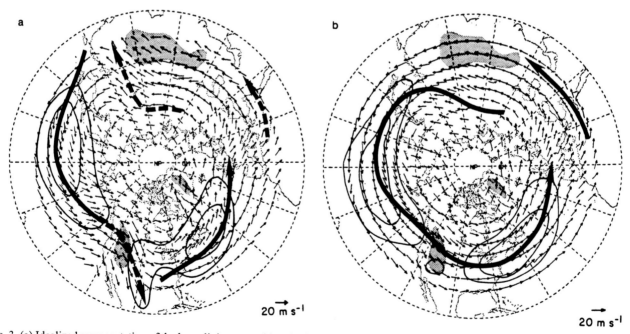

Fig. 3. (a) Idealized representation of the baroclinic waveguides: the heavy arrows correspond to the axes of the bands of high teleconnectivity, where the high-frequency fluctuations are most wavelike; the closed loops correspond to the 40 m (outer) and 50 m (inner) contours transcribed from the standard deviation of the highpass-filtered 1000 hPa height field; and the arrows are the 1000 hPa phase propagation vectors (scale at lower right). The dashed heavy arrows denote the less-pronounced portions of the baroclinic waveguides (Figure 19a of Wallace et al. 1988). (b) As in Fig. 3a but for the highpass-filtered 500 hPa height field. The closed loops represent the 50 m (outer) and 60 m (inner) contours (Wallace et al. 1988, Fig. 19b).

Comparable studies of closed cyclones and anticyclones in the middle troposphere, although not as abundant as the studies of surface systems mentioned above, have been conducted for the Northern Hemisphere by Parker et al. (1989) and Bell and Bosart (1989). Closed cyclones frequently occur in the storm track entrance regions of eastern North America and eastern Asia poleward of the main baroclinic zone and over southwestern North America and southern Europe (in the vicinity of the Alps) equatorward of the main baroclinic zone. In these latter regions lysis and genesis maxima nearly coincide, suggesting that the life cycles of the closed cyclones are strongly influenced by topography. Individual case and composite studies (see, e.g., Colucci 1985; Colucci and Davenport 1987; Bell and Keyser 1993; Bell and Bosart 1993, 1994) suggest that closed cyclone formation equatorward of the polar jet is strongly associated with upstream anticyclogenesis and a shortening of the distance between the upstream ridge and the downstream trough in the northwesterly flow whereas closed cyclone formation poleward of the polar jet is related to vigorous planetary-scale downstream ridge amplification.

It is also interesting that both regions of closed cyclone frequency maxima equatorward of the polar jet occur downstream of regions of large 500 hPa geopotential height variability (Wallace et al. 1988, their Fig. 1) and in an area where the baroclinic waveguide tends to shift equatorward (Fig. 3). This shift is particularly robust in the case of southern Europe where the waveguide appears "fractured" (meridional discontinuity) over a 60 degree longitude band. A similar shift would probably also be apparent over southwestern North America at 300 hPa and 200 hPa, given that the subtropical jet entrance region is frequently in this location during the cooler half of the year. It is hypothesized that "seams" in the baroclinic waveguides downstream of geopotential height variance maxima are particularly well conducive to closed cyclone formation as ridge development occurs poleward and immediately upstream of equatorward cyclone development in a region of large-scale flow deformation coincident with the waveguide seams. Put another way, cutoff cyclones and anticyclones are more likely because the mean flow is weaker in these regions.

4. Self-Development: The Road to Quasigeostrophic Theory

An observationally based research effort was conducted by Sutcliffe and collaborators (see, e.g., Sutcliffe 1939, 1947; Durst and Sutcliffe 1938; Sutcliffe and Forsdyke 1950) in the United Kingdom to diagnose and quantify the dynamical basis of cyclone life cycles. They were motivated by the desire to understand (and eventually forecast) the vertical structure and associated cloud and precipitation patterns in cyclones. A cornerstone of this approach was the use of the concept of thermal vorticity and the advection of vorticity by the thermal wind (Sutcliffe and Forsdyke 1950) to deduce atmospheric thermal (thickness) structures and to diagnose the "development" (defined as the difference between the upper-level and lower-level divergence) of cyclones and anticyclones. As an example, Fig. 4 (taken from Fig. 24 of Sutcliffe and Forsdyke 1950) shows very nicely a schematic "thickness jet" model in which the authors indicate that ascent (descent) is favored in the equatorward (poleward) confluent entrance and the poleward (equatorward) diffluent exit regions of the thickness jet. This work, in which the presence of synoptic scale ascending and descending vertical motion is linked to the advection of vorticity by the thermal wind, might (arguably) be called the "invention" of the quasigeostrophic theory. (The observationally based approach of Sutcliffe and collaborators to the development of quasigeostrophic theory occurred mostly independently of the much more well known parallel theoretical effort by, for example, Charney (1947), Eady (1949), Charney and Stern (1962) and Phillips (1963) that culminated in the formal development of the quasigeostrophic theory.) A review of the theoretical ideas governing extratropical cyclone development, including quasigeostrophic theory, can be found in Hoskins (1990). The work of Sutcliffe and colleagues also anticipated a similar description of four-cell vertical motion patterns in a straight jet model as put forth by Beebe and Bates (1955) in a study of the release of convective instability and used subsequently by many authors to help understand the dynamical importance of jet stream configurations to the evolution of vertical circulations associated with cyclones and

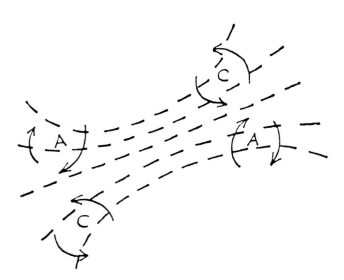

FIG. 4. The thermal jet complex. The development regions favor the existence of depressions in the C regions and anticyclones in the A regions, which tend to distort the thermal pattern by advection and maintain or develop the jet structure. This is the hyperbolic frontogenetic complex which is self-developing. (Sutcliffe and Forsdyke 1950, Fig. 24).

anticyclones and convective weather disturbances. It should also be remarked that Sutcliffe's advance was motivated by the practical necessity of providing improved short-term (one-to-two day) weather forecasts in support of military operations during World War II. In this regard he was one of the truly outstanding early pioneers because of his insightful recognition of the importance of linking the needs of the operational community to progress in understanding the dynamical basis of observed cyclone structures and cyclone life cycles.

The work by Sutcliffe and colleagues stimulated Petterssen and collaborators (see, e.g., Petterssen 1955; Petterssen et al. 1962; Petterssen and Smebye 1971) to expand and quantify the "self-development" concept (cyclone development at sea level begins wherever and whenever an area of cyclonic vorticity advection aloft overspreads a surface baroclinic zone) to a range of continental and oceanic cyclone investigations that included the application of the quasigeostrophic omega equation to diagnose the vertical motion field. Petterssen et al. (1962) and Petterssen and Smebye (1971) also introduced the concept of dual cyclone development paths which they called type "A" and type "B" disturbances. The type A category was assigned to amplifying frontal wave disturbances in which atmospheric development appeared to originate from the bottom upward with little, if any, evidence for a predecessor disturbance aloft. Type B disturbances were characterized by a well defined predecessor vorticity maximum aloft that triggered subsequent surface cyclogenesis as it crossed an old low-level baroclinic zone. The problem with this categorization scheme is that the evidence for type A cyclone development has been tenuous at best as there almost always appears to be some incipient disturbance aloft that can be related to the surface development (see, e.g., Sanders 1986). As evidence for this statement, note that the example used by Petterssen and Smebye (1971) as somewhat illustrative of a type A development had a type B signature as well (see their Figs. 15–16). Similarly, Petterssen et al. (1962) wrote about oceanic cyclone development under a relatively straight and featureless 300 hPa flow (their Fig. 12), but a careful inspection of their figure reveals the likely presence of geostrophic shear vorticity in their "uniform" current aloft. It is quite conceivable that had the relatively abundant jet aircraft observations of today been available for analysis of their cases, the authors might have drawn a different interpretation about featureless 300 hPa geopotential features because the characteristic shear vorticity couplet signature in the vicinity of jet streams would have been better revealed. Instead of attempting to distinguish between type A versus type B cyclonic developments, perhaps it would be more revealing to recognize that type B developments come in many "flavors" representing a full spectrum of predecessor disturbances.

5. Three-Dimensional Structure of Cyclones and Anticyclones

An important aspect of the radiosonde era that began in the late 1940s and continued into the 1960s was the proliferation of synoptic studies designed to elucidate the three-dimensional structure of troughs and ridges in the middle and upper troposphere and their role in controlling the behavior of mobile cyclones and anticyclones. Studies by, for example, Palmén (1949), Hsieh (1949), Peltonen (1963), Omoto (1966), and Krishnamurti (1968) were central to uncovering the life cycle of cutoff cyclones and anticyclones and related features such as cold air outbreaks into lower latitudes associated with the formation of a cold cutoff cyclone. Palmén (1949; see his Fig. 4) showed that the formation of a cutoff cyclone is associated with a deepening diffluent trough in which the largest cyclonic vorticity (mostly in the form of shear) lies upstream of the trough axis in a strong jet stream. As cyclonic vorticity accumulates in the trough axis in association with cold air advection into the trough axis below 500 hPa, the midtropospheric trough intensifies. (Newton (1990) and Newton and Newton (1994) provide illustrative examples from several Palmén papers of trough intensification and cutoff cyclone development from a three-dimensional viewpoint.) Characteristic cutoff cyclone signatures include: (1) a locally depressed tropopause coincident with a dome of cold air (uplifted isentropic surfaces) in the troposphere, and (2) a relatively warm and depressed stratosphere above the tropospheric cold dome. The resultant cyclonic circulation, strongest near the tropopause, follows immediately by thermal wind considerations. A schematic illustration of surface cyclogenesis associated with the arrival of an upper-level disturbance over a low-level baroclinic zone that results in 500 hPa cutoff cyclone development is shown in Fig. 5, taken from Fig. 11.3 of Palmén and Newton (1969). The evolution of the surface and upper-level patterns is consistent with the self-development concepts discussed in Section 4. Alternatively, the cutoff cyclone can be viewed as a positive potential vorticity (PV) anomaly that has broken off from the main PV reservoir farther poleward (see, e.g., Kleinschmidt 1950, 1957; Hoskins et al. 1985; Thorpe 1985) with the reverse situation being true for cutoff anticyclones. This observation, the essence of "PV thinking" as quantified by Hoskins et al. (1985), will be revisited in Section 9 in conjunction with specific examples of cyclone evolution.

Krishnamurti (1968) first employed a general balance equation model to diagnose a case of extratropical cyclogenesis of the type previously studied by Palmén and colleagues. He chose a case of a deepening diffluent trough in northwesterly flow over North America east of the Rockies for his case study. Cyclogenesis originated in a lee trough east of the Rockies beneath northwesterly flow aloft. As this intensifying trough crossed the Rockies the surface cyclone began to deepen substantially as it moved away from the mountains

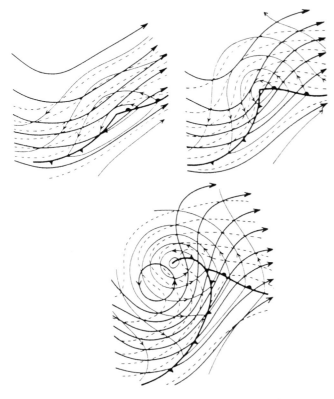

Fɪɢ. 5. Schematic 500 hPa contours (heavy solid lines), 1000 hPa contours (thin lines) and 1000-500 hPa thickness (dashed lines) illustrating the self-development process (Palmén and Newton 1969, Fig. 11.3).

and became situated beneath the forward side (east of the trough axis) of the trough. Krishnamurti (1968) showed how differential sinking motion in the northwesterly flow with the largest subsidence toward the warmer air at mid levels of the atmosphere was crucial to the generation of large values of cyclonic vorticity on the cyclonic shear side of the jet at upper levels. The vorticity thus generated was advected southeastward into the base of the intensifying trough, eventually leading to the formation of a closed cyclone center. An important result from his work was that much of the observed cross-contour flow toward higher heights in the upper troposphere in association with the cyclogenesis could be attributed to the nondivergent wind.

In October 1954 Hurricane Hazel devastated the Carolinas region of the United States. Subsequent to landfall the storm became situated beneath the forward side of a very strong large-amplitude baroclinic trough approaching from the west. Hazel was transformed into an intense extratropical cyclone that moved poleward and westward and resulted in widespread damage from high winds and flooding over portions of the Appalachian mountains and southern Canada. In a seminal scientific contribution Palmén (1958) provided evidence that the transformation of Hurricane Hazel into an extratropical storm was strongly related to the vertical circulations associated with the advancing synoptic-scale trough,

and he speculated that the substantial amount of latent heat release in the tropical air that was swept into the larger-scale circulation must have been crucial to the extraordinary intensity of the transformed cyclone. Anthes (1990) provided confirmation of Palmén's speculation in a numerical simulation of this historic storm. Without latent heat release the cyclone central pressure in the simulated storm (Anthes 1990, Table 12.4) was approximately 25 hPa higher than in the run incorporating latent heat release.

Palmén (1958) also provided the primary motivation for the author and one of his former graduate students, Geoffrey DiMego, to research the transformation of Hurricane Agnes into an extratropical storm over the eastern United States in June 1972. DiMego and Bosart (1982a,b) showed that the ambient cyclonic vorticity and deep moisture surrounding Agnes subsequent to landfall was crucial to the transformation and regeneration process when it was acted upon by strong low-level convergence beneath the updraft region of an unusually strong polar trough advancing from the west. A comparable Southern Hemisphere investigation by Sinclair (1993) showed that ascent in the equatorward entrance region of a propagating subtropical jet helped to sustain the transformation of a tropical cyclone to an extratropical disturbance as tropical moisture was swept poleward beneath the updraft region.

Palmén (1958) also used the Hazel case to calculate kinetic energy and water vapor budgets from which he deduced that extratropical cyclones must play a very important role in the general circulation of the atmosphere. He was particularly intrigued by the generation of a very strong jet stream on the forward side of the aforementioned large-amplitude trough in the 400–200 hPa layer. The existence of this jet stream was deduced from the comparatively few available radiosonde wind observations in the upper troposphere (many stations lost their balloons in the strong winds) combined with geostrophic wind estimates inferred from the more numerous mass field observations. As noted by Palmén (1958), "the kinetic energy produced by the sinking cold air and ascending warm air in the extratropical cyclone Hazel was to a very large extent available for export out of the source region. This export occurred . . . between 400 and 200 hPa in the form of an upper jet stream leaving the active disturbance." Palmén's (1958) demonstration of the outward kinetic energy flux in the upper troposphere away from Hazel and its importance in the maintenance of the upper-level westerlies in middle latitudes is also completely consistent with the recent renewed theoretical interest in the phenomena of downstream development (see, e.g., Orlanski and Chang 1993, 1995; Chang 1993; Lee and Held 1993.

In another landmark scientific contribution, Newton and Palmén (1963) documented the kinematic structure of an intense large-amplitude trough over eastern North America and an equally impressive large-amplitude ridge over the central Atlantic ocean. The author remembers being fascinated by this paper when he read it as a graduate student because the real data example clearly illustrated the important differences between the geostrophic and gradient winds in the upper troposphere, particularly in the Atlantic ridge where the flow aloft was supergeostrophic but consistent with gradient wind balance limitations in a quasi-stationary large-amplitude ridge. Equally noteworthy was the presence of a geostrophic vorticity maxima in the base of the 500 hPa trough while the observed vorticity maxima first appeared on the back side (west of the trough axis) of the trough and then shifted to the forward side of the trough as a transient disturbance rounded the base of the long wave trough. The Newton and Palmén (1963) case study was especially important in illustrating that jet streaks, consisting mostly of shear vorticity in the northwesterly flow behind the trough axis, typically reform on the forward side of the trough as opposed to simple advection around the base of the long wave trough. Taken together, the innovative aspects of the Palmén (1958), Newton and Palmén (1963), and Krishnamurti (1968) papers lay in their emphasis on quantitatively diagnosing the three-dimensional airflow through developing extratropical cyclones and the budget of kinetic energy by a variety of suitable means.

6. The Isentropic View and Air Mass Exchange Concepts

The emphasis on examining the evolution of cyclones in three dimensions as cited above was also at the core of viewing development in isentropic coordinates as originally pioneered by Rossby and collaborators (1937a,b), Montgomery (1937), Wexler and Namias (1938), and Namias (1938, 1939). A crucial assumption in most of these papers is that time-to-space conversion techniques can be used to "create" observations at intermediate times between radiosonde launches to facilitate the preparation of isentropic trajectories necessary for the Lagrangian perspective on storm development. A clear weakness of this approach, however, is that the evolution of highly time-dependent secondary circulations associated with rapidly developing (or weakening) weather systems may be masked by the required steady-state assumption for atmospheric flow between the times of available radiosonde datasets. However, in the comparatively few case studies of nondeepening cyclones (see, e.g., Iskenderian 1988) in the literature, the time-to-space conversion technique has been useful in understanding the mechanism of precipitation growth and concentration near the cyclone center and along and ahead of the surface warm front.

The use of isentropic analysis as a tool to uncover the time-dependent behavior of air movement through middle latitude cyclones received an unexpectedly large boost in the 1960s from atomic bomb fallout studies. For example,

Danielsen and colleagues (Danielsen 1964, 1966, 1967, 1968; Danielsen and Bleck 1967), whose early research was supported by the Atomic Energy Commission, and Shapiro (1974, 1980) pioneered the use of research aircraft observations and conventional datasets to study the structure of middle-latitude cyclones and upper-level fronts using quasi-conservative tracers such as PV, ozone, and selected radioactive isotopes to map air movement on isentropic surfaces. These authors took advantage of earlier work by Reed and Sanders (1953), Newton (1954), Reed (1955), and Staley (1960), among others, that had established the general validity and significance of using PV conservation as a diagnostic tool in synoptic-dynamic studies of upper-level frontogenesis.

To some extent Danielsen's (1964, 1966) scientific contribution to the understanding of airflow through cyclonic disturbances was somewhat undervalued because he published many of his detailed findings in unrefereed scientific reports. Danielsen used his superb manual analysis skills to combine special radiosonde datasets with aircraft observations to prepare vertically consistent three-dimensional analyses of meteorological fields on isentropic surfaces for selected cyclonic disturbances in the early 1960s. These analyses were then manually gridded and digitized for the purpose of diagnosing quantities such as PV. Isentropic trajectories were computed manually using the methodology described in Danielsen (1961). The care in which these analyses were prepared enabled Danielsen to produce a reliable set of isentropic trajectories at 6 or 12 h intervals over 36–48 h periods that were noteworthy for quasi conservatism of PV (Starr and Neiburger 1940), ozone, and radioactive isotope concentration. Danielsen's trajectories provided convincing evidence that air parcels in the upper troposphere and lower stratosphere in the northwesterly flow behind the cyclone descended abruptly with time-averaged vertical motions of upwards of $10\ cm\ s^{-1}$ such that stratospheric air was extruded deep into the lower and middle troposphere. Figure 6, taken from Fig. 6 of Danielsen (1964), illustrates this process in a schematic diagram of the three-dimensional air movement through a developing midlatitude cyclone.

A careful inspection of the plotted maps and analyses in Danielsen (1964, 1966) also reveals that the quasi-PV conservation along the isentropic trajectories enabled him to deduce the existence of the "dry slot" and the characteristic comma cloud shape in cyclonic storms of middle latitudes prior to the widespread availability of satellite imagery that showed such signatures to be commonplace. Furthermore, Danielsen's analyses established that the air parcels comprising the dry slot were very dry because of a previous history of subsidence, often from the lower stratosphere, and that individual air parcels near the tip of the dry slot were beginning to ascend again on the forward side of the trough. Danielsen noted how this dry air could contribute to the generation of deep conditional instability as it was swept

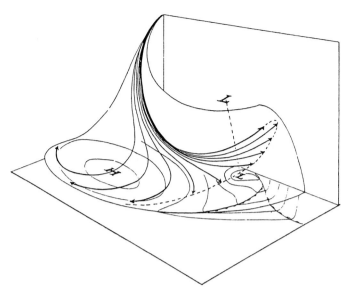

FIG. 6. Trajectories of extruded stratospheric air (Danielsen 1964, Fig. 6).

downstream above the deep warm front cloud (see also Carr and Millard 1985).

7. Explosive Cyclogenesis

A renewed interest in scientific issues related to the cyclogenesis process was sparked by the seminal paper of Sanders and Gyakum (1980) on the synoptic-dynamic climatology of the "bomb" (explosively deepening cyclone). They found that explosively deepening cyclones are primarily a cool season marine phenomenon, are generally confined to the western portions of the Atlantic and Pacific ocean basins, occur preferentially in the vicinity of the Gulf Stream and the Kuroshio Current, and are highly baroclinic systems that are subject to potentially large amounts of diabatic heating. Confirmation and extension of the Sanders and Gyakum (1980) climatology have come in subsequent studies by, for example, Roebber (1984), Rogers and Bosart (1986), Gyakum et al. (1989), Roebber (1989a), and Chen et al. (1992).

The degree of modulation of cyclogenesis by latent heat release has been debated ever since Espy (1841) stressed the importance of latent heat release in cyclones and Margules (1903) advanced the concept of available potential energy. Palmén and Holopainen (1962) stressed the importance of latent heat release in augmenting the intensity of vertical circulations accompanying cyclogenesis. Danard (1964) made a similar point and he emphasized that the impact of latent heat release in augmenting dry adiabatic vertical motions was a function of the environmental static stability.

Johnson (1970), building on the modern concept of available potential energy advanced by Lorenz (1955), constructed an isentropic framework for quantitatively evaluat-

ing available potential energy. His results showed the importance of diabatic heating in general, and latent heat release in particular, as a local source of available potential energy in the presence of enhanced baroclinic zones (e.g., jet streak regions). Krishnamurti (1968) and Johnson and Downey (1976) diagnosed how widespread latent heat release associated with cyclogenesis accelerated surface development and resulted in the formation of a deep vortex in the mid-troposphere. Tracton (1973) hypothesized that one possible reason that early operational NMC models frequently failed to predict the timing and intensity of cyclogenesis when deep convection was present near the cyclone center was that they were unable to replicate the bulk effects of cumulus convection. Tracton further hypothesized that latent heat release associated with deep convection near the cyclone center acted to intensify the ridge aloft over and downstream of the cyclone center. As a result he argued that cyclonic vorticity advection was augmented over the cyclone center immediately ahead of the advancing upstream trough aloft, a view consistent with the original self-development ideas of Sutcliffe and Forsdyke (1950).

The publication of the Sanders and Gyakum (1980) climatology also triggered an avalanche of explosive cyclogenesis studies that included such "infamous" storms as, for example, the Cleveland Superbomb (CSB) of January 1978 (see, e.g., Burrows et al. 1979; Salmon and Smith 1980; and Hakim et al. 1995, 1996); the QEII storm of September 1978 (see, e.g., Anthes et al. 1983; Gyakum 1983a,b; Uccellini 1986; Gyakum 1986, 1991; Manobianco et al. 1992); the Presidents' Day storm of February 1979 (see, e.g., Bosart 1981; Bosart and Lin 1984; Uccellini et al. 1984, 1985, 1987; Whittaker et al. 1988); the eastern Pacific storm of November 1981 (see, e.g., Reed and Albright 1986; Kuo and Reed 1988); the Ocean Ranger storm of February 1982 (see, e.g., Kuo et al. 1991, 1992); the Megalopolitan storm of February 1983 (see, e.g., Sanders and Bosart 1985a,b); the "Great Storm" of October 1987 (see, e.g., Hoskins and Berrisford 1988; Shutts 1990); various ERICA (Experiment on Rapidly Intensifying Cyclones over the Atlantic) storms (see, e.g., Neiman and Shapiro 1993; Neiman et al. 1993; Reed et al. 1993a,b); and the "Storm-of-the-Century" of March 1993 (SS'93) over eastern North America (see, e.g., Caplan 1995; Kocin et al. 1995; Bosart et al. 1996; Schultz et al. 1997; Dickinson et al. 1997). Points of scientific contention in these and other investigations included the issues raised previously as well as questions as to what extent data limitations and uncertainties may have masked attempts to identify the physical processes responsible for the cyclogenesis.

As an example, Petterssen et al. (1962) computed oceanic sensible and latent heat fluxes for a variety of Atlantic cyclone events. They observed that these fluxes maximized in the advancing cold air mass behind the cold front and the surface cyclone and that in general the fluxes acted to damp the lower tropospheric environmental thickness wave, a situation unfavorable for cyclogenesis. The opposite situation occurred in the Presidents' Day storm as Bosart (1981) and Bosart and Lin (1984) demonstrated that oceanic sensible and latent heat fluxes maximized in the southwestern quadrant of an arctic anticyclone. These fluxes were of critical importance in warming, moistening and destabilizing the coastal marine boundary layer, thereby leading to the development of an intense coastal front along which the initial cyclogenesis commenced with explosive development following as a potent trough aloft (PV anomaly) reached the eastern coast of North America. In the view of Bosart (1981) and Bosart and Lin (1984), the antecedent growth of baroclinicity, low-level vorticity, and reduction in atmospheric static stability in the lower troposphere "preconditioned" the atmosphere for future rapid cyclonic vorticity growth as the updraft region and attendant low-level convergence field associated with the advancing upper-level trough reached the coast.

This interpretation of the potential importance of a pre-existing cyclonic vorticity maximum to explosive cyclogenesis is also consistent with the results from DiMego and Bosart (1982a,b) on the reintensification of Hurricane Agnes and its ultimate transformation into an extratropical cyclone and is in agreement with the original ideas on the subject expressed by Palmén and Newton (1969). A similar conclusion was reached by Gyakum (1991), who showed that the growth of a small-scale cyclone along an intense low-level frontal zone provided the antecedent vorticity for later rapid vorticity growth in the QEII storm. More generally, Gyakum et al. (1992), in a study of 794 North Pacific cyclones over nine cold seasons, showed that the "sample of cases with a history of existence prior to maximum deepening is characterized by significantly larger maximum deepening rates," a clear indication of the importance of "preconditioning" as defined above. Similarly, Roebber (1993) argued that the long tail of explosively deepening cyclones in his earlier statistical studies (Roebber, 1984, 1989a) "may be a statistical manifestation of the more robust preconditioning of the cyclone environment in oceanic regions."

The issue of the relative importance of upper-level versus lower-level dynamical and thermodynamical processes (including oceanic sensible and latent heat fluxes) on cyclogenesis has been strenuously debated over the years. Representative papers by, for example, Reed (1979), Hoskins et al. (1985), Sanders (1986, 1987), Tsou et al. (1987), Mullen and Baumhefner (1988), Pauley and Smith (1988), Martin et al. (1989), Roebber (1989b), Nuss and Kamikawa (1990), Shutts (1990), Velden and Mills (1990), Bond and Shapiro (1991), Davis and Emanuel (1991), Kuo et al. (1991a,b), Reed and Simmons (1991), Rogers and Bosart (1991), Gyakum et al. (1992), Lupo et al. (1992), Wash et al. (1992), Bullock and Gyakum (1993), Neiman and Shapiro (1993), Reed et al. (1993a,b), Roebber (1993), Hakim et al. (1995, 1996), Bosart et al. (1995, 1996), Stoelinga (1996), and Dickinson et al. (1997) on these and related topics have helped to both clarify

and sharpen existing scientific controversies and to provide testable scientific hypotheses for reconciling differing viewpoints. Accordingly, the findings from which a scientific consensus may be emerging include: (1) that the role of oceanic heat and moisture fluxes in cyclogenesis in general is time, regional, and life cycle dependent for a wide spectrum of extratropical cyclones (including polar lows and baroclinic developments that transform into warm-core tropical storms), (2) that numerical studies of explosive marine cyclogenesis in particular suggest that the oceanic sensible and latent heat fluxes are most crucial early in the life cycle of the cyclone and that much of the controversy in the literature as to their potential importance could be traced to the choice of model initialization time with respect to the phase of the cyclone life cycle, (3) that latent heat release processes probably dominate dry baroclinic dynamics processes in the most explosive marine cyclogenesis cases and in many weaker frontal waves arising along quasistationary baroclinic zones, (4) that extratropical cyclogenesis needs to be viewed as a nonlinear interaction between ordinary baroclinic dynamics and diabatic processes, and (5) that extratropical cyclones exhibit a rich spectrum of three-dimensional structures that are tied to the configuration and evolution of the synoptic-scale flow and are modified regionally and locally.

8. Planetary-Scale Flow Modulation by Orography

Although not mentioned specifically in the previous paragraph, the configuration of the large-scale flow also figures prominently in the evolution and characteristic cloud-shield signatures of extratropical cyclones and anticyclones. For example, Namias and Clapp (1949) and Sawyer (1956) demonstrated the frontogenetical nature of large-scale confluent flow and the implied generation of cloud and precipitation by ascending motions on the equatorward side of the confluent jet-entrance region. Large-scale confluent flow in the entrance region of the great Asiatic jet stream plays an important role in focusing frontogenesis and small-scale cyclogenesis along east-west oriented bands several thousand kilometers long in association with the Mei-Yu and Baiu rainy seasons in China and Japan, respectively, in spring and early summer (see, e.g., Matsumoto et al. 1971; Ninomiya and Akiyama 1971; Huang et al. 1976; Ninomiya 1984; Chen and Chang 1980; Chen and Chi 1980; Kuo and Anthes 1982; Chen 1983; Kato 1985; Ma and Bosart 1987; Kuo et al. 1988; and Wang et al. 1993). During the cooler half of the year, synoptic disturbances forming within the large-scale confluent flow region over eastern Asia can become quite robust (see, e.g., Chen and Dell'Osso 1987) in response to synoptic-scale forcing mechanisms operating over a comparatively widespread region (Bullock and Gyakum 1993). These large-scale confluent flow patterns are also a prominent feature of the baroclinic waveguides (Fig. 3) of Wallace et al. (1988).

Sanders (1988) constructed a nine-year climatological study of 500 hPa trough birth and death regions around the Northern Hemisphere. Figure 7, taken from Fig. 6 of his paper, suggests that on the average troughs are born (die) preferentially in northwesterly (southwesterly) flow regions downstream (upstream) of prominent mountainous regions. Although Sanders's results are open to challenge (he traced daily trough evolutions on the basis of the configuration/ curvature of the 552 dam geopotential height contour as opposed to tracking individual vorticity centers) because his relatively simple analysis methodology may result in trough "loss" over mountain barriers and "rediscovery" downstream, his findings are in accord with synoptic experience. Frequently one finds large-scale confluence in northwesterly flow in characteristic trough birth regions, as noted by Sanders et al. (1991). More generally, the results from Sanders (1988) clearly establish the long-lived nature of transient disturbances in the middle-latitude westerlies and their importance in triggering surface cyclogenesis, especially in northwesterly flow situations downstream of prominent mountain barriers such as the Rockies (see, e.g., Newton 1956; Hovanec and Horn 1975; Locatelli et al. 1989; Hartjenstein and Bleck 1991; and Keshishian et al. 1994) and Alps (Buzzi and Tibaldi 1978; McGinley 1982; and the review paper by Tibaldi et al. 1990). Finally, Lefevre and Nielsen-Gammon (1995) confirmed the principal findings of Sanders (1988) by means of a superior objective mobile tracking method.

In particular, the initial blocking of low-level cold air by a mountain barrier such as the Alps can lead to a situation where a jet streak crossing the mountain barrier becomes unbalanced (see, e.g., Mattocks and Bleck 1986, their Figs. 1–2) in that there is excess vertical wind shear for the weaker thermal gradient downwind of the barrier so that vigorous deep vertical circulations arise in response to the requirement for appreciable geostrophic adjustment. The mountains can also play an important role in the evolution of frontal structure associated with the developing lee cyclones, so much so that as the cyclone develops and eventually moves away from the mountains the life cycle of the cyclone and its associated fronts through the occlusion stage may depart significantly from that envisioned in the idealized NCM. As noted in Keshishian et al. (1994) the degree of disagreement with the classical NCM is strongly a function of the configuration of the synoptic-scale flow and its orientation with respect to the mountains. This process is illustrated schematically in Fig. 8, taken from Fig. 37 of Keshishian et al. (1994), at two typical stages in the life cycle of a lee cyclone.

9. The Potential Vorticity Perspective

The pioneering use of PV as a quasi-conservative tracer of atmospheric motion on isentropic surfaces (quasiconservative

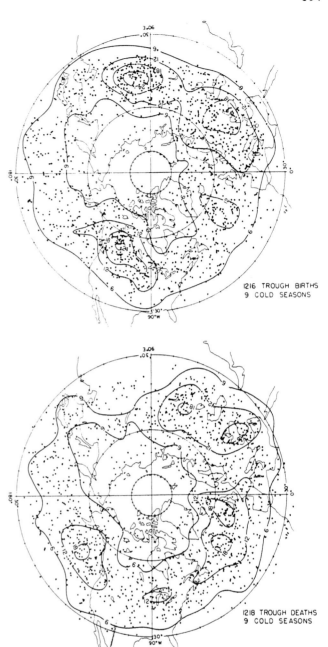

FIG. 7. Positions of mobile trough (a) origins, (b) terminations, and (c) excess of origins over terminations during the nine cold seasons examined. The solid lines are contours of smoothed normalized frequency in areas bounded by 10 degrees of latitude and longitude. See text (Sanders 1988, Fig. 8).

in the sense that data limitations and analysis uncertainties would preclude finding full PV conservation even under pure adiabatic and frictionless conditions) to diagnose life cycles of upper-level fronts and cyclones by, for example, Bleck (1973), Danielsen (1964, 1968), Kleinschmidt (1957), Reed and Sanders (1953), and Staley (1960) among others helped to stimulate the use of PV as a dynamical tool for prognostic and diagnostic research purposes. Bleck (1973, 1974) first

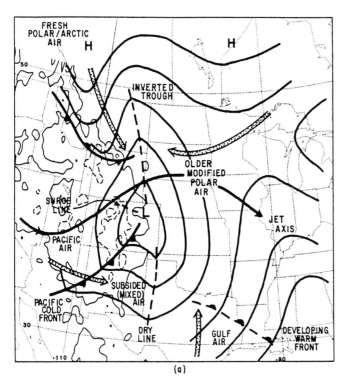

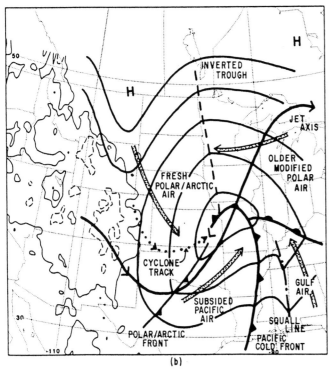

FIG. 8. Schematic of cyclogenesis east of the Rockies when an inverted trough is present for (a) the initial time and (b) some later (~24 h) time. Solid lines denote mean sea level isobars. Principal frontal boundaries are denoted by conventional symbols with the inverted trough appearing as a dashed line. Approximate position of the jet stream is shown by the heavier solid line. Approximate track of the cyclone is shown by the dotted line in (b). The squall line is indicated by a dash-dot line. Estimated surface air trajectories within the labeled air masses are denoted by the hatched arrows (Keshishian et al. 1994, Fig. 37).

conducted numerical forecasting experiments based upon PV conservation to demonstrate that tropopause lowering and folding associated with deepening troughs was a characteristic signature of incipient downstream cyclogenesis. More generally, Hoskins et al. (1985) conceptualized and quantified "PV thinking" as a basis for dynamical studies of atmospheric motions on a wide variety of scales. They accomplished this task by demonstrating that a knowledge of the PV distribution throughout the atmosphere combined with the specification of the boundary potential temperature could be used to deduce unambiguously the three-dimensional mass field in the atmosphere subject to the application of a balanced flow constraint. The publication of this seminal paper (see also the summary paper by Hoskins 1990) more than a decade ago has triggered a new avalanche of PV-based observational, numerical, and theoretical studies, many examples of which can be found elsewhere in the Bergen Symposium preprint volumes, this book, and in the summary of the Cyclone Workshop proceedings held at Val Morin, Quebec, Canada, in October 1992, presented by Bleck et al. (1993).

An attraction of "PV thinking" is that synoptic development can be viewed as the interaction of a predecessor disturbance (PV anomaly) on the dynamical tropopause (defined as a surface of constant PV) with a comparable disturbance near the ground (where a warm region can be viewed as equivalent to a positive PV anomaly on the bottom boundary) within the quasigeostrophic framework. This dynamical interpretation of "PV thinking" is consistent with the original Sutcliffe-Petterssen self-development ideas described in Section 4 and is also in accord with quasigeostrophic theory since the quasigeostrophic height tendency equation is equivalent to the quasigeostrophic potential vorticity equation. It also appeals to many theoreticians and modelers because, subject to a number of restrictions, the governing dynamics of atmospheric motions are "simplified" within the context of the Eady (1949) model in which atmospheric development is interpreted in terms of the mutual interaction of waves propagating along the tropopause and the surface for a uniform PV (typically some small positive value–Northern Hemisphere case) troposphere. The "stampede" to "PV thinking" has not gone unnoticed by some scientists, however. Frederick Sanders has challenged everyone (see Bleck et al. 1993) to demonstrate convincingly what new ideas about cyclone development and cyclone life cycles have been learned from the PV approach that were unknown to Sutcliffe-Petterssen self-development and traditional quasigeostrophic theory aficionados. A viewpoint expressed here is that PV thinking has at least two distinct advantages: (1) comparison of theory with observations is facilitated, and (2) the effects of diabatic and frictional processes can be quantified. Note, however, that despite the physically attractive linkage of the Hoskins et al. (1985) PV inversion technique and partitioning to the dynamics of the Eady model, the partitioning is arbitrary. Other partitionings are possible and there is considerable debate in the literature as to which partitioning technique is "best" (see, e.g., Holopainen and Kaurola 1991).

Illustrative examples of the use of "PV thinking" to conduct diagnostic analyses of cyclone behavior can be found, for example, in papers by Hoskins et al. (1985), Boyle and Bosart (1986), Uccellini et al. (1987), Whittaker et al. (1988), Bleck (1990), Uccellini (1990), Davis and Emanuel (1991), Holopainen and Kaurola (1991), Davis (1992), Reed et al. (1992), Black and Dole (1993), Davis et al. (1993, 1996), Bosart and Lackmann (1995), Hakim et al. (1995, 1996) and Nielsen-Gammon (1995, 1996). Collectively, these and other papers make the point that: (1) predecessor disturbances (positive PV anomalies) are invariably present along the tropopause immediately upstream of the region where the lower-level cyclonic disturbance will form, and (2) the shape, height, and orientation of the tropopause with respect to the surface is of crucial importance to the overall evolution and intensity of cyclones and anticyclones. Hirschberg and Fritsch (1991, their Fig. 25) presented an alternative viewpoint by illustrating schematically how thermal variations along the 200 hPa surface (associated with tropopause undulations) can be related directly to sea level pressure falls and rises accompanying cyclogenesis. More recently, Davis and Emanuel (1991), Davis (1992), Black and Dole (1993), Davis et al. (1993, 1996), and Neilsen-Gammon (1996), among others, have attempted to quantify cyclogenesis on scales from 1000 to 10,000 km by attempting to isolate the dynamical contributions of upper- and lower-level PV anomalies using PV inversion methods. Use of this strategy, patterned after the Eady (1949) problem, enables them to quantitatively compare and contrast the contribution of transient individual tropopause and bottom boundary disturbances to individual cyclone life cycles.

The application of PV thinking to real data cases can be illustrated by means of dynamic tropopause maps that take advantage of the order of magnitude increase in PV that is observed across the tropopause away from tropical latitudes. Earlier research on upper-level fronts (see, e.g., Reed and Sanders 1953; Reed 1955; Reed and Danielsen 1959) established the value of using PV conservation and the observed large vertical PV gradient across the tropopause to diagnose the three-dimensional evolution of frontal circulations. On the basis of a careful analysis of original radiosonde records that preserved evidence of small-scale thermal stratifications (stable laminae) in the atmosphere, Danielsen (1959) concluded that the tropopause must be viewed as an isentropic pathway (and not as a material surface) along which mixing of stratospheric and tropospheric air could take place and across which the strong vertical gradient of static stability required a similar large gradient of PV. The concept of the

dynamic tropopause is built upon these earlier papers and Danielsen's subsequent contributions (discussed in Section 6) that established the crucial role of migratory cyclones in stratospheric-tropospheric exchange processes. The seminal paper by Hoskins et al. (1985) reignited interest in PV thinking from a dynamical perspective and was followed by the application of dynamic tropopause maps to real data cases (see, e.g., Hoskins and Berrisford 1988). Briefly, the dynamic tropopause is defined on the basis of a surface of constant PV (typically 1–2 PV units (PVU), where 1 PVU = 1×10^{-6} m^2 K kg^{-1} s^{-1}) that lies within the observed large vertical gradient of PV that characteristically marks the tropopause poleward of 20 degrees latitude in either hemisphere.

To first approximation (frictionless, adiabatic motion), conservation of PV means that PV contours can be advected by the observed winds on isentropic surfaces. Equivalently, potential temperature contours can be advected by the winds on surfaces of constant PV (dynamic tropopause). Dynamical significance can then be attached to: (1) the advection of potential temperature on the dynamic tropopause (an implicit measure of vertical motion in a sheared environment with ascent and descent favored where there is advection of lower and higher values of potential temperature as is observed to occur ahead of migratory troughs and ridges, respectively); (2) the magnitude of the potential temperature gradient on the dynamic tropopause (a large potential temperature gradient is indicative of a steeply sloping tropopause such as might be found in the vicinity of a jet stream disturbance); (3) the height of the dynamic tropopause above the surface (synoptic-scale troughs and ridges are marked by typically low and high dynamic tropopause heights and potential temperatures, respectively); and (4) the interaction of transient disturbances on the dynamic tropopause with low-level thermal and moisture boundaries. Dynamic tropopause maps, together with conventional surface or lower-level maps, can then be used to qualitatively understand cyclone life cycles. The dynamic tropopause mapping procedure is conceptually analogous to the Eady (1949) model representation of the atmosphere with the major difference that the dynamic tropopause is free to undulate.

These physical concepts are now applied to the 12–14 March 1993 Superstorm (SS'93) over eastern North America. Figures 9a,b,d,e,g,h,j,k show a series of dynamic tropopause maps with superimposed potential temperature and pressure contours and plotted winds for 0000 UTC 12–14 March 1993, all based upon global grids obtained from the NMC Medium Range Forecast (MRF) model. Corresponding maps of mean sea-level pressure, 850 hPa equivalent potential temperature and mean (850–700 hPa layer) PV are displayed in Figs. 9c,f,i,l. For comparison purposes, 24 h forecast MRF dynamic tropopause and sea-level pressure maps verifying 0000 UTC 13 March are also shown in Fig. 9.

Inspection of Fig. 9 reveals that the initial cyclogenesis

over the western Gulf of Mexico subsequent to 0000 UTC 12 March 1993 occurs as an area of lower (higher) potential temperature (pressure) is advected over the Gulf of Mexico at the leading edge of a short-wave trough crossing northern Mexico (Figs. 9a,b,c) and that explosive cyclogenesis ensues as a second trough, also marked by the advection of lower (higher) potential temperature (pressure) and much colder air overall on the dynamic tropopause, pushes southeastward from Wyoming and Montana to Oklahoma and Kansas and begins to interact with the Gulf of Mexico trough in the 24 h ending 0000 UTC 13 March 1993 (Figs. 9d,e,f). The surface cyclone is near maximum intensity (central pressure ~962 hPa) along the Atlantic coast at 0000 UTC 14 March 1993 in response to the interaction and merger of both of these predecessor troughs (Figs. 9g,h,i). An "inverted comma" (or "coat hook") signature in the dynamic tropopause potential temperature and pressure fields (Figs. 9g,h) marks the end of the explosive cyclogenesis period (see also Hakim et al. 1995) and the surface cyclone is centered near the tip of the "inverted comma" and lies immediately downstream of a region of large lower (higher) potential temperature (pressure) advection on the dynamic tropopause at this time.

The dynamic tropopause (viewed from above) descends to almost 800 hPa immediately upstream of the surface cyclone near the time of maximum cyclone intensity at 0000 UTC 14 March 1993 (Fig. 9h). Given that the mean PV in the 850–700 hPa layer increases to over 2 PVU by 0000 UTC 14 March (Fig. 9i), primarily in response to diabatic heating associated with latent heat release (not shown) in the cyclone environment, the extraordinary low dynamic tropopause likely represents an amalgamation of the descending high PV air from the upper troposphere with the rapidly growing 850–700 hPa PV maximum. The poleward expansion of potential temperatures (pressures) >350 K (200 hPa) on the dynamic tropopause ahead of the rapidly developing cyclone (Figs. 9a,b,d,e,g,h) cannot be explained on the basis of horizontal advection and must be associated with diabatic heating processes resulting in the nonconservation of PV (widespread convection and stratiform precipitation is observed in the PV nonconservation region). An outcome of the warming and lifting of the dynamic tropopause on the east side of the developing trough (associated with diabatic heating processes) is the creation of an intense jet that extends through the downstream ridge with speeds in excess of 80 m s^{-1}. A comparison of the 24 h MRF forecast fields of potential temperature and pressure on the dynamic tropopause (Figs. 9j,k) with the MRF initialized fields for 0000 UTC 13 March 1993 (Figs. 9d,e) reveals that the MRF fails to capture the observed warming and lifting of the dynamic tropopause associated with the equatorward PV anomaly. Likewise, the intensity of the cyclogenesis over the Gulf of Mexico and growth of mean PV in the 850–700 hPa layer is underestimated (Figs. 9f,l) in advance of the equatorward PV anomaly. The 24 h MRF forecast underprediction of cyclogenesis sug-

gests a possible MRF model problem with the accurate simulation of diabatic heating processes associated with convection. A more comprehensive discussion of the role of diabatic processes in this storm is contained in Dickinson et al. (1997).

10. Conceptual Models of Idealized Cyclones: Sensitivity to Large-Scale Flow Configuration

Bergeron (1937) compared the surface temperature contrast between the air ahead of the warm front and the air behind the cold front to distinguish between warm and cold frontal occlusions as envisioned in the idealized NCM described in Section 2. Although the idealized NCM has come under increasing criticism recently as being unrepresentative of the spectrum of observed cyclone life cycles (see, e.g., Mass 1991; Uccellini et al. 1992; and Sanders and Doswell 1995), similar criticism exists in the earlier literature. For example, Palmén (1951a) stated, "there is no doubt that many occluded cyclones on surface maps have never gone through a real process of occlusion, although they show the same characteristic structure as really occluded polar-front perturbations." An illustrative example would be the formation of a lee trough downwind of the Rockies accompanying lee cyclogenesis. Typically, the lee trough is marked by a thermal ridge in the 1000–500 hPa thickness field. This thermal ridge differs little in appearance from a similar axis of warm air delineating a typical occluded front. Schultz and Mass (1993) examined 25 previously published case studies of occluding cyclones and found very little definitive evidence for the existence of a cold occlusion structure during the mature stage of the cyclone life cycle. Also, in their detailed numerical simulation (see also Mass and Schultz 1993) of an intense North American cyclone they found evidence of a warm-type occlusion that differed significantly in structure and evolution from that envisioned by the idealized NCM.

At issue is the applicability of an idealized conceptual model of cyclone life cycles to real-world cyclones that

---

FIG. 9. (facing page) (a): Potential temperature (solid contours every 10 K) and winds (m s^{-1} with one pennant, full barb and half barb denoting 25 m s^{-1}, 5 m s^{-1}, and 2.5 m s^{-1}, respectively) on the dynamic tropopause defined by the 1.5 PVU surface (1 PVU = 10^{-6} K m^2 kg^{-1} s^{-1}) for 0000 UTC 12 March 1993; (b): as in (a) except for pressure (solid contours every 100 hPa) and winds on the dynamic tropopause; (c): as in (a) except for mean sea-level pressure (solid contours every 4 hPa) with superimposed 850–700 hPa layer mean PV (shaded according to the scale every 0.5 PVU beginning at 0.5 PVU) and 850 hPa equivalent potential temperature (dashed contours every 5K); (d), (e), (f), as in (a), (b), (c) except for 0000 UTC 13 March 1993; (g), (h), (i) as in (a), (b), (c) except for 0000 UTC 14 March 1993; (j), (k), (l) as in (a), (b), (c) except for 24 h MRF forecasts verifying 0000 UTC 13 March 1993.

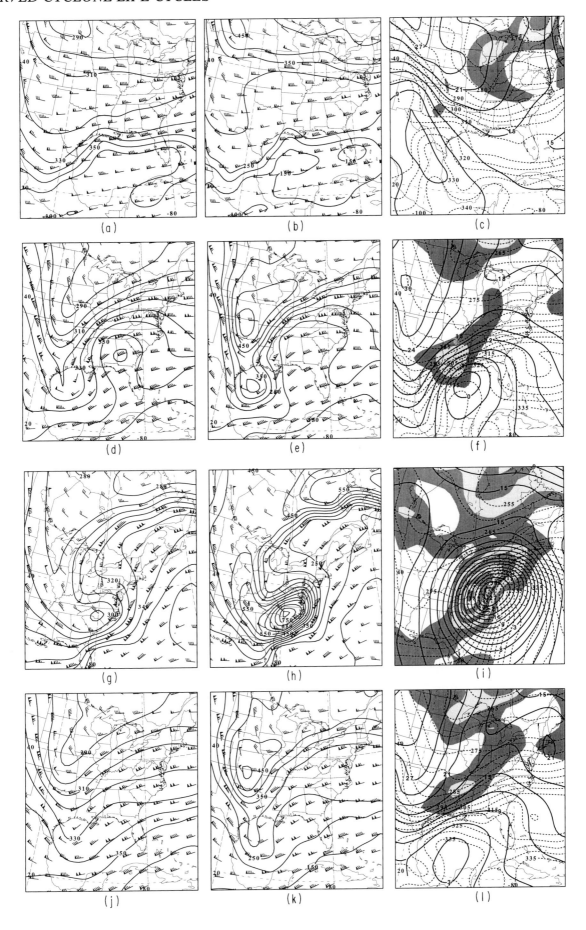

exhibit a rich spectrum of behavior as a function of the mutual interaction of large-scale and synoptic scale circulations that in turn are modulated regionally and locally by an additional spectrum of physiographic factors. Rossby and Willett (1948) suggested that the aspect ratio (zonal to meridional elongation) of individual transient cyclones was strongly dependent upon the amplitude of the large-scale flow. Willett (1949) and Namias (1950) first demonstrated that atmospheric blocking phenomena (see, e.g., Rex 1950a,b) could be related to a quasiperiodic vacillation of the large-scale flow. The flow vacillation was defined by an "index cycle" (a measure of the strength of the westerlies) and could also be associated with cyclogenesis frequency and storm track location. Namias (1978) extended the original ideas of large-scale flow vacillation to include multiscale influences and interactions of the atmosphere and ocean that he believed could help explain the highly anomalous circulation over North America and vicinity during the infamous winter of 1976-77.

More recently, Black and Dole (1993) used PV inversion techniques to investigate the origin and maintenance of persistent cyclonic flow anomalies over the North Pacific in winter. Black and Dole (1993) found that the cyclonic development "in several respects resembles a classical Petterssen Type B development (predecessor disturbance aloft) but occurs on a scale that is much larger than for typical synoptic scale cyclogenesis." Colucci (1985), Colucci and Davenport (1987), Sanders and Davis (1988), Bell and Bosart (1989) and Bullock and Gyakum (1993), among others, have also demonstrated that the configuration of the large-scale circulation does impact, and in turn is impacted by, synoptic-scale flow evolution associated with cyclogenesis and anticyclogenesis. These often complex planetary-scale and synoptic-scale interactions come in a variety of "flavors" that can make it difficult to isolate and classify characteristic flow interaction signatures on the basis of a few idealized conceptual models alone.

Young and Grant (1995) and Evans et al. (1994) used satellite imagery to classify extratropical cyclogenesis over the North Atlantic Ocean (see also the references in these papers to earlier studies). The results from these (and other) studies support the conjecture that there may be characteristic large-scale flow patterns associated with distinct types of cyclogenesis as determined by particular cloud signatures seen in the satellite imagery. An important lesson from the Young and Grant (1995) and Evans et al. (1994) papers is that characteristic observed satellite signatures associated with cyclone life cycles are manifestations of synoptic-scale circulations that owe their existence to fundamental dynamical and thermodynamical processes as modulated by local and regional geographic features. The different conceptual models of cyclone life cycles derived from satellite imagery shown in both papers reflects the structure of the large-scale flow as measured by wave amplitude, wavelength, jet con-figuration, environmental static stability, trough tilt, and the degree of confluence or diffluence in the background flow. These emerging research findings suggest that preference be given to the interpretation of characteristic satellite-derived cyclone signatures in terms of known dynamical and kinematic principles as a future basis for cyclone and anticyclone life cycle studies.

An additional benefit of such a strategy is that it would facilitate the growing communication and cooperation between the observational, theoretical, and numerical modeling communities in addressing cyclone/anticyclone life cycle studies in a wide variety of research studies. For example, over the last 10–15 years researchers have made increasing use of idealized three-dimensional primitive equation models of varying complexity to investigate the life cycle of an idealized baroclinic wave cyclone (see, most recently, e.g., Polavarapu and Peltier 1990; Hines and Mechoso 1993; Schär and Wernli 1993; Thorncroft et al. 1993 and references therein). These studies along with the investigations of, for example, Farrell (1989) and Montgomery and Farrell (1993) illustrate that theoretical progress in understanding the time-dependent behavior of cyclones and anticyclones is contingent upon moving away from purely normal mode type studies to include: (1) a more realistic representation of predecessor disturbances (PV anomalies) initially (initial value problem), (2) the addition of barotropic shear to the background baroclinic flow, and (3) allowing downstream and upstream wave energy dispersion in conjunction with localized disturbances inserted into an unstable background baroclinic flow.

The studies cited above also reveal the existence of important signatures in the simulations that differ significantly from the NCM: (1) fractured cold fronts south of the cyclone center, and (2) bent-back warm fronts west of the cyclone center with a warm air seclusion region. The results of these modeling studies combined with research aircraft and other mesoscale observations of developing marine cyclones prompted Shapiro and Keyser (1990) and Neiman and Shapiro (1993) to produce a new schematic figure illustrating the life cycle of a marine cyclone (shown in Fig. 10.27 of Shapiro and Keyser 1990, which should be compared to the classical NCM schematic shown in Fig. 10.12 of the same paper). Recent numerical simulations of some explosively developing marine cyclones using real data (see, e.g., Kuo et al. 1992; Reed et al. 1992) have yielded bent-back warm front and fractured cold front signatures that are qualitatively in accord with the Shapiro and Keyser (1990) conceptual model. An idealized simulation of baroclinic wave cyclone development by Hines and Mechoso (1993) suggests that the existence of fractured and bent-back frontal structure is critically dependent upon the degree of model friction and is more likely to occur with values typical of oceanic as opposed to continental conditions. Although the extent to which the proposed Shapiro and Keyser (1990) model is representative of a wider spectrum of ocean cyclones remains to be determined, it is of critical importance that continuing studies be conducted to firm up our knowledge and understanding of the dynamical reasons for the existence of three-dimensional structural variability in cyclones and anticyclones.

As an example, Thorncroft et al. (1993) show that the evolution of an idealized baroclinic wave cyclone life cycle is sensitive to the addition of barotropic shear (a very modest 6×10^{-6} s^{-1} between 20°N and 50°N) to the baroclinic background flow (compare their Figs. 5–11 for details and their schematic Fig. 12). In the case without barotropic shear the initial wave develops poleward of the main baroclinic zone and wraps up cyclonically. Subsequently, a southern extension of the wave moves equatorward of the jet, fractures from the main wave, and becomes associated with a cutoff cyclone in low latitudes as a ridge builds poleward of the cutoff. The addition of cyclonic barotropic shear to the background flow enables the growing wave to develop a stronger cyclonic circulation while living out its life cycle poleward of the jet with no indication of additional development equatorward of the jet. Although direct observational verification of these theoretical findings is lacking, it is of interest to note that the formation of closed cyclones equatorward of the jet in the vicinity of the Alps and southwestern United States described by Bell and Bosart (1994) bears some resemblance to the first pattern described above by Thorncroft et al. (1993) whereas eastern United States closed cyclones forming poleward of the jet in the same paper bear some characteristics to the cyclonic barotropic shear case of Thorncroft et al. (1993).

11. Trough Fracture, Trough Merger and Cyclone Development

Synoptic experience shows that existing numerical prediction models frequently have difficulty simulating trough fracture and merger processes properly. Here the term trough fracture refers to the observed tendency of an individual vorticity maximum to split (fracture) into two distinct centers. Similarly, trough merger can be associated with the amalgamation of initially separate vorticity maxima into one disturbance. Trough fracture, for example, is illustrated in the theoretical study of cyclone life cycles using idealized baroclinic models by Thorncroft et al. (1993, see their schematic Figs. 12a, 13b). A case study of a prominent case of unforecast trough merger over eastern North America appears in Lai and Bosart (1988). Gaza and Bosart (1990) conducted a limited climatological analysis of trough merger (defined to be the amalgamation of initially distinct vorticity maxima at 500 mb) over North America. They found that

trough merger was concentrated over central and eastern North America east of the Rockies and was frequently associated with the development of a negatively tilted trough (one oriented northwest-southeast) and explosive cyclogenesis.

Maps of trough fracture and merger frequencies for the Northern Hemisphere as taken from the 30+ year objective climatological study of Dean and Bosart (1996) using twice daily gridded 500 hPa analyses appear in Figs. 10 and 11, respectively. A fracture origination point is located objectively at the center of the vorticity maximum immediately preceding the map showing two separate vorticity centers. A reverse definition is used for trough merger cases. Accordingly, the locations of trough fracture points will lie poleward of the positions of the cutoff cyclones that represent the end stage of the fracture process. Comparison of Figs. 10 and 11 illustrates the following: (1) both fracture and merger events maximize in the vicinity of the storm tracks; (2) merger events lie somewhat poleward of fracture events; (3) merger events avoid the major mountainous areas of North America, consistent with the earlier results of Gaza and Bosart (1990); (4) the equatorward shift of merger events over the Mediterranean region is consistent with the "seam" in the baroclinic waveguide of Wallace et al. (1988) discussed previously; and (5) fracture regions cluster preferentially over extreme eastern Asia, the east-central North Pacific Ocean, extreme southeastern Canada and the western Atlantic, and over extreme western Europe.

A typical trough fracture and merger scenario over North America might have a life cycle of ~10 days. It often begins with a synoptic-scale trough in the eastern Pacific Ocean approaching the western coast of North America. Typically, the trough will elongate meridionally as it approaches land with the northern part moving onshore across the Canadian or American Rockies as the southern part first slows down and stretches equatorward before eventually forming a cutoff cyclone over southwestern North America (see, e.g., Bell and Bosart 1994). Subsequently, the cutoff cyclone may spin harmlessly in place for several days before it begins to move slowly eastward across northern Mexico and the southern United States in response to the approach of another trough from upstream. As the cutoff cyclone begins to overspread the Gulf of Mexico, it is able to tap a source of increasingly warm, moist and unstable air. This new environment is very favorable for cyclogenesis as Gulf air is ingested into the updraft region ahead of the cyclonic circulation center aloft. The resulting cyclone typically moves poleward and eastward and the associated closed vortex aloft opens up and reattaches itself (merges) to the main belt of polar westerlies. In exceptional cases the southern cutoff cyclone may merge with a separate, and very cold, northern system and spectacular cyclogenesis can result, as was observed for the CSB (Hakim et al. 1995, 1996) and SS'93 events (Caplan 1995;

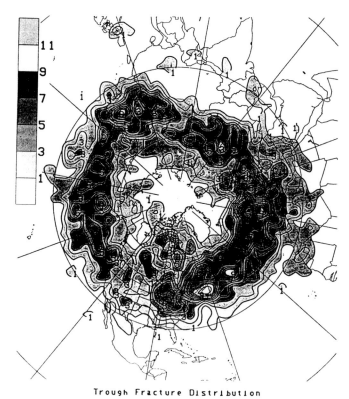

Trough Fracture Distribution

FIG. 10. Total number of trough fracture events occurring from 1 September 1957 to 31 May 1989. The field has been smoothed with a standard two-dimensional five point smoother (1-2-1 in both directions) prior to plotting. Shading contour interval every two events beginning with one event (Dean and Bosart 1996, Fig. 5).

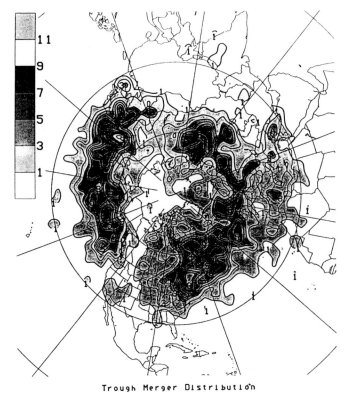

Trough Merger Distribution

FIG. 11. As in Fig. 10 except for trough merger events (Dean and Bosart 1996, Fig. 8).

Kocin et al. 1995; Bosart et al. 1996; Dickinson et al. 1997). The SS'93 event was especially noteworthy because the northern PV anomaly involved in the explosive trough merger originated with the Hudson Bay arctic vortex three weeks before the event (Bosart et al. 1996). Hakim et al. (1995) adopted vortex-vortex interaction techniques (see, e.g., Takayabu 1991 and Guinn and Schubert 1993) to PV inversion methodology to isolate the dynamical effects of the individual baroclinic vortices and the baroclinic background flow for the CSB case. Trough merger was accomplished in this case by the cyclonic rotation of the individual baroclinic vortices as they mutually interacted with one another within a confluent background flow.

A crucial aspect of whether trough fracture and merger can occur is the configuration of the large-scale flow and its evolution in time. Bosart and Bartlo (1991) demonstrated that the development of Hurricane Diana east of Florida in September 1984 occurred in sequential steps beginning with a trough fracture. As an upper-level synoptic-scale trough moved across eastern North America the southern end of the trough elongated and lagged behind with a cutoff cyclone forming over the northeastern Gulf of Mexico. Meanwhile, the sea-level pressure rose over the western Atlantic behind the departing northern trough, setting up a strong northeasterly fetch of relatively cool and dry air over the warm waters of the Gulf Stream where surface sensible and latent heat fluxes in excess of $1000 \, \mathrm{W \, m^{-2}}$ warmed, moistened, and destabilized this air. Cyclogenesis began along the old frontal zone east of Florida as the PV anomaly marking the fractured cutoff cyclone crossed Florida. Subsequently, the strengthening low-level cyclonic circulation became warm core as convective heating short-circuited the isentropic uplift ahead of the cold dome associated with the PV anomaly aloft, eradicating the PV anomaly and allowing for the creation of a negative PV anomaly along the tropopause as Hurricane Diana was born. Montgomery and Farrell (1993) were able to replicate many of the observed results for the Diana case in a theoretical study of tropical cyclone formation initiated in response to the interaction of an upper-level PV anomaly with a preexisting low-level disturbance.

It is stressed that cyclonic disturbances of higher, middle, and lower latitudes share common physical development attributes that are modulated by the configuration of the large-scale flow, the availability (or lack thereof) of warm, moist unstable air, and the influence of nearby regional and local geographic features. Accordingly, individual cyclone and anticyclone life cycles owe their existence and structural evolution to a rich spectrum of dynamical and thermodynamical processes. These processes cannot easily be labeled as tropical, midlatitude, or polar. Future progress in gaining knowledge and understanding of observed cyclone and anticyclone life cycles can be accelerated only by thinking more broadly about the nature of the physical mechanisms involved and the dynamical and thermodynamical processes responsible for the growth of vertical circulations on a variety of time and space scales.

A final illustrative example of trough merger is the Ertel PV signature shown in Fig. 12 (taken from Fig. 9 of Bosart and Lackmann 1995) in conjunction with the landfall of Hurricane David along the southeast coast of the United States on 1200 UTC 4 September 1979 and its subsequent (and noteworthy) inland reintensification along the eastern slopes of the Appalachian mountains. Unlike the transformation of Hurricane Hazel (Palmén 1958; Anthes 1990) and Hurricane Agnes (DiMego and Bosart 1982a,b) into strong extratropical storms in highly baroclinic environments, the redevelopment of David occurred through a sequential and mutual interaction of polar, middle latitude, and tropical PV anomalies. Over the 72 h period illustrated in Fig. 12 (24 h snapshots) the tropical PV anomaly marking David moved northeastward along the eastern coast of North America and eventually reattached itself to the main band of westerlies. Reintensification was well under way by 1200 UTC 6 September (Fig. 12c) as positive PV advection aloft increased over the low-level circulation center in response to the combined eastward advection of low PV air equatorward of the main belt of westerlies and the diabatic generation of low PV air north and east of David. Meanwhile, a weak PV "bump," initially located to the west of a quasicircular polar PV anomaly centered over northern Canada at 1200 UTC 4 September, rotated equatorward and began to merge with a Pacific PV anomaly moving inland across western North America from the North Pacific PV reservoir by 1200 UTC 5 September. Subsequently, these two PV anomalies completed their merger by 1200 UTC 6 September, accompanied by the southward extension of the westerlies over eastern North America. This new PV anomaly continued to rotate eastward and equatorward as it began to interact with the tropical (David) PV anomaly. In this case, however, the interaction consisted of the tropical PV anomaly getting swept into the main flow ahead of the newly minted polar-Pacific PV anomaly. Clearly, it is necessary to view synoptic-scale cyclone life cycles such as occurred in the David case in terms of: (1) the lateral interaction of PV anomalies along the tropopause, (2) the vertical interaction of PV anomalies on the tropopause with surface-based PV anomalies, and (3) the mutual interaction of interior PV anomalies driven mostly by latent heat release (David) with distant PV anomalies both laterally and vertically.

12. Downstream Development and Cyclogenesis as an Initial Value Problem

In addition to the idealized cyclone life cycle studies mentioned above, the theoretical community has shown renewed interest in the downstream development ideas first articulated by Namias and Clapp (1944), Cressman (1948) and Hövmöller (1949) in conjunction with North Pacific cyclogenesis and van Loon (1965) for Southern Hemisphere circulations. The initial theoretical interpretation of the downstream development observations is contained in the classical barotropic wave energy dispersion ideas presented by Rossby (1945) and Yeh (1949). Simmons and Hoskins (1979) extended the barotropic energy dispersion approach to include the response of a baroclinically unstable atmosphere to a localized initial perturbation by means of numerical experiments using idealized primitive equation models. Although Simmons and Hoskins (1979) found similar downstream energy dispersion behavior as for a classical barotropic atmosphere, they also found evidence for upstream development to be clustered over a finite longitude band. More recently, papers by, for example, Lee and Held (1993), Chang (1993), Chang and Orlanski (1993), Orlanski and Chang (1993, 1995), and Swanson and Pierrehumbert (1994) have shown that baroclinic eddies growing in regions of large baroclinicity can transfer energy well downstream, triggering new eddy growth in regions of much weaker overall baroclinicity. As noted by Chang and Orlanski (1993), in an idealized primitive equation model study of the dynamics of a storm track, baroclinic conversion does correlate with a "region of maximum baroclinicity but downstream radiation of energy through ageostrophic geopotential fluxes acts as a trigger for the development and maintenance of eddy activity over less baroclinic regions, extending regions of eddy activity downstream from the region of high baroclinicity."

Chang (1993) quantified these numerical findings by preparing one-point lag correlation maps of the 300 hPa meridional wind for seven winter seasons averaged over the 30–60°N belt. He concluded (see his Fig. 3) that "wave trains exhibit characteristics of downstream development with successive perturbations developing toward the downstream side of existing perturbations" (schematically summarized in his Fig. 13). Chang (1993) also suggested that the failure to find downstream development in earlier statistical studies (see, e.g., Lim and Wallace 1991) was an artifact of the time-filtering procedures employed to process the original observations. Support for these findings was provided by Swanson and Pierrehumbert (1994) who used an inviscid nonlinear model to simulate the life cycle of wave packets that initially emanated from a longitudinally confined disturbance region in a baroclinically unstable flow. Both upstream and downstream wave propagation in the form of traveling synoptic disturbances was observed, the former being damped and associated with a strengthening barotropic jet and the latter being associated with growing "leading edge" waves (see their schematic Fig. 15) that propagated well downstream from the region of the original disturbance. Swanson and Pierrehumbert (1994) noted that the development and propa-

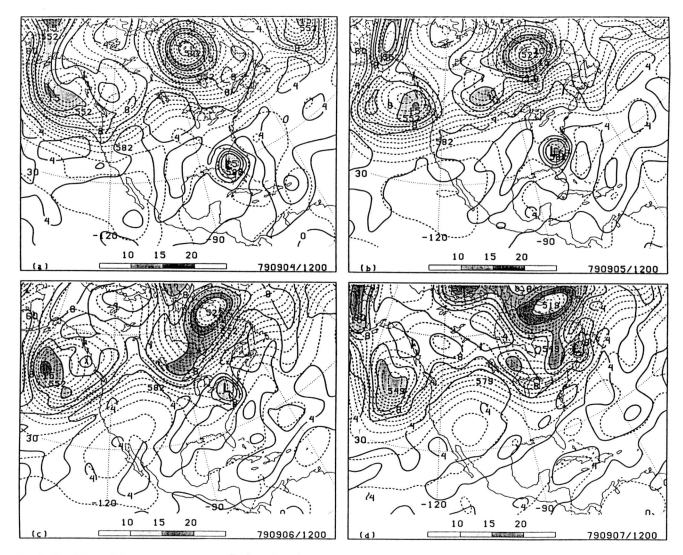

FIG. 12. Ertel PV (solid contours every 2×10^{-7} m^2 K s^{-1} kg^{-1}, where 10 of these units corresponds to the first shaded contour on the map) and 500 hPa geopotential height (dashed contours every 30 dam) for 1200 UTC 4 (a), 5 (b), 6 (c), and 7 (d) September 1979. The Ertel PV was computed using ECMWF observed 500 hPa winds for the vorticity, the 700 and 400 hPa winds for the isobaric coordinate conversion term, and the 700 and 400 hPa potential temperatures for the stability term (Bosart and Lackmann 1995, Fig. 9).

gation of the leading "edge waves" was consistent with the findings of Chang and Orlanski (1993) as to the importance of downstream development associated with ageostrophic geopotential fluxes.

Recent case studies of individual and composite storms have produced mixed evidence in support of the downstream development findings arising from statistical and energetics studies. Nielsen-Gammon (1995) and Nielsen-Gammon and Lefevre (1996) demonstrated that trough development over eastern North America was consistent with downstream development using PV concepts. Lackmann et al. (1996) examined planetary- and synoptic-scale composites of explosive cyclogenesis over the western North Atlantic ocean in winter. Although they found that troughs and ridges tended to amplify downstream from the eastern Pacific across North

America to the western Atlantic, consistent with expectation for downstream development, prior to the explosive cyclogenesis in the Atlantic, they also noted that the quasi-constant amplitude and quasi-stationary phase of the upstream anomalies over western North America were suggestive of an important orographic signal. At issue, however, is whether downstream development ideas are fully applicable to smaller-scale coherent disturbances (PV anomalies) in the upper troposphere that exhibit more vortex-like than wave-like behavior (and which have similar Rossby wave group and phase velocities; Lackmann et al. 1996). Bosart et al. (1996) demonstrated that the March 1993 superstorm over eastern North America could be attributed to the lateral and vertical interaction of long-lived smaller-scale PV anomalies, and they challenged the general applicability of downstream development concepts in this case. Finally, a study of upper-tropospheric cyclogenetic precursors by Lackmann et al. (1997) has uncovered the evidence that confluence downstream of the characteristic western North American ridge contributes to the intensification of a midtropospheric jet-front downshear over interior North America. Cold advection along the flow and cross-stream differential vertical motion contribute to cyclonic vorticity production and frontogenesis as the precursor disturbance approaches the downstream trough. The relevance of this important mesoscale process to classical synoptic-scale downstream development concepts needs to be more fully addressed.

These papers and other related investigations not specifically mentioned illustrate the continuing and growing theoretical interest in solving initial value problems. Such studies, which depart significantly from traditional "normal mode thinking," represent one attempt to reconcile theory with observation and help to reinforce the original self-development concepts articulated by Sutcliffe and Petterssen and summarized in Section 4. Simultaneously, these exciting new theoretical and numerical studies open up many new potential research avenues for observational exploration. At issue, for example, is how storm tracks are maintained, what determines the zonal extent of the storm tracks, and what is the origin of the transient disturbances that propagate along these storm tracks. Hoskins and Valdes (1990) demonstrated that diabatic processes over the western Atlantic Ocean associated with atmospheric latent heat release and surface sensible and latent heat fluxes were of critical importance to regenerating the large-scale baroclinicity depleted by amplifying eddies in the storm track entrance region. However, as noted by Chang and Orlanski (1993), the regions of maximum wintertime eddy activity over the Atlantic Ocean are frequently located well downstream of the maximum diabatic heating region that is centered near the mean position of the Gulf Stream, raising the likelihood that other more subtle dynamical processes responsible for determining the extent and characteristics of the storm tracks remain to be determined.

Another interesting aspect of storm track behavior was raised by Nakamura (1992) when he suggested that midwinter suppression of baroclinic wave activity in the Pacific (as defined by appropriately filtered 250 hPa geopotential height data, his Fig. 1) is negatively correlated with wind speeds in the jet stream when the maximum winds exceed ~ 45 m s^{-1}. The absence of a similarly robust result for the Atlantic Ocean may be an illustration of fundamental dynamical differences between the Atlantic and Pacific storm tracks. It is also possible that the Nakamura result could be an artifact of statistical data processing procedures as Roebber (1994, personal communication) has indicated he can find little evidence to support a reduced frequency of midwinter cyclone crossings along 150°E from 10 years of winter season cyclone data originally developed by Gyakum et al. (1989).

Furthermore, since Nakamura (1992) examined only 250 hPa geopotential data, it is possible that the level of maximum eddy activity was situated below 250 hPa over the Pacific storm track in midwinter when the height of the tropopause is typically lowest.

This writer is hopeful that a coordinated observational, theoretical, and numerical attack over the next decade designed to increase our knowledge and understanding of the dynamical processes associated with storm track origin, structure, and evolution will yield important additional dividends, including a better appreciation of the role played by regional and local physiographic features, modulated by seasonally dependent continental scale differential heating, in controlling individual cyclone and anticyclone life cycles and associated weather regime behavior. It is further anticipated that the development and deployment of new observational technologies such as Doppler radars and remote sensing systems and the ongoing global reanalysis effort by operational centers such as the United States National Meteorological Center (NMC) and the European Centre for Medium Range Weather Forecasts (ECMWF), combined with the theoretical breakaway from the normal mode "straightjacket," will accelerate the process by creating even wider scientific opportunities for young people joining the field.

Acknowledgments

The author gratefully acknowledges continued research support from the United States National Science Foundation under grants ATM-9114598, ATM-9120331, and ATM-9413012, and the many contributions of SUNY/Albany students, past and present, to the research effort. He especially would like to thank Dr. Ronald C. Taylor for several enlightening discussions of meteorological history.

References

Anthes, R. A., 1990: Advances in the understanding and prediction of cyclone development with limited-area fine-mesh models. *Extratropical Cyclones, The Erik Palmén Memorial Volume*, C. W. Newton and E. O. Holopainen, Eds., Amer. Meteor. Soc., 221–253.

——, Y.-H. Kuo, and J. R. Gyakum, 1983: Numerical simulations of a case of explosive marine cyclogenesis. *Mon. Wea. Rev.,* **107**, 963–984.

Beebe, R. G., and F. C. Bates, 1955: A mechanism for assisting in the release of convective instability. *Mon. Wea. Rev.,* **83**, 1–10.

Bell, G. D., and L. F. Bosart, 1989: A 15 year climatology of Northern Hemisphere 500 mb closed cyclone and anticyclone centers. *Mon. Wea. Rev.,* **117**, 2142–2163.

——, and ——, 1993: A case study diagnosis of the formation of an upper level closed cyclonic circulation over the eastern United States. *Mon. Wea. Rev.,* **121**, 1635–1655.

——, and ——, 1994: Midtropospheric closed cyclone formation over the southwestern United States, the Eastern United States, and the Alps. *Mon. Wea. Rev.,* **122**, 791–813.

————, and D. Keyser, 1993: Shear and curvature vorticity and potential-vorticity interchanges: Interpretation and application to a cutoff cyclone event. *Mon. Wea. Rev.*, **121**, 76–102.

Bengtsson, L., 1990: Advances in numerical prediction of the atmospheric circulation in the extratropics. *Extratropical Cyclones, The Erik Palmén Memorial Volume*, C. W. Newton and E. O. Holopainen, Eds., Amer. Meteor. Soc., 193–220.

Berbery, E. H., and C. S. Vera, 1996: Characteristics of the Southern Hemisphere winter storm track with filtered and unfiltered data. *J. Atmos. Sci.*, **53**, 468–481.

Bergeron, T., 1937: On the physics of fronts. *Bull. Amer. Meteor. Soc.*, **18**, 265–275.

————, 1959: Methods in scientific weather analysis and forecasting. An outline in the history of ideas and hints at a program. *The Atmosphere and the Sea in Motion*, B. Bolin, Ed., Rockefeller Institute Press, 440–474.

Bjerknes, J., 1919: On the structure of moving cyclones. *Geofys. Publ.*, **1**, 1–8.

———— 1937: Theorie der aussertropischen Zyklonenbildung. *Meteor. Z.*, **54**, 462–466.

————, and H. Solberg, 1922: Life cycle of cyclones and the polar front theory of atmospheric circulation. *Geofys. Publ.*, **3**(1), 1–18.

————, and E. Palmén, 1937: Investigations of selected European cyclones by means of serial ascents. *Geofys. Publ.*, **12**, 1–62.

————, and J. Holmboe, 1944: On the theory of cyclones. *J. Meteor.*, **1**, 1–22.

Black, R. X., and R. M. Dole, 1993: The dynamics of large-scale cyclogenesis over the north Pacific ocean. *J. Atmos. Sci.*, **50**, 421–442.

Blackmon, M. L., 1976: A climatological study of the 500 mb geopotential height of the Northern Hemisphere. *J. Atmos. Sci.*, **34**, 1607–1623.

Bleck, R., 1973: Numerical forecasting experiments based on the conservation of potential vorticity on isentropic surfaces. *J. Appl. Meteor.*, **12**, 737–752.

————, 1974: Short-range prediction in isentropic coordinates with filtered and unfiltered numerical models. *Mon. Wea. Rev.*, **102**, 813–829.

————, 1990: Depiction of upper/lower vortex interaction associated with extratropical cyclogenesis. *Mon. Wea. Rev.*, **118**, 573–585.

————, and Coauthors, 1993: Eighth cyclone workshop scientific summary, Val Morin, Quebec, Canada, 12–16 October 1992. *Bull. Amer. Meteor. Soc.*, **74**, 1361–1373.

Bluestein, H., 1992: *Synoptic-Dynamic Meteorology in Midlatitudes*, Vol. I: *Principles of Kinematics and Dynamics*, Oxford University Press, 431 pp.

————, 1993: *Synoptic-Dynamic Meteorology in Midlatitudes*, Vol. II: *Observations and Theory of Weather Systems*, Oxford University Press, 594 pp.

Bond, N. A., and M. A. Shapiro, 1991: Polar lows over the Gulf of Alaska in conditions of reverse shear. *Mon. Wea. Rev.*, **119**, 551–572.

Bosart, L. F., 1981: The Presidents' Day snowstorm of 18–19 February 1979: A subsynoptic-scale event. *Mon. Wea. Rev.*, **109**, 1542–1566.

————, and S. C. Lin, 1984: A diagnostic analysis of the Presidents' Day storm of February 1979. *Mon. Wea. Rev.*, **112**, 2148–2177.

————, and J. A. Bartlo, 1991: Tropical storm formation in a baroclinic environment. *Mon. Wea. Rev.*, **119**, 1979–2013.

————, and G. M. Lackmann, 1995: Post-landfall tropical cyclone reintensification in a weakly-baroclinic environment: A case study of Hurricane David (September 1979). *Mon. Wea. Rev.*, **123**, 3268–3291.

————, C.-C. Lai, and E. Rogers, 1995: Incipient explosive marine cyclogenesis: Coastal development. *Tellus*, **47A**, 1–29.

————, G. J. Hakim, K. R. Tyle, M. A. Bedrick, W. E. Bracken, M. J. Dickinson, and D. M. Schultz, 1996: Large-scale antecedent conditions associated with the 12–14 March 1993 cyclone ("Superstorm '93") over eastern North America. *Mon. Wea. Rev.*, **124**, 1865–1891.

Boyle, J. S., and L. F. Bosart, 1986: Cyclone–anticyclone couplets over North America. Part II: Analysis of a major cyclone event over the eastern United States. *Mon. Wea. Rev.*, **114**, 2432–2465.

Browning, K. A., 1990: Organization of clouds and precipitation in extratropical cyclones. *Extratropical Cyclones, The Erik Palmén Memorial Volume*, C. W. Newton and E. O. Holopainen, Eds., Amer. Meteor. Soc., 129–153.

Bullock, T. A., and J. R. Gyakum, 1993: A diagnostic study of cyclogenesis in the western North Pacific ocean. *Mon. Wea. Rev.*, **121**, 65–75.

Burrows, W. R., R. A. Triedl, and R. G. Lawford, 1979: The southern Ontario blizzard of January 26 and 27, 1978. *Atmos.-Oceans*, **17**, 306–320.

Buzzi, A., and S. Tibaldi, 1978: Cyclogenesis in the lee of the Alps: A case study. *Quart. J. Roy. Meteor. Soc.*, **104**, 271–287.

Caplan, P. M., 1995: The 12–14 March 1993 Superstorm: Performance of the NMC global medium-range model. *Bull. Amer. Meteor. Soc.*, **76**, 201–212.

Carleton, A. M., 1981: Monthly variability of satellite derived cyclonic activity for the Southern Hemisphere winter. *J. Climatol.*, **1**, 21–38.

————, 1985: Satellite climatological aspects of the "polar low" and "instant occlusion". *Tellus*, **37A**, 433–450.

Carr, F. H., and J. P. Millard, 1985: A composite study of comma clouds and their association with severe weather over the Great Plains. *Mon. Wea. Rev.*, **113**, 370–387.

Chang, E. K. M., 1993: Downstream development of baroclinic waves as inferred from regression analysis. *J. Atmos. Sci.*, **50**, 2038–2053.

————, and I. Orlanski, 1993: On the dynamics of a storm track. *J. Atmos. Sci.*, **50**, 999–1015.

Charney, J. G., 1947: The dynamics of long waves in a baroclinic westerly current. *J. Meteor.*, **4**, 125–162.

————, and M. E. Stern, 1962: On the stability of internal baroclinic jets in a rotation atmosphere. *J. Atmos., Sci.*, **19**, 159–172.

Chen, G. T.-J., 1983: Observational aspects of the Mei-Yu system in subtropical China. *J. Meteor. Soc. Japan*, **61**, 306–312.

————, and C.-P. Chang, 1980: The structure and vorticity budget of an early summer monsoon trough (Mei-Yu) over southeastern China and Japan. *Mon. Wea. Rev.*, **108**, 942–953.

————, and S.-S. Chi, 1980: On the frequency and speed of Mei-Yu fronts over southern China and the adjacent areas. *Paper Meteor. Res.*, **3**, 31–42.

Chen, S.-J., and L. Dell'Osso, 1987: A numerical case study of east Asian coastal cyclogenesis. *Mon. Wea. Rev.*, **115**, 477–487.

————, Y.-H. Kuo, P.-Z. Zhang, and Q.-F. Bai, 1991: Synoptic climatology of cyclogenesis over East Asia, 1958–1987. *Mon. Wea. Rev.*, **119**, 1407–1418.

————, ————, ————, and ————, 1992: Climatology of explosive cyclones off the east Asian coast. *Mon. Wea. Rev.*, **120**, 3029–3035.

Clarke, R. H., 1954: Frontal analysis over the southern ocean. *Aust. Meteor. Mag.*, **5**, 33–54.

Colucci, S. J., 1985: Explosive cyclogenesis and large-scale circulation changes: Implications for atmospheric blocking. *J. Atmos. Sci.*, **42**, 2701–2717.

——, and J. C. Davenport, 1987: Rapid surface anticyclogenesis: Synoptic climatology and attendant large-scale circulation changes. *Mon. Wea. Rev.*, **115**, 822–836.

Cressman, G. P., 1948: On the forecasting of long waves in the upper westerlies. *J. Meteor.*, **5**, 44–57.

Danard, M. B., 1964: On the influence of released latent heat on cyclone development. *J. Appl. Meteor.*, **3**, 27–37.

Danielsen, E. F., 1959: The laminar structure of the atmosphere and its relation to the concept of a tropopause. *Arch. Meteor., Geophys. Bioclimatol.*, **A11**, 294–332.

——, 1961: Trajectories: Isobaric, isentropic and actual. *J. Meteor.*, **18**, 479–486.

——, 1964: Project Springfield Report. Defense Atomic Support Agency, DASA 1517, 97 pp. [NTIS AD-607980].

——, 1966: Research in four-dimensional diagnosis of cyclonic storm cloud systems. The Pennsylvania State University, Scientific Report 1, AFCRL 66–30, 1–53.

——, 1967: Transport and diffusion of stratospheric radioactivity based on synoptic hemispheric analyses of potential vorticity. The Pennsylvania State University, Final Scientific Report part 3, NYO-3317-3, 1–91.

——, 1968: Stratospheric-tropospheric exchange based on radioactivity, ozone and potential vorticity. *J. Atmos. Sci.*, **25**, 502–518.

——, and R. Bleck, 1967: Research in four-dimensional diagnosis of cyclonic storm cloud systems. The Pennsylvania State University, Final Scientific Report, AFCRL-67-0617, 1–96.

Davis, C. A., 1992: A potential-vorticity diagnosis of the importance of initial structure and condensational heating in observed extratropical cyclogenesis. *Mon. Wea. Rev.*, **120**, 2409–2428.

——, and K. A. Emanuel, 1991: Potential vorticity diagnostics of cyclogenesis. *Mon. Wea. Rev.*, **119**, 1929–1953.

——, M. T. Stoelinda, and Y.-H. Kuo, 1993: The integrated effect of condensation in numerical simulations of extratropical cyclogenesis. *Mon. Wea. Rev.*, **121**, 2309–2330.

——, E. D. Grell, and M. A. Shapiro, 1996: The balanced dynamical nature of a rapidly intensifying oceanic cyclone. *Mon. Wea. Rev.*, **124**, 3–26.

Dean, D. B., and L. F. Bosart, 1996: Northern Hemisphere 500 hPa trough merger and fracture: A climatology and case study. *Mon. Wea. Rev.*, **124**, 2644–2671.

Dickinson, M. J., L. F. Bosart, W. E. Bracken, G. J. Hakim, D. M. Schultz, M. A. Bedrock, and K. R. Tyle, 1997: The March 1993 Superstorm: Incipient phase synoptic- and convective-scale flow interaction and model performance. *Mon. Wea. Rev.*, **125**, 3041–3072.

DiMego, G. J., and L. F. Bosart, 1982a: The transformation of tropical storm Agnes into an extratropical cyclone. Part I: The observed fields and vertical motion computations. *Mon. Wea. Rev.*, **110**, 385–411.

——, and ——, 1982b: The transformation of tropical storm Agnes into an extratropical cyclone. Part II: Moisture, vorticity and kinetic energy budgets. *Mon. Wea. Rev.*, **110**, 412–433.

Durst, C. S., and R. C. Sutcliffe, 1938: The importance of vertical motion in the development of tropical revolving storms. *Quart J. Roy. Meteor. Soc.*, **64**, 75–84.

Eady, E. J., 1949: Long waves and cyclone waves. *Tellus*, **1**, 33–52.

Eliassen, A., 1990: Transverse circulations in frontal zones. *Extratropical Cyclones, The Erik Palmén Memorial Volume*, C. W. Newton and E. O. Holopainen, Eds., Amer. Meteor. Soc., 27–45.

Espy, J. P., 1841: *Philosophy of Storms*, Little and Brown, 552 pp.

Evans, M. S., D. Keyser, L. F. Bosart, and G. M. Lackmann, 1994: A satellite-derived classification scheme for rapid maritime cyclogenesis. *Mon. Wea. Rev.*, **122**, 1381–1416.

Farrell, B. F., 1989: Optimal excitation of baroclinic waves. *J. Atmos. Sci.*, **46**, 1193–1206.

Gan, M. A., and V. B. Rao, 1991: Surface cyclogenesis over South America. *Mon. Wea. Rev.*, **119**, 1293–1302.

Gaza, R. S., and L. F. Bosart, 1990: Trough merger characteristics over North America. *Wea. Forecasting*, **5**, 314–331.

Guinn, T. A., and W. H. Schubert, 1993: Hurricane spiral bands. *J. Atmos. Sci.*, **50**, 3380–3504.

Gyakum, J. R., 1983a: On the evolution of the QE II storm. I: Synoptic aspects. *Mon. Wea. Rev.*, **111**, 1137–1155.

——, 1983b: On the evolution of the QE II storm. II: Dynamic and thermodynamic structure. *Mon. Wea. Rev.*, **111**, 1156–1173.

——, 1986: Experiments in temperature and precipitation forecasting for Illinois. *Wea. Forecasting*, **1**, 77–88.

——, 1991: Meteorological precursors to the explosive intensification of the QE II Storm. *Mon. Wea. Rev.*, **119**, 1105–1131.

——, J. R. Anderson, R. H. Grumm, and E. L. Gruner, 1989: North Pacific cold-season surface cyclone activity: 1975–1983. *Mon. Wea. Rev.*, **117**, 1141–1155.

——, P. J. Roebber, and T. A. Bullock, 1992: The role of antecedent surface vorticity development as a conditioning process in explosive cyclone intensification. *Mon. Wea. Rev.*, **120**, 1465–1489.

Hakim, G. J., L. F. Bosart, and D. Keyser, 1995: The Ohio valley wave-merger cyclogenesis event of 25–26 January 1978. Part 1: Observations. *Mon. Wea. Rev.*, **123**, 2663–2692.

——, D. Keyser, and L. F. Bosart, 1996: The Ohio valley wave-merger cyclogenesis event of 25–26 January 1978. Part II: Diagnosis using quasigeostrophic potential vorticity inversion. *Mon. Wea. Rev.*, **124**, 2176–2205.

Hanson, H. P., and B. Long, 1985: Climatology of cyclogenesis over the east China sea. *Mon. Wea. Rev.*, **113**, 697–707.

Hartjenstein, G., and R. Bleck, 1991: Factors affecting cold-air outbreaks east of the Rocky Mountains. *Mon. Wea. Rev.*, **119**, 2280–2292.

Hines, K. M., and C. R. Mechoso, 1993: Influence of surface drag on the evolution of fronts. *Mon. Wea. Rev.*, **121**, 1152–1175.

Hirschberg, P. A., and J. M. Fritsch, 1991: Tropopause undulations and the development of extratropical cyclones. Part I: Overview and observations from a cyclone event. *Mon. Wea. Rev.*, **119**, 496–517.

Holopainen, E., and J. Kaurola, 1991: Decomposing the atmospheric flow using potential vorticity framework. *J. Atmos. Sci.*, **48**, 2614–2625.

Holton, J. R., 1992: *An Introduction to Dynamic Meteorology*, 3d ed., Academic Press, 511 pp.

Hoskins, B. J., 1990: Theory of extratropical cyclones. *Extratropical Cyclones, The Erik Palmén Memorial Volume*, C. W. Newton and E. O. Holopainen, Eds., Amer. Meteor. Soc., 63–80.

——, and P. Berrisford, 1988: A potential vorticity perspective of the storm of 15–16 October 1987. *Weather*, **23**, 122–129.

——, and P. J. Valdes, 1990: On the existence of storm-tracks. *J. Atmos. Sci.*, **47**, 1854–1864.

——, M. E. McIntyre, and A. W. Robertson, 1985: On the use and significance of isentropic potential vorticity maps. *Quart. J. Roy. Meteor. Soc.*, **111**, 877–946.

Hovanec, R. D., and L. H. Horn, 1975: Static stability and the 300 mb isotach field in the Colorado cyclogenetic area. *Mon. Wea. Rev.*, **103**, 628–638.

Hövmöller, E., 1949: The trough-and-ridge diagram. *Tellus*, **1**, 62–66.

Hsieh, Y.-P., 1949: An investigation of a selected cold vortex over North America. *J. Meteor.*, **6**, 401–410.

Huang, S.-S., Y.-B. Lin, and T.-J. Wei, 1976: The genesis and development of cyclone waves in the Yangtze and Huaihe rivers and its associated heavy rain. *Sci. Atmos. Sinica*, (1).

Iskenderian, H., 1988: Three-dimensional airflow and precipitation structure in a nondeepening cyclone. *Wea. Forecasting*, **3**, 18–32.

Johnson, D. R., 1970: The available potential energy of storms. *J. Atmos. Sci.*, **27**, 727–741.

——, and W. K. Downey, 1976: The absolute angular momentum budget of an extratropical cyclone: Quasi-Langrangian diagnostics—3. *Mon. Wea. Rev.*, **104**, 3–14.

Kato, K., 1985: On the abrupt change in the structure of the Baiu front over the China continent in late May of 1979. *J. Meteor. Soc. Japan*, **63**, 20–35.

Kennedy, S. E., and P. J. Smith, 1983: On the release of eddy available potential energy in an extratropical cyclone system. *Mon. Wea. Rev.*, **111**, 745–755.

Keshishian, L. G., L. F. Bosart, and E. W. Bracken, 1994: Inverted troughs and cyclogenesis over interior North America: A limited regional climatology and case studies. *Mon. Wea. Rev.*, **122**, 565–607.

Keyser, D., and L. W. Uccellini, 1987: Regional models: Emerging research tools for synoptic meteorologists. *Bull. Amer. Meteor. Soc.*, **68**, 306–320.

Klein, W. H., 1957: Principal tracks and mean frequencies of cyclones and anticyclones in the Northern Hemisphere. Res. Pap. No. 40, U.S. Weather Bureau. U.S. Government Printing Office, 60 pp.

——, 1958: The frequency of cyclones and anticyclones in relation to the mean circulation. *J. Meteor.*, **15**, 98–102.

Kleinschmidt, E., 1950: On the structure and origin of cyclones (Part 1). *Meteor. Rundsch.*, **3**, 1–6.

——, 1957: Cyclones and anticyclones. *Dynamic Meteorology, Handbuch der Physik*, **48**, S. Flügge, Ed., Springer-Verlag, 112–154.

Kocin, P. J., P. N. Schumacher, R. F. Morales Jr., and L. W. Uccellini, 1995: Overview of the 12–14 March 1993 superstorm. *Bull. Amer. Meteor. Soc.*, **76**, 165–182.

Krishnamurti, T. N., 1968: A study of a developing wave cyclone. *Mon. Wea. Rev.*, **96**, 208–217.

Kuo, Y.-H., and R. A. Anthes, 1982: Numerical simulation of a Mei-Yu system over southeastern Asia. *Paper Meteor. Res.*, **5**, 15–36.

——, and R. J. Reed, 1988: Numerical simulation of an explosively deepening cyclone in the eastern Pacific. *Mon. Wea. Rev.*, **116**, 2081–2105.

——, L. Cheng, and J.-W. Bao, 1988: Numerical simulation of the 1981 Sichuan flood. Part I: Evolution of a mesoscale southwest vortex. *Mon. Wea. Rev.*, **116**, 2481–2504.

——, M. A. Shapiro, and E. G. Donall, 1991a: Interaction of baroclinic and diabatic processes in numerical simulations of a rapidly developing marine cyclone. *Mon. Wea. Rev.*, **119**, 368–384.

——, R. J. Reed, and S. Low-Nam, 1991b: Effects of surface energy fluxes during the early development and rapid intensification stages of seven explosive cyclones in the western Atlantic. *Mon. Wea. Rev.*, **119**, 457–476.

——, ——, and ——, 1992: Thermal structure and airflow in a model simulation of an occluded marine cyclone. *Mon. Wea. Rev.*, **120**, 2280–2297.

Lackmann, G. M, L. F. Bosart, and D. Keyser, 1996: Planetary- and synoptic-scale characteristics of explosive wintertime cyclogenesis over the western north Atlantic ocean. *Mon. Wea. Rev.*, **124**, 2672–2702.

——, D. Keyser, and L. F. Bosart, 1997: A characteristic life cycle of upper-tropospheric cyclogenetic precursors during the experiment on Rapidly Intensifying Cyclones over the Atlantic (ERICA). *Mon. Wea. Rev.*, **125**, 2729–2758.

Lai, C.-C., and L. F. Bosart, 1988: A case study of trough merger in split westerly flow. *Mon. Wea. Rev.*, **116**, 1838–1856.

Lee, S., and I. M. Held, 1993: Baroclinic wave packets in models and observations. *J. Atmos. Sci.*, **50**, 1413–1428.

Lefevre, R. J., and J. W. Nielsen-Gammon, 1995: An objective climatology of mobile troughs in the Northern Hemisphere. *Tellus*, **47A**, 638–655.

Lim, G. H., and J. M. Wallace, 1991: Structure and evolution of baroclinic waves as inferred from regression analysis. *J. Atmos. Sci.*, **48**, 1718–1732.

Locatelli, J. D., J. M. Sienkiewicz, and P. V. Hobbs, 1989: Organization and structure of clouds and precipitation on the Mid-Atlantic coast of the United States. Part I: Synoptic evolution of a frontal system from the Rockies to the Atlantic coast. *J. Atmos. Sci.*, **46**, 1327–1348.

Lorenz, E. N., 1955: Available potential energy and the maintenance of the general circulation. *Tellus*, **7**, 157–167.

Lupo, A. R., P. J. Smith, and P. Zwack, 1992: A diagnosis of the explosive development of two extratropical cyclones. *Mon. Wea. Rev.*, **120**, 1490–1523.

Ma, K.-Y., and L. F. Bosart, 1987: A synoptic overview of a heavy rain event in southern China. *Wea. Forecasting*, **2**, 90–112.

Manobianco, J., L. W. Uccellini, K. F. Brill, and Y.-H. Kuo, 1992: The impact of dynamic data assimilation on the numerical simulations of the QE II cyclone and an analysis of the jet streak influencing the precyclogenetic environment. *Mon. Wea. Rev.*, **120**, 1973–1996.

Margules, M., 1903: Uber die Energie der Stürme. Jahrb. Zentralanst. Meteor., 1–26 (Transl. C. Abbe, 1910: *The Mechanics of the Earth's Atmosphere*. 3rd Coll., Smithsonian Inst., Washington, DC, 553–595.).

Martin, J. E., J. D. Locatelli, and P. V. Hobbs, 1993: Organization and structure of clouds and precipitation on the mid-Atlantic coast of the United States. Part VI: The synoptic evolution of a deep tropospheric frontal circulation and attendant cyclogenesis. *Mon. Wea. Rev.*, **121**, 1299–1316.

Mass, C. F., 1991: Synoptic frontal analysis: Time for a reassessment? *Bull. Amer. Meteor. Soc.*, **72**, 348–363.

——, and D. M. Schultz, 1993: The structure and evolution of a simulated, midlatitude cyclone over land. *Mon. Wea. Rev.*, **121**, 889–917.

——, W. J. Steenburgh, and D. M. Schultz, 1991: Diurnal surface-pressure variations over the continental United States and the influence of sea level reduction. *Mon. Wea. Rev.*, **119**, 2814–2830.

Matsumoto, S., K. Ninomiya, and S. Yoshizumi, 1971: Characteristic features of "Baiu" front associated with heavy rainfall. *J. Meteor. Soc. Japan*, **49**, 267–281.

Mattocks, C., and R. Bleck, 1986: Jet streak dynamics and geostrophic adjustment processes during the initial stage of lee cyclogenesis. *Mon. Wea. Rev.*, **114**, 2033–2056.

McGinley, J. A., 1982: A diagnosis of Alpine lee cyclogenesis. *Mon. Wea. Rev.*, **110**, 1271–1287.

Montgomery, M. T., and B. Farrell, 1993: Tropical cyclone formation. *J. Atmos. Sci.*, **50**, 285–332.

Montgomery, R. B., 1937: A suggested method for representing gradient flow in isentropic surfaces. *Bull. Amer. Meteor. Soc.*, **18**, 210–212.

Mullen, S. L., and D. P. Baumhefner, 1988: Sensitivity of numerical simulations of explosive oceanic cyclogenesis to changes in physical parameterizations. *Mon. Wea. Rev.*, **116**, 2289–2329.

Nakamura, H., 1992: Midwinter suppression of baroclinic wave activity in the Pacific. *J. Atmos. Sci.*, **49**, 1629–1642.

Namias, J., 1938: Thunderstorm forecasting with the aid of isentropic charts. *Bull. Amer. Meteor. Soc.*, **19**, 1–14.

——, 1939: The use of isentropic analysis in short term forecasting. *J. Aero. Sci.*, **6**, 295–298.

——, 1950: The index cycle and its role in the general circulation. *J. Meteor.*, **7**, 130–139.

——, 1978: Multiple causes of the North American abnormal winter 1976/77. *Mon. Wea. Rev.*, **106**, 279–295.

——, 1983: The history of polar front and air mass concepts in the United States—An eyewitness account. *Bull. Amer. Meteor. Soc.*, **64**, 734–755.

——, and P. F. Clapp, 1944: Studies of the motion and development of long waves in the westerlies. *J. Meteor.*, **1**, 57–77.

——, and ——, 1949: Confluence theory of the high tropospheric jet stream. *J. Meteor.*, **6**, 330–336.

Neiman, P. J., and M. A. Shapiro, 1993: The life cycle of an extratropical marine cyclone. Part I: Frontal cyclone evolution and thermodynamic air–sea interaction. *Mon. Wea. Rev.*, **121**, 2153–2176.

——, ——, and L. S. Fedor, 1993: The life cycle of an extratropical marine cyclone. Part II: Mesoscale structure and diagnostics. *Mon. Wea. Rev.*, **121**, 2177–2199.

Newton, C. W., 1954: Frontogenesis and frontolysis as a three-dimensional process. *J. Meteor.*, **11**, 449–461.

——, 1956: Mechanisms of circulation change during lee cyclogenesis. *J. Meteor.*, **11**, 528–539.

——, 1990: Erik Palmén's contributions to the development of cyclone concepts. *Extratropical Cyclones, The Erik Palmén Memorial Volume*, C. W. Newton and E. O. Holopainen, Eds., Amer. Meteor. Soc., 1–18.

——, and E. Palmén, 1963: Kinematic and thermal properties of a large-amplitude wave in the westerlies. *Tellus*, **15**, 99–119.

——, and H. R. Newton, 1994: The Bergen school concepts come to America. *The Life Cycles of Extratropical Cyclones. Vol. I*, Bergen Symposium, 27 June–1 July 1994, 22–31.

Nielsen-Gammon, J. W., 1995: Dynamical conceptual models of upper-level mobile trough formation: Comparison and application. *Tellus*, **47A**, 705–721.

——, and R. J. Lefevre, 1996: Piecewise tendency diagnosis of dynamical processes governing the development of an upper-tropospheric mobile trough. *J. Atmos. Sci.*, **53**, 3120–3142.

Ninomiya, K., 1984: Characteristics of Baiu front as a predominant subtropical front in the summer Northern Hemisphere. *J. Meteor. Soc. Japan*, **62**, 880–893.

——, and K. Akiyama, 1971: The development of the medium-scale disturbance in the Baiu front. *J. Meteor. Soc. Japan*, **49**, 663–677.

Nuss, W. A., and S. I. Kamikawa, 1990: Dynamics and boundary layer processes in two Asian cyclones. *Mon. Wea. Rev.*, **118**, 755–771.

Omoto, Y., 1966: On the structure of an intense upper cyclone (Part 1). *J. Meteor. Soc. Japan*, **44**, 320–340.

Orlanski, I., and E. K. M. Chang, 1993: Ageostrophic geopotential fluxes in downstream and upstream development of baroclinic waves. *J. Atmos. Sci.*, **50**, 212–225.

——, and ——, 1995: Stages in the energetics of a baroclinic system. *Tellus*, **47A**, 605–628.

Palmén, E., 1931: Die Beziehung zweischen troposphärischen und stratosphärischen Temperatur- und Luftdruckschwankungen. (Über die Natur der sog. premären und sekundären Druckwellen). *Beitr. Phys. Atmos.*, **17**, 102–116.

——, 1949: On the origin and structure of high-level cyclones south of the maximum westerlies. *Tellus*, **1**, 22–31.

——, 1951a: The aerology of extratropical disturbances. *Compendium of Meteorology*, T. F. Malone, Ed., Amer. Meteor. Soc., 599–620.

——, 1951b: The role of atmospheric disturbances in the general circulation (Symons Memorial Lecture). *Quart. J. Roy. Meteor. Soc.*, **77**, 337–354.

——, 1958: Vertical circulation and release of kinetic energy during the development of hurricane Hazel into an extratropical cyclone. *Tellus*, **10**, 1–23.

——, and E. O. Holopainen, 1962: Divergence, vertical velocity and conversion between potential and kinetic energy in an extratropical disturbance. *Geophysica*, **8**, 89–113.

——, and C. W. Newton, 1969: *Atmospheric Circulation Systems: Their Structure and Physical Interpretation.* Academic Press, 603 pp.

Palmer, C. W., 1942: Synoptic analysis over the southern oceans. Prof. Note No. 1, New Zealand Meteorological Office, Wellington, 38 pp.

Parker, S. S., J. T. Hawes, S. J. Colucci, and B. P. Hayden, 1989: Climatology of 500 mb cyclones and anticyclones, 1950–1985. *Mon. Wea. Rev.*, **117**, 558–570.

Pauley, P. M., and P. J. Smith, 1988: Direct and indirect effects of latent heat release on a synoptic-scale wave system. *Mon. Wea. Rev.*, **116**, 1209–1235.

Peltonen, T., 1963: A case study of an intense upper cyclone over eastern and northern Europe in November 1959. *Geophysica*, **8**, 225–251.

Petterssen, S., 1955: A general survey of factors influencing development at sea-level. *J. Meteor.*, **12**, 36–42.

——, 1956: *Weather Analysis and Forecasting.* Vol. I, 2d ed. McGraw Hill, 428 pp.

——, and S. J. Smebye, 1971: On the development of extratropical cyclones. *Quart. J. Roy. Meteor. Soc.*, **97**, 457–482.

——, D. L. Bradbury, and K. Pedersen, 1962: The Norwegian cyclone models in relation to heat and cold sources. *Geofys. Publ.*, **24**, 243–280.

Phillips, N. A., 1963: Geostrophic motion. *Rev. Geophys.*, **1**, 123–176.

Polavarapu, S. M., and W. R. Peltier, 1990: The structure and nonlinear evolution of synoptic scale cyclones: Life cycle simulations with a cloud-scale model. *J. Atmos. Sci.*, **47**, 2645–2672.

Reed, R. J., 1955: A study of a characteristic type of upper-level frontogenesis. *J. Meteor.*, **12**, 542–552.

——, 1979: Cyclogenesis in polar air streams. *Mon. Wea. Rev.*, **107**, 38–52.

——, 1990: Advances in knowledge and understanding of extratropical cyclones during the past quarter century: An overview. *Extratropical Cyclones, The Erik Palmén Memorial Volume*, C. W. Newton and E. O. Holopainen, Eds., Amer. Meteor. Soc., 27–45.

——, and F. Sanders, 1953: An investigation of the development of a mid-tropospheric frontal zone and its associated vorticity field. *J. Meteor.*, **10**, 338–349.

——, and E. F. Danielsen, 1959: Fronts in the vicinity of the tropopause. *Archiv. Meteorol., Geophys. Bioclimatol.*, **A11**, 1–17.

——, and M. D. Albright, 1986: A case study of explosive cyclogenesis in the Eastern Pacific. *Mon. Wea. Rev.*, **112**, 2297–2319.

——, and A. J. Simmons, 1991: Numerical simulation of an explosively deepening cyclone over the North Atlantic that was unaffected by concurrent surface energy fluxes. *Wea. Forecasting*, **6**, 117–122.

——, M. T. Stoelinga, and Y.-H. Kuo, 1992: A model-aided study of the origin and evolution of the anomalously high potential vorticity in the inner region of a rapidly deepening marine cyclone. *Mon. Wea. Rev.*, **120**, 893–913.

——, G. A. Grell, and Y.-H. Kuo, 1993a: The ERICA IOP5 Storm. Part I: Analysis and simulation. *Mon. Wea. Rev.*, **121**, 1577–1594.

——, ——, and ——, 1993b: The ERICA IOP5 Storm. Part II: Sensitivity tests and further diagnosis based on model output. *Mon. Wea. Rev.*, **121**, 1595–1612.

Rex, D. F., 1950a: Blocking action in the middle- troposphere and its effects on regional climate. I: An aerological study of blocking. *Tellus*, **2**, 196–211.

——, 1950b: Blocking action in the middle troposphere and its effects on regional climate. II: The climatology of blocking action. *Tellus*, **2**, 275–301.

Roebber, P. J., 1984: Statistical analysis and updated climatology of explosive cyclones. *Mon. Wea. Rev.*, **112**, 1577–1589.

——, 1989a: On the statistical analysis of cyclone deepening rates. *Mon. Wea. Rev.*, **117**, 2293–2298.

——, 1989b: The role of heat and moisture fluxes associated with large-scale ocean current meanders in maritime cyclogenesis. *Mon. Wea. Rev.*, **117**, 1676–1694.

——, 1993: A diagnostic case study of self-development as an antecedent conditioning process in explosive cyclogenesis. *Mon. Wea. Rev.*, **121**, 976–1006.

Rogers, E., and L. F. Bosart, 1986: An investigation of explosively deepening oceanic cyclones. *Mon. Wea. Rev.*, **114**, 702–718.

——, and ——, 1991: A diagnostic study of two intense oceanic cyclones. *Mon. Wea. Rev.*, **119**, 967–997.

Rossby, C.-G., 1939: Relation between variations in the intensity of the zonal circulation of the atmosphere and the displacements of the semipermanent centers of action. *J. Mar. Res.*, **2**, 38–55.

——, 1940: Planetary flow patterns in the atmosphere. *Quart. J. Roy. Meteor. Soc.*, **66** (Suppl.), 68–87.

——, 1945: On the propagation of frequencies and energy in certain types of oceanic and atmospheric waves. *J. Meteor.*, **2**, 187–204.

——, and Coauthors, 1937a: Aerological evidence of large-scale mixing in the atmosphere. Trans. Amer. Geophys. Union, 19th annual meeting.

——, and Coauthors, 1937b: Isentropic analysis. *Bull. Amer. Meteor. Soc.*, **18**, 201–209.

——, and H. C. Willett, 1948: The circulation of the upper troposphere and lower stratosphere. *Science*, **108**, 643–652.

Salmon, E. W., and P. J. Smith, 1980: A synoptic analysis of the 25–26 January 1978 blizzard cyclone in the central United States. *Bull. Amer. Meteor. Soc.*, **61**, 453–460.

Sanders, F., 1986: Explosive cyclogenesis over the west-central North Atlantic Ocean, 1981–84. Part I: Composite structure and mean behavior. *Mon. Wea. Rev.*, **114**, 1781–1794.

——, 1987: A study of 500 mb vorticity maxima crossing the east coast of North American and associated surface cyclogenesis. *Wea. Forecasting*, **2**, 70–83.

——, 1988: Life history of mobile troughs in the upper westerlies. *Mon. Wea. Rev.*, **116**, 2629–2648.

——, and J. R. Gyakum, 1980: Synoptic-dynamic climatology of the "bomb." *Mon. Wea. Rev.*, **108**, 1589–1606.

——, and L. F. Bosart, 1985a: Mesoscale structure in the megalopolitan snowstorm of 11–12 February 1983. Part I: Frontogenetical forcing and symmetric instability. *J. Atmos. Sci.*, **42**, 1050–1061.

——, and ——, 1985b: Mesoscale structure in the megalopolitan snowstorm of 11–12 February 1983. Part II: Doppler radar study of the New England snowband. *J. Atmos. Sci.*, **42**, 1398–1407.

——, and C. A. Davis, 1988: Patterns of thickness anomaly for explosive cyclogenesis over the west-central North Atlantic Ocean. *Mon. Wea. Rev.*, **116**, 2725–2730.

——, and C. A. Doswell III, 1995: A case for detailed surface analysis. *Bull. Amer. Meteor. Soc.*, **76**, 505–521.

——, L. F. Bosart, and C.-C. Lai, 1991: Initiation and evolution of an intense upper-level front. *Mon. Wea. Rev.*, **119**, 1337–1367.

Satyamurty, P., C. D. C. Ferreira, and M. A. Gan, 1990: Cyclonic vortices over South America. *Tellus*, **42A**, 194–201.

Sawyer, J. S., 1956: The vertical circulation at meteorological fronts and its relation to frontogenesis. *Proc. Roy. Soc. London*, **A234**, 346–362.

Schär, C. J., and H. Wernli, 1993: Structure and evolution of an isolated semi-geostrophic cyclone. *Quart. J. Roy. Meteor. Soc.*, **119**, 57–90.

Schultz, D. M., and C. F. Mass, 1993: The occlusion process in a midlatitude cyclone over land. *Mon. Wea. Rev.*, **121**, 918–940.

——, W. E. Bracken, L. F. Bosart, G. J. Hakim, M. A. Bedrick, M. J. Dickinson, and K. R. Tyle, 1997: The 1993 Superstorm cold surge: Frontal structure, gap flow, and tropical impact. *Mon. Wea. Rev.*, **125**, 5–39.

Shapiro, M. A., 1974: A multiple structured frontal zone-jet stream system as revealed by meteorologically instrumented aircraft. *Mon. Wea. Rev.*, **102**, 244–253.

——, 1980: Turbulent mixing within tropopause folds as a mechanism for the exchange of chemical constituents between the stratosphere and troposphere. *J. Atmos. Sci.*, **37**, 994–1004.

——, and D. Keyser, 1990: Fronts, jet streams and the tropopause. *Extratropical Cyclones, The Erik Palmén Memorial Volume*, C. W. Newton and E. O. Holopainen, Eds., Amer. Meteor. Soc., 167–191.

Shaw, N., 1923: *The Air and Its Ways: The Rede Lecture (1921) in the University of Cambridge, with other Contributions to Meteorology for Schools and Colleges.* The University Press, Cambridge, 237 pp.

——, and R. G. K. Lempfert, 1906: Life History of surface air currents. Meteorol. Committee, His Majesty's Stationery Office, 174.

Shutts, G. J., 1990: Dynamical aspects of the October storm 1987: A study of a successful finemesh simulation. *Quart. J. Roy. Meteor. Soc.*, **116**, 1315–1347.

Simmons, A. J., and B. J. Hoskins, 1979: The downstream and upstream development of unstable baroclinic waves. *J. Atmos. Sci.*, **36**, 1239–1254.

Sinclair, M. R., 1993: Synoptic-scale diagnosis of the extratropical transition of a southwest Pacific tropical cyclone. *Mon. Wea. Rev.*, **121**, 941–960.

——, 1994: An objective cyclone climatology for the Southern Hemisphere. *Mon. Wea. Rev.*, **122**, 2239–2256.

——, 1995: A climatology of cyclogenesis for the Southern Hemisphere. *Mon. Wea. Rev.*, **123**, 1601–1619.

——, 1996: A climatology of anticyclones and blocking for the Southern Hemisphere. *Mon. Wea. Rev.*, **124**, 245–263.

Staley, D. O., 1960: Evaluation of potential vorticity changes near the tropopause and the related vertical motions, vertical advection of vorticity and transfer of radioactive debris from stratosphere to troposphere. *J. Meteor.*, **17**, 591–620.

Starr, V. P., and M. Neiburger, 1940: Potential vorticity as a conservative property. *J. Mar. Res.*, **3**, 202–210.

Stoelinga, M. T., 1996: A potential vorticity-based study of the role of diabatic heating and friction in a numerically simulated baroclinic cyclone. *Mon. Wea. Rev.*, **124**, 849–874.

Sutcliffe, R. C., 1939: Cyclonic and anticyclonic development. *Quart. J. Roy. Meteor. Soc.*, **65**, 518–524.

——, 1947: A contribution to the problem of development. *Quart. J. Roy. Meteor. Soc.*, **73**, 370–383.

——, and A. G. Forsdyke, 1950: The theory and use of upper air thickness patterns in forecasting. *Quart. J. Roy. Meteor. Soc.*, **76**, 189–217.

Swanson, K., and R. T. Pierrehumbert, 1994: Nonlinear wave packet evolution on a baroclinically unstable jet. *J. Atmos. Sci.*, **51**, 384–396.

Takayabu, I., 1991: "The coupling development": An efficient mechanism for the development of extratropical cyclones. *J . Meteor. Soc. Japan*, **69**, 609–628.

Taljaard, J. J., 1972: Synoptic meteorology of the Southern Hemisphere. *Meteor. Monogr.*, No. 13, 139–214.

——, W. Schmidt, and H. van Loon, 1961: Frontal analysis with application to the Southern Hemisphere. *Notos*, **10**, 25–58.

Thorncroft, C. D., B. J. Hoskins, and M. E. McIntyre, 1993: Two paradigms of baroclinic-wave life-cycle behavior. *Quart. J. Roy. Meteor. Soc.*, **119**, 17–56.

Thorpe, A. J., 1985: Diagnosis of balanced vortex structure using potential vorticity. *J. Atmos. Sci*, **42**, 397–406.

Tibaldi, S., A. Buzzi, and A. Speranza, 1990: Orographic cyclogenesis. *Extratropical Cyclones, The Erik Palmén Memorial Volume*, C. W. Newton and E. O. Holopainen, Eds., Amer. Meteor. Soc., 107–127.

Tracton, M. S., 1973: The role of cumulus convection in the development of extratropical cyclones. *Mon. Wea. Rev.*, **101**, 573–593.

Trenberth, K. E., 1981: Observed Southern Hemisphere eddy statistics at 500 mb: Frequency and spatial dependence. *J. Atmos. Sci.*, **38**, 2585–2605.

——, 1991: Storm tracks in the Southern Hemisphere. *J. Atmos. Sci*, **48**, 2159–2178.

Tsou, C.-H., P. J. Smith, and P. M. Pauley, 1987: A comparison of adiabatic and diabatic forcing in an intense extratropical cyclone system. *Mon. Wea. Rev.*, **115**, 763–786.

Uccellini, L. W., 1986: The possible influence of upstream upperlevel baroclinic processes on the development of the QEII storm. *Mon. Wea. Rev.*, **114**, 1019–1027.

——, 1990: Processes contributing to the rapid development of extratropical cyclones. *Extratropical Cyclones, The Erik Palmén Memorial Volume*, C. W. Newton and E. O. Holopainen, Eds., Amer. Meteor. Soc., 81–105.

——, P. J. Kocin, R. A. Petersen, C. H. Wash, and K. F. Brill, 1984: The Presidents' Day cyclone of 18–19 February 1979: Synoptic overview and analysis of the subtropical jet streak influencing the precyclogenetic period. *Mon. Wea. Rev.*, **112**, 31–55.

——, D. Keyser, K. F. Brill, and C. H. Wash, 1985: The Presidents'

Day cyclone of 18–19 February, 1979: The influence of upstream trough amplification and associated tropopause folding on rapid cyclogenesis. *Mon. Wea. Rev.*, **113**, 962–988.

——, R. A. Petersen, K. F. Brill, P. J. Kocin, and J. J. Tuccillo, 1987: Synergistic interactions between an upper-level jet streak and diabatic processes that influence the development of a low-level jet and a secondary coastal cyclone. *Mon. Wea. Rev.*, **115**, 2227–2261.

——, S. F. Corfidi, N. W. Junker, P. J. Kocin, and D. A. Olson, 1992: Report on the Surface Analysis Workshop held at the National Meteorological Center 25–28 March 1991. *Bull. Amer. Meteor. Soc.*, **73**, 459–472.

van Loon, H., 1965: A climatological study of the atmospheric circulation in the Southern Hemisphere during IGY. Part I: 1 July 1957–31 March 1958. *J. Appl. Meteor.*, **4**, 479–491.

Velden, C. S., and G. A. Mills, 1990: Diagnosis of upper-level processes influencing an unusually intense extratropical cyclone over southeast Australia. *Wea. Forecasting*, **5**, 449–482.

Wallace, J. M., G.-H. Lim, and M. L. Blackmon, 1988: Relationships between cyclone tracks, anticyclone tracks and baroclinic waveguides. *J. Atmos. Sci.*, **45**, 439–462.

Wang, W., Y.-H. Kuo, and T. T. Warner, 1993: A diabatically driven mesoscale vortex in the lee of the Tibetan Plateau. *Mon. Wea. Rev.*, **121**, 2542–2561.

Wash, C. H., R. A. Hale, P. H. Dobos, and E. J. Wright, 1992: Study of explosive and nonexplosive cyclogenesis during FGGE. *Mon. Wea. Rev.*, **120**, 40–51.

Wexler, H., and J. Namias, 1938: Mean monthly isentropic charts and their relation to departures of summer rainfall. Transactions, American Geophysical Union, 19th Annual Meeting, 164–170.

Whittaker, J. S., L. W. Uccellini, and K. F. Brill, 1988: A model-based diagnostic study of the rapid development phase of the Presidents' Day cyclone. *Mon. Wea. Rev.*, **116**, 2337–2365.

Whittaker, L. M., and L. H. Horn, 1981: Geographical and seasonal distribution of North American cyclogenesis, 1958–1977. *Mon. Wea. Rev.*, **109**, 2312–2322.

——, and ——, 1984: Northern Hemisphere extratropical cyclone activity for four mid-season months. *J. Climatol.*, **4**, 297–310.

Willett, H. C., 1949: Long-period fluctuations of the general circulation of the atmosphere. *J. Meteor.*, **6**, 34–50.

Yeh, T.-C., 1949: On energy dispersion in the atmosphere. *J. Meteor.*, **6**, 1–16.

Young, M. V., and J. R. Grant, 1995: Depressions in midlatitudes. *Images in Weather Forecasting*, M. J. Bader, G. S. Forbes, J. R. Grant, R. B. E. Lilley, and A. J. Waters, Eds., Cambridge University Press, 206–212.

Zishka, K. M., and P. J. Smith, 1980: The climatology of cyclones and anticyclones over North America and surrounding ocean environs for January and July, 1950–77. *Mon. Wea. Rev.*, **108**, 387–401.

Theories of Frontogenesis

HUW C. DAVIES

Institute for Atmospheric Science, ETH, Zürich, Switzerland

Abstract

A review is presented of theoretical frontogenesis studies. The framework for the study is that of quasi- and semigeostrophic dynamics, and the approach is to consider idealized models that incorporate some presumed quintessential frontogenetic process(es). The focus of the study is the temporal development and nature of surface fronts.

Insight is sought first from consideration of two archetypal settings for two-dimensional frontal formation—the Bergeron configuration for deformation-induced frontogenesis and the Eliassen configuration for horizontal shear-induced frontogenesis. These two settings serve to illustrate the role of the ageostrophic circulation in frontal scale-contraction and the key features and distinction between prototype cold and warm fronts. They also form a platform for the exploration of the factors that can limit the scale-contraction and generate unbalanced flow, and the examination of frictional and diabatic effects.

The concurrent development of fronts and cyclones is studied in the extended Eady-setting of baroclinic waves growing on a jet-like basic flow. Simulations reproduce variants of the classic λ-shaped configuration for the surface frontal palette that include elongated cold fronts, warm fronts that arch around the center of the depression, and the fracture of the cold front in the vicinity of the cyclone. Consideration is also given to the frontal palette associated with paradigmatic precursor patterns for cyclogenesis. Diagnostic analyses based predominantly on the potential vorticity perspective illuminate the role of ambient lateral shear and the precursor flow pattern in determining the nature of the frontal palette.

1. Introduction

Fronts in the form of elongated (~1000 km long and ~100 km wide) bands of enhanced baroclinicity are distinctive and key features of surface weather patterns. The terms "frontogen-esis" and "frontolysis" were introduced to signify, respectively, the dynamical development and decay of such fronts and were coined by a member of the Bergen school (Bergeron 1928). There were three salient ingredients to Bergeron's landmark study. First, there was the recognition that fronts were not immutable but underwent a life cycle. Thus the study directed attention to the role of the larger-scale flow in generating fronts and was a counterpoint to the Bergen school's earlier portrayal of midlatitude weather patterns in terms of the distortion and convolution of the pre-existing polar-front. Second, there was the identification and the theoretical consideration of the possible frontogenetic role of deformation in frontogenesis. Bergeron's study constitutes a forerunner of a class of theoretical frontogenesis studies that, based upon idealized and readily analyzable mathematical models that incorporate some presumed quintessential frontogenetic feature(s), seek to advance our understanding/ insight of the natural phenomena. Third, Bergeron explored briefly the implications of the linkage between deformation and frontogenesis to the climatology of surface fronts.

This review attempts to follow the same pattern. It seeks to distill the knowledge and understanding of frontal dynamics that has been gained from idealized mathematical models, and to relate the results to the observed range of fronts. Particular consideration will be given to the factors that determine the temporal development, nature, and classification of *surface* fronts. Detailed consideration of observational and diagnostic aspects of frontogenesis can be found in the contributions to this book of Browning, Keyser, Shapiro, and Thorpe.

2. Observational and Theoretical Backcloth

2.1 Some Salient Observational Characteristics

Surface fronts are regarded as slender transition zones possessing certain salient flow characteristics. These include the aforementioned enhancement of the horizontal baroclinicity ($\nabla_h \theta$) relative to the background value, a cyclonic wind shift, and a kink in the isobars across the zone—consistent with the existence of a locally intensified vertical component of relative vorticity (ζ), significant prefrontal bands of strong relative airflow (so-called frontal airstreams), and substantial bands of convergence and cloud activity. Cold fronts are usually elongated and often match this standard description whereas warm fronts tend to be stubbier, less distinct at the surface, and have a shallower slope aloft.

Both cold and warm fronts are depicted on surface synoptic charts as either the main components of an incipient wave disturbance growing within a frontal zone or, in combination with their occluded progeny, as the subsynoptic embroidery of a maturing or matured low pressure system. An incipient frontal palette accompanying cyclogenesis exhibits a $\wedge$ shape in the Northern Hemisphere with the apex at the cyclone's center and the southwestern- and southeastern-directed segments corresponding to the cold and warm fronts. Thereafter the pattern frequently evolves to λ-shape with the cold front appearing to overtake the warm front to form an "occlusion" in the neighborhood of the cyclone. However, there is considerable variation from event to event, and particularly noteworthy are the following variants: the frequent absence of a distinct warm frontal structure with merely the presence of an amorphous band of "weather" in its presumed vicinity; the nonoccurrence of an occlusion process; a fracture of the original baroclinic zone such that the cold front is dislocated from the cyclone; and the occurrence of a strong warm front that arches around the poleward sector of the low.

There are also strong regional differences in the form of the prevalent frontal types. For example, synoptic experience suggests that warm fronts tend to occur less frequently in the vicinity of Australia, whereas they are the most frequent and distinctive frontal type over northeastern Canada.

2.2. A Theoretical Framework

First consider the theoretical rudiments of frontogenesis. A natural starting point is to examine the kinematics accompanying the generation of intense thermal gradients. Vector and scalar forms for the Lagrangian development of a fluid parcel's potential temperature gradient ($\nabla\theta$) are,

$$\frac{D}{Dt}(\nabla\theta) = -(\nabla\theta \cdot \nabla)\mathbf{v} - (\nabla\theta \wedge \varpi) + \nabla E, \quad (2.1)$$

and

$$\frac{D}{Dt}\frac{1}{2}(\nabla\theta)^2 = -(\nabla\theta)\cdot(\nabla\theta\cdot\nabla)\mathbf{v} + (\nabla\theta)\cdot(\nabla E), \quad (2.2a)$$

$$= -(\nabla\theta)^2(\partial v / \partial n) + (\nabla\theta)\cdot(\nabla E), \quad (2.2b)$$

Here ϖ is the vector vorticity, E is a term proportional to the diabatic heating rate, and v is the velocity component in the n-direction aligned perpendicular to the isentropes along

the baroclinicity vector. (Lalaurette et al. (1994) record the tensor equivalent of the foregoing relationships.)

Equation (2.1) indicates that, following a fluid parcel, the three-dimensional baroclinicity ($\nabla\theta$) can be modified in three ways: a tendency for "baroclinicity lines" to move *against* the flow (cf. the equation for a material line $D\mathbf{a}/Dt = \{\mathbf{a}\cdot\nabla\}\mathbf{v}$) involving both tilting and compression; reorientation by the ambient vorticity; and generation/destruction due to spatial variations of diabatic effects. Likewise, Eq. (2.2) emphasizes that a decrease of v along the baroclinicity vector connotes frontogenesis.

There are three significant caveats attached to the foregoing general formulae. First, surface frontogenesis is usually linked to the horizontal component ($\nabla_h\theta$) of the baroclinicity. Hence it is appropriate to note that at a flat surface the adiabatic form of Eq. (2.2) reduces to (Petterssen 1936),

$$\frac{D}{Dt}|\nabla_h\theta| = \frac{1}{2}|\nabla_h\theta|\{D\cos 2\delta - (\nabla\cdot\mathbf{v}_h)\}, \quad (2.3)$$

where D is the resultant deformation and δ is the local angle between the dilatation axis and the surface isentrope. This vector invariant form of the equation serves to emphasize the role of deformation and horizontal divergence, and moreover the latter term disappears in the geostrophic limit.

Second, in a framework moving with the front, the along-front wind component tends to dominate the across-front component. This hampers direct inference of the frontogenetic tendency using Eq. 2.2 because this involves estimating the spatial gradient ($\partial v/\partial n$) of the across-front flow. In the geostrophic limit this difficulty can be circumvented because at the surface the equivalent of Eq.(2.1) takes the form (Hoskins et al. 1978),

$$\frac{D}{Dt}(\nabla_h\theta) = \mathbf{Q} + \nabla_h E, \quad (2.4a)$$

where

$$\mathbf{Q} = -|\nabla_h\theta|\{\mathbf{k} \wedge (\partial\mathbf{v}_g / \partial s)\}, \quad (2.4b)$$

with (s, n) prescribing a local Cartesian framework such that s is aligned along an isentrope and n points toward low thermal values. Thus in this formulation the frontogenetic forcing $\mathbf{Q}$ is expressed in terms of the product of ($\nabla_h\theta$) and the along-front variation of the geostrophic velocity ($\mathbf{v}_g$), and a useful qualitative appraisal of the frontogenetic tendency can be ascertained directly by a visual inspection of the geopotential and thermal fields on a standard low-level synoptic chart. The vector $\mathbf{Q}$ can also be recast in natural coordinates (Keyser et al. 1988) or decomposed into its invariant components (Schär and Wernli 1993).

The third caveat pertains to all the foregoing formulas.

Each provides a kinematic index of the frontogenesis experienced by fluid parcels that are within but also possibly traversing across a frontal zone, as opposed to representing a dynamic frontogenesis index for the front itself.

Consideration of frontal dynamics requires an internally consistent, self-contained, and appropriate theoretical framework. The focus here is on the quintessence of frontogenesis, and this can be explored using the quasigeostrophic set of equations (QG) for an incompressible and inviscid fluid on an f-plane. The set takes the compact form expressed by the conservation, in the absence of diabatic effects, of quasigeostrophic potential vorticity (q) in the interior,

$$\left(\frac{D}{Dt}\right)_{QG}(q) = (f/N^2)(g/\theta_0)\partial E/\partial z , \quad (2.5a)$$

and potential temperature at rigid horizontal boundaries,

$$\left(\frac{D}{Dt}\right)_{QG}(\theta) = E \quad (2.5b)$$

at, say, $z = 0$, where

$$\left(\frac{D}{Dt}\right)_{QG} = \{\partial/\partial t + u_g\partial/\partial x + v_g\partial/\partial y\} \quad (2.5c)$$

is the advective derivative following the geostrophic flow $\mathbf{v}_g = [-\partial\psi/\partial y, \partial\psi/\partial x]$. The potential vorticity (q) is given by,

$$\left(\frac{D}{Dt}\right)_{QG} = \{\partial/\partial t + u_g\partial/\partial x + v_g\partial/\partial y\} , \quad (2.5d)$$

and here the potential temperature (θ),

$$(g/\theta_0)\theta = f(\partial\psi/\partial z) , \quad (2.5e)$$

refers to the deviation from a basic state defined by a reference value θ_0 plus an ambient stratification N^2 (here assumed for simplicity to be uniform).

In this framework key frontal parameters—the vertical component of vorticity (ζ), the vertical velocity (w), and the baroclinicity ($\nabla_h\theta$) at the surface—are interrelated via the equations (Davies and Wernli 1997)

$$\left(\frac{D}{Dt}\right)_{QG}(\zeta) = f(\partial w/\partial z) \quad (2.6)$$

$$\{N^2(\nabla_h^2 w) + f^2(\partial^2 w/\partial z^2)\} = (g/\theta_0)\{2(\nabla_h \cdot \mathbf{Q}) + \nabla_h^2 E\} \quad (2.7)$$

$$\left(\frac{D}{Dt}\right)_{QG}(\nabla_h\theta) = \mathbf{Q} + \nabla_h(E) \text{ at } z = 0. \quad (2.8)$$

Hence a suitable distribution of ageostrophic forcing (i.e., the $\mathbf{Q}$ field) and/or diabatic heating (the E-field) within a frontal zone could concomitantly enhance the ambient thermal gradient (Eq. 2.8), induce low-level ascent (Eq. 2.7), and thereby increase the vorticity (2.6).

The basic QG set (2.5) is itself internally consistent and self-contained provided the diabatic heating rate is expressible in terms of the geostrophic field. However, in the present context the appropriateness of the set is called into question by the intrinsic subsynoptic space and time scales of observed frontogenesis events. Major shortcomings are linked to the strength of the ageostrophic circulation (and with it the neglect of tilting effects in (2.8)), the amplitude of the relative vorticity (ζ), and the horizontal variation of the stratification. The ageostrophic circulation can be significant and, although partially prescribed by (2.7), it is not taken into account in the advection of q and θ (2.5a,b). Likewise ζ attains values comparable to f, but ($\zeta + f$) is replaced by f in the term (on the right-hand side of (2.6)) for the production of vorticity due to horizontal convergence.

These shortcomings are to a measure circumvented in the semigeostrophic system (SG). It is based on the observation that frontal flow is quasi-balanced with anisotropic synoptic along-front and meso across-front scales, and that on a time-scale comparable to f^{-1} the fluid parcels move predominantly along rather than traversing substantially across the frontal zone. The resulting inference is that the essence of two-dimensional frontogenesis is captured by assuming geostrophy of the along-front flow component, that is, the semigeostrophic approximation (Eliassen 1962). A natural extension to three-dimensions is the so-called geostrophic momentum system (Hoskins 1975, 1982) wherein the geostrophic momentum is advected by the sum of the two-dimensional geostrophic flow ($\mathbf{v}_g$) and the three-dimensional geostrophically forced ageostrophic flow ($\mathbf{v}_{ag}$), that is,

$$\left(\frac{D}{Dt}\right)_{SG}(\mathbf{v}_g) = \left[(\partial/\partial t) + \{(\mathbf{v}_g + \mathbf{v}_{ag}) \cdot \nabla\}\right](\mathbf{v}_g) \quad (2.9)$$

Here and hereafter the terms semigeostrophic and geostrophic momentum are deemed synonymous. Hoskins (1975) further showed that, subject to the transform,

$$[X, Y, Z, T] = [\{x + (v_g/f)\}, \{y - (u_g/f)\}, z, t](2.10)$$

the SG set assumes a form similar to that of the QG set, and indeed the two are isomorphic for the case of uniform interior potential vorticity.

The philosophy adopted here is to seek insight and qualitative interpretations using the QG and SG sets but bearing in mind their limitations.

3. Frontogenesis in Two Dimensions

Quasi two-dimensional models of frontogenesis are idealizations geared specifically to elucidating the nature of the process in the presence of certain simple archetypal **Q**-field configurations. Two basic configurations for frontogenesis (assumed to be aligned along the y-coordinate) are illustrated in Fig. 1, and they represent: (i) a horizontal deformation field, $[u_d = -\alpha x, v_d = \alpha y]$, acting on a two-dimensional thermal distribution $\theta = \theta(x,z)$; and (ii) horizontal shear $v = v(x,z)$ aligned transverse to an ambient field of uniform baroclinicity $U = \Lambda z$. These two configurations correspond, respectively, to those postulated by Bergeron (1928) and Eliassen (1959, 1990). At the surface they represent, respectively, the two forcing terms in the x-component of (2.1), i.e.,

$$\frac{D}{Dt}(\partial\theta/\partial x) = -\{(\partial\theta/\partial x)(\partial u/\partial x) + (\partial\theta/\partial y)(\partial v/\partial x)\} \quad (3.1)$$

$$\text{at } z = 0$$

The time development of these two configurations has been the subject of intensive study for the limit case of the *adiabatic* and *inviscid* flow of a medium of *uniform potential vorticity* (e.g., Stone 1966; Williams 1967, 1968; Hoskins 1971; Hoskins and Bretherton 1972; Blumen 1981; Bannon 1984; Davies and Müller 1988).

In this limit the SG set takes the form of Eq. (2.5) with (x, z, t) replaced by (X, Z, T), so that the equation for the potential vorticity departure away from its uniform background value is

$$\{\psi_{XX} + (f/N)^2 \psi_{ZZ}\} = 0, \quad (3.2)$$

and the thermal boundary equation is a reduced form of Eq. (2.5b). Also the cross-front ageostrophic circulation can be deduced from the Sawyer-Eliassen equation (Sawyer 1956; Eliassen 1962) for the streamfunction (χ),

$$(N/f)^2 \chi_{XX} + \chi_{ZZ} = \Im. \quad (3.3)$$

This is a two-dimensional refined form of Eq. (2.8) with $\Im$ denoting the geostrophic forcing field and the ageostrophic flow components given by $(u', w) = [-(\partial\chi/\partial z), (\partial\chi/\partial x)]$. Specific solutions of this system (i.e., Eqs. (3.2), (2.5b), and (3.3)) can be derived in geostrophic space, $[X, Z, T]$, and then displayed in physical space using the reverse of the transform (2.10).

The approach adopted here is to first consider separately surface frontogenesis for the two foregoing configurations, and then to comment upon their limitations and the modifying effect of cloud diabatic heating and surface friction.

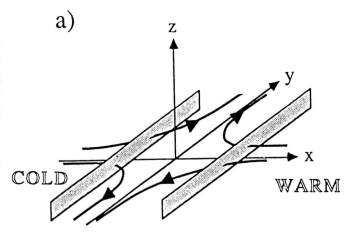

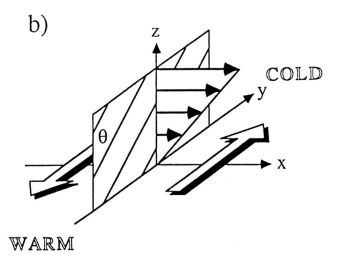

FIG. 1. Schematic of the basic states for (a) the Bergeron and (b) the Eliassen archetypal settings corresponding respectively to deformation-induced and horizontal shear–induced frontogenesis.

3.1 Deformation-Induced Frontogenesis

For illustrative purposes consider a semi-infinite domain ($z > 0$) with a flow setting comprising a pure horizontal deformation field, $(u_d = -\alpha x, v_d = \alpha y)$, and an initial thermal distribution composed of a reference potential temperature θ_0, uniform stratification N, and a spatial variation $\theta = \theta(X, Z)$.

Surface frontogenesis accomplished by the deformation field alone—that is, merely by the quasigeostrophic flow (or equivalently by the dynamics in geostrophic space)—is described by the following reduced form of Eq. (3.1),

$$\left(\frac{D}{Dt}\right)_{SG}(\partial\theta/\partial X) = (\alpha)(\partial\theta/\partial X) \text{ at } Z = 0.$$

Hence at the surface a fluid parcel's baroclinicity would increase such that after a lapse-time T,

$$\{\partial \theta / \partial X\}_{T=T} = \{\partial \theta / \partial X\}_{T=0} \exp\{\alpha T\},$$

and it would require an infinite time to develop a frontal discontinuity.

The essence of the SG dynamics can be inferred as follows. For this configuration the geostrophic forcing field $\Im$ takes the form,

$$\Im = +2(\alpha / f^2)(g / \theta_0)(\partial \theta / \partial X)$$
$$= +2(\alpha / f)(\partial v / \partial Z)$$

The solution of Eq. (3.3) in the half space $Z > 0$ is given by (Davies and Müller 1988):

$$\chi = (\alpha / f) z \, v + \text{const.}$$

so that

$$u' = -(\alpha / f)(v + z \, \partial v / \partial z), \text{ and } w = (\alpha / f)(z \, \partial v / \partial x). \quad (3.4a)$$

It follows that at the surface,

$$(\partial u' / \partial x) = -(\alpha / f)(\partial v / \partial x). \quad (3.4b)$$

Hence the surface relative vorticity component ($\partial v / \partial x$) and surface horizontal baroclinicity ($\partial \theta / \partial x$) satisfy the equations (see Eqs. (2.9) and (3.1)),

$$\left(\frac{D}{Dt}\right)_{SG} (\partial v / \partial x) = (\alpha / f)(\partial v / \partial x)(\partial v / \partial x + f), \text{ at } Z = 0,$$
$$(3.5)$$

$$\left(\frac{D}{Dt}\right)_{SG} (\partial \theta / \partial x) = (\alpha / f)(\partial \theta / \partial x)(\partial v / \partial x + f), \text{ at } Z = 0,$$
$$(3.6)$$

and

$$\left(\frac{D}{Dt}\right)_{SG} [(\partial v / \partial x) / (\partial \theta / \partial x)] = 0 \text{ at } Z = 0. \quad (3.7)$$

It follows, since the ageostrophic surface convergence remains in phase with the surface vorticity (3.4b), that the nonlinear vortex-stretching effect (the term on the right-hand side of (3.5)) and **Q**-forcing (the term on the right-hand side of (3.6)) continually increase the growth rates of the vorticity and baroclinicity of all surface fluid parcels with positive vorticity. Thus there is a sustained and increasing rate of frontogenesis, and the two frontal parameters become discontinuous after a lapse time t_c,

$$(\alpha t_c) = \ln\left\{1 + (f / v_x)_{\min(z=0, t=0)}\right\}. \quad (3.8)$$

An example of deformation-induced frontogenesis is displayed in Fig. 2. It shows in physical space the solution whose thermal development in geostrophic space (X, Z, T) is given by

$$\theta(X, Z, T) = (\Delta \theta)(X^*) / \{(X^*)^2 + (Z^*)^2\} \text{ for } Z > 0 \quad (3.9)$$

where
$X^* = (X/L) \exp(\alpha T)$ and $Z^* = [1 + (N/f)^2 (Z/L) \exp(\alpha T)]$.

In effect an initial thermal transition zone of amplitude ($\Delta \theta$) and width $2L$ (see Fig. 2a) undergoes a scale contraction (Fig. 2b) and attains an infinite value of surface baroclinicity (and vorticity) after a time $t = t_c$,

$$(\alpha t_c) = \ln\{(8/3\sqrt{3})R_0^{-1}\} \quad (3.10)$$

at the location $x = x_c$,

$$(x_c / L) = (9/8)R_0\{1 + \Xi / \sqrt{3}L\} \quad (3.11)$$

Here $R_0 = V/(fL)$ is a Rossby Number based upon a measure (V) of the along-front velocity, $V = [g(\Delta \theta)/(N\theta_0)]$, and Ξ refers to an initial displacement of the θ-field relative to the dilatation axis. In the approach to this scale collapse the vertical slope of the baroclinicity maximum (see Fig. 2c) can be shown to approach the value $\sim \sqrt{3}(f/N)$. The direct ageostrophic circulation, with surface convergence focused at the vorticity maximum and ascent aligned along this baroclinicity axis, is evident in Fig. 2d.

The development is governed by the two dimensionless parameters—R_0 and (Ξ/L). Higher values of R_0—that is, stronger fronts and/or a less stable ambient atmosphere (cf. Bannon 1984)—induce a more rapid frontogenesis. For typical atmospheric values, $t_c \sim 3$ days. This is not particularly rapid because it is comparable to the time scale of cyclogenesis and, moreover, our idealized model assumes that the deformation remains optimally aligned to the baroclinicity throughout the frontal contraction. Again, the relative displacement (Ξ/L) modifies the propagation speed and the final location x_c of the baroclinicity maximum, but not the internal frontal structure. Hence the direction of frontal movement, determined by the sign of the factor $\{1 + \Xi / \sqrt{3}L\}$—see Eq. (3.11)—is not a dynamical indicator of the strength (strong/weak) or type (cold/warm) of the front.

Ancillary inferences can also be drawn from Eq. (3.5) regarding frontal predictability and the development of multiple fronts. First, from Eq. (3.5) it can be deduced that errors in the vorticity field will amplify rapidly (decay) in regions of positive (negative) relative vorticity. In turn, this implies that

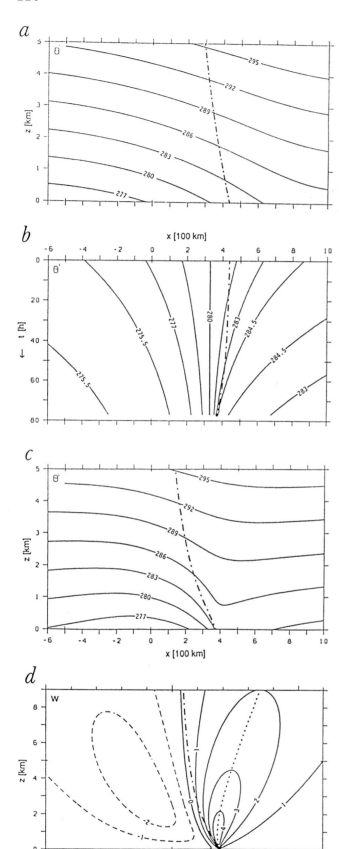

for this idealized model, setting accurate prediction of frontogenesis requires an accurate specification of the flow field within and in the vicinity of the frontal zone. Second, a surface vorticity maximum is always linked to the same fluid parcel, and each local maximum will undergo frontogenesis. Thus the presence of vorticity/thermal maxima within the zone will lead to a multiple frontal structure, and this bolsters the suggestion (Hoskins et al. 1984) that frontal rainbands may be caused by mesoscale inhomogeneities in the potential temperature field prior to the onset of frontogenesis.

It is apparent from the foregoing that frontogenesis induced by a pure deformation field, although clearly an overt idealization of the observed phenomenon, serves as an excellent theoretical model insofar as it readily yields a rich variety of transparent and conceptual insights. There are also occasions when it replicates the evolution of an observed event (see, e.g., Ostdiek and Blumen 1995). However, the amplitude (~1 cm s^{-1}) of the vertical velocity induced in this setting is significantly less than that of observed frontal ascent, and this suggests that other effects, in particular diabatic and surface frictional effects, are seminal to this aspect of frontogenesis.

3.2 Horizontal Shear-Induced Frontogenesis

In this configuration a horizontal shear $v = v(x,z)$ is aligned transverse to an ambient field of uniform baroclinicity with its associated shear, $U = \Lambda z$. The rate of surface frontogenesis is now prescribed by the other reduced form of Eq. (3.1), viz.

$$\frac{D}{Dt}(\partial\theta / \partial x) = -\{(\partial\theta / \partial y)(\partial v / \partial x)\} \text{ at } z = 0, \quad (3.12)$$

that is, the baroclinicity of a fluid parcel will increase if the vorticity and baroclinicity are suitably correlated.

In the SG limit, the equations governing this system are the expression for the PV (i.e., Eq. (3.2)),

$$\{\psi_{XX} + (f / N)^2 \psi_{ZZ}\} = 0, \quad (3.13)$$

and the thermal development on the bounding surface(s)—the reduced form of Eq. (2.5b),

---

FIG. 2. Features of a semigeostrophic realization of deformation-induced frontogenesis. The panels show (a) the cross-section of the initial isentropic distribution, (b) the space-time development of the surface isentropes, and (c) the cross-sectional distribution of the isentropic distribution and (d) the vertical velocity field shortly before frontal collapse. The dash-dot curves refer to the baroclinicity maximum, and the dotted line in panel (d) demarks the vorticity maximum. Units are deg K for the isentropes and mm s^{-1} for the vertical velocity (adapted from Davies and Müller 1988).

$\{\partial / \partial T + u\,\partial / \partial X\}(g / \theta_0)(\theta) - (\Lambda f)v = 0$ at $Z = 0$. (3.14)

Also the forcing term in the Sawyer-Eliassen equation (3.3) now takes the form

$$\Im = -2(\Lambda / N^2)(g / \theta_0)(\partial\theta / \partial Z) . \quad (3.15)$$

Consider in turn frontogenesis in (1) vertically semi-infinite and (2) doubly bounded domains. For Case (1) the dynamics of the linear flow system (Eqs. (3.13)–(3.14)) in *the semi-infinite domain* ($Z > 0$) can be viewed as a composite of noninteracting surface thermal waves each of the form

$$\theta(X, Z) = B\sin\{kX - \omega T\}\exp[-(\mu Z)] \quad (3.16)$$

with $\omega = (\Lambda f / N)$, and $\mu = (Nk/f)$, that is, vertically evanescent waves of constant amplitude (B) that propagate along the surface ($Z = 0$) in the positive x-direction with a phase-speed inversely proportional to their wavenumber (k). The corresponding group velocity is identically zero, and thus a packet of such waves will undergo a periodic temporal evolution with a period $T = 2\pi(N/\Lambda f)$, ~60 hours. Transient frontogenesis and frontolysis can occur as the baroclinicity of the individual waves experience constructive and destructive interference (Müller et al. 1989), and the accompanying ageostrophic circulation is given by the solution (see Müller et al. 1989) of Sawyer-Eliassen equation (3.15), viz.

$$\chi = -(\Lambda / N^2)(g / \theta_0)z\theta .$$

Figure 3 shows an example of such a cycle. A surface band of cold air is first replaced by a warm band advected from the south, and thereafter it is restored after the flow direction reverses (Fig. 3a,b). Frontal features evolve rapidly on the western edge of the thermal anomalies and attain maturity at ($T/4$) and ($3T/4$). The cross-sectional structure at ($T/4$) reveals features (see Fig. 3d–f) that bear comparison with typical warm fronts—a frontal region with a slope ~(f/N), a cold conveyor belt (Bjerknes 1919; Browning 1990) ahead of a comparatively weak surface front, and ascent through the frontal zone (cf. Eliassen 1962; Keyser and Pecnick 1987).

Case (2), *the bounded domain* with rigid horizontal boundaries at, say, $Z = 0$ and d, is the Eady paradigm for baroclinic instability. For this case it is customary to examine the finite-time development of wave-like perturbations, and by analogy with Eq. (3.16) the system's dynamics in SG space comprise the interaction of two edge waves of the form (Davies and Bishop 1994),

$$\theta_B = B(\sinh\mu d)^{-1}\{\sinh[\mu(d - Z)]\}\sin(kX + \varepsilon_B) \quad (3.17a)$$

$$\theta_T = T(\sinh\mu d)^{-1}\{\sinh(\mu Z)\}\sin(kX + \varepsilon_T) \quad (3.17b)$$

with $\mu = (N/f)k$, the variables (B, ε_B) denoting respectively the time-dependent amplitude and phase of the lower boundary wave, and (T, ε_T) representing the corresponding variables for the upper boundary wave. The temporal development is governed by the time-scale $r = (\Lambda f/N)$ and a geometric factor $\alpha = \{\cosh(\mu d) - (\mu d/2)\sinh(\mu d)\}$.

From the so-called potential vorticity perspective (Hoskins et al. 1985), a warm anomaly at the lower (upper) boundary constitutes a positive (negative) PV anomaly with a δ-function vertical structure. The interaction of the two pseudo-PV surface waves arises from the advection of one boundary wave by the component of the flow field attributable to the other. This interaction can (i) enhance the wave-amplitude provided the relative phase, $\varepsilon = (\varepsilon_T - \varepsilon_B)$, is negative (i.e., a westward slope to the pseudo-PV wave anomalies from the lower to the upper boundary), and (ii) counter the tendency of each wave to propagate against the ambient flow at its level. Synchronized development (i.e., $T = B$) will prevail if the waves are initially of equal amplitude, and for such a development the long wavelength waves (corresponding to $[1 - \alpha^2] > 0$) will amplify and evolve to a phase-locked state ($\varepsilon = $ const.). The special case with ab initio phase-locking corresponds to the exponentially growing Eady normal modes with a growth rate (σ_E),

$$\sigma_E = r\{(f / N)k\,\mathrm{cosech}(\mu d)\}(1 - \alpha^2)^{1/2} \text{ with } \cos\varepsilon = \alpha .$$

The ageostrophic circulation distorts these waves and leads to a scale-collapse at the boundaries after a time t_c (Blumen 1981) given by

$$(\sigma_E t_c) = \ln\left\{1 + (f / v_x)_{\min_{t=0}}\right\}.$$

Figure 4a shows the thermal structure of such a system in physical space immediately prior to the scale-collapse. It is reminiscent of a surface cold front maturing on the western flank of the warm air, but there is a notable lack of a corresponding warm front on the eastern flank. Figure 4b is a schematic of the location of the corresponding pseudo-PV boundary anomalies with an indication of their accompanying signals in the near-field vorticity and isentropic displacement. Recall that these signals are also present but with reduced amplitude in the far field.

This PV perspective prompts a sequence of inferences that derive from the westward slope of the pseudo-PV axes with height. This alignment is a necessary concomitant of growth with the upper- and lower-level anomalies configured to enhance one another, but it also implies that the incipient cold front near A and the "expected" warm front near B are overlain by positive and negative PV-anomalies, respectively. Thus in the vicinity of A the effect is to (i) increase the

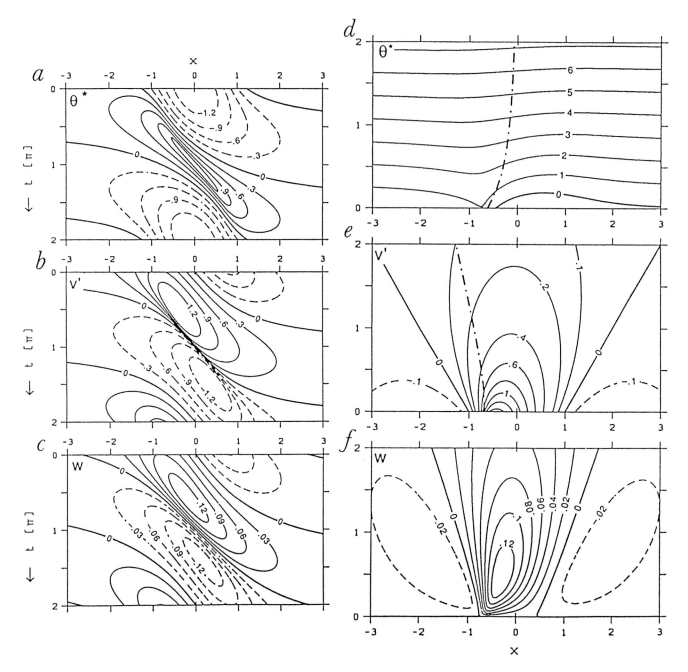

FIG. 3. Features of a semigeostrophic realization of a horizontal shear-induced frontogenesis in a semi-infinite domain. Left-hand panels show the space-time development of (a) the isentropes, (b) the along-front wind component at the surface, and (c) the vertical wind field at height of 2.5 km. Panels (d), (e), and (f) depict the cross-sectional distribution of the corresponding variables after a quarter of the cycle. The dash-dot curves refer to the baroclinicity maximum, and the parameter settings are such that unit values of (θ, v, w) correspond respectively to (20 K, 20 m s^{-1}, and 0.15 m s^{-1}). The domain of the cross-section is (3000 × 10) km (adapted from Müller et al. 1989).

surface vorticity, (ii) reduce the ambient stratification, and (iii) increase the slope of the front. (The reverse pertains near B). The first two effects serve to promote frontogenesis at A and inhibit it at B (see Eqs. (2.6), (2.7)). In effect the vertical alignment of the pseudo-PV anomalies establishes a dynamical distinction between the incipient cold and warm fronts.

This PV-based distinction bears comparison with other dynamical classifications of frontal types. The westward slope of the PV axes with height is accompanied by a forward tilt of the thermal wave and the latter's contribution to the difference in the frontal slope was already noted by Hoskins and Heckley (1981). Again the PV-based distinction leads directly to the deduction of enhanced (diminished) low-level vorticity at the cold (warm) front, and this contrasts with the

(a)

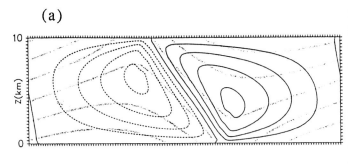

(b)

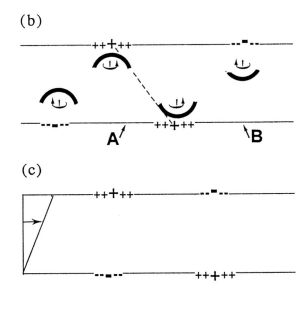

(c)

(d)

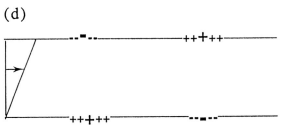

FIG. 4. Features of a semigeostrophic realization of horizontal shear-induced frontogenesis in a vertically bounded domain. The setting is that of the nonlinear normal mode for the two-dimensional Eady model. Panel (a) shows the isentropes (shaded) and the along-front velocity field (continuous and dashed lines) shortly before frontal collapse (Snyder et al. 1993). Panel (b) is a schematic of the pseudo-PV anomalies on the bounding surfaces together with an indication of their near-field vorticity/circulation and isentropic displacement. Dashed line indicates the alignment of the pseudo-PV anomalies. Panels (c) and (d) depict two alternative configurations of initially localized PV distributions.

more customary interpretation—low-level frontal convergence results in cyclonic shear (2.6) and this shear's contribution to the **Q**-field is frontogenetic for a cold front and frontolytic for a warm front. The PV interpretation also distinguishes between the contribution of the in situ and far-

field PV anomalies to the low-level **Q**-field. The former is indicative of a tendency for an eastward propagation of the surface edge-wave yielding frontogenesis in the warm sector and frontolysis in the cold, whereas the latter signals frontogenesis at the cold front.

Examples of case-study comparisons of the Eady model frontogenesis with observed events include those of Blumen (1980), Ogura and Portis (1982), and Reeder and Smith (1986). The caveat regarding the weakness of the vertical velocity field applies in this case again, but nevertheless the degree of similitude with the observed events is surprising considering the overt simplicity of the model.

Alternative initial alignments of the pseudo-PV anomalies on the bounding surfaces—for example, the two examples schematized in Figs. 4c,d—result in notably different flow evolutions. Consideration of the far-field contribution of the anomalies and their advection by the background shear suggests rapid cyclo- and frontogenesis for the configuration of Fig. 4c as it transits toward the phase slope of growing Eady modes. In contrast incipient frontolysis can be expected for the case of Fig. 4d. These two cases bear comparison with, and shed light on, the differing form of evolution associated with the so-called cold and warm advection cases studied by Keyser and Pecnick (1987).

3.3 Other Two-Dimensional Settings

A natural extension of the foregoing is to consider the combined effect of deformation and horizontal shear in inducing frontogenesis (Keyser and Pecnick 1987). Bishop (1993) derived analytical solutions for this generalized setting in the limit of uniform potential vorticity. His results indicate that the presence of deformation oriented so as to increase the baroclinicity modifies the Eady-type instability such that smaller-scale waves can evolve to form fronts.

Another setting for shear-induced frontogenesis is that of a uniformly sheared flow in a semi-infinite domain with an isolated interior PV anomaly (Thorncroft and Hoskins 1990; Davies and Bishop 1994). In the linear limit the interior anomaly is merely advected with the background flow at its level while continually forcing surface pseudo-PV/thermal waves (cf. Eq. (3.16)). For illustration consider an anomaly comprising a δ-function vertical structure located at a height h and with a bell-shape horizontal distribution. Surface waves generated with a wavenumber $k \sim (f/Nh)$ propagate with a phase-velocity $\sim \Lambda f/Nk$ that is comparable to that of the advecting velocity at level h, i.e., $U_{z=h} = \Lambda h$. Such waves will experience a quasimonotonic forcing and can grow almost resonantly. Figure 5 shows the space-time evolution of the surface potential temperature field in response to such a PV-anomaly. This frontogenetic mechanism is also germane to some dynamical aspects discussed in Section 3.5 and again in Section 5.

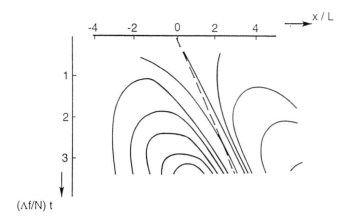

$(\Lambda f/N) t$

FIG. 5. The quasigeostrophic space-time evolution of the surface isentropes in response to a localized interior PV-anomaly initially implanted at $(x, z) = (0, h)$ in a uniform baroclinic shear within a semi-infinite domain. The anomaly has a δ-function structure in the vertical and a bell-shape, that is, $[1 + (x/L)^2]^{-1}$, in the horizontal. The long-dashed line denotes the location of the center of the PV anomaly at $z = h$ as it is advected with the background baroclinic flow field. To ensure quasi-monochromatic forcing the half width L and the height h have been set such that $L \sim 1.5 \, (N/f)h$. (Note also that $(\Lambda f/N) t = 2.5$ is of the order of 1 day, and that for this linear problem the constant increments of the isentropic isolines are arbitrary.)

3.4 Limits on the Frontogenesis

The foregoing theoretical examples illustrated the scale-collapse of a front in a finite time. This growth in the baroclinicity might be curtailed by the reorientation or reduction in amplitude of the deformation associated with the large-scale field, unbalanced effects excluded from the semigeostrophic system, three-dimensional effects, and diffusive effects.

A modification of the deformation field on the time scale of frontal development is certainly possible since as noted earlier, the frontogenesis time scale is linked to that of the evolution of the synoptic-scale flow. Likewise, for horizontal shear-induce frontogenesis the finite width of an ambient baroclinic zone will limit the surface thermal contrast that can evolve across the front.

From the standpoint of the limitations of the SG system itself, studies indicate that the two-dimensional semi-geostrophic flow developments discussed earlier remain valid until just prior to frontal collapse (e.g., Williams 1967; Hoskins and Bretherton 1972; Davies and Müller 1984; Snyder et al. 1993; Nakamura 1994), and even then only break down in a shallow narrow band. For example, for the deformation-induced frontogenesis discussed in Section 3.1, this band is of order 1 km wide and 30 m deep (Fig. 6a).

One theoretical approach to the study of the post-collapse phase is to presume that the SG equations remain valid and to allow extrusions of the boundary pseudo-PV into the

interior (e.g., Cullen and Purser 1984; Koshyk and Cho 1992). Such an extrusion forms a sloping frontal line of discontinuity comprising an interior δ-function of potential vorticity (Fig. 6b). In this supra-balanced development the intrusion from the surface is accompanied by an upwelling of the low-level isentropes and hence in enhanced ascent ahead of the front.

More generally the development of spatially rich flow features in the transition toward a scale collapse can result in the forcing of unbalanced flow, an enhanced role for diffusive effects, and the development of balanced flow structures that are potentially unstable. Here we give consideration to each of these effects.

3.4.1 Generation of unbalanced flow

The generation of unbalanced flow in the region of impending flow breakdown could both modify the in situ balanced flow and possibly influence the far-field by the emission of inertia-buoyancy waves. Equilibration might prevail if the balanced flow-tendency for local scale contraction was either countered by diffusive effects or matched by an adjustment of energy and momentum induced by the wave emission. The former possibility has a counterpart in the development of shock or bore-like phenomena, whereas the latter has a parallel with the turbulent generation of aerodynamic noise (Lighthill 1952) and is also analogous to the breakdown of balanced flow over topography (Trüb and Davies 1995).

For equilibration via wave emission an assessment or interpretation of the nature of wave response can be sought by decomposing the flow into balanced and residual flow components and evaluating the space-time structure of the balanced field's forcing of residual flow (Ley and Peltier 1978; Snyder et al. 1993).

Explicit consideration of the dynamics of the breakdown using primitive equation numerical models requires very fine spatial and temporal resolution to accurately capture the nature of the ensuing motion without generating spurious buoyancy waves (see the discussions in Pecnick and Keyser 1989; Volkert and Bishop 1990; Snyder et al. 1993; Gall et al. 1988). The simulations of Snyder et al. (1993) show evidence of wave activity in the approach to breakdown, and indicate that in the post-collapse phase, when dissipative effects are strong, the waves become stationary relative to the front and modify the prefrontal ascent (Fig. 6c). An earlier suggestion (Orlanski and Ross 1984) that the residual flow could induce a negative feedback upon the surface divergence and so curtail the frontogenesis has not received corroboration (see Levy and Bretherton 1987; Garner 1989).

3.4.2 Diffusive effects

Diffusive effects can directly and significantly counter the scale contraction once the front attains a very fine spatial

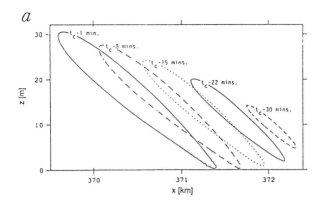

a

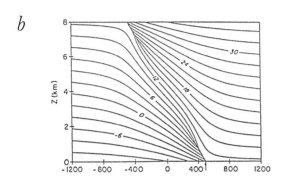

b

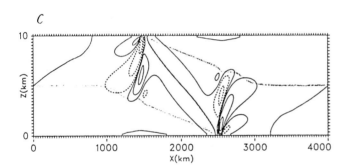

c

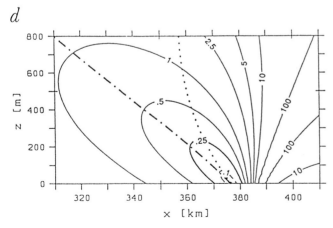

d

scale. In addition surface processes condition the potential vorticity in the planetary boundary layer and therefore air extruded into the interior will form PV anomalies (Nakamura and Held 1989; Cooper et al. 1992). Simulations of the two-dimensional Eady wave (Nakamura and Held 1989; Nakamura 1994) indicate an equilibration of the development as the PV anomalies from the upper (artificial) rigid boundary and the surface are vertically realigned by the background flow.

3.4.3 Unstable flow configurations

An instability of the balanced flow in the approach to a scale-collapse would ensue if the evolving flow was unstable to unbalanced two-dimensional or (balanced or unbalanced) three-dimensional perturbations. To explore these possibilities consider the semigeostrophic potential vorticity (q_{SG}),

$$q_{SG} = \{(\zeta + f)(\partial\theta/\partial z) - (\partial v/\partial z)(\partial\theta/\partial x) + (\partial u/\partial z)(\partial\theta/\partial y)\}$$

This can be rewritten, using the thermal wind relationship, in the form

$$(g/\theta_0)f\,q_{SG} = f^2 N^2\{(1 + \zeta/f) - Ri\} \quad (3.18)$$

or equivalently,

$$(g/\theta_0)f\,q_{SG} = S^2\{(1 + \zeta/f)Ri - 1\}, \quad (3.19)$$

where the Richardson number (Ri) and the baroclinicity measure (S) are defined as

$$Ri = N^2[(\partial u/\partial z)^2 + (\partial v/\partial z)^2], \text{ and}$$
$$S^2 = (g/\theta_0)^2[(\partial\theta/\partial x)^2 + (\partial\theta/\partial y)^2]$$

Note further that the potential vorticity of the ambient atmosphere is usually positive and is such that

$$(g/\theta_0)f\,q_{SG} \approx f^2 N^2 \quad (3.20)$$

because typically both (ζ/f) and $Ri \ll 1$.

FIG. 6. Illustrations of some features related to frontal scale collapse. Panel (a): Error evolution of the SG set during the deformation-induced frontogenesis corresponding to Fig. 2 (from Davies and Müller 1988). The error is defined as the ratio of the neglected inertia terms to the Coriolis term in the across-front momentum equation. Isolines demark the shape of the $10(\alpha/f)^2$ error measure, and t_c refers to the time of the scale collapse. (Spatial scale of the domain is ~2.5 km × 30 m.) Panel (b): Isentropic distribution following an extrusion of the pseudo-PV anomalies of the bounding surfaces into the interior in the post-collapse phase (from Koshyk and Cho 1992). Panel (c): Inertia-buoyancy waves emitted in a PE simulation at an advanced stage of the Eady two-dimensional setting of horizontal shear-induced frontogenesis (from Snyder et al. 1993). Shown is the nonbalanced component of the vertical velocity field. Panel (d): The Ri distribution two hours prior to scale-collapse for the case of deformation-induced frontogenesis corresponding to Fig. 2 (from Müller 1989).

Conservation of q_{SG} during the inviscid, dry frontogeneses discussed earlier implies that

$$\{(1 + \zeta / f) - Ri\} > 0. \qquad (3.21)$$

This inequality is also the necessary condition for stability to ageostrophic, two-dimensional (x,z) perturbations of the balanced flow if the latter is assumed to be time-independent.

Again note that an increase of ζ and S during frontogenesis induces a concomitant decrease in a fluid parcel's Richardson number (see Eq. (3.19)). It has been argued that domains characterized by $Ri < 1$ can sustain turbulence, whereas the criterion $Ri < 1/4$ is linked to the Kelvin-Helmholz (K–H) type of instability—at least in flows that are less laterally structured than frontal zones. It follows from Eqs. (3.19) and (3.20) that the evolution of the Richardson number in semigeostrophic flow is prescribed by the relationship,

$$Ri \approx \left\{1 + (f^2 N^2)_{t=0} / S^2\right\} / \left\{1 + \zeta / f\right\},$$

and, when the baroclinicity S assumes large values characteristic of frontal zones, this reduces further to

$$Ri \approx 1 / \{1 + \zeta / f\} \qquad (3.22)$$

Thus the critical Richardson number $(Ri \sim 1/4)$ is attained for vorticity values of $\zeta / f \sim 3$, and such values for the vorticity can develop significantly prior to frontal collapse. For example, Fig. 6d shows the Richardson number structure two hours before the scale collapse for the case of deformation-induced frontogenesis considered earlier. The $Ri < 1/4$ domain is ~ 15 km $\times 250$ m, and thus the onset and occurrence of a K-H instability could place a limit on the frontal width and baroclinicity. For this particular case the surface vorticity and baroclinicity are related and the presumed critical across-frontal length scale $(\Delta x)_{crit}$ is scaled by the relationship,

$$(\Delta x)_{crit} \sim \tfrac{1}{3}(g / \theta_0)[\Delta\theta / (f N)]_{t=0} \qquad (3.23)$$

where $\Delta\theta$ and N represent, respectively, the initial values of the potential temperature difference across the broad frontal zone and the ambient stratification. A caveat to the foregoing discussion is that it remains to detect or explicitly model this small-scale three-dimensional process within a zone undergoing strong frontogenesis.

Note also that a banded interior PV anomaly formed by extrusion of anomalous positive potential vorticity from the planetary boundary layer will form a maximum of PV in the interior and such a distribution is a potential seat for a three-dimensional balanced-flow instability. Diabatic effects (see Section 3.5) can also give rise to interior PV extrema.

3.5   Cloud-Diabatic and Surface Friction Effects

The nature of convection triggered within a low-level frontal convergence zone will be highly dependent upon the vertical thermodynamic structure. An organized ensemble of such convective elements could feed back directly or indirectly upon the front. The feedback will be indirect if the organization takes the form of one or more squall lines propagating ahead of the front and thereby conditioning the pre-frontal environment. It can be direct if there is a space and time-scale matching of the ensemble and the front—cf. an elongated nontransient mesoscale cloud band aligned almost parallel to and propagating with the surface front. Such interaction could be sustained and compatible with balanced-flow dynamics, cf. the postulate of quasi balance for some mesoscale convective systems (Raymond 1992; Davis and Weisman 1994; Parker and Thorpe 1995).

It is therefore instructive to examine the form of the convection-front interaction(s) from a balanced-flow and PV perspective. The initial location and width of the convective region within an evolving front would be linked to the low-level convergence pattern and the domain demarked by fluid parcels first attaining their lifting condensation level. In general the latter will be significantly less than the width of the ascending region itself. This decreased width coupled with an effective decrease in the vertical stratification within the convective region will favor (Eq. (2.7)) a narrower and stronger region of ascent. Within the region of organized convection the diabatic heating will yield a PV tendency that is positive beneath the level of maximum heating and negative above (see Eq. (2.5a)). The realization of a positive low-level PV anomaly would be accompanied at lower levels by (i) cyclonic vorticity and (ii) an elevation of the isentropes. In general signal (i) will be favored by a deep narrow diabatic domain (cf. deep penetrative convection) and signal (ii) by a shallower broader distribution (cf. layered stratocumulus).

The foregoing features can influence the frontogenesis in several ways. For the case of deformation-induced frontogenesis, prior to the onset of convection, the low-level ascent and surface convergence is centered at the vorticity maximum ahead of the surface front (Eqs. (3.4a, b)). Thus, signal (i) will increase the rate of vorticity production (Eq. (2.6)) and signal (ii) will promote additional ascent. The latter can lead to a positive feedback with the additional low-level ascent enhancing the moisture flux convergence into the cloud region. The isentropic and perturbed PV fields (Fig. 7) derived from a SG simulation of deformation-induced frontogenesis in the presence of parameterized stratiform clouds (Chan and Cho 1991) shows the surface front and the low-level PV anomaly suitably located to sustain and to be enhanced by these effects.

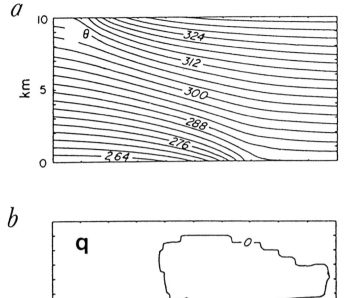

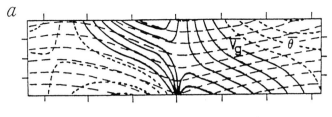

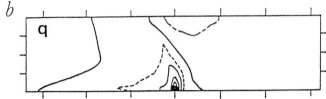

FIG. 8. An SG simulation of the Eady setting with a parameterization of slant-wise convection. Panel (a) shows the potential temperature (dashed isolines) and along-front velocity fields (solid isolines), and panel (b) the perturbed PV field (Joly and Thorpe 1989).

FIG. 7. An SG simulation of deformation-induced frontogenesis with a parameterization of stratiform clouds. The upper and lower panels show, respectively, the potential temperature field and the perturbed PV field (Chan and Cho 1991).

For the case of horizontal shear-induced frontogenesis prior to the onset of convection, the low-level convergence is located in the warm sector ahead of the surface front. In addition to the aforementioned interactions there are now effects linked to the ambient along-front baroclinicity. To illustrate this, consider the influence and evolution of a diabatically induced low-level PV anomaly. The accompanying cyclonic vorticity will augment the surface frontogenesis. There is also the possibility (see Section 3.3) that as this anomaly is advected eastward by the ambient flow it will remain suitably phased relative to the propagating anomaly in the surface baroclinicity, and thereby instigate and be sustained by further diabatic heating. Figure 8 is an example of the structure that evolves within an SG model of horizontal shear-induced frontogenesis that includes a parameterization of slant-wise convection.

In general, numerical modeling studies of cloud diabatic effects for the deformation case (Mak and Bannon 1984; Thorpe and Emanuel 1985; Chan and Cho 1991) and the horizontal shear case (Emanuel et al. 1987; Knight and Hobbs 1988; Joly and Thorpe 1989; Whitaker and Davis 1994) do show narrower and more intense updrafts, but the scale, structure, and location(s) of the updrafts are sensitively dependent upon the cloud parameterization schemes.

It was noted earlier that diffusive effects within the boundary layer can counter frontal scale contraction and lead to interior PV anomalies. In addition surface frictional effects can modify the rate and direction of frontal propagation within the boundary layer, and the strength and width of the low-level convergence zone (Blumen 1990). Detailed consideration of these effects for a range of boundary layer structures has yet to be undertaken. Simulations of the archetypal frontogenesis configurations undertaken with numerical models that include a representation of the turbulent boundary layer do produce a narrower and more intense updraft (Keyser and Anthes 1982, 1986; Reeder and Smith 1986) and significant interior PV anomalies (Cooper et al. 1992). An important caveat to these studies is our limited understanding of the turbulent PBL response when forced by highly structured and rapidly varying mesoscale free-atmosphere flow.

4. Frontogenesis in an Idealized Three-Dimensional Setting

Fronts and cyclones develop concurrently in the atmosphere and are interrelated both in terms of their origin and dynamics. This co-development is studied here in the classical idealized setting of baroclinic instability.

Consider the extended Eady model comprising an incompressible flow of uniform potential vorticity sandwiched between two rigid horizontal boundaries on an f-plane (see, e.g., Hoskins and West 1979; Davies et al. 1991). The basic state is assigned the form of a two-dimensional jet-like baroclinic flow that is characteristic of the zonally averaged midlatitude troposphere. As before, the dynamics of perturbations to this state can be viewed in terms of upper- and lower-boundary pseudo-PV waves propagating on the laterally confined baroclinicity at these surfaces. In contrast to the two-dimensional setting, the Q-vector frontogenetic forcing field now develops internally during the flow evolution, and moreover the vorticity field can rotate the evolving surface frontal zone into, or out of, alignment with the dilatation axis of the prevailing deformation field (cf. Eqs. (2.1), (2.3)).

4.1 Evolution Within a Symmetric Basic State

Figure 9 shows an example of the finite-amplitude structure of the most unstable normal-mode of a symmetric jet-like flow in the SG, adiabatic, and inviscid limits. To the south of the evolving surface cyclone the presence of a distinct thermal gradient and significant wind shift and vorticity signatures are all indicative of strong frontogenesis. This cold front's thermal gradient is weaker in the vicinity of the cyclone itself, but to the southwest it extends around the base of the anticyclone. There is also some evidence of a weak, short warm front that bends outward and eastward from the cyclone center.

At the surface the cold front's resemblance to the counterpart two-dimensional fronts (see Sections 3.1, 3.2) is striking. It has narrow collocated bands of vorticity and convergence ahead of the baroclinic zone, indications of a strong deformation field with its dilatation axis aligned along the front, and active frontogenesis along a substantial portion of the frontal band. For the bent-back warm front it is the lack of correspondence with the archetypes of Section 3 that is striking. Far from the cyclone the dominant feature is the broad and strong signature in the $(\nabla_h \cdot \mathbf{Q})$ field, whereas the surface baroclinicity and vorticity signatures attain significant amplitude only in the vicinity of the cyclone. A tentative inference, which is explored later, is that this warm front is an intrinsically three-dimensional feature.

4.2 Evolution for an Asymmetric Basic State

The foregoing frontal features reflect a symmetry constraint imposed upon the flow development by the combination of the SG system on an f-plane and the symmetric basic state (Davies et al. 1991). This symmetry does not in general prevail for more complex systems (the PE equations), different geometrical settings (the β-effect or spherical geometry), or asymmetric basic states (e.g., the presence of lateral barotropic shear or internal PV gradients).

In essence, the symmetry state of the SG system can be viewed as a bifurcation point in parameter space with the nature of the ensuing fronts differing radically in response to the sign of the asymmetry. The PE system on an f-plane

favors the earlier, stronger, and spatially more extended development of the warm front (see, e.g., Takayabu 1986; Keyser et al. 1989; Polavarapu and Peltier 1990; Snyder et al. 1991; Tremblay 1992; Lalaurette et al. 1994; Rotunno et al. 1994). On the other hand, the PE system in spherical geometry tends to induce stronger and more elongated cold frontal features (see, e.g., Thorncroft and Hoskins 1990; Hines and Mechoso 1991; Whitaker and Snyder 1992). These dichotomous responses can be replicated in the SG f-plane system by the simple expedient of adding to the symmetric basic state a weak uniform positive (negative) barotropic shear to favor surface warm (cold) frontogenesis with upper-level vortex-like (band-like) pseudo-PV features (Davies et al. 1991).

There are indications that lateral shear is indeed the dominant symmetry-breaking factor. For example, the starkly differentiated response to lateral shear in the SG f-plane system is paralleled in the PE f-plane system (Wernli et al. 1998), and are integral to the two paradigms of baroclinic-wave life-cycle behavior identified in simulations with the PE spherical geometry system (see Thorncroft et al. 1993; Simmons, this volume). In light of the foregoing we focus here on the two forms of response engendered by the addition of weak barotropic shear.

Figures 10 and 11 illustrate the characteristics of the two forms in the f-plane SG setting. For the "cyclonic" case of positive shear (Fig. 10) the thermal pattern of the cold front is severely disrupted to the immediate southeast of the cyclone (cf. the phenomenon of frontal fracture), and there is evidence of a bent-back warm frontal feature. The frontal palette takes on a distinctive λ-shape—particularly in the $(\nabla_h \cdot \mathbf{Q})$ field, and aloft at the lid a vorticity band extends northwestward from a vortex structure that is located almost above the surface cyclone. In the "anticyclonic" case of negative lateral shear (Fig. 11) the patterns of vorticity and $(\nabla_h \cdot \mathbf{Q})$ directly link the cyclone and cold front, whereas the warm front is weaker. At the lid the southward protrusion of cold air is now bordered by frontal/vorticity features and the leading edge is aligned parallel to, but displaced NW of, the surface cold front.

The ambient lateral shear, which in both cases amplifies considerably during the flow development, clearly exerts a profound influence upon the strength and structure, and the distinctiveness of the ensuing fronts. In relation to the strength and structure note that positive (negative) lateral shear can be decomposed (Fig. 12a) into a positive (negative) rotation plus a deformation field with a dilatation axis aligned NW-SE (NE-SW). Thus in the positive (negative) lateral shear case the incipient warm (cold) front is favorably oriented to benefit from the additional deformation. This is consistent with the lateral shear's contribution to the frontogenesis index (Eq. (2.2)). The shear's contribution to the vertical velocity, via the $(\nabla_h \cdot \mathbf{Q})$ field (Eq. (2.7)), can be

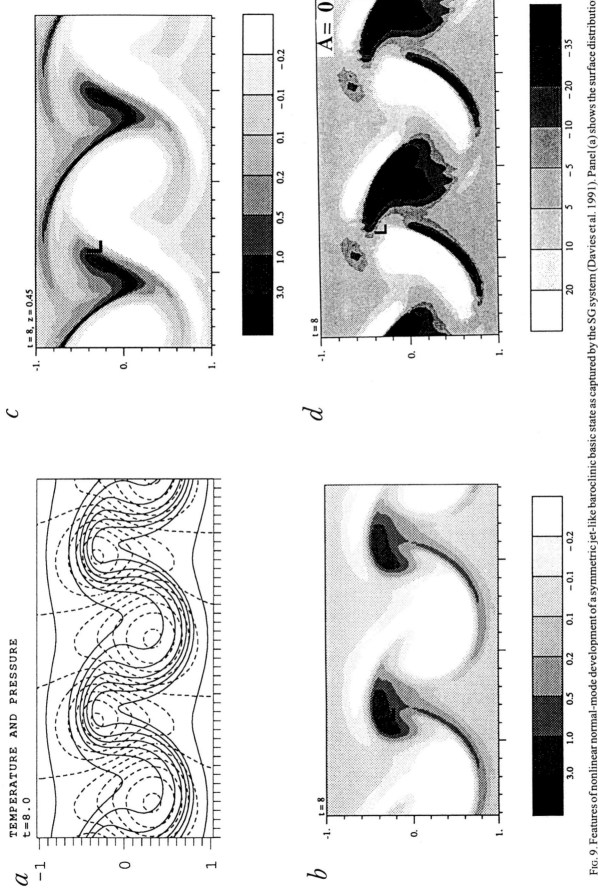

FIG. 9. Features of nonlinear normal-mode development of a symmetric jet-like baroclinic basic state as captured by the SG system (Davies et al. 1991). Panel (a) shows the surface distribution of the isentropes (solid lines) and stream function (dashed lines). Panels (b) and (c) show respectively the surface and upper-boundary vorticity patterns, and panel (d) the surface quasi-geostrophic ($\nabla_h \cdot \mathbf{Q}$) field.

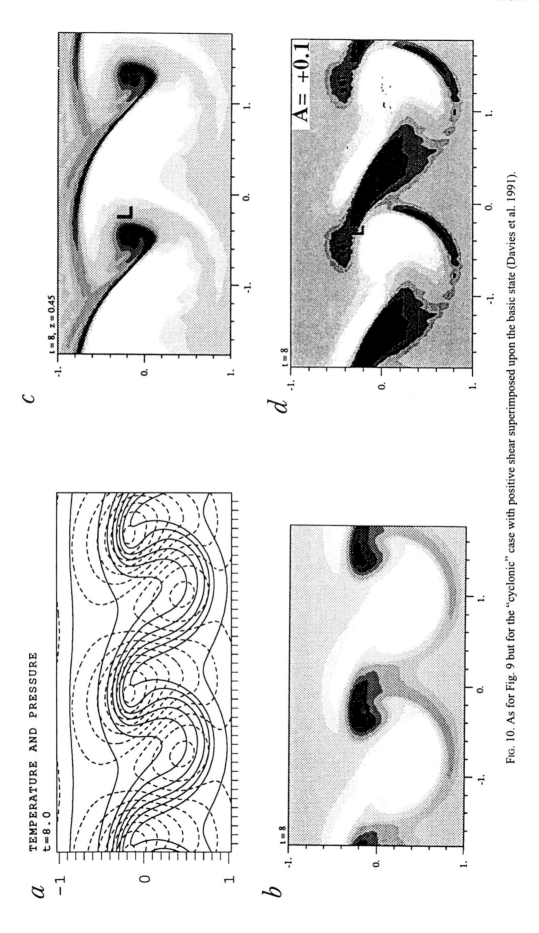

FIG. 10. As for Fig. 9 but for the "cyclonic" case with positive shear superimposed upon the basic state (Davies et al. 1991).

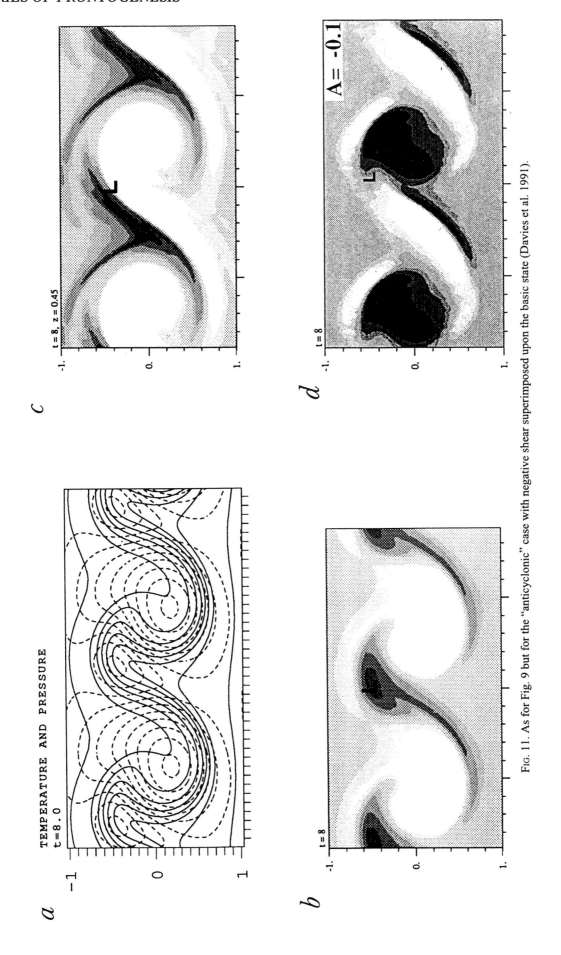

Fig. 11. As for Fig. 9 but for the "anticyclonic" case with negative shear superimposed upon the basic state (Davies et al. 1991).

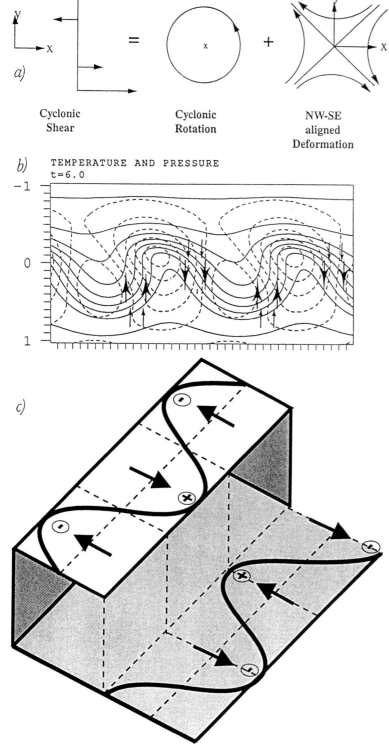

Cyclonic
Shear

Cyclonic
Rotation

NW-SE
aligned
Deformation

b)   TEMPERATURE AND PRESSURE
     t=6.0

c)

Fig. 12. Illustration of features influencing the nature of frontal development. Panel (a): the decomposition of a positive lateral shear into its invariant components—that is, a rotation and a NW-SE aligned deformation field. Panel (b): the dominant Q-vector signal attributable to positive ambient lateral shear (indicated by the arrows) within an incipient baroclinic wave development. Panel (c) the displacement and subsequent locations of the upper- and lower-boundary pseudo-PV anomalies for a basic state with the upper baroclinic zone located northward of the surface zone. (Note the approach of the upper and lower level positive pseudo-PV anomalies.)

inferred qualitatively from the schematic of the shear-related Q-vectors of Fig. 12b. They suggest that for the cyclonic shear case frontogenesis is extended southeastward at the warm front, and curtailed in the vicinity of the cyclone at the cold front. The reverse applies for the case of anticyclonic shear.

In relation to the distinction between frontal types a notable feature is the disparity of the lateral-shear–induced changes to the $(\nabla_h \cdot Q)$ and vorticity patterns at both the warm and cold front (Figs. 10 and 11). Yet these two parameters are strongly coupled via Eqs. (2.7) and (2.8). To interpret this difference note that, in the presence of strong surface convergence and associated vorticity production, a weak (strong) relative airflow along the front will result in the vorticity being realized in situ (downstream). Now consider the contributions to this low-level relative air-flow from a PV perspective, that is, in terms of the contributions attributable to the thermal anomalies on the two bounding surfaces. For the case of positive lateral-shear the contributions of the surface and upper-level thermal anomalies counter one another significantly at the cold front, whereas the upper-level contribution is comparatively weak at the warm front. This is consistent with the appearance (nonappearance) of an in situ band of vorticity along the cold (warm) front. (The location of the upper- and lower-level anomalies can be inferred qualitatively from the upper-level vorticity signatures in Fig. 10.) This dependence of the nature of the frontogenesis upon the relative alignment of the upper- and lower-level pseudo-PV anomalies resembles and extends the dynamical distinction drawn earlier between two-dimensional cold and warm fronts.

At the forefront of the preceding discussion has been the diagnosis of the influence of the ambient lateral shear. The dynamics of the shear's development itself is more subtle. The shear renders the basic jet flow asymmetric and this induces a lateral shift in the relative locations of the upper and lower boundary thermal waves. During the nonlinear development this lateral shift for the cyclonic case (say) results in the positive pseudo-PV anomalies on the upper and lower boundaries approaching, interacting, and almost overlying one another, whereas the negative anomalies move apart. (The initial approach is schematized in Fig. 12c and the result of the subsequent evolution is evident in Fig. 10.) This relative movement of the anomalies can in turn account for the amplification of the shear during the evolution.

4.3 Influence of Some Other Factors

The frontal patterns in Figs. 10 and 11 resemble many of the structural features of observed atmospheric fronts, but diabatic and boundary-layer processes can again modify the nature of the frontogenesis. For the simulated cold fronts their structural and dynamical resemblance to the two-dimensional counterparts suggests that their diabatic modification might also be somewhat similar. For the warm fronts their intrinsically three-dimensional structure with bands of strong front-relative flow and low-level convergence suggests that their dynamics might be sensitive to frictional and diabatic effects. These inferences are supported by numerical simulations performed with PE models. For instance, the inclusion of a bulk-aerodynamic representation of the surface drag (Hines and Mechoso 1993) reduces the strength of the front-relative flow in the warm-frontal region and decreases that front's length. Likewise the inclusion of a simple convective parameterization scheme (Whitaker and Davis 1994) yields low-level PV bands in the frontal convergence zones, and the subsequent advection of the potential vorticity at low-level along the bent-back portion of the warm front contributes directly to a more rapid and intense cyclogenesis.

The shear-related cyclonic and anticyclonic paradigms of development (Sections 4.1 and 4.2) might also be triggered by other physical variations that induce an asymmetry of the basic-state jet flow. Possible examples are anomalous shifts in SST patterns, for example, a lateral shift of the lower boundary baroclinic zone relative to the upper zone or the presence of a finer-scaled warm band embedded within the surface zone. In the latter case numerical simulations indicate that the nature of the response is sensitive to the warm band's location with elements of the cyclonic paradigm resulting if the band is near the center of the surface zone .

4.4 Comments on Upper-Level Fronts

In this study the focus throughout has been on surface frontogenesis. However distinctive front-like features also develop at the tropopause and are generally associated with V-shaped intrusions of stratospheric air down to tropospheric levels. (See Shapiro and Keyser 1990 for a recent overview.)

The theoretical examination of upper-level frontal features has paralleled that for surface fronts. Studies include idealized models of two-dimensional deformation-induced frontogenesis ranging from a simple two-PV-layer representation of the troposphere-stratosphere interface (Hoskins 1971; Hoskins and Bretherton 1972) to a continuous PV representation (Moore 1993; Saute 1993). Likewise there have been detailed studies of the front-like upper-level features that develop within three-dimensional baroclinic waves growing on a jet-like basic state (e.g., Hines and Mechoso 1991; Thorncroft et al. 1993; Bush and Peltier 1994; Rotunno et al. 1994). From an interpretative standpoint it is germane to record that the starkly different upper-level patterns (Figs. 10 and 11) that evolve in a simple flow setting (*f*-plane, rigid lid) bear a marked resemblance to the patterns that evolve in a more general setting (spherical geometry, tropopause structure) and that were termed paradigmatic by Thorncroft et al. (1993).

Here we comment briefly upon tropopause-level versus surface frontogenesis, and, following Davies and Rossa (1998), confine our remarks primarily to PV-related aspects. At the surface a frontal development usually signifies the dynamical juxtaposition of air masses with distinctively different spatial origin and thermal characteristics. It is therefore appropriate to devise a dynamical index for frontogenesis in terms of the evolution of the surface baroclinicity (cf. Eqs. (2.1)–(2.4)). In contrast, astride the tropopause there are pre-existing air masses with starkly different characteristics, and thus formulating an index for upper-level frontogenesis is more problematic. For example, consider the traditional indices based upon potential temperature (Eqs. (2.1) and (2.2)). In the free atmosphere these indices are influenced by the effect of tilting due to differential vertical advection, and their simple reduced forms, valid on a flat surface (Eqs. (2.3) and (2.4)), are inapplicable. Moreover, within a direct frontal circulation the tilting can substantially offset the in situ effects of deformation and/or horizontal shear, and thereby render the diagnosis of frontogenesis a subtle assessment of counteracting effects (see, e.g., Keyser and Shapiro 1986; Hines and Mechoso 1991; Shapiro and Keyser 1990; Rotunno et al. 1994).

From a potential vorticity perspective the free-atmosphere analog of the baroclinicity at the surface is the PV gradient on a θ-surface. Likewise an analog of potential temperature frontogenesis at the surface would be the sharpening of an interior PV gradient on a θ-surface. Moreover, it is at the midlatitude tropopause-break that this PV gradient is maximum, and indeed Reed and Danielsen (1960) identified an upper-level front as a quasi discontinuity of potential vorticity on a θ-surface. It follows that the free-atmosphere dynamical counterpart of the traditional indices for surface frontogenesis (i.e., Eqs. (2.2) and (2.4)) are

$$\left(\frac{D}{Dt}\right)_\theta \frac{1}{2}(\nabla_\theta q)^2 = -(\nabla_\theta q)^2(\partial v / \partial n) \qquad (4.1)$$

$$\left(\frac{D}{Dt}\right)_\theta (\nabla_\theta q) = |\nabla_\theta q|\{\mathbf{k} \times (\partial \mathbf{v}_g / \partial s)\} \qquad (4.2)$$

where q is the potential vorticity, and the θ subscript refers to derivatives along a θ-surface.

These PV-related measures circumvent consideration of the aforementioned counteracting effects, and enable a qualitative appraisal of the frontogenesis to be made by inspection of the potential vorticity and wind field distributions on a tropopause-transcending isentropic chart (cf. discussion following Eq. (2.4)). In addition many of the deductions and interpretations advanced earlier (Section 4.2) to account for the characteristics/types of surface fronts can be reformulated in terms of PV patterns on an isentropic chart and applied to upper-level fronts. (As an aside we note that these frontogenesis measures cannot be applied to the rudimentary but standard two PV-layer representation of the tropopause used for model studies of frontogenesis/tropopause folding. This is either a deficiency of the measures outlined here or an indication of the inappropriateness of the model as an idealization of free-atmosphere frontogenesis.)

5. The Frontal Palette and Cyclogenesis Types

Consideration is now given to the nature of the frontogenesis and the frontal palette associated with presumed paradigmatic precursor patterns for cyclogenesis. Classifications of surface cyclogenesis exist that are based upon a synthesis of synoptic (e.g., Petterssen and Smeybe 1971), satellite (e.g., Smigielski and Ellrod 1985; Weldon and Holmes 1991), or composite (Bader et al. 1995) data. These schemes identify two broad classes of surface development, and here we comment briefly upon the nature of the frontal palette that can accompany these classes.

5.1   Upper-Level Induced Development

The first type is the response to the passage of an upper-level trough (or equivalently a positive PV anomaly) toward a surface baroclinic zone with further refinements to distinguish between a northwest-southeast and northeast-southwest alignment of an elongated upper-level anomaly. This first type can be viewed as an initial-value counterpart of baroclinic instability with a finite-amplitude initial perturbation, and the refinements relate to the incipient structures of the cyclonic and anticyclonic paradigms discussed in Sections 4.1 and 4.2.

A theoretically based approach to the study of this type is to implant a finite-amplitude upper-level anomaly upon a jet-like baroclinic basic state, and to examine the resulting flow evolution (e.g., Takayabu 1991; Schär and Wernli 1993). A key initial feature is the anomaly's distortion of the surface baroclinic zone to form a dipole of negative-positive thermal anomalies aligned along the zone. For a prescribed basic state the nature of the subsequent development will be a function of the location, strength, and structure of the PV anomaly.

Conformity with observations would usually place the anomaly over, or to the cold side of, the surface baroclinic zone. An anomaly located directly over the zone would favor (see Eq. (2.2b), and Keyser et al. 1988) surface cold and warm frontogenesis, respectively, to the southwest and northeast of

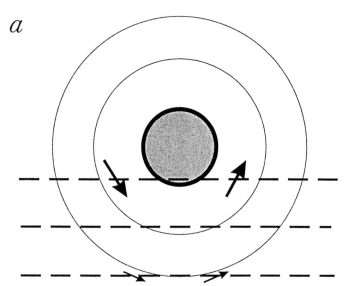

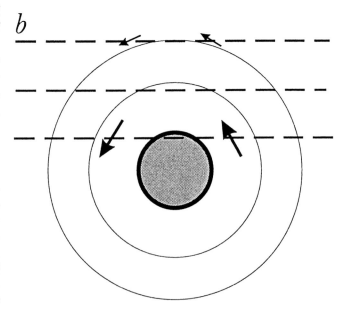

FIG. 13. Schematics of features associated with a positive PV-anomaly (shaded feature) located (a) poleward and (b) equatorward of a low-level baroclinic zone (dashed lines denoting the surface isentropes). The arrows signify the contribution of the PV-anomaly to the flow within the baroclinic zone.

the anomaly. Likewise, it can be inferred from the Q-vector diagnostic of Eq. 2.4b that an anomaly located to the cold side of the baroclinic zone (see Fig. 13a) would contribute to frontogenesis (frontolysis) to its southwest (southeast). Also, the anomaly's contribution to the along-front flow counters that attributable to the surface baroclinic zone and thereby helps promote the in situ realization of the vorticity.

The subsequent evolution is influenced by a range of effects including the dispersion of the initial surface dipole along the low-level zone of enhanced baroclinicity, the analog process on the upper-level PV gradient, and the distortion and advection of the original PV anomaly by the ambient flow. Growth of the initial surface perturbations will again be enhanced in this case if they remain suitably phased with the translating upper-level anomaly (see Section 3.5 and Fig. 5).

5.2 PV-Band Instability

The second type is the evolution of wave-like disturbances on a mature frontal zone. In recent years this type has been linked to the instability of the low-level prefrontal band of diabatically produced potential vorticity produced during moist frontogenesis. Growth to finite amplitude of unstable waves on this band can lead to the band's break up into a train of isolated segments (Schär and Davies 1990; Malardel et al. 1993). Each segment now constitutes a low-level PV anomaly located on the warm side of the surface baroclinic zone (see Fig. 13b). In this case the simple qualitative considerations of Section 5.1 lead to the reversed inferences of frontogenetic (frontolytic) tendencies on the front to the northeast (northwest) of the segment, and these inferences are sustained by the results of numerical simulations (Schär and Davies 1990; Malardel et al. 1993).

5.3 Other Configurations

In general the two of the key cyclogenetic features of the ambient atmosphere—the surface baroclinic zone and the zone of strong PV gradient on troposphere-transcending potential temperature surfaces—are not co-aligned. In such situations the predilection for a surface disturbance to propagate along the surface baroclinic zone can result in a change in its location relative to a pre-existing upper-level anomaly. Shapiro (this volume) draws attention to such an effect. For example, consider a shallow system propagating northeastward from the subtropics along a surface baroclinic zone aligned with the North-Atlantic SST pattern. The system would in effect move away from the influence of the strong tropopause-level PV gradient of the subtropical jet stream and come closer to the meandering and anomaly-rich features of the midlatitude jet. In this process the surface disturbance would experience a marked change in its ambient acrossfront shear and, in light of the foregoing, be susceptible to a modification of its frontal palette.

6. Further Remarks

Research on fronts and frontogenesis has both forecastoriented and explanatory-related objectives. It is instructive to view the development and current trends in the theoretical study of frontogenesis through the lens of these objectives.

In the development of the research the contributions of the Bergen school were seminal. Their method of synoptic analysis delivered a crisp, physically based portrayal of the frontal palette of midlatitude weather systems on surface charts. It provided meteorologists with a readily assimilated tool capable of representing key frontal features, pinpointing probable areas of precipitation, and tracing the day-to-day evolution of frontal weather patterns (Davies 1997).

The Bergen school method prompted primarily forecastoriented studies. The comparative success and utility of the method long discouraged the development and use of alternative analysis techniques. On the one hand, the basis for the analysis method was accepted and attempts were made to elaborate the procedure (see, e.g., Chromow 1940, Willett 1944), whereas, on the other hand, the forecaster often pragmatically tailored the results of applying the technique to mask rather than highlight incongruities with the prevailing weather features (Sutcliffe 1952). Neither approach was undergirded by a significant theoretical base, and Sawyer (1964) remonstrated (and demonstrated) that the increase in available observations, technological advances, and better understanding of atmospheric dynamics did make "it possible to display succinctly on one chart several features of atmospheric structure which are not adequately described by conventional frontal analysis." This refrain has been repeated sporadically ever since. For example, Mass (1991), assessing the issue from a synoptician's standpoint, provides a detailed critique of the currently perceived limitations of the original frontal analysis technique.

Bergeron's exploration of frontogenesis (Bergeron 1928) was predominantly explanatory in character, as have been the subsequent theoretical studies surveyed in the present study. Seminal to advance in this field was the formulation of the semigeostrophic (SG) system of equations (Eliassen 1959, 1962; Hoskins 1971, 1975; Hoskins and Bretherton 1972). The foundational studies (Section 3) based principally upon the SG system isolated the key mechanisms for inducing frontogenesis, illustrated the central role of the geostrophically forced ageostrophic circulation in generating rapid frontogenesis, established a dynamical basis for the distinction between cold and warm fronts, pinpointed the processes limiting frontal scale-contraction, and helped clarify the role of frictional and diabatic effects. Studies of frontogenesis in the idealized setting of growing baroclinic waves (Section 4) indicated that the range of simulated patterns include several realistic features, and that the realized frontal palette is sensitive to the geometry, the nature of the flow system, and the ambient shear. These studies provide a platform for understanding the frontal palette that accompanies observed events of cyclogenesis (Section 5).

In relation to forecast-oriented objectives, the theoretical studies of frontogenesis have served, for example, to demonstrate the value and limitations of the primitive equations; to indicate the two-way interscale linkages between fronts, cyclones, and the larger-scale flow; to underline the significance of these linkages to the occurrence and nature of frontogenetic events; and to provide physically meaningful diagnostic tools for analyzing the observed phenomena.

In the future, theoretical studies can help in the development of insightful way(s) of examining both analysis and forecast fields. This challenge is heightened by the ability to view synoptic and frontal development with time loop displays from one (or more) three-dimensional perspectives. Future theoretical (and observational) research areas can be highlighted by the limitations and capabilities of the current operational forecasting suites. For example rapid fronto- and cyclogenesis remains a major forecasting problem. Also the ability to perform ensemble forecasts with even high resolution mesoscale limited-area models raises the intriguing issue of the degree of deterministic space-time predictability of fronts. These possible future research directions are linked to both forecast and explanatory science, and as such follow the pathway exemplified by the Bergen school.

References

Bader, M. J., G. S. Forbes, J. R. Grant, R. B. E. Lilley, and A. J. Waters, Eds., 1995: *Images in Weather Forecasting.* Cambridge University Press, 499 pp.

Bannon, P. R., 1984 : Effects of stratification on surface frontogenesis: Warm and cold fronts. *J. Atmos. Sci.*, **41**, 2021–2026.

Bergeron, T., 1928. Über die dreidimensional verknüpfende Wetteranalyse. *Geofys. Publ.*, **5**(6), 111.

Bishop, C. H., 1993: On the behaviour of baroclinic waves undergoing horizontal deformation . I: The "RT" phase diagram. *Quart. J. Roy. Meteor. Soc.*, **119**, 221–240.

Bjerknes, J., 1919: On the structure of moving cyclones. *Geofys.Publ.*, **1**(2), 8.

Blumen, W., 1980: A comparison between the Hoskins–Bretherton model of frontogenesis and the analysis of an intense surface frontal zone. *J. Atmos. Sci.*, **37**, 64–77.

———, 1981: The geostrophic coordinate transformation. *J. Atmos. Sci.*, **38**, 1100 –1105.

———, 1990: A semigeostrophic Eady-wave frontal model incorporating momentum diffusion: Part I: Model and solutions. *J. Atmos. Sci.*, **47**, 2890–2902.

Browning, K. A., 1990: Organization of clouds and precipitation in extratropical cyclones. *Extratropical Cyclones, The Erik Palmén Memorial Volume*, C. W. Newton and E. O. Holopainen, Eds., Amer. Meteor. Soc.

Chan, D., and H-R. Cho, 1991: The dynamics of moist frontogenesis in a semigeostrophic model. *Atmos.-Ocean*, **29**, 85–101.

Chromow, S. P., 1940: *Einfürung in die synoptische Wetteranalyse*, Springer, 532 pp.

Cooper, I. M., A. J. Thorpe, and C. H. Bishop, 1992: The role of diffusive effects on potential vorticity in fronts. *Quart. J. Roy. Meteor. Soc.*, **118**, 629–647.

Cullen, M. J. P., and R. J. Purser, 1984: An extended Lagrangian theory of semigeostrophic frontogenesis. *J. Atmos. Sci.*, **41**, 1477–1497.

Davies, H. C., 1997: Emergence of the mainstream cyclogenesis theories. *Meteor. Z.*, N.F. **6**, 261–274.

———, and C. H. Bishop, 1994: Eady edge waves and rapid development. *J. Atmos. Sci.*, **51**, 1930–1946.

———, and J. C. Müller, 1988. Detailed description of deformation-induced semigeostrophic frontogenesis. *Quart. J. Roy. Meteor. Soc.*, **114**, 1201–1219.

———, and A. Rossa, 1998: PV-frontogenesis and upper-tropospheric fronts. *Mon. Wea. Rev.*, **126**, 1528–1539.

———, and H. Wernli, 1997: On studying the structure of synoptic systems. *Meteor. Appl.*, **4**, 365–374.

———, C. Schär, and H. Wernli, 1991: The palette of fronts and cyclones within a baroclinic wave-development. *J. Atmos. Sci.*, **48**, 1666–1689.

Davis, C. A., and M. L. Weisman, 1994: Balanced dynamics of mesoscale vortices produced in simulated convective systems. *J. Atmos. Sci.*, **51**, 2005–2030.

Eliassen, A., 1959: On the formation of fronts in the atmosphere. *The Rossby Memorial Volume* , Rockefeller Institute Press, 277–287.

———, 1962: On the vertical circulation in frontal zones. *Geofys. Publ.* **24**, 147–160.

———, 1990: Transverse circulations in frontal zones. *Extratropical Cyclones, The Erik Palmén Memorial Volume*, C. W. Newton and E. O. Holopainen, Eds., Amer. Meteor. Soc.

Emanuel, K. A., M. Fantini, and A. J. Thorpe, 1987: Baroclinic instability in an environment of small stability to slantwise moist convection. Part I: Two-dimensional models. *J. Atmos. Sci.*, **44**, 1559–1573.

Gall, R., R. T. Williams, and T. L. Clark, 1988: Gravity waves generated during frontogenesis. *J. Atmos. Sci.*, **46**, 2205–2219.

Garner, S. T., 1989 : Fully Lagrangian numerical solutions of unbalanced frontogenesis and frontal collapse. *J. Atmos. Sci.*, **46**, 717–739.

Hines, K. M., and C. R. Mechoso, 1993: Influence of surface drag on the evolution of fronts. *Mon. Wea. Rev.*, **121**, 1152–1175.

———, and ———, 1991: Frontogenesis processes in the middle and upper troposphere. *Mon. Wea. Rev.*, **119**, 1225–1241.

Hoskins, B. J., 1971: Atmospheric frontogenesis models: Some solutions. *Quart. J. Roy. Meteor. Soc.* **97**, 139–153.

———, 1975: The geostrophic momentum approximation and the semigeostrophic equations. *J. Atmos. Sci.*, **32**, 233–242.

———, 1982: The mathematical theory of frontogenesis. *Ann. Rev. Fluid Mech.*, **14**, 131–151.

———, and F. P. Bretherton, 1972: Atmospheric frontogenesis models: Mathematical formulation and solutions. *J. Atmos. Sci.*, **29**, 11–37.

———, and W. A. Heckley, 1981: Cold and warm fronts in baroclinic waves. *Quart. J. Roy. Meteor. Soc.*, **107**, 79–90.

———, and N. V. West, 1979: Baroclinic waves and frontogenesis. Part II: Uniform potential vorticity jet flows. *J. Atmos. Sci.*, **36**, 1663–1680.

———, I. Draghici, and H. C. Davies, 1978: A new look at the ω-equation. *Quart. J. Roy. Meteor. Soc.*, **104**, 31–38.

———, E. C. Neto, and H.-R. Cho, 1984: The formation of multiple fronts. *Quart. J. Roy. Meteor. Soc.*, **110**, 881–896.

———, M. E. McIntyre, and A.W. Robertson, 1985: On the use and significance of isentropic potential vorticity maps. *Quart. J. Roy. Meteor. Soc.*, **111**, 877–946.

Joly, A., and A. J. Thorpe, 1989: Warm and occluded fronts in two-dimensional moist baroclinic instability. *Quart. J. Roy. Meteor. Soc.*, **115**, 513–534.

Keyser, D., and R. A. Anthes, 1982: The influence of planetary boundary layer physics on frontal structure in the Hoskins–

Bretherton horizontal shear model. *J. Atmos. Sci.*, **39**, 1783–1802.

——, and ——, 1986: Comments on "Frontogenesis in a moist semigeostrophic model." *J. Atmos. Sci.*, **43**, 1051–1054.

——, and M. A. Shapiro, 1986: A review of the structure and dynamics of upper-level frontal zones. *Mon. Wea. Rev.*, **114**, 452–499.

——, and M. J. Pecnick, 1987: The effect of along-front temperature variation in a two-dimensional primitive equation model of surface frontogenesis. *J. Atmos. Sci.*, **44**, 577–604.

——, M. J. Reeder, and R. J. Reed, 1988: A generalization of Petterssen's frontogenesis function and its relation to the forcing of vertical motion. *Mon. Wea. Rev.*, **116**, 762–780.

——, B. D. Schmidt, and D. G. Duffy, 1989: A technique for representing three-dimensional vertical circulations in baroclinic disturbances. *Mon. Wea. Rev.*, **117**, 2463–2494.

Knight, D. J., and P. V. Hobbs, 1988: The mesoscale and microscale structure and organization of clouds and precipitation in midlatitude cyclones. Part XV: A numerical study of frontogenesis and cold-frontal rainbands. *J. Atmos. Sci.*, **45**, 915–930.

Koshyk, J. N., and H.-R. Cho, 1992: Dynamics of a mature front in a uniform potential vorticity semigeostrophic model. *J. Atmos. Sci.*, **49**, 497–510.

Lalaurette, F., C. Fischer, and J.-P. Cammas, 1994: Location and interaction of upper- and lower-tropospheric adiabatic frontogenesis. *Mon. Wea. Rev.*, **122**, 2004–2021.

Levy, G., and C. S. Bretherton, 1987: On a theory of evolution of surface cold fronts. *J. Atmos. Sci.*, **44**, 3413–3418.

Lighthill, M. J., 1952: On sound generated aerodynamically. I. General Theory. *Proc. Roy. Soc. A*, **211**, 564–587.

Malardel, S., A. Joly, F. Courbet, and P. Courtier, 1993: Nonlinear evolution of ordinary frontal waves induced by low level potential vorticity anomalies. *Quart. J. Roy. Meteor. Soc.*, **119**, 681–714.

Mak, M., and P. R. Bannon, 1984: Frontogenesis in a moist semigeostrophic model. *J. Atmos. Sci.*, **41**, 3485–3500.

Mass, C. F., 1991: Synoptic analysis: Time for a reassessment. *Bull. Amer. Meteor. Soc.*, **72**, 348–363.

Moore, G. W. K., 1993: The development of tropopause folds in two-dimensional models of frontogenesis. *J. Atmos. Sci.*, **50**, 2321–2334.

Müller, J., 1989: Semi-geostrophische Entwicklung von Fronten und Störungen in der unteren Atmosphäre. Ph.D. dissertation No. 8878, ETH Zürich, 179 pp.

——, H. C. Davies, and Ch. Schär, 1989: An unsung mechanism for development. *J. Atmos. Sci.*, **46**, 3666–3672.

Nakamura, N., 1994: Nonlinear equilibriation of two-dimensional Eady waves: Simulations with viscous geostrophic momentum equations. *J. Atmos. Sci.*, **51**, 1023–1035.

——, and I. M. Held, 1989: Nonlinear equilibriation of two-dimensional Eady waves. *J. Atmos. Sci.*, **46**, 3055–3064.

Ogura, Y., and D. Portis, 1982: Structure of the cold front observed in SESAME-AVE III and its comparison with the Hoskins–Bretherton frontogenesis model. *J. Atmos. Sci.*, **39**, 2773–2792.

Orlanski, I., and B. B. Ross, 1984: The evolution of an observed cold front. Part II: Mesoscale dynamics. *J. Atmos. Sci.*, **41**, 1669–1703.

Ostdiek, V., and W. Blumen, 1995: Deformation frontogenesis: Observations and theory. *J. Atmos. Sci.*, **39**, 1783–1801.

Parker, D.J., and A.J. Thorpe, 1995: Conditional convective heating in a baroclinic atmosphere: A model of convective frontogenesis. *J. Atmos. Sci.*, **52**, 1699–1711.

Pecnick, M. J., and D. Keyser, 1989: The effect of spatial resolution on the simulation of upper-tropospheric frontogenesis using a sigma-coordinate primitive equation model. *Meteor. Atmos. Phys.*, **40**, 137–149.

Petterssen, S., 1936: A contribution to the theory of frontogenesis. *Geofys. Publ.*, **11**, 1–27.

——, and S. J. Smeybe, 1971: On the development of extratropical cyclones. *Quart. J. Roy. Meteor. Soc.*, **97**, 457–482.

Polavarapu, S. M., and W. R. Peltier, 1990: The structure and nonlinear evolution of synoptic scale cyclones: Life cycle simulations with a cloud scale model. *J. Atmos. Sci.*, **47**, 2645–2672.

Raymond, D. J., 1992: Nonlinear balance and potential-vorticity thinking at large Rossby number. *Quart. J. Roy. Meteor. Soc.*, **118**, 987–1105.

Reed, R. J., and E. F. Danielsen, 1960: Fronts in the vicinity of the tropopause. *Arch. Meteor. Geophys. Biokl. A*, **11**, 1–17.

Reeder, M. J., and R. K. Smith, 1986: A comparison between frontogenesis in the two-dimensional Eady model of baroclinic instability and summertime cold fronts in the Australian region. *Quart. J. Roy. Meteor. Soc.*, **112**, 293–313.

Rotunno, R., W. C. Skamarock, and C. Snyder, 1994: An analysis of frontogenesis in numerical simulations of baroclinic waves. *J. Atmos. Sci.*, **51**, 3373–3398.

Saute, M., 1993: The influence of tropospheric static stability on upper-level frontogenesis. *Tellus*, **45A**, 159–167.

Sawyer, J. S., 1956: The vertical circulation at meteorological fronts and its relation to frontogenesis. *Proc. Roy. Soc. A*, **234**, 346–362.

——, 1964: Meteorological analysis—A challenge for the future. *Quart. J. Roy. Meteor. Soc.*, **90**, 227–247.

Schär, C., and H. C. Davies, 1990: An instability of mature cold fronts. *J. Atmos. Sci.*, **47**, 929–950.

——, and H. Wernli, 1993: Structure and evolution of an isolated semigeostrophic cyclone. *Quart. J. Roy. Meteor. Soc.*, **119**, 57–90.

Shapiro, M. A., and D. Keyser, 1990: On the structure and dynamics of fronts, jet streams and the tropopause. *Extratropical Cyclones, The Erik Palmén Memorial Volume*, C. W. Newton and E. O. Holopainen, Eds., Amer. Meteor. Soc.

Smigielski, F. J., and G. P. Ellrod, 1985: Surface cyclogenesis as indicated by satellite imagery. NOAA (T.M.NESDIS 9), 30 pp.

Snyder, C., W. C. Skamarock, and R. Rotunno, 1991: A comparison of primitive-equation and semi-geostrophic simulations of baroclinic waves. *J. Atmos. Sci.*, **48**, 2179–2194.

——, ——, and ——, 1993: Frontal dynamics near and following frontal collapse. *J. Atmos. Sci.*, **50**, 3194–3211.

Stone, P. H., 1966: Frontogenesis by horizontal wind deformation fields. *J. Atmos. Sci.*, **23**, 455–465.

Sutcliffe, R. C., 1952: Principles of synoptic weather forecasting. *Quart. J. Roy. Meteor. Soc.*, **78**, 291–320.

Takayabu, I., 1986: Roles of the horizontal advection on the formation of surface fronts and on occlusion of a cyclone developing in the baroclinic westerly jet. *J. Meteor. Soc. Japan*, **64**, 329–345.

——, 1991: "Coupling development": An efficient mechanism for the development of extratropical cyclones. *J. Meteor. Soc. Japan*, **69**, 609–628.

Thorncroft, C., and B. J. Hoskins, 1990: Frontal cyclogenesis. *J. Atmos. Sci.*, **47**, 2317–2336.

——, ——, and M. E. McIntyre, 1993: Two paradigms of baroclinic–wave life-cycle behaviour. *Quart. J. Roy. Meteor. Soc.*, **119**, 17–56.

Thorpe, A. J., and K. A. Emanuel, 1985: Frontogenesis in the presence of small stability to slantwise convection. *J. Atmos. Sci.*, **42**, 1809–1824.

Tremblay, A., 1992: Structure and development of mesoscale baroclinic waves in a nonhydrostatic numerical model. *Mon. Wea. Rev.,* **120,** 463–481.

Trüb, J., and H. C. Davies, 1995: Flow over a mesoscale ridge: Pathways to regime transition. *Tellus,* **47A,** 502–524.

Volkert, H., and C. H. Bishop, 1990: The semigeostrophic Eady problem as a testbed for numerical simulations of frontogenesis. *Tellus,* **42A,** 202–207.

Weldon, R. B., and S. J. Holmes, 1991: Water vapour imagery interpretation and applications to weather analysis and forecasting. NOAA Tech. Rep. NESDIS 57, 213 pp.

Wernli, H., R. Fehlmann, and D. Luthi, 1998: The effect of barotropic shear on upper-level induced cyclogenesis: Semigeostrophic and primitive equation numerical simulations. *J. Atmos. Sci.,* **55,** 2080–2094.

Whitaker, J. S., and C. A. Davis, 1994: Cyclogenesis in a saturated environment. *J. Atmos. Sci.,* **51,** 889–907.

———, and ———, 1992 : The effects of spherical geometry on the evolution of baroclinic waves. *J. Atmos. Sci.,* **50,** 597–612.

Willett, H. C., 1944: *Descriptive Meteorology.* Academic Press.

Williams, R. T., 1967: Atmospheric frontogenesis: A numerical experiment. *J. Atmos. Sci.,* **24,** 627–641.

———, 1968: A note on quasi-geostrophic frontogenesis. *J. Atmos. Sci.,* **25,** 1157–1159.

On the Representation and Diagnosis of Frontal Circulations in Two and Three Dimensions

DANIEL KEYSER

Department of Earth and Atmospheric Sciences, The University at Albany/State University of New York, Albany, New York, USA

1. Introduction

Ageostrophic circulations play a significant role in the kinematics and dynamics of a variety of synoptic- and mesoscale weather systems found within extratropical baroclinic wave regimes. Here the ageostrophic circulation is defined as (V_{ag}, ω), where the horizontal and vertical "branches" are linked through mass continuity, which may be expressed as $\nabla_p \cdot V_{ag} + \partial \omega / \partial p = 0$, provided that the geostrophic wind, V_g, is defined in terms of constant Coriolis parameter, f_0. Ageostrophic circulations contribute to: (1) modulation of the structure of cloud and precipitation systems; (2) cyclogenesis through vortex stretching; (3) frontogenesis through positive feedbacks with the geostrophic flow involving horizontal scale contractions and tilting; and (4) energy transformations involving cross-contour ageostrophic flow, ageostrophic geopotential fluxes, and correlations between vertical velocity, ω, and potential temperature, θ (i.e., direct/indirect circulations). Diagnostic studies of ageostrophic circulations offer a means both to quantify predictions of conceptual models of the structure of synoptic- and mesoscale weather systems, such as cyclones and fronts, and to identify signatures of dynamical processes operative in theoretical models of these systems.

In view of recent interest within the synoptic community in applying potential-vorticity concepts to the diagnosis of baroclinic development and frontogenesis that has followed the publication of the influential review by Hoskins et al. (1985), one might question the need for continuing to emphasize ageostrophic circulations in their own right. In the adiabatic, frictionless, quasigeostrophic (QG) framework to be adopted in the present chapter, ageostrophic circulations are implicit, and one can model and interpret the dynamics of cyclogenesis without considering them directly. Nevertheless, it is recognized that ageostrophic circulations need to be treated explicitly in the QG baroclinic development problem when vertical-motion-dependent processes, such as latent heat release, are included. Furthermore, dynamical feedbacks involving ageostrophic circulations play a fundamental role in the development of both surface and upper-level fronts, as is demonstrated in the now-classic study by Hoskins and Bretherton (1972), which is posed in terms of the semigeostrophic (SG) equations. In this generalization of the QG equations, ageostrophic circulations are essential to the dynamics of the scale-contraction process and to the realistic replication of frontal structure. These considerations serve as a reminder that interpretations deriving from "potential-vorticity thinking" need not be seen as superseding those deriving from analysis of ageostrophic circulations—rather each diagnostic framework may be understood to yield complementary perspectives on the dynamics of balanced-flow phenomena such as cyclones and fronts.

There is extensive observational and theoretical support for a close correspondence between ageostrophic circulations and frontal regions situated within extratropical cyclones and baroclinic wave disturbances. Visible and infrared satellite imagery regularly reveal elongated cloud bands in the vicinity of frontal zones (e.g., Browning 1990, Fig. 8.4), and water-vapor imagery often exhibits dark streaks indicative of deep dry layers in the middle and upper troposphere (e.g., Muller and Fuelberg 1990; Appenzeller and Davies 1992). The two-dimensional Sawyer-Eliassen circulation equation (Sawyer 1956; Eliassen 1962) and the Q-vector form of the three-dimensional QG omega equation (Hoskins et al. 1978; Hoskins and Pedder 1980) relate the vertical motion and the divergent part of the ageostrophic wind to the frontogenetical action of the geostrophic wind on the potential temperature field. This relationship yields the interpretation that the forcing of ageostrophic circulations is concentrated in frontal zones, underscoring the major role that these features play in the dynamics of extratropical baroclinic flows.

In two dimensions, where ageostrophic circulations $((u_{ag}, \omega);$ linked through mass continuity, $\partial u_{ag}/\partial x + \partial \omega/\partial p = 0)$ are confined to the cross-front or transverse vertical (x, p) plane, the representation and diagnosis of ageostrophic circulations in frontal zones in terms of the Sawyer-Eliassen equation is sufficiently well-established to have taken a prominent place in review articles (e.g., Keyser and Shapiro 1986, Section 3b; Eliassen 1990) and in textbooks concerned with synoptic/dynamic meteorology (e.g., Carlson 1991, Section 14.4; Bluestein 1993, Section 2.5.2). In contrast, progress in three dimensions does not appear to be as advanced. A variety of approaches for representing ageostrophic circulations appear in the literature, serving as the basis for well-known conceptual models of frontal and jet-streak archetypes, but the kinematic properties of these approaches and their relationship to each other and to the two-dimensional case is not always apparent. Although real-data applications of the Sawyer-Eliassen equation can be found in the literature (e.g., Shapiro 1981; Keyser and Carlson 1984; Uccellini et al. 1985; Lagouvardos and Kotroni 1995), the appropriateness of such applications may be questioned because the assumption that ageostrophic circulations are two-dimensional (i.e., confined to the cross-front plane) is not generally satisfied in realistic flows, especially when such flows are strongly curved. Finally, observational application of dynamical equations for three-dimensional ageostrophic circulations, such as the generalization of the Sawyer-Eliassen equation proposed by Hoskins and Draghici (1977), appears to be limited; the only case-study application of this equation of which the author is aware is that of Bosart and Lin (1984, Section 9). This state of affairs motivates the present review.

The representation and diagnosis of ageostrophic circulations will be reviewed and illustrated with examples involving frontal zones and jet streaks (localized wind speed maxima within jet streams). A primary objective of this review is to suggest a coherent framework for diagnosing and interpreting three-dimensional ageostrophic circulations that follows naturally from two-dimensional considerations. In order to minimize complexity and to avoid excessive detail, diagnostic approaches will be formulated in terms of the adiabatic, frictionless QG equations applicable to an f-plane in the absence of topography. Furthermore, the complicating effects of Earth geometry and lateral boundary conditions will be disregarded through the adoption of tangent-plane Cartesian coordinates and the assumption of localized disturbances on an unbounded domain or of periodic disturbances in x and y. Although the foregoing abstractions appear to leave little room for realism, the resulting mathematical simplification is deemed appropriate for the conceptual flavor of this review.

In the following section, various representations of ageostrophic circulations are identified from a survey of selected observational studies and conceptual models of fronts and jet streaks. In Section 3, the two-dimensional case is considered, starting with the kinematic properties of ageostrophic circulations and concluding with a brief review of the Sawyer-Eliassen equation. This analysis provides the basis for the three-dimensional case, which is the subject of Section 4. Concluding remarks are presented in Section 5.

2. Representations of Frontal Circulations

Here and throughout this chapter, the terms "frontal circulation" and "vertical circulation" (as opposed to "ageostrophic circulation") shall be used somewhat loosely to refer to the vertical motion field and an associated horizontal wind field (consistent with mass continuity) in the vicinity of fronts and jet streaks. Furthermore, "cross-front" (x, p) and "along-front" (y, p) refer respectively to circulations in vertical planes oriented normal and parallel to the long axis of a frontal zone, defined to be a highly anisotropic region of concentrated horizontal gradients of potential temperature and strong vertical wind shears (consistent with thermal-wind balance), as well as large cyclonic vorticity and enhanced static stability. When the terms "cross-jet" and "along-jet" are used, it is understood that these orientations correspond closely to their frontal counterparts.

2.1 Natural-coordinate Representations

This approach consists of using the natural-coordinate form of the vector-momentum equation to infer ageostrophic flow in the cross- and along-jet directions:

$$v_{agn} = \frac{1}{f} \frac{dV}{dt}, \qquad (2.1a)$$

$$v_{ags} = -\frac{K_t V^2}{f}, \qquad (2.1b)$$

where v_{agn} and v_{ags} are the components of the ageostrophic wind in a "right-handed" natural-coordinate system, such that $\mathbf{n} = \mathbf{k} \times \mathbf{s}$ $(\mathbf{s} = V/V)$; V is the speed of the horizontal wind, V; f is the Coriolis parameter (not necessarily constant); d/dt $(= \partial/\partial t + V \cdot \nabla_p + \omega \partial/\partial p)$ is the Lagrangian rate of change following the three-dimensional flow; and K_t is parcel-trajectory curvature. (For further details on the natural-coordinate equations of motion, see Holton (1992, Section 3.2.1) and Bell and Keyser (1993, Section 2).) In traditional applications of (2.1a,b), patterns of v_{agn} and v_{ags} are related to wind-speed and flow-curvature configurations characteristic of jet streaks and baroclinic waves, respectively. Vertical motions are inferred qualitatively from contributions to the horizontal divergence from the ageostrophic wind components in the cross- and along-jet directions, yielding vertical circulations in these respective directions. Inference of vertical motions typically assumes a "two-layer" atmosphere, such that the ageostrophic winds and horizontal divergence are maximized near the surface and tropopause, and the vertical motion is of one sign and maximized in the middle troposphere. In applications of (2.1a,b), the jet-front system often is taken to be steady state, and it is assumed that vertical advections may be neglected (which is justifiable at jet level, i.e., the level of maximum wind (LMW)). Assuming a steady-state system and neglecting vertical advections allows Lagrangian rates of change to be approximated in terms of advection by the horizontal flow relative to the motion of the jet-front system.

A variant of this approach is to apply the vorticity equation at the LMW, allowing neglect of the vertical-advection and tilting terms. This yields the following expression for the horizontal divergence:

$$\nabla_p \bullet \mathbf{V} = -(\zeta_p + f)^{-1} \frac{d}{dt_h}(\zeta_p + f), \quad (2.2)$$

where ζ_p is the vertical component of relative vorticity in the pressure-coordinate system and $d/dt_h (= \partial/\partial t + V \bullet \nabla_p)$ is the Lagrangian rate of change following the horizontal flow. As before, (2.2) often is applied assuming a steady-state jet-front system. In order to isolate "components" of the horizontal divergence in the cross- and along-jet directions, the relative vorticity on the right-hand side of (2.2) may be partitioned into shear and curvature components (e.g., Beebe and Bates 1955).

Perhaps the best-known conceptual model based on applications of (2.1a,b) and (2.2) is the classic four-quadrant model of vertical motion associated with straight jet streaks. (See Uccellini (1990, Section 6.3.1) for a comprehensive review and pertinent citations concerning this conceptual model.) For a sufficiently straight jet streak (in the sense that $|K_t| << f/V$; refer to (2.1b)), it may be assumed that the along-jet flow is approximately geostrophic. The ageostrophic flow is thus confined to transverse vertical planes and is described by (2.1a); thermodynamically direct and indirect transverse ageostrophic circulations are inferred in the entrance and exit regions of the jet streak, respectively (Figs. 1a,b). In terms of (2.2), the same pattern may be inferred from the advection of relative vorticity by the jet-relative flow at the LMW (Fig. 1c). This model may be generalized to include the effects of flow curvature by applying (2.1b) to a curved jet stream (Fig. 2). It is noteworthy with respect to the historical contributions of the Bergen school to the topic of the present chapter that the schematics in Figs. 1 and 2 derive from papers by Bjerknes (1951) and Bjerknes and Holmboe (1944), respectively. Applications of the so-called "natural-coordinate" approach of inferring ageostrophic circulations in upper-

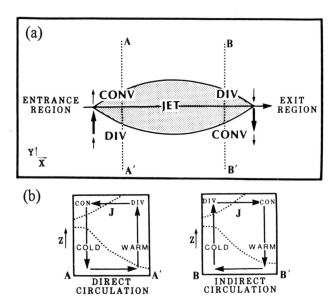

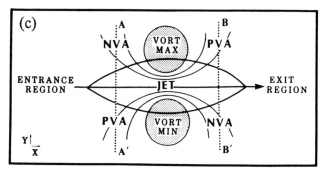

FIG. 1. Schematic illustration of ageostrophic circulations and vorticity patterns for a straight jet streak: (a) transverse ageostrophic wind components and associated patterns of convergence (CONV) and divergence (DIV) in the entrance and exit regions at the level of maximum wind; (b) transverse ageostrophic circulations in the entrance (cross section AA′) and exit (cross section BB′) regions of the jet streak depicted in (a), along with schematic isentropes (dotted lines) and location of jet core (J); (c) relative vorticity and associated advection patterns, with NVA and PVA indicating anticyclonic (negative) and cyclonic (positive) relative vorticity advection, respectively (Uccellini and Kocin 1987; also presented in Kocin and Uccellini 1990 and Uccellini 1990).

level jet-front systems are found in the diagnostic studies by Uccellini and Johnson (1979), Shapiro and Kennedy (1981), Bluestein and Thomas (1984), and Cammas and Ramond (1989).

The natural-coordinate approach of inferring frontal circulations may be considered empirical in the following respects: (1) the momentum/vorticity equations are used without simultaneous consideration of the thermodynamic equation, (2) the steady-state assumption is invoked, and (3) vertical motion is inferred from the spatial distribution of horizontal divergence at jet level through qualitative appeal to the continuity equation. Nevertheless, this approach introduces the notion that horizontal divergence and vertical

(a)

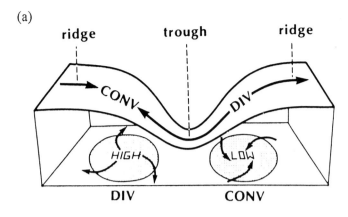

(b)

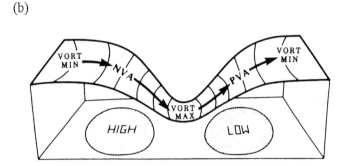

Fig. 2. Schematic illustration of ageostrophic wind and vorticity patterns for a curved jet stream: (a) along-flow ageostrophic wind and associated patterns of convergence (CONV) and divergence (DIV) at the level of maximum wind; (b) relative vorticity and associated advection patterns, with NVA and PVA indicating anticyclonic (negative) and cyclonic (positive) relative vorticity advection, respectively (Kocin and Uccellini 1990 and Uccellini 1990).

motion, although fundamental kinematic quantities, may be partitioned meaningfully to isolate vertical circulations respectively associated with jet-front systems and with the baroclinic waves in which these systems are embedded. The notion of partitioned vertical motions and vertical circulations will be reexamined in Section 2.3.

2.2. Transverse Vertical Circulations

The next method of representing frontal circulations consists of projecting particular components of the horizontal wind onto the cross-front plane in conjunction with the vertical motion. This representation is displayed as a vector (vertical part scaled appropriately) on a vertical cross section oriented in the cross-front plane. The particular transverse horizontal wind components to be discussed in this subsection are: total (u), system-relative ($u - c$) (where c is the translation speed of the frontal system (assumed to coincide with the cross-front direction)), and ageostrophic (u_{ag}).

The (u, ω) and ($u - c$, ω) representations often are used to provide an indication of vertical displacement in the cross-front plane, a quantity which is a better discriminator of the formation and presence of clouds than instantaneous vertical motion (e.g., Schär and Wernli 1993; Wernli 1995). In the case where the flow (translating frontal system) is steady-state and two-dimensional, that is, invariant in the along-front direction, streamlines of (u, ω) (($u - c$, ω)) correspond to projections of trajectories with respect to the Earth (with respect to the translating frontal system) onto the cross-front plane, from which vertical displacements may be assessed. Furthermore, in the two-dimensional limit, these representations satisfy the continuity equation in the cross-front plane (i.e., there are no sources/sinks of mass in the y direction, consistent with $\partial v/\partial y = 0$).

Examples of these representations are given for a simulated upper-level front by Newton and Trevisan (1984; reproduced here in Fig. 3) and for an analyzed surface cold front by Eliassen (1966, 1990; reproduced here in Fig. 9). In both cases, the frontal zones coincide closely with a confluent asymptote in the flow in the cross-front plane (Figs. 3c and 9c), a signature that may be characteristic of these particular representations. It is of interest that conceptual models of frontal archetypes based on principles of relative-flow isentropic analysis (Green et al. 1966; Carlson 1991, Sections 12.2 and 12.3), such as forward- and rearward-sloping cold fronts (e.g., Browning 1990, Section 8.3), make use of depictions of front-relative flow (in map and cross-section form), and thus may be viewed as an example of this type of representation of the vertical circulation.

In interpreting streamline patterns of (u, ω) in cross-section form, it is convenient to exploit the thermodynamic equation:

$$V_2 \bullet \nabla_y \theta = -\frac{\partial \theta}{\partial t} - v \frac{\partial \theta}{\partial y} + \frac{H}{\pi c_p}, \qquad (2.3)$$

where $V_2 = u\mathbf{i} - \omega\mathbf{k}$, $\nabla_y = \mathbf{i}\partial/\partial x - \mathbf{k}\partial/\partial p$ is the gradient operator in the (x, p) plane, H is the heating rate, and $\pi = (p/p_0)^{R/c_p}$. Eq. (2.3) reveals that if the flow is adiabatic and locally steady, cross-isentrope flow in the (x, p) plane indicates the sense and relative magnitude of the along-front potential temperature advection. Provided that these assumptions pertain to the case shown in Fig. 3, cold-air advection is evident in the cross section depicting the upper-level frontal zone (note the orientation of the streamlines relative to the isentropes in Fig. 3c), consistent with the sense of the geostrophic potential temperature advection in the portion of the frontal zone upstream of the 500 hPa trough axis (Fig. 3a). If the flow is steady following the frontal zone, the motion of which is assumed to be in the x direction (such that $\partial\theta/\partial t = -c\partial\theta/\partial x$), (2.3) pertains with u replaced by $u - c$ on the left-hand side and without $\partial\theta/\partial t$ on the right-hand side.

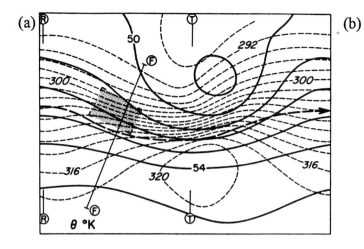

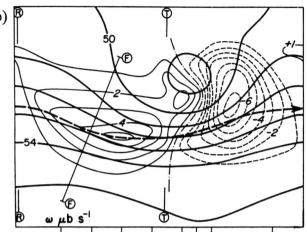

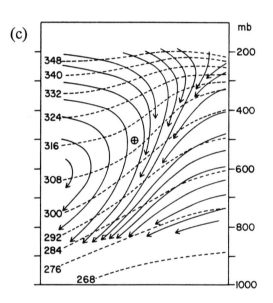

FIG. 3. Depiction of structure of and vertical circulations in an upper-level jet-front system derived from integration of a baroclinically unstable wave to finite amplitude in an adiabatic, β-plane, primitive-equation channel model: (a) 500 hPa geopotential height (contour interval 100 m, 50 denotes 5000 m; thick solid), potential temperature (contour interval 2 K, dashed), and jet-stream axis (thick dashed arrow); (b) as in (a), except for pressure-coordinate vertical velocity (contour interval 1 μb s^{-1}; thin solid (positive), dot-dashed (zero), dashed (negative)) replacing potential temperature; (c) streamlines of total transverse circulation (u, ω) and isentropes (contour interval 8 K, dashed) for the vertical section, FF, crossing the northwesterly flow inflection, shown in (a) and (b). Circled plus sign indicates location of 500 hPa jet stream (Newton and Trevisan 1984).

In summary, the (u, ω) and ($u - c$, ω) representations of frontal circulations are convenient proxies for vertical displacement provided that the steady-state and two-dimensional assumptions are applicable. Even in cases where the flow is three-dimensional, these representations can provide an indication of the along-front potential temperature advection, provided that the flow may be considered adiabatic. In such cases, it should be recognized that these representations do not satisfy mass continuity in the cross-front plane.

The third representation of frontal circulations is the transverse component of the ageostrophic wind and the total vertical motion (u_{ag}, ω). In the two-dimensional limit and for constant Coriolis parameter, this representation satisfies continuity in the cross-front plane and forms closed circulations. Furthermore, in this limit, the (u_{ag}, ω) representation coincides with the secondary circulation (consisting of the divergent wind in the cross-front plane and the total vertical motion) for straight frontal zones, assumed implicitly in the synthesis of the four-quadrant model of jet-streak vertical circulations (Section 2.1) and explicitly in the derivation of the Sawyer-Eliassen equation (Section 3). For these reasons, this representation has appeared prominently in observational studies of vertical circulations in the vicinity of jet-front systems (e.g., Keyser et al. 1978; Uccellini et al. 1985; Uccellini and Kocin 1987; Hakim and Uccellini 1992). In the three-dimensional case, this representation does not satisfy continuity in the cross-front plane ($\partial v_{ag}/\partial y \neq 0$).

2.3 Partitioned Vertical Circulations

In order to circumvent the problem that continuity is not satisfied in any particular vertical plane when the flow is three-dimensional, it is advantageous to express the vertical motion in terms of a vector streamfunction (Hoskins and Draghici 1977):

$$\omega = \nabla_p \bullet \psi , \qquad (2.4a)$$

which, to satisfy mass continuity, requires that

$$V = -\frac{\partial \psi}{\partial p}. \qquad (2.4b)$$

It may be confirmed readily that this representation also satisfies mass continuity independently in the cross- and along-front vertical planes, where it is understood that $\omega_x = \partial \psi_x / \partial x$ and $\omega_y = \partial \psi_y / \partial y$ (here subscripts denote components, not partial derivatives). A schematic illustration of the orientation of ψ in relation to the horizontal and vertical branches of the circulation is given in Fig. 4. The reader may refer ahead to Fig. 15 for an example of the ψ-vector representation involving the divergent ageostrophic wind in a channel-model simulation of a finite-amplitude baroclinic wave similar to that shown in Fig. 3. An illustration of the decomposition of the ω field into cross-front (ω_x) and along-front (ω_y) components follows in Fig. 16 for a cross section traversing the midtropospheric frontal zone in the model simulation.

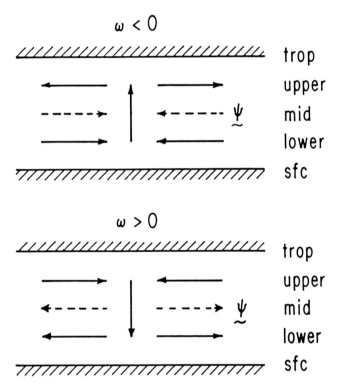

FIG. 4. Schematic illustrating relationship between ψ, ω, and V, as given by (2.4a,b) for a "two-layer" vertical circulation. Top and bottom panels depict cases of rising and sinking motion, respectively. The vertical velocity (vertical arrows) is assumed to vanish at the surface (sfc) and tropopause (trop), and to attain maximum magnitude in the middle troposphere (mid). The horizontal branches of the circulation (solid horizontal arrows) are depicted at pressure levels in the upper and lower troposphere; ψ (dashed horizontal arrows) is shown in the middle troposphere at the level where $|\omega|$ reaches a maximum. The schematic shows that ψ points toward (away from) areas of ascent (descent) and that the lower-level (upper-level) flow is parallel (antiparallel) to ψ, regardless of the sign of ω. This relationship is the same as that between Q, ω, and V_{ag} (see Hoskins and Pedder 1980) (Keyser et al. 1989).

The vector-streamfunction representation has been applied using various definitions of the horizontal wind component: total (V: Sanders and Bosart 1985), ageostrophic (V_{ag}: Hoskins and Draghici 1977; Xu 1990; Xu and Keyser 1993), and divergent[1] (V_d: Keyser et al. 1989, 1992a; Reeder et al. 1991; Loughe et al. 1995). Keyser et al. (1989) point out that unless the definition based on the divergent wind is adopted, the possibility exists for internal cancellation between ω_x and ω_y. Such cancellation is possible because the total horizontal and ageostrophic winds possess a rotational component, which must cancel internally in the continuity equation (i.e., $\partial u_r / \partial x = -\partial v_r / \partial y$; cancellation between ω_x and ω_y occurs when these respective contributions to the horizontal divergence are integrated vertically from either the lower or upper boundary, p_b or p_t, at one of which it is assumed that $\omega_x = \omega_y = 0$). Finally, because ω_x and ω_y do not vanish in general at both p_b and p_t for the alternatives involving the total and ageostrophic horizontal winds (implying internal cancellation within the horizontal divergence), the respective circulations in the cross- and along-front planes are not necessarily closed. Closed circulations do result, however, for the alternative involving the divergent wind.

3. Diagnosis of Frontal Circulations: Two-dimensional Case

Here the Sawyer-Eliassen equation is described briefly to provide the conceptual basis for its extension to three dimensions, to be discussed in Section 4. In two dimensions, the representations (u_{ag}, ω) and (u_d, ω) are identical, so in keeping with past practice, the former will be adopted in the remainder of this section.

The following discussion applies to straight fronts subject to the approximation of cross-front geostrophic balance (e.g., Hoskins and Bretherton 1972), so that the along-front wind is replaced by its geostrophic counterpart and the ageostrophic wind is confined to the cross-front plane. Under these circumstances, the continuity equation reduces to $\partial u_{ag} / \partial x + \partial \omega / \partial p = 0$ and (2.4a,b) to $\omega = \partial \psi / \partial x$ and $u_{ag} = -\partial \psi / \partial p$. (Note that here the subscript x is dropped from ω and ψ for convenience.) This definition implies clockwise (counterclockwise) circulations around minima (maxima) in ψ when looking toward the positive along-front direction (i.e., increasing y). These circulations are thermodynamically indirect (direct) provided that the cross-front coordinate, x, increases toward warmer air. The assumption that ω vanishes

---

[1]Here and in the remainder of this chapter, it is understood that a two-component partition (i.e., rotational/divergent, equivalent to nondivergent/irrotational) of the horizontal wind in terms of the Helmholtz theorem is defined unambiguously, a consequence of the idealized lateral boundary conditions assumed in Section 1.

at p_b and p_t guarantees that ageostrophic circulations are closed in the cross-front plane.

In the adiabatic, frictionless case, the Sawyer-Eliassen equation relates the transverse ageostrophic circulation (u_{ag}, ω) to frontogenetical configurations of the geopotential field, which is assumed to be known. Technically speaking, the geopotential is "balanced" in the sense that it is derived from some form of potential-vorticity inversion (e.g., Hoskins et al. 1985), so that it is free of the influence of high-frequency phenomena such as gravity waves and convection. Given the balanced geopotential and suitable boundary conditions on the ageostrophic streamfunction, and provided that an ellipticity condition is satisfied (i.e., positive reference-state static stability in the QG case), the ageostrophic streamfunction may be determined uniquely from the geopotential field.

Hoskins et al. (1978) articulated the now well-known principle that the frontogenetical action of the geostrophic flow tends to disrupt thermal wind balance. From this perspective, ageostrophic circulations are required to counteract this disruptive tendency, and thus to restore thermal wind balance. For the cross-front direction, it follows that

$$\frac{d}{dt_g}\left(f_0 \frac{\partial v_g}{\partial p}\right) = hQ_x - f_0^2 \frac{\partial u_{ag}}{\partial p}, \quad (3.1a)$$

$$\frac{d}{dt_g}\left(-h\frac{\partial\theta}{\partial x}\right) = -hQ_x - \sigma\frac{\partial\omega}{\partial x}. \quad (3.1b)$$

In (3.1a,b), d/dt_g $(= \partial/\partial t + V_g \cdot \nabla_p)$ is the Lagrangian rate of change following the geostrophic flow; h $(= (\rho\theta)^{-1})$ is a function of pressure, expressed equivalently as $h = (R/p_0)\times (p_0/p)^{c_v/c_p}$; σ $(= -hd\Theta/dp)$ is the static-stability coefficient; Θ is a pressure-dependent reference profile of potential temperature; and Q_x is the cross-front component of the Q vector:

$$Q_x = -\frac{\partial u_g}{\partial x}\frac{\partial\theta}{\partial x} - \frac{\partial v_g}{\partial x}\frac{\partial\theta}{\partial y}, \quad (3.2)$$

where the first and second terms describe frontogenesis due to confluence (Fig. 5) and horizontal shear (Fig. 6), respectively. For subsequent reference, the Q vector is given by

$$Q = -\left(\frac{\partial V_g}{\partial x} \cdot \nabla_p\theta\right)i - \left(\frac{\partial V_g}{\partial y} \cdot \nabla_p\theta\right)j. \quad (3.3)$$

For background material on the Q vector, the reader is referred to Hoskins et al. (1978), Hoskins and Pedder (1980), Carlson (1991, Section 14.7), and Holton (1992, Section 6.4.2).

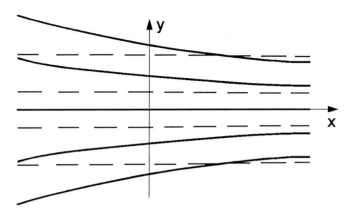

FIG. 5. Schematic illustrating frontogenetical confluence effect. Thin solid lines are streamlines of the geostrophic flow and dashed lines are isentropes. Along- and cross-front coordinates transform to those in this chapter as follows: $x \to y$ and $y \to -x$, where the first entry refers to the coordinate indicated in the figure and the second to its counterpart in the chapter (adapted from Eliassen 1990).

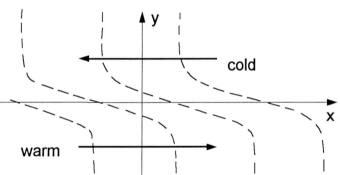

FIG. 6. Schematic illustrating frontogenetical horizontal shear effect. Arrows indicate sense of the geostrophic flow and dashed lines are isentropes. Along- and cross-front coordinates transform to those in the chapter as in Fig. 5 (adapted from Eliassen 1990).

Subtracting (3.1b) from (3.1a) and rearranging terms yields

$$\sigma\frac{\partial\omega}{\partial x} - f_0^2\frac{\partial u_{ag}}{\partial p} = -2hQ_x, \quad (3.4a)$$

which becomes, with the use of the definitions of (u_{ag}, ω) in terms of ψ given in the second paragraph of this section,

$$\sigma\frac{\partial^2\psi}{\partial x^2} + f_0^2\frac{\partial^2\psi}{\partial p^2} = -2hQ_x. \quad (3.4b)$$

Eq. (3.4b) is the QG version of the Sawyer-Eliassen equation.

For completeness, the SG version of the Sawyer-Eliassen equation is now presented without derivation:

$$\sigma_{loc}\frac{\partial^2\psi}{\partial x^2} - \left(f_0\frac{\partial v_g}{\partial p} - h\frac{\partial\theta}{\partial x}\right)\frac{\partial^2\psi}{\partial x\partial p} + f_0\left(f_0 + \frac{\partial v_g}{\partial x}\right)\frac{\partial^2\psi}{\partial p^2}$$

$$+ \left(\mu h\frac{\partial\theta}{\partial x}\right)\frac{\partial\psi}{\partial x} = -2hQ_x\,,$$

$$(3.5)$$

where σ_{loc} $(= -h\partial\theta/\partial p)$ is the local counterpart of the static-stability coefficient introduced in connection with (3.1a,b) and the coefficient μ $(= d\ln h/dp)$ equals $-(c_v/c_p)p^{-1}$. The form of the Q-vector forcing in (3.5) is identical to that in (3.4b), but the operator on the left-hand side is more general, incorporating the effects of local static stability, baroclinicity, and inertial stability. For background on the derivation of (3.5), along with its mathematical properties (e.g., the ellipticity condition and Green's function for the operator on ψ) and physical interpretation, the reader is referred to Eliassen

(1962), as well as to the reviews of the Sawyer-Eliassen equation cited in Section 1.

Patterns of the ageostrophic circulation streamfunction, ψ, for idealized distributions of confluence and horizontal shear forcing are presented in Figs. 7a and 8a, respectively, and for an observed cold front in Fig. 9b. (Figures 7b and 8b portray the front-relative transverse circulation $(u - c, \omega)$ (see Section 2.2) corresponding to Figs. 7a and 8a, respectively.) Figure 10 displays hypothetical configurations of transverse ageostrophic circulations in which the individual patterns attributed to the exit region of an upper-level jet streak and to a surface frontal zone and prefrontal low-level jet are unfavorably (Fig. 10b) and favorably (Fig. 10d) aligned in the vertical to contribute to the release of convective instability. This latter example, which contains elements of the natural-coordinate conceptual model of ageostrophic circulations associated with a straight jet streak (Figs. 1a,b), reveals the generality of the Sawyer-Eliassen equation in allowing qualitative inference of vertical circulations in situations where the

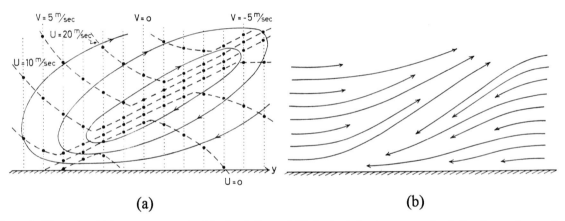

(a)   (b)

FIG. 7. Schematic illustrating (a) transverse ageostrophic circulation and (b) front-relative transverse circulation for confluence frontogenesis depicted in Fig. 5. In (a), dashed lines are isotachs of along-front geostrophic wind (indicated by U); dotted lines are isotachs of cross-front geostrophic wind (indicated by V); and solid lines are streamfunction for ageostrophic circulation. In (b), solid lines are streamlines of front-relative transverse circulation. Along- and cross-front coordinates (wind components) transform to those in this chapter as follows: $x \rightarrow y\,(U \rightarrow v_g)$ and $y \rightarrow -x\,(V \rightarrow -u_g)$, where the first entry refers to the coordinate (component) in the figure and the second to its counterpart in the chapter (Eliassen 1962).

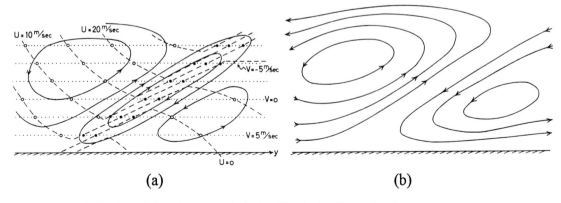

(a)   (b)

FIG. 8. As in Fig. 7, except for horizontal shear frontogenesis depicted in Fig. 6 (Eliassen 1962).

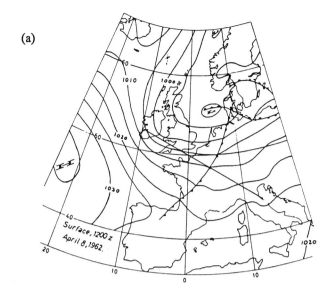

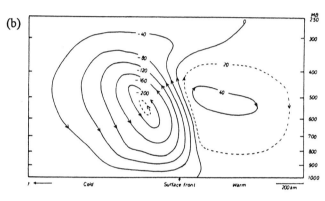

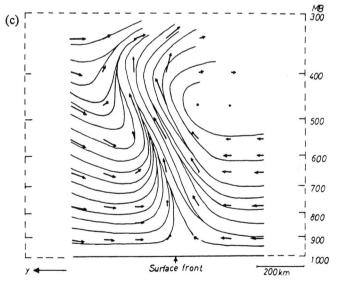

FIG. 9. Calculation of transverse circulation for an analyzed cold front over western Europe at 1200 UTC 8 April 1962: (a) conventional surface analysis, indicating northwest-southeast cross section to be depicted in subsequent panels; (b) streamfunction for transverse ageostrophic circulation (contour interval 40×10^3 Pa m s^{-1}) calculated from an SG version of the Sawyer-Eliassen equation (similar to (3.5))

assumption of a two-layer atmosphere (see Section 2.1) is not expected to apply. Further discussion on the vertical coupling of ageostrophic circulations in relation to the development of severe convective storms may be found in Uccellini and Johnson (1979), who analyzed and conceptualized this process using the natural-coordinate approach, and in Carlson (1991, Section 15.5) and Bluestein (1993, Section 2.8.4).

The Sawyer-Eliassen equation in either its QG or SG forms may be viewed as an advance relative to the momentum/vorticity equations (Section 2.1) for the following reasons: (1) both the momentum and thermodynamic equations are considered simultaneously; (2) there are no assumptions on the behavior of the local tendency term (unsteady flows and systems can be described); (3) the vertical circulation is described by a linear, elliptic partial differential equation, allowing straightforward interpretation of the ageostrophic circulation as a forced response to frontogenesis; and (4) by virtue of the linearity of the operator on ψ, the vertical circulation may be partitioned *termwise* into individual contributions associated with confluence and horizontal shear. Nevertheless, *piecewise* partitioning of ψ among geopotential distributions within various subregions of a domain, as is possible with QG potential-vorticity inversion, is rendered problematic by the nonlinear form of Q_x with respect to geopotential.

4. Diagnosis of Frontal Circulations: Three-dimensional Case

In three-dimensional flows, the ageostrophic wind contains a rotational part, the components of which may be oriented in both the cross- and along-front directions. Thus, a distinction needs to be made between ageostrophic (V_{ag}, ω) and divergent (V_d, ω) circulations. For the assumption of constant Coriolis parameter, the geostrophic wind is nondivergent, so that the divergent part of the wind is ageostrophic. As a reminder of this property, V_d will be denoted as V_{agd}.

4.1 Ageostrophic Circulation

Hoskins and Draghici (1977) are credited with the first generalization of the Sawyer-Eliassen equation to three dimensions, the derivation of which is performed in geostrophic coordinates (e.g., Hoskins 1975). The interested reader is referred to Xu (1990) for a thorough and comprehensive treatment of the generalized Sawyer-Eliassen equation in

incorporating a simple parameterization of latent heat release; (c) streamlines of front-relative transverse circulation, with arrows indicating parcel displacements over a 1.5 h period. Cross-front coordinate in (b) and (c) transforms as in Figs. 5–8; sign of streamfunction in (b) should be reversed to be consistent with the definition used in this chapter (Eliassen 1966, 1990; based on work originally by Todsen 1964).

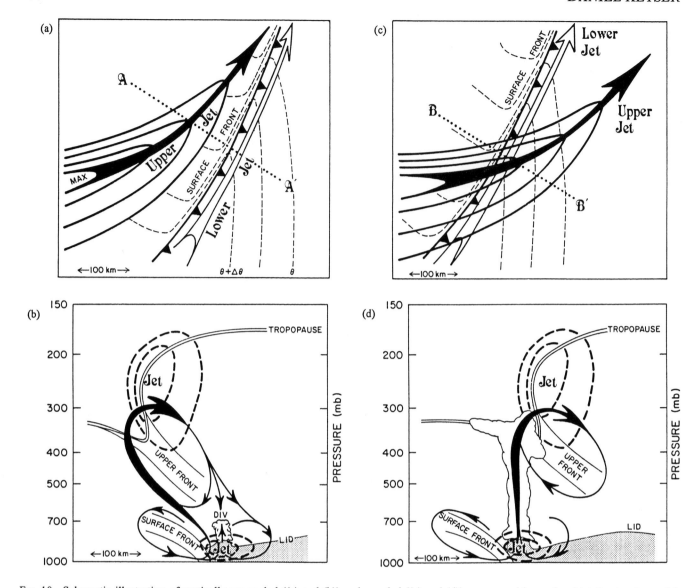

FIG. 10. Schematic illustration of vertically uncoupled ((a) and (b)) and coupled ((c) and (d)) upper- and lower-level jet-front systems: (a) horizontal map showing upper-level jet-exit region (isotachs, heavy solid lines; solid arrow, jet axis) aligned along and displaced toward cold side of surface frontal zone (isentropes, dashed lines; cold front, conventional symbols) and low-level jet (open arrow, jet axis); (b) cross section along AA′ indicated in (a) depicting upper- and lower-level jets (isotachs, thick dashed lines), upper-level and surface frontal zones (bounded by thin solid lines), tropopause (double solid lines), moist boundary layer (stippled) capped by lid, and streamlines of transverse ageostrophic circulation (heavy arrows, strength of circulation proportional to width); (c) as in (a), except for upper-level jet-exit region aligned across surface frontal zone; (d) as in (b), except for cross section along BB′ indicated in (c) (Shapiro 1982).

both physical and geostrophic coordinates. A complementary development of the material in this subsection is given by Xu (1992), wherein the three-dimensional ageostrophic circulation is analyzed in terms of the three-dimensional ageostrophic vorticity vector, the geostrophic forcing of which is referred to as the "*C* vector." Consistent with previous sections, the present development is outlined for the QG system. Combining (3.4a) with its counterpart applicable to the (y, p) plane, which results from equating expressions for $d(f_0 \partial u_g/\partial p)/dt_g$ and $d(h\partial\theta/\partial y)/dt_g$ (not shown), yields the vector analogue to (3.4a):

$$\sigma\nabla_p\omega - f_0^2\frac{\partial V_{ag}}{\partial p} = -2h\boldsymbol{Q}. \qquad (4.1)$$

Substitution of the form of (2.4a,b) applicable to (V_{ag}, ω) into (4.1) and use of the vector identity, $\nabla_p(\nabla_p\cdot\psi_{ag}) = \nabla_p^2\psi_{ag} + \nabla_p\times(\nabla_p\times\psi_{ag})$, leads to the following expression:

$$\sigma\left[\nabla_p^2\psi_{ag} + \nabla_p\times(\nabla_p\times\psi_{ag})\right] + f_0^2\frac{\partial^2\psi_{ag}}{\partial p^2} = -2h\boldsymbol{Q}.$$

$$(4.2)$$

Assuming that the total vertical motion and its components, ω_x and ω_y, vanish at p_b leads to the lower boundary condition, $\psi_{ag}(p_b) = 0$. Consistent with considerations in Section 2.3, a homogeneous Dirichlet boundary condition for ψ_{ag} at p_t does not apply in general because of the possibility of internal cancellation between ω_x and ω_y at p_t, owing to the presence of V_{agr}. As discussed by Xu and Keyser (1993), adoption of $\psi_{ag}(p_t) = 0$ yields the so-called "baroclinic" part of the ageostrophic wind, V_{ag}^{bc} (i.e., the deviation of V_{ag} from its vertical average in the layer bounded by p_b and p_t, satisfying the condition, $\int_{p_t}^{p_b} V_{ag}^{bc} dp = 0$), and the total vertical motion. A complete solution for the ageostrophic wind requires determining its leftover portion, the so-called "barotropic" part, V_{ag}^{bt} $[= (p_b - p_t)^{-1} \int_{p_t}^{p_b} V_{ag} dp]$. The barotropic part of the ageostrophic wind is rotational, which may be verified by integrating the continuity equation between p_b and p_t and recalling that ω is assumed to vanish at these boundaries. Because the barotropic part of the ageostrophic wind is rotational, its solution requires consideration of the QG divergence equation (4.10a,b), to be discussed subsequently.

Interpretation of (4.2) is facilitated by extracting its cross-front component, which may be compared with (3.4b). The resulting expression is:

$$\sigma \frac{\partial^2 \psi_{agx}}{\partial x^2} + f_0^2 \frac{\partial^2 \psi_{agx}}{\partial p^2} = -2hQ_x - \sigma \frac{\partial \omega_y}{\partial x}. \quad (4.3)$$

Thus, the diagnostic equation for the component of the ageostrophic circulation in the cross-front plane is identical in form to its two-dimensional counterpart (3.4b), except for the addition of a coupling term involving the vertical motion in the along-front plane (note that the second term on the right-hand side of (4.3) may be rewritten as $-\sigma \partial^2 \psi_{agy}/\partial x \partial y$) and the necessity of considering a nonhomogeneous boundary condition for ψ_{agx} at p_t. This so-called coupling term may be interpreted physically as a forcing of ψ_{agx} due to differential adiabatic warming/cooling in the cross-front plane induced by the ageostrophic circulation in the along-front plane. The rationale for viewing these potential temperature changes as thermal forcing is that they are associated with a process external to the ageostrophic circulation in the cross-front plane.

4.2. Divergent/Rotational Ageostrophic Circulations

It is possible to solve for the divergent ageostrophic circulation (V_{agd}, ω) using either the approach reviewed in the previous section or the QG omega equation. For completeness, a solution for the rotational ageostrophic wind, V_{agr}, is presented as well.

Following Eliassen (1984), the vector streamfunction defined in (2.4a,b) may be expressed in terms of a potential function, χ, such that

$$\psi_{agd} = -\nabla_p \chi, \quad (4.4)$$

which implies the following relations:

$$\omega = -\nabla_p^2 \chi, \quad (4.5a)$$

$$V_{agd} = \nabla_p \frac{\partial \chi}{\partial p}. \quad (4.5b)$$

Substituting (4.5a,b) into (4.1), and applying the identity immediately above (4.2) to a vector of the form $\nabla_p \chi$ to verify that $\nabla_p(\nabla_p^2 \chi) = \nabla_p^2(\nabla_p \chi)$, yields

$$\sigma \nabla_p^2 \psi_{agd} + f_0^2 \frac{\partial^2 \psi_{agd}}{\partial p^2} = -2hQ + f_0^2 \frac{\partial V_{agr}}{\partial p}, \quad (4.6)$$

which may be taken a step further through application of a two-component (rotational/divergent) Helmholtz decomposition. Keyser et al. (1992a, Section 2b) justify the applicability of the two-component partition for the choices of lateral boundary conditions assumed for this presentation, which eliminates the harmonic component (e.g., Lynch 1989) from consideration. The resulting equations are:

$$\sigma \nabla_p^2 \psi_{agd} + f_0^2 \frac{\partial^2 \psi_{agd}}{\partial p^2} = -2hQ_d, \quad (4.7a)$$

$$f_0^2 \frac{\partial^2 \psi_{agr}}{\partial p^2} = -2hQ_r, \quad (4.7b)$$

where in (4.7b) use has been made of the relationship, $V_{agr} = -\partial \psi_{agr}/\partial p$, which may be verified by recalling (2.4b). Alternatively, (4.7a,b) may be derived directly from (4.2), where it is noted from the vector identity given immediately above (4.2) that $\nabla_p(\nabla_p \cdot \psi_{agd}) = \nabla_p^2 \psi_{agd}$ and $\nabla_p(\nabla_p \cdot \psi_{agr}) = 0$.

In (4.7a), homogeneous Dirichlet boundary conditions apply at p_b and p_t for ψ_{agd}. This may be verified from (4.5a) for the case of $\omega = 0$ and noting that $\omega_x = -\partial^2 \chi/\partial x^2$ and $\omega_y = -\partial^2 \chi/\partial y^2$. Solution of (4.6) requires knowledge of the rotational ageostrophic wind, which may be determined from (4.7b) or from the QG divergence equation (4.10a,b). Because solution of (4.7b) is dependent upon knowledge of V_{agr} at one level, this form is probably more of conceptual than practical interest.

The projection of (4.6) onto the cross-front plane is

$$\sigma \frac{\partial^2 \psi_{agdx}}{\partial x^2} + f_0^2 \frac{\partial^2 \psi_{agdx}}{\partial p^2} = -2hQ_x + f_0^2 \frac{\partial u_{agr}}{\partial p} - \sigma \frac{\partial \omega_y}{\partial x}. \quad (4.8)$$

where $\omega_y = \partial \psi_{agdy}/\partial y$. Comparing (4.3) with (4.8) shows that, aside from differing boundary conditions on ψ_{agdx} and

ψ_{agx} at p_t, defining the frontal circulation in terms of (V_{agd}, ω) results in an additional forcing term involving the vertical shear of the rotational ageostrophic wind in the cross-front plane. This term may be interpreted as a momentum forcing, as its role is to contribute to the rate of change of the vertical shear of the along-front geostrophic wind, $f_0 \partial v_g / \partial p$. As in the case of the thermal forcing term, which appears in both (4.3) and (4.8), the momentum forcing is viewed as external to the divergent ageostrophic circulation in the cross-front plane because this term consists of the rotational part of the ageostrophic wind. Despite the equivalent form of the thermal forcing terms on the right-hand sides of (4.3) and (4.8), it should be recognized that they are not necessarily equivalent numerically in any given case, since values of ω_y generally will differ between the "ageostrophic" and "divergent" definitions of ψ. In the two-dimensional limit, where the horizontal component of the ageostrophic circulation is divergent and the circulation is confined to the cross-front plane, (4.8) reduces to (3.4b).

An alternative method of determining (V_{agd}, ω) consists of solving the QG omega equation, which may be derived by taking the divergence ($\nabla_p \cdot$) of (4.1):

$$\sigma \nabla_p^2 \omega + f_0^2 \frac{\partial^2 \omega}{\partial p^2} = -2h(\nabla_p \cdot \boldsymbol{Q}). \qquad (4.9)$$

Given ω from (4.9), Eq. (4.5a) may be inverted for χ, from which V_{agd} may be determined using (4.5b).

The rotational ageostrophic wind, V_{agr}, is determined by inverting a Poisson equation relating a streamfunction to the ageostrophic vorticity, ζ_{ag}, which is given by the QG divergence equation:

$$\zeta_{ag} = -\frac{2}{f_0} J_{xy}(u_g, v_g), \qquad (4.10a)$$

or, equivalently,

$$\zeta_{ag} = \frac{1}{2f_0}\left(E_g^2 - \zeta_g^2\right). \qquad (4.10b)$$

In (4.10a), J_{xy} is the Jacobian operator applicable to the (x, y) plane; in (4.10b), E_g is the geostrophic resultant deformation and ζ_g the geostrophic relative vorticity. For completeness, it is noted that taking the curl ($\boldsymbol{k} \cdot \nabla_p \times$) of (4.1) results in

$$f_0^2 \frac{\partial \zeta_{ag}}{\partial p} = 2h\boldsymbol{k} \cdot \left(\nabla_p \times \boldsymbol{Q}\right). \qquad (4.11)$$

Vertical differentiation of (4.10a,b) yields a result consistent with (4.11). The particular forms given for ζ_{ag} (4.10b) and $\partial \zeta_{ag} / \partial p$ (4.11) are discussed and interpreted by Xu (1990,

Section 3b; 1992, Section 2) and by Hoskins and Pedder (1980, Appendix B), respectively.

4.3. Total or Divergent/Rotational Ageostrophic Circulation: Which Definition?

A question to be resolved before considering applications of the diagnostic methodologies presented in Sections 4.1 and 4.2 is an appropriate definition of a frontal circulation in three dimensions. The present view of the author is that the (V_{agd}, ω) alternative is preferable for several reasons, despite complications introduced by the Helmholtz decomposition that arise from the need for lateral boundary conditions on limited-area domains and from the nonlocal nature of the decomposition on both limited-area and global domains. As pointed out by Sardeshmukh and Hoskins (1987), the Helmholtz decomposition is "nonlocal" in the following sense: if a vector field is modified within a subregion of a domain then the rotational and divergent parts will differ outside of the subregion, where the vector field is unchanged, not only within the subregion where the vector field is modified. This property is the result of the inversion of elliptic operators to determine the rotational and divergent parts of a vector field, a process that extends the influence of the vorticity and divergence at a point or small region to remote (i.e., far-field) portions of a domain.

The first argument in favor of the (V_{agd}, ω) alternative is the elimination of the risk of internal cancellation between the cross- and along-front components of ω that arises when the ageostrophic wind possesses a rotational part, although it should be noted that, to the author's knowledge, the practical significance of this problem with the (V_{ag}, ω) alternative has not been investigated. Second, the (V_{agd}, ω) alternative is the natural extension of the two-dimensional definition of a frontal circulation, where (u_{ag}, ω) corresponds to the divergent circulation. A potential shortcoming is that the projection of (V_{agd}, ω) onto the cross-section plane does not reveal local parcel accelerations through cross-contour ageostrophic flow in the vicinity of jet streaks, which may be dominated in realistic situations by the rotational ageostrophic wind (e.g., Krishnamurti 1968b, p. 212). The dominance of the rotational relative to the divergent ageostrophic wind is to be expected in balanced flows, where the magnitude of the vorticity of the ageostrophic wind is greater than the magnitude of its divergence ($|\zeta_{ag}| > |\nabla_p \cdot V_{ag}|$), a condition that is consistent with small Rossby number. In response to the concern that the divergent ageostrophic wind does not reveal local parcel accelerations, it is of interest to point out that, globally, conversions between available potential and kinetic energy through cross-contour ageostrophic flow reduce to consideration of the cross-contour component of the divergent ageostrophic flow, as the rotational contribution integrates to zero (Chen and Wiin-Nielsen 1976; Chen et al. 1978; Blackburn 1985).

Third, and consistent with the recognition that in the two-dimensional limit (u_{ag}, ω) corresponds to the divergent circulation, it may be shown that the principle of preservation of thermal wind balance (3.1a,b) generalizes to the preservation of thermal vorticity balance. The three-dimensional counterparts of (3.1a,b) are:

$$\frac{d}{dt_g}\left(f_0\frac{\partial\zeta_g}{\partial p}\right) = -f_0\,E_g^2\frac{\partial\alpha_g}{\partial p} + h(\nabla_p\bullet\boldsymbol{Q})$$

$$-f_0^2\frac{\partial}{\partial p}(\nabla_p\bullet\boldsymbol{V}_{agd}), \qquad (4.12a)$$

$$\frac{d}{dt_g}\left(-h\nabla_p^2\theta\right) = -f_0\,E_g^2\frac{\partial\alpha_g}{\partial p} - h(\nabla_p\bullet\boldsymbol{Q}) - \sigma\nabla_p^2\omega, \qquad (4.12b)$$

where $\alpha_g\left[=\frac{1}{2}\tan^{-1}(E_{g2}/E_{g1})\right]$ is the angle with respect to the x axis of the dilatation axis of the geostrophic flow, and E_{g1} ($= \partial u_g/\partial x - \partial v_g/\partial y$) and E_{g2} ($= \partial v_g/\partial x + \partial u_g/\partial y$) are the components of the geostrophic deformation. Eqs. (4.12a,b), which may be combined to form the QG omega equation (4.9), reveal that the Q-vector forcing of the omega equation tends to disrupt thermal vorticity balance and that the divergent circulation (V_{agd}, ω) tends to restore it, analogous to the interpretation of (3.1a,b) applicable to the two-dimensional case.

4.4. Application of Diagnostic Methodologies to Elucidate the Dynamics of Upper-level Frontogenesis

An example of topical interest, where the diagnostic equations developed in Sections 4.1 and 4.2 prove revealing, is interpreting upper-level frontogenesis in terms of "two-dimensional" mechanisms. One such mechanism, referred to by Rotunno et al. (1994) as the "Shapiro effect" (Shapiro 1981; summarized in Carlson 1991, Section 15.8, and in Bluestein 1993, Section 2.6.2), involves the superposition of confluence and horizontal shear forcing in the entrance region of an upper-level jet-front system in the presence of cold-air advection (see Fig. 3a). These forcing terms combine to shift a thermodynamically direct ageostrophic circulation in the cross-front plane toward the warm-air side of the frontal zone, resulting in frontogenetical tilting in the prognostic equations for the cross-front potential temperature gradient and the vertical component of vorticity. It is noted that a thermodynamically direct circulation, as opposed to an indirect circulation, is necessary for upper-level frontogenesis, because the cross-contour ageostrophic flow at jet level is directed toward lower geopotential heights in the former case, allowing the jet to intensify during frontogenesis.

The horizontal and vertical structure of the upper-level jet-front system that was diagnosed by Shapiro (1981) to identify the frontogenetical configuration of the transverse ageostrophic circulation using the Sawyer-Eliassen equation is depicted in Figs. 11 and 12a; the potential temperature field and the along- and cross-front geostrophic wind components

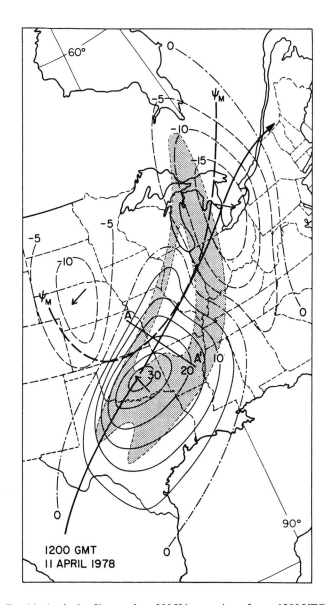

FIG. 11. Analysis of jet streak on 330 K isentropic surface at 1200 UTC 11 April 1978. Jet streak is outlined by stippling of 40–60 m s^{-1} wind speed interval, with isotachs indicated by thin dashed lines. A streamline for the total wind is denoted by a thick solid line; a streamline for the geostrophic wind (Montgomery streamfunction, Ψ_M) tangent to the former within the jet maximum is indicated by a thick dashed line. The cross-contour component of the ageostrophic wind (contour interval 5 m s^{-1}) is shown by thin solid (dashed) lines where directed toward lower (higher) values of Ψ_M; dot-dashed lines correspond to zero isotach of this ageostrophic wind component. Selected orientations of the cross-contour ageostrophic wind are denoted by short arrows. The line labeled AA′ refers to the cross section displayed in Fig. 12 (Shapiro 1981).

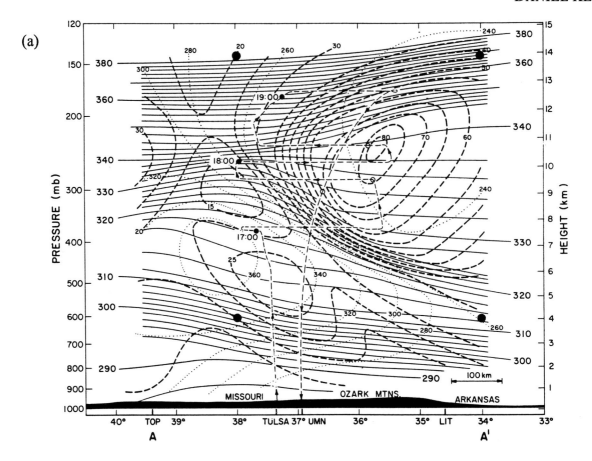

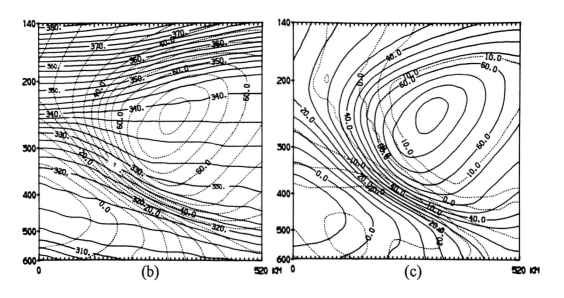

FIG. 12. Cross section for 11 April 1978 along AA′ in Fig. 11: (a) potential temperature (contour interval 2 K, solid), wind speed (contour interval 5 m s⁻¹, dashed), wind direction (contour interval 20°, dotted), flight track (arrowed dashed line; solid (open) circles indicate full (half) hours in UTC), and corner points for computational domain (520 km × 140–600 hPa) in (b) and (c) (large solid dots); (b) along-front geostrophic wind component (U; contour interval 5 m s⁻¹, dotted) and potential temperature (contour interval 2.5 K, solid); (c) U repeated from (b) (contour interval 5 m s⁻¹, solid) and cross-front geostrophic wind component (V; contour interval 5 m s⁻¹, dotted). In (b) and (c), along- and cross-front coordinates (wind components) transform to those in this chapter as follows: $x \to y (U \to v_g)$ and $-y \to x (V \to -u_g)$, where the first entry refers to the coordinate (component) in the figure and the second to its counterpart in the chapter (Shapiro 1981).

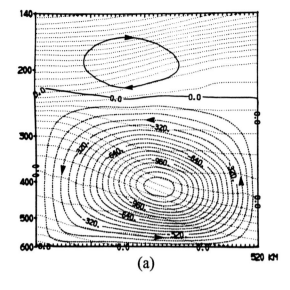

(a)

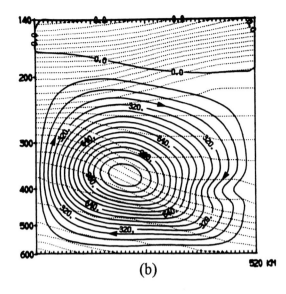

(b)

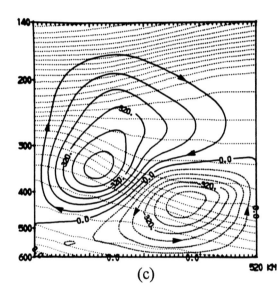

(c)

FIG. 13. Transverse ageostrophic circulation streamfunction (contour interval 80×10^2 Pa m s^{-1}) calculated from an SG version of the Sawyer-Eliassen equation (similar to (3.5)) for the computational domain in Figs. 12b,c. In order to reconcile the definition of the streamfunction used in this figure with that in the chapter, solid (dashed) contours should be interpreted as negative (positive). Background field is potential temperature (dotted) repeated from Fig. 12b: (a) streamfunction diagnosed from confluence term of Q_x (3.2); (b) as in (a), except for horizontal shear; (c) sum of (a) and (b) (Shapiro 1981).

used to solve for the transverse ageostrophic circulation are shown in Figs. 12b,c. The configuration of the cross-front geostrophic wind component shows a pattern of sloping isotachs in the vicinity of the frontal zone, such that the zone lies in a region of confluence. Furthermore, the vertical shear of the cross-front geostrophic wind component (directed toward the cold-air side of the frontal zone with decreasing pressure) is such that warmer air is located into the page, implying cold-air advection by the upper-level jet. This pattern of confluence and differential cold-air advection by the upper-level jet results in thermodynamically direct and indirect transverse ageostrophic circulations due to confluence (Fig. 13a) and horizontal shear (Fig. 13b), respectively, the sum of which (Fig. 13c) consists of a pair of indirect and direct cells within the frontal zone. The total circulation places subsidence within and to the warm-air side of the frontal zone, establishing the conditions necessary for

frontogenetical tilting terms in the equations for the cross-front component of the potential temperature gradient and the vertical component of vorticity.

In retrospect, the diagnosis presented by Shapiro (1981) resolved a paradox existing at that time in the conceptual understanding of upper-level frontogenesis. Earlier studies, dating back to Reed and Sanders (1953), Newton (1954), and Reed (1955) (see Keyser and Shapiro (1986, Section 2d) for additional discussion and references), emphasize the importance of tilting associated with a cross-front gradient of vertical motion such that subsidence is maximized on the warm-air side of a developing upper-level frontal zone in the presence of confluent northwesterly flow upstream of the axis of a synoptic-scale trough (see Figs. 3a,b). In terms of the Sawyer-Eliassen equation, such a confluent-flow pattern would be expected to yield a direct transverse ageostrophic circulation (as in Fig. 13a), which supports tilting in a *frontolytical* sense. At the time of publication of Shapiro (1981), it was widely held (e.g., Buzzi et al. 1981) that the only way to explain the maximum subsidence on the warm-air side of an upper-level frontal zone was to account for the along-front ageostrophic circulation (Fig. 2a). The recognition that the horizontal shear term can modify the direct transverse ageostrophic circulation in the proper sense to account for observed upper-level frontogenesis resolved the foregoing paradox without the need to appeal to three-dimensional arguments involving ageostrophic circulations in the along-front direction.

Subsequent two-dimensional modeling of the Shapiro effect by Keyser and Pecnick (1985) and Reeder and Keyser (1988) show that a robust positive feedback can develop between the vorticity and vertical motion fields through the horizontal shear forcing term of (3.2). The question arises as to whether the Shapiro effect may be found in more realistic models of frontogenesis, and eventually in the real atmosphere. Toward this end, Keyser et al. (1992a) applied (4.8) to a developing upper-level front in a channel-model simulation of baroclinic-wave evolution similar to that portrayed in Fig. 3. The results of this application are reproduced here in Figs. 14–19. Figures 14a–c display the midtropospheric geopotential height, potential temperature, and vertical motion patterns (compare with Figs. 3a,b), indicating confluent geostrophic flow upstream of the trough axis in the region of the frontal zone and subsidence maximized on the warm-air side of the latter. Figure 15 illustrates the relationship between ψ_{agd} (Fig. 15a) and χ (Fig. 15c) given by (4.4), as well as that between ψ_{agd} and V_{agd} at lower (Fig. 15b) and upper (Fig. 15d) levels given by (2.4b) and presented schematically in Fig. 4. The midtropospheric ω pattern (Fig. 14c) may be related to ψ_{agd} through (2.4a) and to χ through (4.5a).

Figure 16 presents a decomposition of ω into cross- and along-front components, ω_x and ω_y, along a meridionally oriented vertical section crossing the upper-level frontal zone depicted in Fig. 14 upstream of the trough axis. Comparison of the magnitude of the along-front component of ω (Fig. 16c) with its cross-front counterpart (Fig. 16b) reveals the "three dimensionality" of the vertical circulation (the two-dimensional limit corresponds to increasingly anisotropic patterns of ω elongated in the along-front direction, such that $|\omega_y| \ll |\omega_x|$; the three-dimensional limit corresponds to isotropic or circular patterns of ω, such that $|\omega_y| \sim |\omega_x|$). Although relatively three-dimensional by this definition, there is a distinct difference between the characteristic cross-front scales of ω_x and ω_y, with the former mesoscale and the latter synoptic scale. In addition to reducing the magnitude of the tilting effects associated with the ω_y field relative to the ω_x field, this difference in scale results in the placement of the regions in the ω_x and ω_y fields favorable to frontogenetical tilting (i.e., $\partial\omega_x/\partial x$ and $\partial\omega_y/\partial x < 0$) within the upper-level frontal zone and on the cold-air side of the frontal zone, respectively. These considerations suggest that frontogenetical

tilting is associated predominantly with the cross-front divergent circulation, a finding that is supported by a comparable diagnosis of a simulated upper-level frontal zone by Hines and Mechoso (1991). The link between frontogenetical tilting and the cross-front divergent circulation justifies applying (4.8) to search for the signature of the Shapiro effect in the model simulation depicted in Fig. 14.

Figure 17 presents the ψ_{agdx} field diagnosed from the channel-model simulation using (4.5a) and (4.4) and its counterpart diagnosed using (4.8). The latter qualitatively reproduces the former, although its amplitude is too large, which is to be expected from applying QG vertical-circulation diagnostics to curved flow regimes where the Rossby number is non-negligible (Keyser et al. 1992a). Figure 18 shows the fields required to evaluate the forcing terms in (4.8); of particular relevance to the Shapiro effect is the configuration of the cross-front geostrophic wind component (Fig. 18a). This configuration reveals confluence coincident with the frontal zone through a deep layer of the troposphere and the vertical shear of the cross-front geostrophic wind component directed toward the cold-air side of the frontal zone with decreasing pressure. This sense of the vertical shear is such that colder air is located into the page, implying cold-air advection by the upper-level jet. This geostrophic wind pattern, which qualitatively reproduces that in Shapiro's observational diagnosis (Fig. 12c), yields direct and indirect cross-front divergent circulations forced by confluence (Fig. 19c) and horizontal shear (Fig. 19d), the sum of which consists of juxtaposed direct and indirect circulations (Fig. 19b), in agreement with the corresponding two-dimensional diagnosis in Fig. 13. Comparison of the circulations diagnosed from the forcing terms in (4.8) involving three-dimensional effects (Figs. 19e,f), along with the contribution of (nonhomogeneous) boundary conditions (Fig. 19a), with the circulation diagnosed from the combination of confluence and horizontal shear (Fig. 19b) reveals the dominance of these two-dimensional forcing mechanisms. Thus, the cited diagnostic evidence suggests the presence of the Shapiro effect in the three-dimensional model simulation, although it remains for future research to investigate whether a time-dependent positive feedback involving this effect, as isolated in two-dimensional modeling studies, is actually taking place.

Recently, Rotunno et al. (1994), in a diagnosis of upper-level frontogenesis using a numerical model generically similar to that used by Keyser et al. (1992a), identified forcing for a cross-front vertical circulation consistent with the Shapiro effect, but in the absence of cold-air advection. An interpretation of this discrepancy (R. Rotunno, personal communication) is that the Shapiro effect occurs in their model, but at a later (i.e., finite-amplitude) stage of baroclinic-wave development; their analysis is concerned with the early (i.e., linear) stage of wave evolution, where the phase lag between the midtropospheric temperature and geopotential patterns is too small to establish along-front cold-air advec-

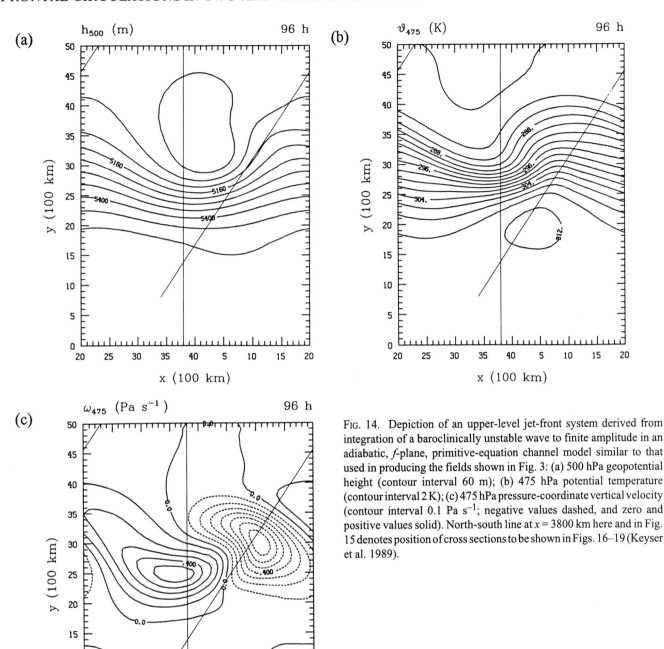

FIG. 14. Depiction of an upper-level jet-front system derived from integration of a baroclinically unstable wave to finite amplitude in an adiabatic, f-plane, primitive-equation channel model similar to that used in producing the fields shown in Fig. 3: (a) 500 hPa geopotential height (contour interval 60 m); (b) 475 hPa potential temperature (contour interval 2 K); (c) 475 hPa pressure-coordinate vertical velocity (contour interval 0.1 Pa s^{-1}; negative values dashed, and zero and positive values solid). North-south line at $x = 3800$ km here and in Fig. 15 denotes position of cross sections to be shown in Figs. 16–19 (Keyser et al. 1989).

tion as required by the Shapiro effect. The mechanism proposed by Rotunno et al. (1994) is hypothesized to involve the confluence term alone, with diffluence on the cold-air side of the frontal zone and confluence on the warm-air side inducing indirect and direct transverse ageostrophic circulations centered on the cold- and warm-air sides of the frontal zone, respectively. These adjacent circulation patterns combine to

yield subsidence and associated frontogenetical tilting within the frontal zone. This configuration of vertical motion strengthens the front, thereby initiating a positive feedback between the cross-front potential temperature gradient and vertical motion through the confluence forcing term of (3.2). It would be of interest to apply (4.3) and/or (4.8) to verify the Rotunno et al. mechanism both at an instant, as has been done by

(a)

ψ_{475} (Pa m s^{-1}), h$_{475}$ (m)   96 h

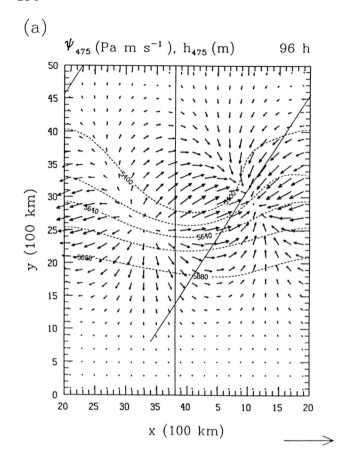

x (100 km)

(b)

$V_{agirr\ 950}$ (m s^{-1}), h$_{950}$ (m)   96 h

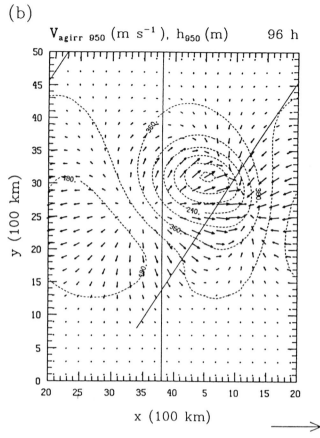

x (100 km)

(c)

χ_{475} (10^9 Pa m^2 s^{-1})   96 h

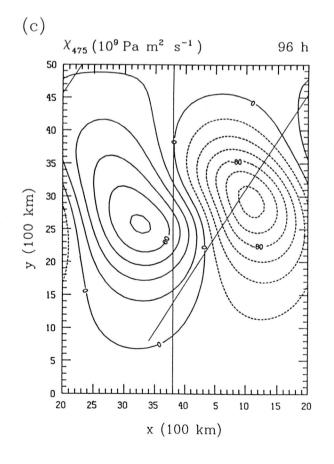

x (100 km)

(d)

$V_{agirr\ 200}$ (m s^{-1}), h$_{200}$ (m)   96 h

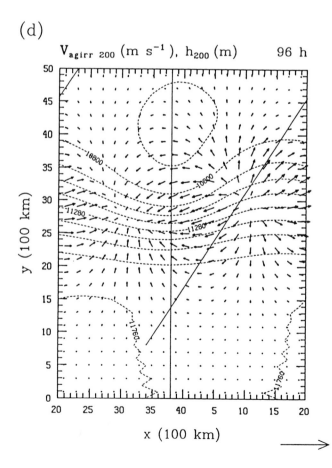

x (100 km)

Fig. 15 (facing page). Patterns of ψ_{agd}, V_{agd}, and χ for the channel-model simulation depicted in Fig. 14: (a) ψ_{agd} (magnitude of reference vector is 60×10^4 Pa m s^{-1}) and geopotential height (contour interval 120 m, dashed) at 475 hPa; (b) V_{agd} (magnitude of reference vector is 35 m s^{-1}) and geopotential height (contour interval 60 m, dashed) at 950 hPa; (c) χ (contour interval 20×10^9 Pa m^2 s^{-1}) at 475 hPa; (d) V_{agd} (magnitude of reference vector is 35 m s^{-1}) and geopotential height (contour interval 120 m, dashed) at 200 hPa (Keyser et al. 1989).

Keyser et al. (1992a) for the Shapiro effect, and over time to confirm whether a positive feedback is operative.

4.5. Extension of Diagnostic Methodologies to Real-data Applications

In order to extend the diagnostic methodologies for both (V_{ag}, ω) (Section 4.1) and (V_{agd}, ω) (Section 4.2) to real data, a number of conceptual and technical issues require resolution. First, diabatic, frictional, and topographical effects should be incorporated into the formulation of the diagnostic equations; the first two are included in the development of the (V_{agd}, ω) alternative by Keyser et al. (1992a). Diabatic and frictional tendencies either may be extracted from a numerical model or parameterized in terms of the vertical motion and other known quantities; for example, an expression for latent heat release due to forced ascent in saturated, conditionally stable flow ($-\partial\theta_e/\partial p > 0$) for the QG system is presented by Whitaker and Davis (1994), adapted from a formulation proposed by Emanuel et al. (1987). Probably the simplest representation of surface friction is Ekman pumping (e.g., Carlson 1991, Sections 3.1 and 9.1), where friction enters as a lower boundary condition on the vertical motion (proportional to geostrophic relative vorticity), rather than explicitly as tendencies in the momentum equations. Orography also may enter as a lower boundary condition on vertical motion through the assumption that this quantity consists of the upslope/downslope component of the flow implied by the orientation of the geostrophic wind relative to terrain contours (e.g., Carlson 1991, Section 9.2). The next level of realism, adapting the frontal-circulation equations to a domain with a pressure- or height-dependent lower boundary, appears to be feasible through adoption of eta- or sigma-type coordinates, although the increase in mathematical and computational complexity is likely to be considerable.

Earth geometry may be incorporated readily, either through reformulation of the equations in spherical or conformal-map coordinate systems (e.g., Haltiner and Williams 1980, Sections 1.7 and 1.8). Assuming that regional (as opposed to global) domains will be used for frontal-circulation diagnosis in the foreseeable future, lateral boundary conditions need to be considered. Xu and Keyser (1993) and Xu and Davies-Jones (1993) propose Dirichlet and Neumann conditions, respectively, for the (V_{ag}, ω) alternative; extend-

ing their analyses to the (V_{agd}, ω) alternative should be straightforward. The two-component Helmholtz partition used to decompose (4.6) into (4.7a,b) is not generally applicable to limited-area domains; the three-component approach discussed by Lynch (1989) or the "domain-independent" approach involving the use of free-space Green's functions proposed by Bishop (1996) may facilitate the desired decomposition in such domains.

In the author's opinion, the greatest challenge lies in generalizing the diagnostic methodologies for three-dimensional frontal circulations to intermediate models beyond QG (e.g., McWilliams and Gent 1980). An obvious approach is to consider the SG circulation equations, either in physical or in geostrophic-coordinate space (Hoskins and Draghici 1977; Xu 1990). The first alternative is realized at the cost of losing the simple mathematical form characteristic of QG, raising the specter of diminishing returns. In the second alternative, the simple mathematical form of QG is preserved, but physical interpretation is complicated by the use of a stretched coordinate (e.g., fundamental quantities such as the Q vector are not invariant under the transformation between physical and geostrophic-coordinate space). And to complicate matters further, considerations by Snyder et al. (1991) suggest that in realistically curved flows the SG equations are not formally any more accurate than the QG equations. The nonlinear balance equations are a viable option for curved flows; it remains for future research to ascertain whether tractable circulation equations are derivable for this system, as well as whether those that have been proposed for related intermediate dynamical models (e.g., Davies-Jones 1991; Xu 1994) prove practical for diagnostic applications involving real data.

A straightforward alternative is to avoid the use of circulation equations altogether by combining appropriate higher-order omega equations with the definition of the vector streamfunction given in (2.4a,b). For example, one might solve a nonlinear-balance-type of omega equation (e.g., Krishnamurti 1968a; Eliassen 1984; Iversen and Nordeng 1984; Davis and Emanuel 1991) and then infer the associated streamfunction for the divergent circulation using (4.5a) and (4.4), allowing directional (i.e., cross- and along-front) partitioning of the circulation. Although explicit links to the two-dimensional confluence and horizontal shear forcing effects do not appear to be possible, if physically meaningful termwise partitions of the forcing of the suggested omega equations could be developed, then this approach may be of interest. The rotational ageostrophic wind could be determined using a truncated form of the divergence equation, analogous to (4.10a,b) for the QG case.

As an example of an application of the above procedure, it is proposed that vertical circulations associated with jet streaks and baroclinic wave disturbances could be "isolated" through consideration of a shear/curvature partition of the relative vorticity in Eliassen's (1984) omega equation and a partition of a generalized Q vector (where the geostrophic wind in (3.3) is replaced by its nondivergent counterpart) into cross- and along-isentrope components in Davies-Jones' (1991) omega equation. This suggestion derives from promising applications by Barnes and Colman (1993, 1994) of the Q vector partitioned into cross- and along-isentrope components (Keyser et al. 1988; Keyser et al. 1992b; Kurz 1992) to isolate the vertical motion patterns associated with upper-level jet streaks and their accompanying short-wave disturbances. The Barnes-Colman approach, which is intended for "quick-look" diagnosis in an operational setting, consists of calculating the divergences of the cross- and along-isentrope components of the Q vector without solving the QG omega equation. In a case of strongly curved flow associated with a midtropospheric cyclone located over the southwestern United States, their approach yields the expected four- and two-cell patterns of vertical-motion forcing associated with jet-streak and wave disturbances (Figs. 1 and 2). Schematics consistent with this statement may be found in Figs. 5b and 4, respectively, of Sanders and Hoskins (1990), which may be viewed as the Q-vector counterparts of the natural-coordinate approach illustrated in Figs. 1 and 2.

Work in progress by the author and colleagues confirms that the four- and two-cell vertical motion signatures attributable to jet streaks and short waves, respectively, may be isolated by solving the QG omega equation with the forcing partitioned into the divergences of the cross- and along-front components of the Q vector. Furthermore, it is hypothesized from consideration of (4.10b) that the rotational ageostrophic wind pattern in a jet streak embedded within a short-wave trough should consist of an anticyclonic circulation centered in the trough (associated with the relative vorticity term in (4.10b)) and cyclonic circulations within the entrance and exit regions of the jet streak (associated with the deformation term in (4.10b)). Whereas the anticyclonic circulation is consistent with upstream-directed ageostrophic winds in the base of the trough (Fig. 2a), the cyclonic circulations are consistent with the orientation of cross-contour ageostrophic winds toward lower and higher geopotential heights in jet-entrance and jet-exit regions (Fig. 1a), respectively. It is of interest that this perspective yields the proper orientation of the rotational ageostrophic wind to account for Lagrangian rates of change of kinetic energy in jet-entrance and jet-exit regions, in accord with the discussion in Section 4.3 regarding the expected dominance of the rotational part of the ageostrophic wind in curved flows.

5. Concluding Remarks

This review surveys the topic of diagnosing ageostrophic circulations in two and three dimensions with a view toward eventual applications involving observed fronts and jet streaks. In the context of the QG equations, the ageostrophic wind and

vertical motion fields may be determined from the balanced geopotential field, provided that appropriate boundary conditions can be specified and that diabatic and frictional effects are known or are amenable to parameterization. In two dimensions, the classic Sawyer-Eliassen equation is sufficient to describe the ageostrophic vertical circulation, which is synonymous with the divergent vertical circulation.

In three dimensions, however, account must be taken of the rotational part of the ageostrophic wind. It is argued that internal cancellation between contraction and stretching in orthogonal directions arising in the presence of rotational flow may render directional partitioning of the vertical motion field problematic. This consideration suggests that a natural generalization of the two-dimensional circulation described by the Sawyer-Eliassen equation (for constant Coriolis parameter) is the divergent part of the ageostrophic wind and vertical motion. Diagnostic equations for a vector streamfunction for alternatives involving the total ageostrophic wind and the rotational/divergent ageostrophic wind are presented and interpreted; their relationship to the two-dimensional Sawyer-Eliassen equation also is considered. It is significant that two-dimensional forcing terms (i.e., confluence and horizontal shear), which are the basis of theoretically derived interpretations of frontogenesis, reappear intact in the three-dimensional equations. The linear nature of the elliptic operators for the circulation streamfunction in the QG case facilitates termwise partitioning of the circulation streamfunction, so that the relative contribution of two-dimensional effects may be evaluated in a three-dimensional context.

The diagnostic methodologies presented in this chapter are illustrated with an example of how the confluence and horizontal shear forcing terms in the Sawyer-Eliassen equation can yield a transverse ageostrophic circulation conducive to upper-level frontogenesis through tilting in the case of an upper-level frontal zone forming in an environment of confluence and cold-air advection upstream of the trough axis of a baroclinic wave. This "two-dimensional" signature, which consists of a thermodynamically direct circulation displaced toward the warm-air side of the frontal zone and a compensating indirect circulation on the cold-air side, appears to carry over to three dimensions, as suggested by the results of diagnosis of a channel-model simulation of baroclinic wave development. Considerations for the eventual application of the diagnostic methodologies to observed fronts and jets also are explored. At this point, it appears that the most vexing problem is identifying well-posed intermediate systems of dynamical equations that retain the conceptual and mathematical simplicity of the QG equations, yet are relatively accurate in situations where the Rossby number is non-negligible, such as in strongly curved flows.

Provided that the question of tractable intermediate dynamical systems can be resolved, application of the diagnos-

(a)

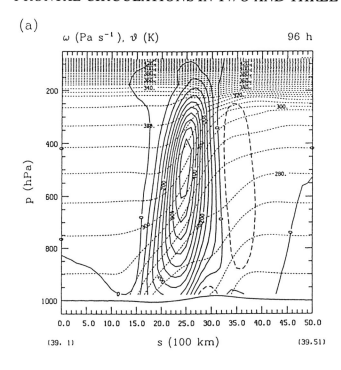

(b)

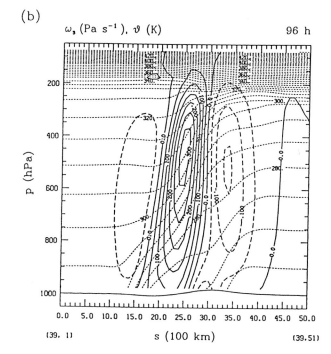

(c)

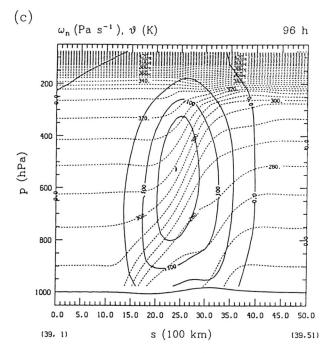

FIG. 16. Cross sections showing total vertical velocity partitioned into cross- and along-front components for the simulated upper-level jet-front system depicted in Fig. 14: (a) total pressure-coordinate vertical velocity, ω; (b) vertical velocity associated with divergent circulation in cross-front direction, ω_s; (c) vertical velocity associated with divergent circulation in along-front direction, ω_n. Contour interval for vertical velocity fields is 0.05 Pa s^{-1}; negative values are long dashed; zero and positive values are solid. Background field is potential temperature (contour interval 5 K, short dashed). Cross sections extend from grid point (39, 1) ($x = 3800$ km, $y = 0$ km) to grid point (39, 51) ($x = 3800$ km, $y = 5000$ km), as indicated on the lower abscissa; distance s along the cross section is in units of 100 km. The nearly horizontal line immediately above the lower abscissa indicates surface pressure. The same conventions are used in Figs. 17–19. Cross- and along-front coordinates (vertical velocity components) transform to those in this chapter as follows: $s \rightarrow x$ ($\omega_s \rightarrow \omega_x$) and $n \rightarrow y$ ($\omega_n \rightarrow \omega_y$), where the first entry refers to the coordinate (component) in the figure and the second to its counterpart in the chapter (Keyser et al. 1989).

tic methodologies discussed in this chapter could become increasingly commonplace, especially as gridded datasets explicitly resolving fronts and jets become routinely available. The availability of such datasets may be anticipated as a result of: (1) continuing interest in mesoscale phenomena and processes as witnessed by a growing list of field programs internationally in recent years; (2) the ongoing operational deployment of wind-profiling and Doppler-radar technologies; (3) the advent of four-dimensional data-assimilation schemes producing dynamically consistent analyses with mesoscale resolution; (4) the ongoing revolution in the networking of computers, allowing real-time data access using the Internet; and (5) recent developments in computing technology that facilitate visualization of complex structures in four-dimensional datasets. The foregoing advances should make it possible to interpret conceptual models describing the structure and evolution of fronts and jets from a dynamical perspective, as well as to reconcile such conceptualizations with a hierarchy of dynamical models of frontogenesis in two and three dimensions.

(a)

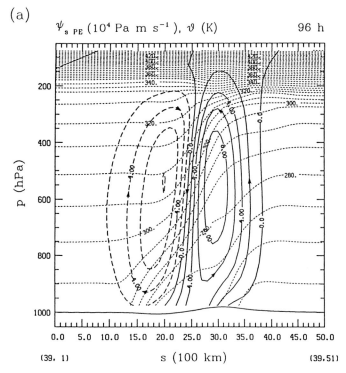

(b)

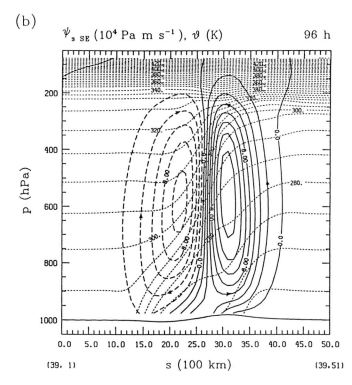

FIG. 17. Cross sections for the simulated upper-level jet-front system depicted in Fig. 14 showing streamfunction for cross-front component of divergent ageostrophic circulation (denoted by ψ_s in the figure and by ψ_{agdx} using notation of chapter (contour interval 2×10^4 Pa m s^{-1})) (a) diagnosed from primitive-equation model and (b) diagnosed from generalized QG Sawyer-Eliassen equation (4.8). Negative values of ψ_s are long dashed; zero and positive values are solid. Background field is potential temperature (contour interval 5 K, short dashed). Cross- and along-front coordinates (s, n) transform as in Fig. 16 (Keyser et al. 1992a).

Acknowledgments

During the preparation of this chapter, the author has benefited from a critical review by H. B. Bluestein and from discussions with H. B. Bluestein, L. F. Bosart, G. J. Hakim, J. C. Jusem, G. M. Lackmann, R. Rotunno, M. A. Shapiro, and L. W. Uccellini. Financial support of this effort has been provided by the National Science Foundation through Grants ATM-9114743 and ATM-9421678 and by the Office of Naval Research through Grant N00014-92-J-1532 awarded to the University at Albany.

References

Appenzeller, C., and H. C. Davies, 1992: Structure of stratospheric intrusions into the troposphere. *Nature*, **358**, 570–572.

Barnes, S. L., and B. R. Colman, 1993: Quasigeostrophic diagnosis of cyclogenesis associated with a cutoff extratropical cyclone–-The Christmas 1987 storm. *Mon. Wea. Rev.*, **121**, 1613–1634.

——, and ——, 1994: Diagnosing an operational numerical model using Q-vector and potential vorticity concepts. *Wea. Forecasting*, **9**, 85–102.

Beebe, R. G., and F. C. Bates, 1955: A mechanism for assisting in the release of convective instability. *Mon. Wea. Rev.*, **83**, 1–10.

Bell, G. D., and D. Keyser, 1993: Shear and curvature vorticity and potential-vorticity interchanges: Interpretation and application to a cutoff cyclone event. *Mon. Wea. Rev.*, **121**, 76–102.

Bishop, C. H., 1996: Domain-independent attribution. Part I: Reconstructing the wind from estimates of vorticity and divergence using free space Green's functions. *J. Atmos. Sci.*, **53**, 241–252.

Bjerknes, J., 1951: Extratropical cyclones. *Compendium of Meteorology*, T. F. Malone, Ed., Amer. Meteor. Soc., 577–598.

——, and J. Holmboe, 1944: On the theory of cyclones. *J. Meteor.*, **1**, 1–22.

Blackburn, M., 1985: Interpretation of ageostrophic winds and implications for jet stream maintenance. *J. Atmos. Sci.*, **42**, 2604–2620.

Bluestein, H. B., 1993: *Synoptic-Dynamic Meteorology in Midlatitudes*. Vol. II, *Observations and Theory of Weather Systems*. Oxford University Press, 594 pp.

——, and K. W. Thomas, 1984: Diagnosis of a jet streak in the vicinity of a severe weather outbreak in the Texas panhandle. *Mon. Wea. Rev.*, **112**, 2499–2520.

Bosart, L. F., and S. C. Lin, 1984: A diagnostic analysis of the Presidents' Day snowstorm of February 1979. *Mon. Wea. Rev.*, **112**, 2148–2177.

Browning, K. A., 1990: Organization of clouds and precipitation in extratropical cyclones. *Extratropical Cyclones, The Erik Palmén Memorial Volume*, C. W. Newton and E. O. Holopainen, Eds., Amer. Meteor. Soc., 129–153.

Buzzi, A., A. Trevisan, and G. Salustri, 1981: Internal frontogenesis: A two-dimensional model in isentropic, semi-geostrophic coordinates. *Mon. Wea. Rev.*, **109**, 1053–1060.

Cammas, J.-P., and D. Ramond, 1989: Analysis and diagnosis of the composition of ageostrophic circulations in jet-front systems. *Mon. Wea. Rev.*, **117**, 2447–2462.

Carlson, T. N., 1991: *Mid-Latitude Weather Systems*. HarperCollins Academic, 507 pp.

Chen, T.-C., and A. C. Wiin-Nielsen, 1976: On the kinetic energy of the divergent and nondivergent flow in the atmosphere. *Tellus*, **28**, 486–498.

(a)
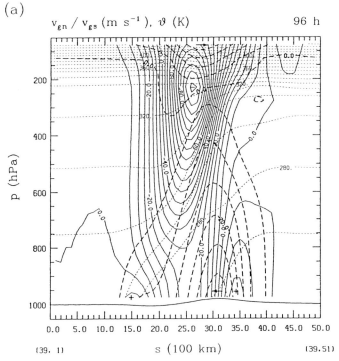

(b)
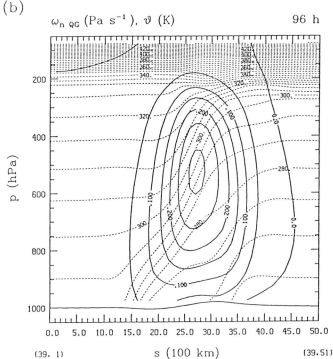

(c)
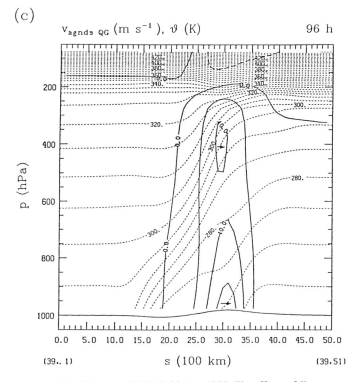

FIG. 18. Cross sections for the simulated upper-level jet-front system depicted in Fig. 14: (a) along-front geostrophic wind component (v_{gn}; contour interval 5 m s^{-1}, solid) and cross-front geostrophic wind component (v_{gs}; contour interval 5 m s^{-1}, long dashed); (b) component of QG-diagnosed vertical velocity associated with divergent circulation in along-front direction (ω_n; contour interval 0.05 Pa s^{-1}, solid); (c) component of QG-diagnosed rotational ageostrophic wind in cross-front direction (v_{agnds}; contour interval 5 m s^{-1}, solid and long dashed). Negative and positive wind components respectively denote flow out of and into cross section (v_{gn}), and flow toward the left (decreasing s) and right (increasing s) in the plane of the cross section (v_{gs} and v_{agnds}). Selected extrema of v_{gn} are indicated by "+" and "−" signs, and of v_{gs} and v_{agnds} by arrows. Background field is potential temperature (contour interval 10 K in (a) and 5 K in (b) and (c), short dashed). Cross- and along-front coordinates (wind components) transform to those in this chapter as follows: $s \rightarrow x$ ($v_{gs} \rightarrow u_g$; $v_{agnds} \rightarrow u_{agr}$) and $n \rightarrow y$ ($v_{gn} \rightarrow v_g$; $\omega_n \rightarrow \omega_y$), where the first entry refers to the coordinate (component) in the figure and the second to its counterpart in the chapter (Keyser et al. 1992a).

——, J. C. Alpert, and T. W. Schlatter, 1978: The effects of divergent and nondivergent winds on the kinetic energy budget of a midlatitude cyclone: A case study. *Mon. Wea. Rev.*, **106**, 458–468.

Davies-Jones, R., 1991: The frontogenetical forcing of secondary circulations. Part I: The duality and generalization of the Q vector. *J. Atmos. Sci.*, **48**, 497–509.

Davis, C. A., and K. A. Emanuel, 1991: Potential vorticity diagnostics of cyclogenesis. *Mon. Wea. Rev.*, **119**, 1929–1953.

Eliassen, A., 1962: On the vertical circulation in frontal zones. *Geofys. Publ.*, **24**(4), 147–160.

——, 1966: Motions of intermediate scale: Fronts and cyclones. *Advances in Earth Science*, P. M. Hurley, Ed., M.I.T. Press, 111–138.

——, 1984: Geostrophy. *Quart. J. Roy. Meteor. Soc.* **110**, 1–12.

——, 1990: Transverse circulations in frontal zones. *Extratropical Cyclones, The Erik Palmén Memorial Volume*, C. W. Newton and E. O. Holopainen, Eds., Amer. Meteor. Soc., 155–165.

Emanuel, K. A., M. Fantini, and A. J. Thorpe, 1987: Baroclinic instability in an environment of small stability to slantwise moist convection. Part I: Two-dimensional models. *J. Atmos. Sci.*, **44**, 1559–1573.

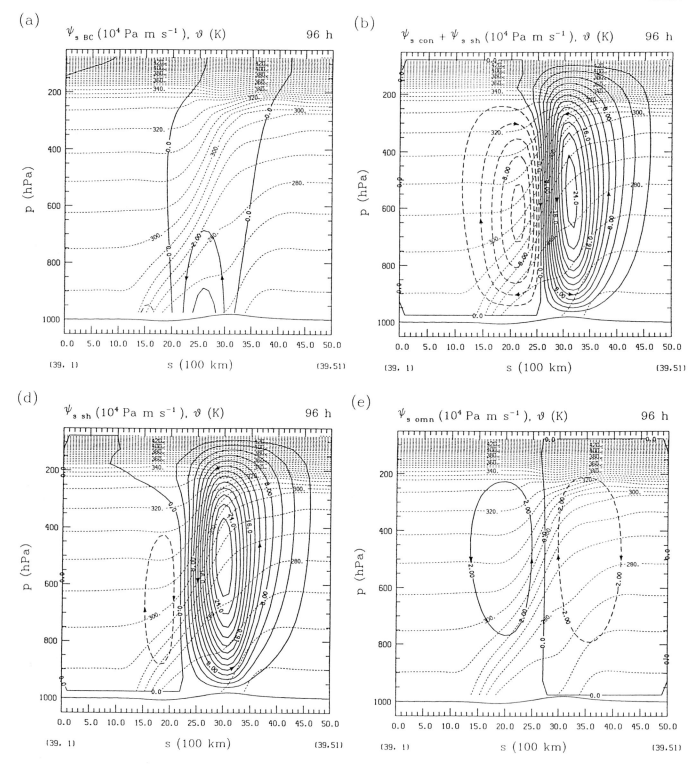

FIG. 19 (this page and facing page). Cross sections for the simulated upper-level jet-front system depicted in Fig. 14 showing ψ_s (ψ_{adgx} using notation of chapter; contour interval 2×10^4 Pa m s^{-1}) diagnosed from generalized QG Sawyer-Eliassen equation (4.8) (see Fig. 17b) partitioned into components attributable to: (a) boundary conditions; (b) sum of confluence and horizontal shear; (c) confluence; (d) horizontal

shear; (e) vertical velocity associated with divergent circulation in along-front direction; (f) rotational ageostrophic wind in cross-front direction. Negative values of ψ_s are long dashed; zero and positive values are solid. Background field is potential temperature (contour interval 5 K, short dashed). Cross- and along-front coordinates (s, n) transform as in Fig. 16 (Keyser et al. 1992a).

(c)

$\psi_{s\ con}\ (10^4\ Pa\ m\ s^{-1}),\ \vartheta\ (K)$ 96 h

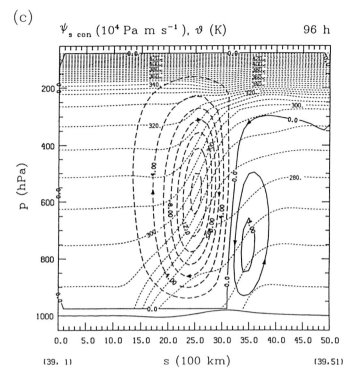

s (100 km)

(f)

$\psi_{s\ vagnds}\ (10^4\ Pa\ m\ s^{-1}),\ \vartheta\ (K)$ 96 h

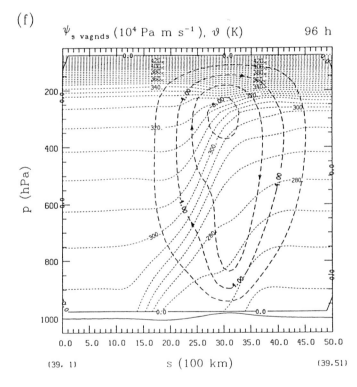

s (100 km)

Green, J. S. A., F. H. Ludlam, and J. F. R. McIlveen, 1966: Isentropic relative-flow analysis and the parcel theory. *Quart. J. Roy. Meteor. Soc.*, **92**, 210–219.

Hakim, G. J., and L. W. Uccellini, 1992: Diagnosing coupled jet-streak circulations for a northern plains snow band from the operational nested-grid model. *Wea. Forecasting*, **7**, 26–48.

Haltiner, G. J., and R. T. Williams, 1980: *Numerical Prediction and Dynamic Meteorology.* 2d ed. Wiley, 477 pp.

Hines, K. M., and C. R. Mechoso, 1991: Frontogenesis processes in the middle and upper troposphere. *Mon. Wea. Rev.*, **119**, 1225–1241.

Holton, J. R., 1992: *An Introduction to Dynamic Meteorology.* 3d ed. International Geophysics Series, Vol. 48, Academic Press, 511 pp.

Hoskins, B. J., 1975: The geostrophic momentum approximation and the semi-geostrophic equations. *J. Atmos. Sci.*, **32**, 233–242.

———, and F. P. Bretherton, 1972: Atmospheric frontogenesis models: Mathematical formulation and solution. *J. Atmos. Sci.*, **29**, 11–37.

———, and I. Draghici, 1977: The forcing of ageostrophic motion according to the semi-geostrophic equations and in an isentropic coordinate model. *J. Atmos. Sci.*, **34**, 1859–1867.

———, and M. A. Pedder, 1980: The diagnosis of middle latitude synoptic development. *Quart. J. Roy. Meteor. Soc.*, **106**, 707–719.

———, I. Draghici, and H. C. Davies, 1978: A new look at the ω-equation. *Quart. J. Roy. Meteor. Soc.*, **104**, 31–38.

———, M. E. McIntyre, and A. W. Robertson, 1985: On the use and significance of isentropic potential vorticity maps. *Quart. J. Roy. Meteor. Soc.*, **111**, 877–946.

Iversen, T., and T. E. Nordeng, 1984: A hierarchy of nonlinear filtered models–Numerical solutions. *Mon. Wea. Rev.*, **112**, 2048–2059.

Keyser, D., and T. N. Carlson, 1984: Transverse ageostrophic circulations associated with elevated mixed layers. *Mon. Wea. Rev.*, **112**, 2465–2478.

———, and M. J. Pecnick, 1985: A two-dimensional primitive equation model of frontogenesis forced by confluence and horizontal shear. *J. Atmos. Sci.*, **42**, 1259–1282.

———, and M. A. Shapiro, 1986: A review of the structure and dynamics of upper-level frontal zones. *Mon. Wea. Rev.*, **114**, 452–499.

———, ———, and D. J. Perkey, 1978: An examination of frontal structure in a fine-mesh primitive equation model for numerical weather prediction. *Mon. Wea. Rev.*, **106**, 1112–1124.

———, M. J. Reeder, and R. J. Reed, 1988: A generalization of Petterssen's frontogenesis function and its relation to the forcing of vertical motion. *Mon. Wea. Rev.*, **116**, 762–780.

———, B. D. Schmidt, and D. G. Duffy, 1989: A technique for representing three-dimensional vertical circulations in baroclinic disturbances. *Mon. Wea. Rev.*, **117**, 2463–2494.

———, ———, and ———, 1992a: Quasigeostrophic diagnosis of three-dimensional ageostrophic circulations in an idealized baroclinic disturbance. *Mon. Wea. Rev.*, **120**, 698–730.

———, ———, and ———, 1992b: Quasigeostrophic vertical motions diagnosed from along- and cross-isentrope components of the Q vector. *Mon. Wea. Rev.*, **120**, 731–741.

Kocin, P. J., and L. W. Uccellini, 1990: *Snowstorms along the Northeastern Coast of the United States: 1955 to 1985.* Meteor. Monogr., No. 44, Amer. Meteor. Soc., 280 pp.

Krishnamurti, T. N., 1968a: A diagnostic balance model for studies of weather systems of low and high latitudes, Rossby number less than 1. *Mon. Wea. Rev.*, **96**, 197–207.

———, 1968b: A study of a developing wave cyclone. *Mon. Wea. Rev.*, **96**, 208–217.

Kurz, M., 1992: Synoptic diagnosis of frontogenetic and cyclogenetic processes. *Meteor. Atmos. Phys.*, **48**, 77–91.

Lagouvardos, K., and V. Kotroni, 1995: Upper-level frontogenesis: Two case studies from the FRONTS 87 experiment. *Mon. Wea. Rev.*, **123**, 1197–1206.

Loughe, A. F., C.-C. Lai, and D. Keyser, 1995: A technique for diagnosing three-dimensional ageostrophic circulations in baroclinic disturbances on limited-area domains. *Mon. Wea. Rev.*, **123**, 1476–1504.

Lynch, P., 1989: Partitioning the wind in a limited domain. *Mon. Wea. Rev.*, **117**, 1492–1500.

McWilliams, J. C., and P. R. Gent, 1980: Intermediate models of planetary circulations in the atmosphere and ocean. *J. Atmos. Sci.*, **37**, 1657–1678.

Muller, B. M., and H. E. Fuelberg, 1990: A simulation and diagnostic study of water vapor image dry bands. *Mon. Wea. Rev.*, **118**, 705–722.

Newton, C. W., 1954: Frontogenesis and frontolysis as a three-dimensional process. *J. Meteor.*, **11**, 449–461.

——, and A. Trevisan, 1984: Clinogenesis and frontogenesis in jet-stream waves. Part II: Channel model numerical experiments. *J. Atmos. Sci.*, **41**, 2735–2755.

Reed, R. J., 1955: A study of a characteristic type of upper-level frontogenesis. *J. Meteor.*, **12**, 226–237.

——, and F. Sanders, 1953: An investigation of the development of a mid-tropospheric frontal zone and its associated vorticity field. *J. Meteor.*, **10**, 338–349.

Reeder, M. J., and D. Keyser, 1988: Balanced and unbalanced upper-level frontogenesis. *J. Atmos. Sci.*, **45**, 3366–3386.

——, ——, and B. D. Schmidt, 1991: Three-dimensional baroclinic instability and summertime frontogenesis in the Australian region. *Quart. J. Roy. Meteor. Soc.*, **117**, 1–28.

Rotunno, R., W. C. Skamarock, and C. Snyder, 1994: An analysis of frontogenesis in numerical simulations of baroclinic waves. *J. Atmos. Sci.*, **51**, 3373–3398.

Sanders, F., and L. F. Bosart, 1985: Mesoscale structure in the megalopolitan snowstorm of 11–12 February 1983. Part I: Frontogenetical forcing and symmetric instability. *J. Atmos. Sci.*, **42**, 1050–1061.

——, and B. J. Hoskins, 1990: An easy method for estimation of Q-vectors from weather maps. *Wea. Forecasting*, **5**, 346–353.

Sardeshmukh, P. D., and B. J. Hoskins, 1987: On the derivation of the divergent flow from the rotational flow: The χ problem. *Quart. J. Roy. Meteor. Soc.*, **113**, 339–360.

Sawyer, J. S., 1956: The vertical circulation at meteorological fronts and its relation to frontogenesis. *Proc. Roy. Soc. London*, **A234**, 346–362.

Schär, C., and H. Wernli, 1993: Structure and evolution of an isolated semi-geostrophic cyclone. *Quart. J. Roy. Meteor. Soc.*, **119**, 57–90.

Shapiro, M. A., 1981: Frontogenesis and geostrophically forced secondary circulations in the vicinity of jet stream-frontal zone systems. *J. Atmos. Sci.*, **38**, 954–973.

——, 1982: Mesoscale weather systems of the central United States. CIRES/NOAA Tech.l Rep., University of Colorado, 78 pp.

——, and P. J. Kennedy, 1981: Research aircraft measurements of jet stream geostrophic and ageostrophic winds. *J. Atmos. Sci.*, **38**, 2642–2652.

Snyder, C., W. C. Skamarock, and R. Rotunno, 1991: A comparison of primitive-equation and semigeostrophic simulations of baroclinic waves. *J. Atmos. Sci.*, **48**, 2179–2194.

Todsen, M., 1964: A study of the vertical circulations in a cold front. Part IV, Final Report, Contract No. AF61(052)–525.

Uccellini, L. W., 1990: Processes contributing to the rapid development of extratropical cyclones. *Extratropical Cyclones, The Erik Palmén Memorial Volume*, C. W. Newton and E. O. Holopainen, Eds., Amer. Meteor. Soc., 81–105.

——, and D. R. Johnson, 1979: The coupling of upper and lower tropospheric jet streaks and implications for the development of severe convective storms. *Mon. Wea. Rev.*, **107**, 682–703.

——, and P. J. Kocin, 1987: The interaction of jet streak circulations during heavy snow events along the east coast of the United States. *Wea. Forecasting*, **1**, 289–308.

——, D. Keyser, K. F. Brill, and C. H. Wash, 1985: The Presidents' Day cyclone of 18–19 February 1979: Influence of upstream trough amplification and associated tropopause folding on rapid cyclogenesis. *Mon. Wea. Rev.*, **113**, 962–988.

Wernli, J. H., 1995: Lagrangian perspective of extratropical cyclogenesis. Dr. Sc. Nat. thesis, Swiss Federal Institute of Technology (ETH), Zürich, Switzerland, Dissertation No. 11016, 157 pp.

Whitaker, J. S., and C. A. Davis, 1994: Cyclogenesis in a saturated environment. *J. Atmos. Sci.*, **51**, 889–907.

Xu, Q., 1990: Cold and warm frontal circulations in an idealized moist semigeostrophic baroclinic wave. *J. Atmos. Sci.*, **47**, 2337–2352.

——, 1992: Ageostrophic pseudovorticity and geostrophic C-vector forcing–A new look at the Q vector in three dimensions. *J. Atmos. Sci.*, **49**, 981–990.

——, 1994: Semibalance model–Connection between geostrophic-type and balanced-type intermediate models. *J. Atmos. Sci.*, **51**, 953–970.

——, and R. Davies-Jones, 1993: Boundary conditions for the psi equations. *Mon. Wea. Rev.*, **121**, 1566–1571.

——, and D. Keyser, 1993: Barotropic and baroclinic ageostrophic winds and completeness of solution for the psi equations. *J. Atmos. Sci.*, **50**, 588–596.

Mesoscale Aspects of Extratropical Cyclones: An Observational Perspective

KEITH A. BROWNING

Joint Centre for Mesoscale Meteorology, University of Reading, Reading, UK

1. Introduction

The understanding of the mesoscale structure of frontal cyclones has, as shown in this review, benefited greatly from special mesoscale observations such as dropwindsondes and other measurements from aircraft. These observations are usually interpreted within the context of numerical weather prediction (NWP) model analyses based on the operational synoptic-scale observations. A few studies have also used high-resolution mesoscale models to assimilate the observations. Another tool for mesoscale analysis—and a major source of inspiration for this review—is imagery. This includes cloud and water-vapor imagery from satellites, especially geostationary satellites, and also precipitation imagery from ground-based radars, especially in networks. By providing the overall pattern of the mesoscale phenomena and an indication of the relationship to the larger-scale forcing, the imagery is invaluable for giving qualitative understanding of the processes at work. Often it provides the only clues as to what is happening on the mesoscale.

There is a whole range of cyclone structures, which differ according to the synoptic-scale setting. The scope of this presentation is restricted to variants of one particular class of cyclone. Another limitation in the scope of this review is a focus on cyclones of oceanic origin in and near northwestern Europe. References are made to observational studies in North America and elsewhere but the specific cases illustrated in detail are drawn mainly from European experience at the eastern end of the North Atlantic storm track. Reference to a self-consistent set of examples portrays more clearly how all the bits fit together into a coherent picture. Despite this restrictive approach, the conceptual framework presented has a broader applicability. The notions of conveyor belts, dry intrusions, and frontal fractures used to unify the observations, although not helpful in every context, do provide a useful framework within which to visualize and interpret three-dimensional mesoscale organization. This review concentrates on cyclones that undergo substantial development. Although weak frontal waves may form solely from low-level instabilities, strongly developing cyclones are usually affected by upper-level forcing of the kind described here.

One of the earliest satellite-based cyclone classification schemes was by Zillman and Price (1972), later adapted by Browning and Hill (1985)—see Fig. 1. (This and all other figures in this review are portrayed from a Northern Hemisphere point of view.) The scheme identifies three types of cyclogenesis according to the distance between the area of enhanced cumulus or comma cloud associated with forcing by an upper-level trough or potential vorticity (PV) anomaly, on the one hand, and a pre-existing polar-front cloud band associated with a baroclinic zone, on the other. When the two remain completely separate, the cyclogenesis is referred to as a comma cloud type. When the clouds associated with the upper-level trough begin as a separate entity and later merge with the frontal cloud band to give a λ-shaped cloud system, the process is referred to as instant occlusion. And when the cyclogenesis occurs close to the polar-front cloud band and involves it dynamically from an early stage, it is referred to as a frontal-wave type development: this type of cyclone is the focus of this review.

A recent satellite-based study of rapidly developing cyclones in the western North Atlantic by Evans et al. (1994) has modified the above classification by adding, as a fourth category, a variant of the instant occlusion, which they refer to as a "left exit" type. Otherwise their study confirms the three categories in the earlier scheme. Evans et al., however, rename the "frontal wave" type as the "emerging cloud head" type. This is in recognition of a characteristic cloud feature that will be described in detail later. Emerging-cloud-head cyclogenesis corresponds to the type of development in the scheme by Weldon (1977), which begins with a so-called

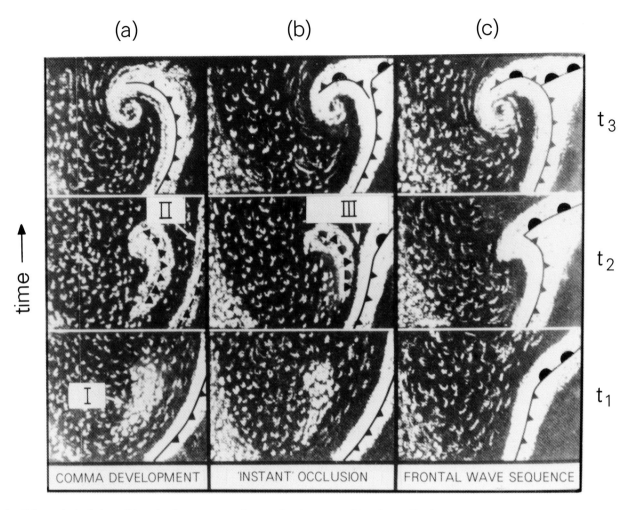

FIG. 1. Schematic depiction of three basic sequences of vortex development evident in satellite imagery: (a) development of a simple comma cloud entirely within the cold air, (b) development of an instant occlusion, (c) development of a frontal wave (adapted from Zillman and Price 1972). Frontal symbols represent the authors' way of representing the various evolution sequences using the tools of conventional frontal analysis. The labels I, II and III, respectively, indicate a region of enhanced convection, a decaying cloud band and a convective cloud band merging with the polar-front cloud band.

baroclinic leaf cloud forming directly over a low-level baroclinic zone.

Within the category of cyclone known as the "frontal wave" or "emerging cloud head" type, a variety of different configurations can occur depending on the synoptic-scale upper-air flow pattern. Young (1994) and Bader et al. (1995) distinguish between cyclones of this kind that develop in the presence of a meridional trough, a flat trough with confluent flow and a flat trough with diffluent flow. This review, however, is concerned not so much with the differences between these subtypes as with the features that are common to all of them, namely the cloud head (Section 3), the dry intrusion (Section 4), and the warm and cold conveyor belts (Section 2).

2. Conceptual Framework

2.1 Frontal Fracture

A conceptual model that provides a useful framework for describing the kind of cyclones discussed in this review is the life-cycle model of Shapiro and Keyser (1990) (Fig. 2), aspects of which were anticipated by McGinnigle et al. (1988). The first stage in this life cycle shows the incipient frontal cyclone. The second and third stages show the development of the so-called frontal fracture in which the main cold front to the south of the cyclone center advances ahead of the center while cold air to the north of the warm front travels rearwards and wraps around the cyclone center. The boundary of the latter flow goes close to the low center to give

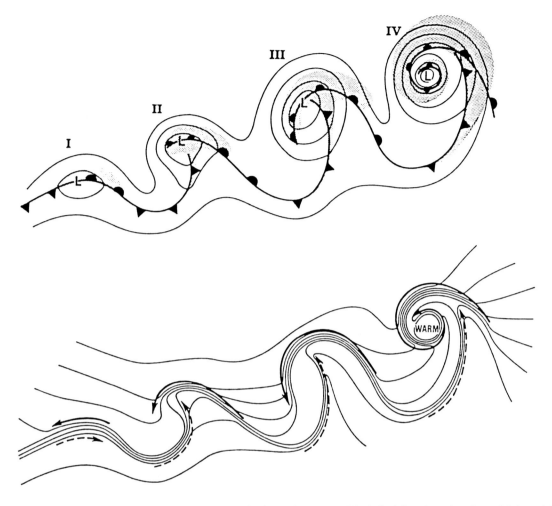

FIG. 2. The life cycle of a marine extratropical frontal cyclone (Shapiro and Keyser 1990): (I) incipient frontal cyclone; (II) frontal fracture; (III) bent-back warm front and frontal T-bone; (IV) warm-core frontal seclusion. Upper panel: sea-level pressure, solid lines; fronts, bold lines; and cloud signature, shaded. Lower panel: temperature, solid lines; cold and warm air currents, solid and dashed arrows, respectively.

what traditional synopticians refer to as a bent-back occlusion. McGinnigle et al. (1988) and Shapiro and Keyser (1990) reanalyzed the bent-back occlusion as a bent-back warm front, although part of it does in fact advance with a component toward the warm air and is properly analyzed as a cold front. This is referred to here simply as a bent-back front. As the bent-back front develops, the temperature gradient in the northern part of the original cold front weakens and air from the warm sector in this region gets drawn toward the low center. Then, as the cyclone matures, the bent-back front begins to encircle the low center and tends to seclude a core of warm air at the center.

The true surface cold front may advance several hundred kilometers ahead of the cyclone center. But often, as in some of the examples illustrated in this review, it advances only 200–300 km ahead of the cyclone center. Given good radar network data, the frontal fracture is perhaps most easily observed in the pattern of precipitation; Fig. 3 shows a typical pattern as seen at Stage 2 of frontal fracture. Poleward of the bent-back front and warm front there is a broad area of widespread precipitation associated with the cloud-head feature discussed in Section 3. Along the main cold front there is a narrower band of rain, which becomes patchy at its northern end where the surface temperature gradient weakens. The rain here is due to an upper-level front, which marks the leading edge of dry air aloft. This upper front, to a first approximation, is often continuous with the main surface cold front, as in Fig. 3. Between this upper-level front and the bent-back front, the capping of the warm moist low-level air by dry air tends to inhibit precipitation just to the warm side of the cyclone center, although, as discussed in Section 6, potential instability in this region may lead to outbreaks of convective showers. To understand all these features we need to visualize the three-dimensional thermodynamic structure and the airflows that give rise to it. A way of doing this is to use the concepts of conveyor belts and the dry intrusion.

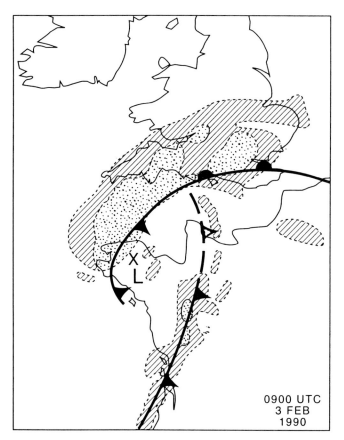

FIG. 3. Rainfall distribution and intensity derived from radar data for a cyclone centered over Brittany. Hatching represents rainfall rates less than or equal to 4 mm h^{-1} and stippling greater than 4 mm h^{-1}. Surface fronts are shown solid; the dashed front is an upper-level feature (Young 1994).

2.2   The Principal Airstreams

The notion of air masses was enshrined in the early Norwegian models of frontal cyclones. To visualize the three-dimensional structure of cyclones we find it helpful to extend the air-mass concept by focusing on airstreams known as conveyor belts. These are defined in a frame of reference moving with the cloud systems they produce. Relative-flow analyses within surfaces of constant wet-bulb potential temperature (θ_w) enable one to build up an overall picture of these cloud-generating conveyor-belt flows, which can then be related to the imagery.

This terminology was first used by Harrold (1973), and ever since the term "warm conveyor belt" (WCB) has been used to describe the flow of air with high θ_w in frontal systems. Ahead of a cold front there is often a strong and more-or-less well-defined tongue of high-θ_w airflow that justifies the name conveyor belt in the sense that the air is conveyed with a strong component along its axis. However, when a frontal wave develops, the wave travels along the front so that, relative to the wave, the flow within the conveyor belt is slower and less like a conveyor belt. As the

cyclone develops, the isentropic surfaces become increasingly distorted with time and it becomes difficult to carry out relative-flow analyses. Thus, Kuo et al. (1992) have argued that the concept is no longer useful in the region of developing cyclones. These difficulties are real but, so long as its limitations are kept in mind, the conveyor-belt paradigm still provides a helpful device for qualitatively visualizing three-dimensional cyclone structure and in particular for providing a framework for interpreting mesoscale structure.

Figure 4 shows the conveyor-belt portrayal of cyclone structure in the framework of frontal fracture. The stage of cyclone development depicted here corresponds to Stage 2 of Shapiro and Keyser's (1990) model. The conveyor-belt concept has been extended beyond that originally proposed by Harrold: thus as many as three conveyor belts are shown in Fig. 4. The three conveyor-belt flows are represented by coded arrows; however, it should be kept in mind that the arrows represent the instantaneous motion of air parcels relative to the system (individual parcels do not have sufficient time to travel along the entire length of each arrow):

- *The primary warm conveyor belt, W1* (after Harrold 1973). This is the main belt of high-θ_w air that rises from the low troposphere to the upper troposphere within the polar-front cloud band. Objective criteria for identifying the warm conveyor belt have been developed and evaluated by Wernli and Rossa (1994). The air in the WCB first travels along the main cold front, often accompanied by a 25 m s^{-1} low-level jet, and then it ascends over the warm front as in Bergeron's (1937) model of an ana-warm front. The WCB should not be regarded as a closed tube of air conveying all air parcels from one end to the other; rather, the WCB flow is augmented by warm low-level air entering it along much of its length (Letestu 1994). Depending on the circumstances, a wave or waves may form on the main cold front (not shown in Fig. 4), in which case parts of the WCB will undergo rearward-sloping ascent over the cold air as it travels along the cold front. Rearward-sloping ascent often occurs along the trailing cold front even in the absence of such a wave.
- *The secondary warm conveyor belt, W2* (after Young et al. 1987). This flow, of limited length in the configuration shown in Fig. 4, is derived from low-level air approaching the cyclone center. It consists of boundary-layer air coming from the lower left-hand side of W1, which peels off toward the cyclone center. This air ascends at the bent-back front into the feature known as the cloud head (see Section 3). The W2 flow is part of the ageostrophic transverse circulation at the exit of an upper-level jet streak (Young 1989), enhanced by frictional turning within the boundary layer. It is responsible for bringing the low-level flow of high-θ_w air into the region of frontal fracture.

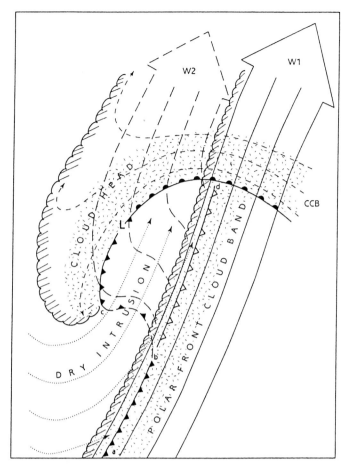

FIG. 4. Structure of a developing extratropical cyclone (confluent-flow type) (based on Browning and Roberts 1994). The cyclone center (L) is traveling toward the top right. The surface warm front is shown conventionally. Part of the bent-back front (cd) is plotted as a cold front with closely spaced frontal symbols. The main surface cold front (ab) is shown similarly. In between the two sharp surface cold fronts (bc) there is a diffuse surface cold front drawn dashed with widely spaced frontal symbols. The cold front drawn with open symbols (bd) is an upper cold θ_w-front (UCF) marking the leading edge of the dry intrusion. Principal airflows, drawn relative to the system, are the main warm conveyor belt (W1) (solid lines), the secondary warm conveyor belt (W2) (long-dash lines), the cold conveyor belt (CCB) (short-dash lines), and the dry intrusion (dotted lines). The cold-air sides of the main cloud features are drawn scalloped: the polar-front cloud band is due to W1, and the cloud head is due to the combined effect of W2 and the CCB. Precipitation (shaded areas) reaches the surface along the left side of W1, near the cold fronts, and above the warm frontal zone; precipitation also falls from the inner parts of the cloud head starting at the bent-back front. The two main areas of precipitation are separated by a dry slot where the shallow W2 flow is capped by the dry intrusion. The dry slot is usually characterized by partial shallow cloudiness.

- *The cold conveyor belt, CCB* (after Carlson 1980). This is an airstream which, relative to the moving cyclone, initially flows rearwards at low levels along the cold side of the warm front and bent-back front. The CCB is often associated with a low-level jet, occasionally exceeding 25 m s^{-1} as, for example, in

the Great Storm of October 1987 (Shutts 1990; Bader et al. 1995). Some of the air at the outer edge of the CCB descends, especially where precipitation falling from above evaporates and cools it, but most of the CCB air eventually rises and fans out within the cloud head as shown in Fig. 4. Whereas W1 and W2 each tend to be characterized by a rather narrow range of θ_w, the CCB is less well defined and may contain a significant range of θ_w-values because it occurs within the warm frontal zone.

A fourth airstream, shown in Fig. 4, is:

- *The dry intrusion* (using the terminology of Rodgers et al. 1985 and Young et al. 1987). This is part of a dry airstream, characterized by low θ_w, that descends from high levels behind the cyclone and approaches the cyclone center at middle levels to give a so-called dry slot in the satellite imagery. This dry slot is best seen in the infrared and water vapor channels. In the dry slot the dry-intrusion air begins to ascend over a shallow moist zone (SMZ) characterized by the high-θ_w air in W2. The right-hand edge of the dry intrusion, where it overruns the moist air, is marked in Fig. 4 as an upper cold front (UCF), strictly an upper cold θ_w-front (see Section 4.1).

The configuration in Fig. 4 is a common one and it forms the basis for much of the discussion in this review. However, as mentioned in the Introduction, the airstreams take on different configurations in different synoptic settings. Figure 4 displays aspects of both Young's (1994) "meridional trough" and his "flat trough confluent flow" subtypes. A distinctly different structure is shown in Fig. 5, corresponding to Young's "diffluent flow" subcategory. Most of the ingredients of Fig. 4 are present in Fig. 5; they are merely reconfigured. The W2 feature, in particular, is more elongated than the W2 flow in Fig. 4. As in Fig. 4, a dry intrusion overruns part of W2 but the overrunning is more extensive. This overrunning process is discussed in Section 4.1.

3. The Cloud Head

The term "cloud head" was coined by Böttger et al. (1975) to describe an entire cloud system, the northern edge of which develops a sharp convex-poleward boundary during rapid cyclogenesis. We adopt a more restrictive definition that confines use of the term to the portion of cloud associated with (and poleward of) the bent-back front. This cloud is often a well-defined entity, commonly separated from the polar-front cloud band by a dry slot. A cloud head often takes on a hook shape as cyclogenesis progresses; Young (1994) refers to this as the cloud hook, but here the term cloud head is used to describe the feature both before and after it develops a hook shape.

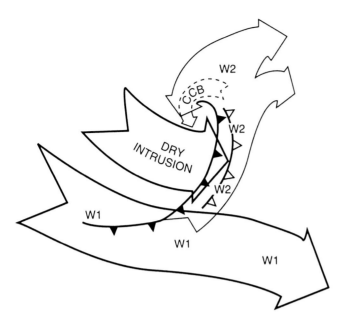

FIG. 5. Conceptual model showing system-relative airflows within a cyclone forming ahead of a strongly diffluent trough. The arrows labeled W1 and W2 are warm conveyor belts. The dashed arrow labeled CCB is a shallow cold conveyor belt, which forms a hook of low cloud around the cyclone center (after Young 1994).

The cloud head is such a characteristic feature that, as mentioned earlier, Evans et al. (1994) in their classification scheme were led to refer to frontal-wave cyclones as emerging-cloud-head cyclones, that is, a cloud head emerging from beneath the polar-front cloud band. Not all cloud heads "emerge," however; sometimes they form alongside (poleward) of the polar-front cloud band. For a more detailed discussion of this topic, see Bader at al. (1995). When a cloud head does emerge from beneath a polar-front cloud band during the course of cyclone development, it does so in the manner shown by the satellite image sequence in Fig. 6. The whiteness of the cloud pattern in this figure is a measure of the height (coldness) of the cloud top and the gray cloud head can be seen emerging in the 0300 UT 13 January image on the left side, to the west of the low center, labeled X. The cloud head grows with time during the rapid deepening of the cyclone, and at the end of this sequence can be seen to have developed into a major hook-shaped feature.

Colored lines superimposed on the cloud images in Fig. 6 represent the ascending portions of the warm-conveyor-belt W1 (red) and cold-conveyor-belt (blue) flows drawn relative to the moving system. (For simplicity the W2 flow is not depicted.) Initially in Fig. 6 there is no cloud head: the red flow predominates and is responsible for the leaf cloud that develops into a sharp-edged feature from an initially ragged area of deep cloud. The red flow originates as a belt of strong winds with high θ_w situated at low levels on the equatorward side of the front. It ascends slantwise within the cloud shield

north of the front, probably producing substantial precipitation in this region. Although the flow relative to the earth is oriented nearly parallel to the front, its orientation takes on the more northerly aspect, as shown, when viewed in a frame of reference traveling with the cloud system. The ascent of the warm (red) flow is part of a thermally direct circulation, and the flow accelerates as it rises, to give a jet maximum (J) at upper levels, consistent with conversion of potential energy into kinetic energy.

The blue CCB flow, which forms as the cyclone continues to develop, originates at low levels ahead of the warm front and gives rise to the cloud head. As the blue flow travels around the northern side of the cyclone center, it ascends and fans out. The cloud shield generated by the blue flow has a characteristic convex-shaped leading edge: hence the name "cloud head." The emergence of the blue flow and its cloud head is associated with a second upper-level jet maximum, which is located just upwind of the dry slot. The ascent of the relatively cold air in the blue flow is part of a thermally indirect circulation in the exit region of this second jet. Thus, the frontal cyclone goes through a characteristic evolution in which the precipitation generated in its earlier stage is associated with ascending warm air (the red WCB flow), while the precipitation generated in the later stages is increasingly due to the ascent of cold air (the blue CCB flow) moistened to some extent by evaporation of precipitation falling from the overrunning warm conveyor belt.

The radar imagery that is superimposed upon the satellite cloud imagery in Fig. 7 shows the typical pattern of precipitation associated with a cloud head. Although this is a different event from that represented in Fig. 6, the situation is similar: a band of high frontal cloud extends from France to the North Sea, with a cloud head over southern Britain north of a cyclone centered near Brittany. The pattern of the cloud-head precipitation resembles that of the cloud itself except that the outer boundary of the surface precipitation is up to 100 km inside the outer boundary of the cloud aloft. This is due to evaporation within dry air that descends beneath the outer edge of the cloud head. The tops of cloud and precipitation are typically at 2 to 3 km at the inner (southern) edge of a cloud head and they rise to 6 or 7 km toward the outer (northwestern) boundary (Fig. 8).

FIG. 6 (facing page). Satellite sequence showing the development of a frontal cyclone. The sequence of mainly 6-hourly infrared images (white represents the coldest/highest clouds) shows the evolution from the early stage of the cyclone, characterized by a "leaf cloud" with a smooth s-shaped edge, to the mature stage with a well-formed "cloud head." Superimposed on the imagery are the positions of the cyclone center (X), 300-mb jet cores (J), surface fronts (black line), edges of cloud leaf and cloud head (scalloped lines), and the principal cloud-producing flows (red and blue) drawn relative to the moving cyclone system (Browning 1994).

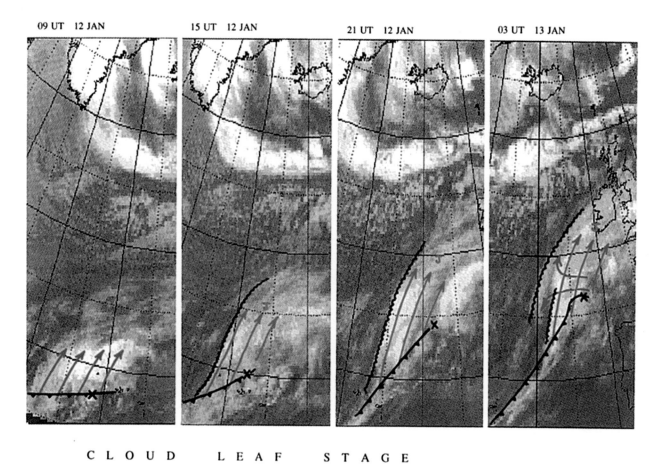

09 UT 12 JAN   15 UT 12 JAN   21 UT 12 JAN   03 UT 13 JAN

C L O U D   L E A F   S T A G E

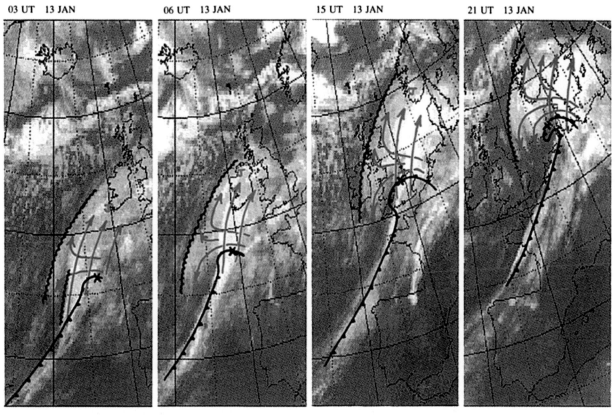

03 UT 13 JAN   06 UT 13 JAN   15 UT 13 JAN   21 UT 13 JAN

C L O U D   H E A D   S T A G E

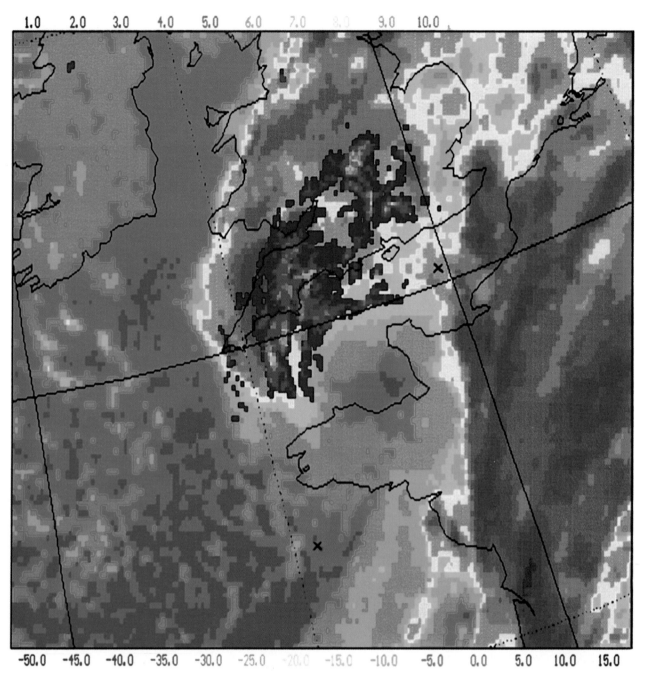

FIG. 7. Radar-network precipitation pattern superimposed upon Meteosat infrared imagery for 0300 UTC 28 April 1992. Rain rates are shown as shades of blue according to the scale (mm h^{-1}) at the top of the figure. Cloud-top temperatures are shown as shades of red through yellow to green (coldest red) according to the scale (°C) at the foot of the figure. The pale- and medium-blue areas are mainly cloud-free. XX denotes the position of the cross-section in Fig. 10 adjusted to take account of the 3-hour time difference. A cloud head covers England and Wales. A dry slot exists over northwest France. The absence of rain beneath the cold cloud to the east of the dry slot is due to the lack of radar data.

A characteristic feature of the inner boundary of many cloud heads is a narrow and rather two-dimensional, although gently curved, rainband associated with a rope-like cloud line with tops at or below 3 km (e.g., Monk and Bader 1988). It can be seen in the cloud and precipitation patterns in Figs. 8 and 7, respectively. It is collocated with the bent-back front and, in terms of its dimensions and sharpness, it resembles narrow cold-frontal rainbands and line convection as reviewed by Browning (1990). Although parts of bent-back fronts do indeed behave as cold fronts, narrow rainbands are sometimes observed to extend into the stationary or warm-frontal parts of the bent-back fronts. In such circumstances the

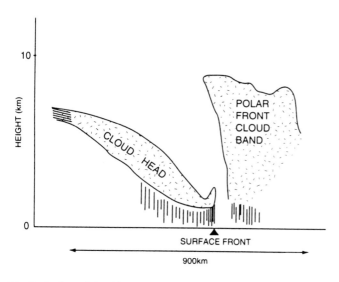

FIG. 8. Outline of cloud boundaries across a rapidly deepening cyclone, inferred from cloud-top temperatures and numerical model thermal structure (after Shutts 1990).

narrow rainbands are unlikely to be due to the density current mechanism proposed for narrow cold-frontal rainbands (e.g., Carbone 1982; Hobbs and Persson 1982).

More complex substructures may also be observed in association with cloud heads. Parallel cloud lines about 10 km apart are occasionally seen during rapid cyclogenesis at the top of the cloud head, extending from the position of the bent-back front and at right angles to it. The cause of these lines is not properly understood. Shutts (1990) has suggested that they may be due to some form of marginally unstable roll instability associated with the intense frontal vertical wind shear within the cloud head. Thunderstorms are also a fairly common occurrence close to the bent-back front, especially at times of rapid cyclogenesis because this is within a region characterized by strong potential instability where the intrusion of dry air with low θ_w overruns warm-conveyor-belt air.

4. The Dry Intrusion

4.1 Upper-Level Origin of the Dry Air and the Overrunning Process

The notion of the dry intrusion as a key part of the cyclone structure responsible for the satellite-detected dry slot has already been introduced in Section 2.2 above, and is embodied in the conceptual model in Fig. 4. The source

of the dry-intrusion air is to be found upwind of the cyclone in the upper troposphere and lower stratosphere in the region of a tropopause fold. The descent of such air in association with tropopause folding is well known following the work of Reed (1955) and Danielsen (1964). Danielsen's three-dimensional visualization is reproduced in Fig. 9, which shows that the dry airstream cannot be approximated as a narrow belt-like flow; rather the upper-level air fans out over a large area as it enters the lower troposphere. Only part of this dry airstream enters the region of the dry slot in the form of a dry intrusion, as shown in Fig. 4. There is some ambiguity over the use of the term dry intrusion: sometimes it is used in a broad sense to describe a large part of the dry airstream as detected in satellite water vapor imagery well upstream of the cyclone. At other times, as here, the term is used specifically to describe the portion of the dry airstream that actually intrudes into the dry slot close to the cyclone center.

The pattern of relative flow in Fig. 9 looks deceptively simple but it has been carefully drawn to reveal several interesting facets of the overall dry airstream, part of which feeds the dry intrusion. Starting on the left side of Fig. 9, some of the dry air, upon reaching the lower troposphere, gets left behind and flows rearward ahead of the warm front associated with the next cyclone in the sequence. Such dry air is often encountered in and beneath a warm frontal zone; it is moistened by the evaporation of precipitation falling into it and it may develop into the next cyclone's cold conveyor belt.

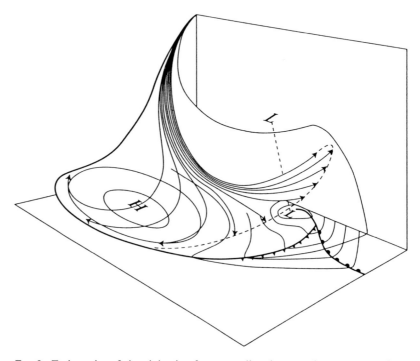

FIG. 9. Trajectories of air originating from a small region near the tropopause, drawn within a curved isentropic surface. The surface is shown sloping from a position near a tropopause fold down to the ground. Trajectories come close to the ground in the left part of the diagram but not in the right-hand part, where they overrun the surface fronts (Danielsen 1964).

In the middle portion of Fig. 9, in the region of the first cyclone's trailing cold front, the dry airstream often approaches the ground as the descending limb of an ana-cold frontal circulation (see Fig. 8.8 of Browning 1990). The portion of the dry airstream that becomes the dry intrusion is the set of five arrows on the right-hand side of Fig. 9, shown overrunning the surface fronts. This air is often very dry because of its history of descent but by the time it gets to this position it is likely to be re-ascending within the dry slot.

Wernli (1995) has identified a diffluent airflow pattern similar to that in Fig. 9 using a new method of objective trajectory analysis. He showed that the parts of the dry airstream that turn cyclonically and enter the dry slot emanate from the stratospheric side of a tropopause fold, whereas the parts of the dry airstream that turn anticyclonically toward the south emanate from the tropospheric side of the tropopause fold.

Figure 10 is a cross section along the axis of the dry intrusion/dry slot shown in Fig. 7. It shows that the actual cold (θ) frontal zone aloft is located well behind (rather than directly beneath) the upper cold θ_w-front. The cold θ-front is a dynamically induced phenomenon affected by differential subsidence of the upper-level dry air. The cold θ_w-front is the true front in airmass terms and is the more significant front in terms of weather. Given sufficient local lifting, the cold θ_w-front also manifests itself as a cold θ-front. Upper cold θ_w-fronts are ubiquitous features of developing cyclones and have been variously described as moisture fronts (Mass and Schultz 1993) and cold fronts aloft (Hobbs et al. 1990), as well as upper cold fronts. There is a tendency in the literature to emphasize the differences between these fronts, but we would prefer to stress the key factor they have in common, namely the role of the dry intrusion in generating what is strictly an upper cold θ_w-front. With this amended terminology, the observations conform to the simple split-front model of Browning and Monk (1982). The word "split" in this model refers to the upper part of the cold θ_w-front running ahead of the surface cold front. The overrunning of low θ_w air behind the upper cold front generates potential instability, which sometimes leads to thunderstorms in this region, as discussed in Section 5.

4.2   The Association of High Potential Vorticity with the Dry Intrusion

Since the early work of Reed and Danielsen it has been known that part of the dry airstream that descends from near tropopause level during tropopause-folding events brings with it values of potential vorticity (PV) characteristic of the lower

stratosphere. The output from operational forecast models shows clearly how, during the development of vigorous cyclones, the dry intrusions sometimes contain air with PV of 1.5 to 2 or more PV units (e.g., Uccellini et al. 1985; Young et al. 1987). A value of about 2 PV units is regarded as defining the dynamical tropopause; broadly speaking, it discriminates the generally high-PV stratospheric air from the generally low-PV tropospheric air. Admittedly, there are regions of high PV generated in situ within the troposphere by diabatic effects, but these are within the moist air and, given NWP model output of sufficient resolution, they can be distinguished from the high-PV air in the dry intrusion. As explained by Hoskins et al. (1985), the overrunning of such high-PV air above a low-level baroclinic zone leads to cyclogenesis and so it comes as no surprise that dry intrusions/dry slots are most in evidence at the stage of rapid cyclogenesis.

The example chosen to illustrate the occurrence of high PV within a dry intrusion corresponds to Young's "diffluent flow" subcategory of frontal cyclones (see Fig. 5). The results for this case, shown in Figs. 11 to 13, reveal a strong mesoscale substructure: a single dry intrusion splits into two smaller ones and each becomes associated with an identifiable dry slot and notable cyclonic events. Figure 11 shows

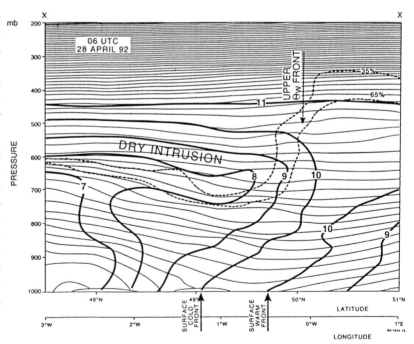

FIG. 10. Vertical section along XX in Fig. 7 showing θ_w (thick solid lines at 1°C intervals), θ (thin solid lines at 1°C intervals), and relative humidity (dashed lines for 65% and 35%), as obtained from a 12-h mesoscale model forecast with assimilation of soundings from a mesoscale array of drop-windsondes (courtesy of S. Ballard and C. Davitt). Note that the surface cold front, marking the leading edge of a diffuse θ_w-gradient is overrun by dry-intrusion air with low θ_w, which gives an upper θ_w-front more than 100 km ahead of the surface cold front. An upper cold θ-frontal zone does not accompany the upper θ_w-front here but, rather, lags behind.

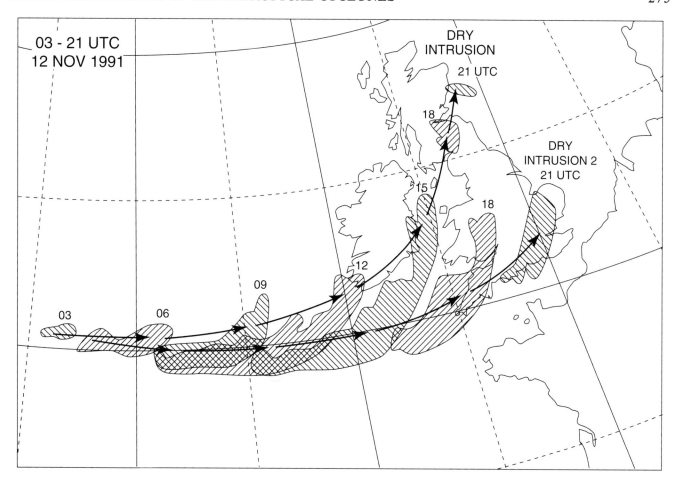

FIG. 11. Tracings of the "dark zone" in Meteosat water vapor imagery at 3-h intervals, showing the tracks of two pockets of dry-intrusion air (Browning and Golding 1995).

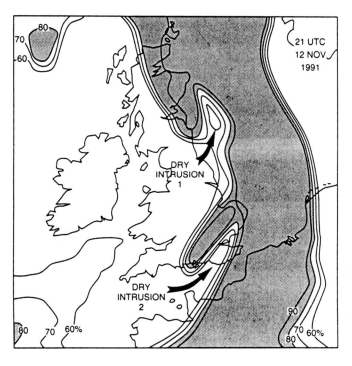

successive 3-hour positions of the dry slot seen in WV imagery. The dry slot became elongated within a strong deformation zone before splitting into two. One of the resulting dry slots then traveled northeastward across Scotland in association with a small rapidly deepening cyclone. The other traveled eastward accompanied by a sharp trough and tornadic squall line. The relative humidity pattern at 3 km (Fig. 12), obtained from a mesoscale model, shows mesoscale dry intrusions associated with both of the satellite-detected dry slots. Time sequences of model diagnostics on an appropriate isentropic surface showed that the region of dry air and collocated high PV associated with one of the dry

FIG. 12 (left). Mesoscale model forecast for the case shown in Fig. 11, at 2100 UTC, initialized from operational limited-area model fields at 1200 UTC, showing relative humidity at 3 km above mean sea level (MSL). Contours are for 60, 70, 80, and 90%. Areas in excess of 80% are shaded. The two dry intrusions seen here are also evident in the satellite imagery sketched in Fig. 11 (Browning and Golding 1995).

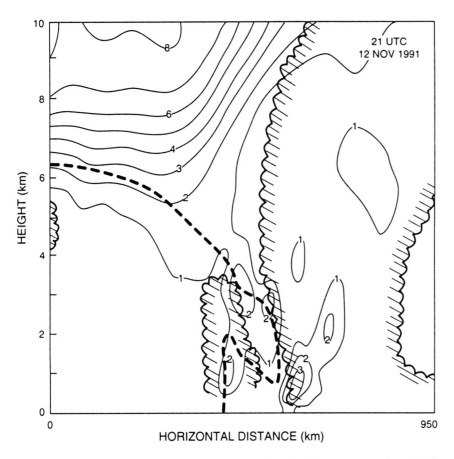

FIG. 13. Forecast cross-section along the south coast of England, from a mesoscale model, for the case shown in Figs. 11 and 12, depicting potential vorticity (solid contours at intervals of 1 PV unit), the $\theta_w = 7C$ isopleth (dashed line), and the cloud boundary (scalloped) (Browning and Golding 1995).

5. Mesoscale Substructure of Surface Cold Fronts in the Region of the Dry Intrusion

This section discusses the kata-cold fronts that occur, often close to the cyclone center, where the low-θ_w dry-intrusion air overruns the warm-conveyor-belt flow W2 as in the split-front model of Browning and Monk (1982). We examine the structure of the surface cold frontal part of the split cold front, and we show that the surface cold front may itself be characterized by multiple structures. The substructure is revealed using a combination of satellite data and soundings from a mesoscale array of about 60 dropwindsondes. The sondes were released during the FRONTS 92 experiment in the cold-frontal region of a developing frontal cyclone over the eastern Atlantic.

The satellite picture in Fig. 14 shows a cloud head separated from a polar-front cloud band by a small dry slot. The distinction between these two cloud features is actually clearer in the infrared image (not shown) from which the tops in the cloud head can be seen to be lower than those in the polar-front cloud band. At this early stage of cyclogenesis the dry intrusion had not yet penetrated fully into the system, and this is why the dry slot was not yet well formed. The discussion here focuses on the region of low-level cloud lines near the dry slot to the southwest of the cloud head. According to the detailed analysis in Browning et al. (1995), they were associated with two surface cold fronts (labeled CF1 and CF2) plus other similar but weaker cold frontal substructure: we now summarize some of their results.

Figure 15 (a, b, and c) shows relative-flow analyses within isentropic surfaces ($\theta = 20C$, $\theta = 10C$, and $\theta_w = 7C$) behind and within the cloud head. The cloud head (stippled) is shown in each of the panels, together with surface positions of CF1 and/or CF2. The right-hand panel (c) shows a flow characterizing the cloud head. The left-hand panels show flow in the dry air just above (top panel (a)) and just beneath (bottom panel (b)) the shallow southwestern end of the cloud head. The north-northeasterly relative flow beneath the cloud head descends to the surface behind CF2. The north-northwesterly relative flow just above the cloud head descends close to the surface at CF1; some of it penetrates to the surface (hatched area along CF1) and some overruns it and begins to re-ascend as a dry intrusion.

intrusions descended from about 3 km to 1 km within 6 hours. The fact that high PV is generally associated with dry intrusions has led Mansfield (1994) to recommend evaluating NWP model performance by comparing model-derived PV patterns with WV imagery.

The vertical section in Fig. 13 shows that the dry intrusion in this case was associated with a narrow filament of high PV extending down from a tropopause fold at 6 km. Parts of the dry intrusion descending below 3 km had a relative humidity less than 30 percent and a PV as high as 2 PV units. The moist air mass into which the dry intrusion was penetrating also had pockets of high PV, owing to the locally generated diabatic effects; however, continuity considerations made it possible to attribute the PV features to the appropriate level of origin. Although low-θ_w air in the dry intrusion was overrunning high-θ_w air in the lowest kilometer and leading to intense convection, Fig. 13 shows that it was also partly undercutting the deep mass of cloudy high-θ_w air associated with the warm conveyor belt. This often happens, in which case the satellite imagery is not capable of detecting the full extent of the dry intrusion.

Fig. 14. Meteosat visible image for the case analyzed in Fig. 15, showing two cold frontal cloud bands, CF1 and CF2, extending southwestward from the tip of a cloud head. The dark zone between CF1 and CF2 is the dry slot (Browning et al. 1995).

Fig. 15. System-relative flows for the case shown in Fig 14, derived from dropwindsondes. The flows represent: (a) dry air in the $\theta = 20C$ surface mainly behind CF1, (b) dry air in the $\theta = 10C$ surface behind CF2, and (c) moist air in the $\theta_w = 7C$ surface within the cloud head. Part of the flow in (c) intrudes between the flows in (a) and (b). Dashed lines show the heights of these surfaces. Dotted lines show relative humidity. Also shown are the position of the cyclone center (L) and extent of moist air (RH > 90%) above 1.5 km associated with the cloud head (stippled shading). The 0.5 km base of a shallow layer of dry air beneath the warm frontal zone is denoted by the dot-dashed isopleth in (c). Another region, along CF1, where dry air penetrated downward as far as 0.5 km is shown in (a) by hatched shading. AB in (c) shows the line of the cross-section in Fig. 16 (Browning et al. 1995).

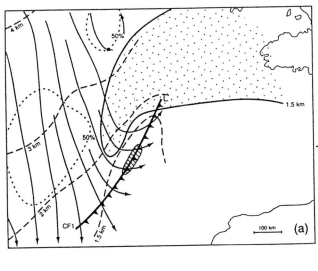

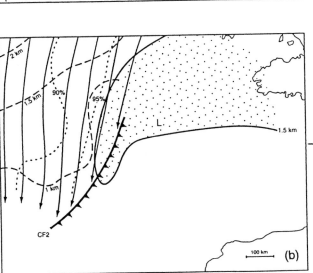

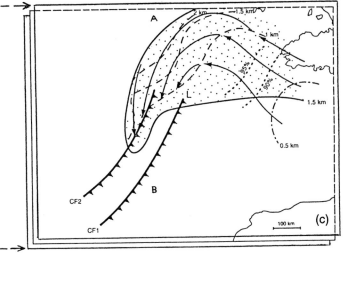

Figure 16 (a and b) shows a vertical section of relative humidity and θ_w constructed from 17 dropsondes along AB in Fig. 15(c). The broad-scale pattern is of dry, low-θ_w air overrunning a moist boundary layer and penetrating toward the surface in places, notably just behind CF1 and CF2. The satellite water vapor imagery showed two zones of dry air plunging downward, as shown schematically by the two bold arrows in Fig. 16(b). The leading edge of one of these corresponds to the $\theta = 10C$ flow undercutting the cloud head; the other corresponds to the $\theta = 20C$ flow overrunning the southwestern tip of the cloud head (cf. Fig. 15(a)). At the top of the boundary layer, mesoscale circulations associated with each of these dry intrusions distort the strong vertical gradient of humidity to produce fingers of dry and moist air (dry axes are shown dashed in Fig. 16 (a)). Actually these fingers were extensive into the plane of the diagram and should be regarded as being interpenetrating dry and moist laminae, with a front-normal slope of 1 in 60. Although the water vapor imagery could resolve only two distinct dry zones, each of these led to a multiply-laminated mesoscale structure, as evidenced by the multiple dashed lines in Fig. 16(a).

Corresponding fine structure existed in the wind field, and Fig. 16(c) shows the lamination in vertical shear of the front-parallel component which, in view of the geometry of these features, can be used as a proxy for (vertical) vorticity. It is possible that these vortex sheets or shear layers were due to preexisting unresolved mesoscale blobbiness in the pattern of vorticity at tropopause level, which was then stretched out into thin laminae by the deformation field as the dry air descended. Alternatively they may have been due to mesoscale instabilities such as conditional symmetric instability developing where the descending dry air encountered the moist boundary-layer air. Figure 16(c) suggests that both mechanisms could have been occurring: The shear layer extending down from the top left corner existed at high levels, where it was entirely within the dry air, as well as extending to lower levels, where it seems to have perturbed the interface between moist and dry air. The other shear layers in Fig. 16(c)

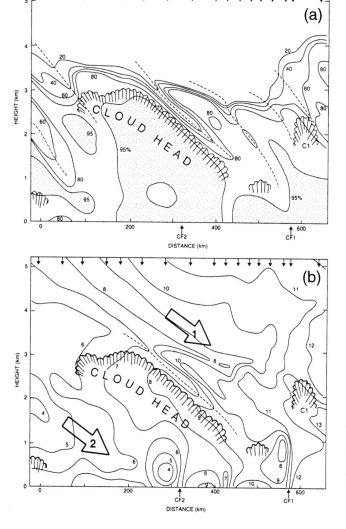

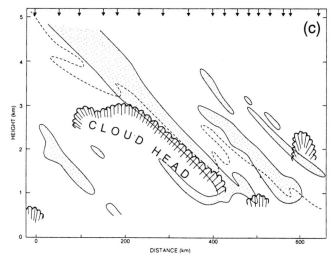

FIG. 16. Cross-section along AB (see Fig. 15(c)) derived from dropwindsondes (locations at tops of figures) showing isopleths of (a) relative humidity (shaded above 95%), (b) wet-bulb potential temperature (isopleths at 1C intervals), and (c) vertical shear in the component of the wind resolved parallel to the cold fronts (stippled laminae show shear greater than 1 ms^{-1} per 100 m). Cloud tops inferred from satellite imagery are shown in all three figures. The two large arrows labeled 1 and 2 in (b) draw attention to the two dry intrusions. The dashed curve in (c) shows a critical level (see text for definition) (Browning et al. 1995).

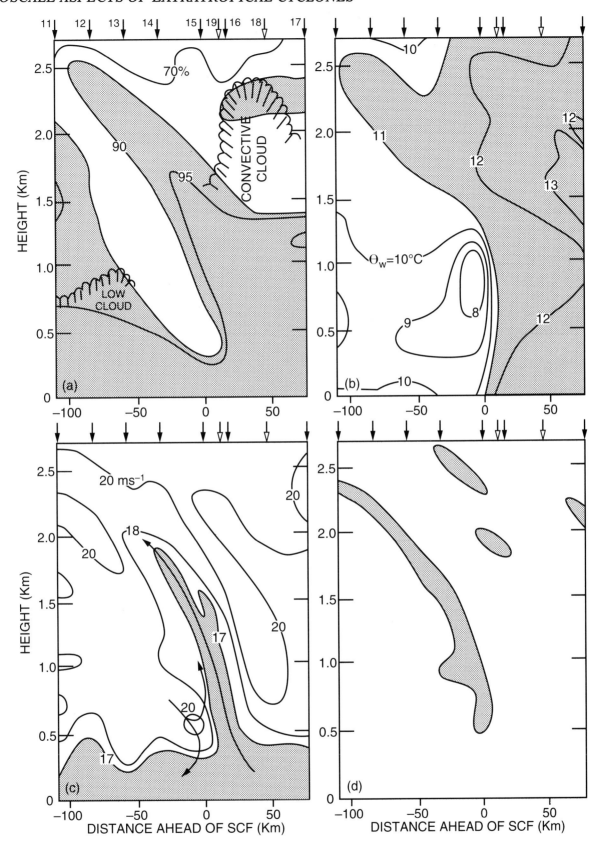

FIG. 17. Cross-section through surface cold front CF1 along the southern part of AB in Fig 15(c) based on 9 dropwindsondes (see arrows at top of diagrams). The four panels show: (a) relative humidity (isopleths at 70, 90, and 95%, shaded above 90%) and cloud tops (derived from satellite), (b) wet-bulb potential temperature, θ_w (isopleths at 1C intervals, shaded above 11C), (c) front-normal wind component (isopleths at 17 m s^{-1}, 18 m s^{-1}, and 20 m s^{-1}, shaded below 17 m s^{-1}; velocity of front was between 17 m s^{-1} and 18 m s^{-1}), and (d) Richardson number less than 0.25 measured over 20 mb layers (Browning 1995).

all appear to be closely confined to the vicinity of the interface between the dry descending air and the moist boundary-layer air. It may be significant that these laminations occur in the vicinity of a critical level (dashed curve in Fig. 16(c)) where the component of the wind resolved along the direction of travel of the cyclone system was equal to the system velocity.

A detailed analysis of dropwindsonde data within a small area at the bottom right-hand corner of Fig. 16 has been carried out to reveal the fine structure associated with one of the surface cold fronts (CF1). Figures 17(a)–(c) show the resulting analyses of (a) relative humidity and cloud, (b) wet-bulb potential temperature and (c) transverse-front wind-component. The shaded areas in each panel draw attention to the narrow lamina of rearward-sloping moist (but mainly subsaturated) ascent just ahead of a corresponding lamina of forward-sloping dry (low-θ_w) descent. Figure 17(d) shows that the interface between the two laminae corresponded to a shear layer characterized by a subcritical Richardson number indicative of ongoing shear-induced turbulent mixing. The surface position of the cold front occurred where this shear layer penetrated to the ground, immediately ahead of which it triggered a band of shallow convective cloud. This convection can be seen in Fig. 17(a) to have risen to a level between 2 km and 2.5 km, where there was a weak flow overtaking the front, which carried the cloud debris ahead of the surface front.

The front in Fig. 17 was a kata-cold front (Bergeron 1937), but this detailed analysis has shown that even at a kata-cold front, with dry-intrusion air overrunning it aloft, there may still be a hint of a small-scale ana-cold frontal circulation. Unlike a fully developed classical ana-cold front, where air is peeled out of the boundary layer and ascends as a major precipitation-producing flow into the middle or upper troposphere, the slantwise ana-cold frontal ascent in Fig. 17 is weak, shallow, and, in this case, largely cloud-free. Without a very dense line of dropwindsondes it would have been unresolvable. Nevertheless, it did exist and its very existence is thought-provoking. The writer used to think of kata- and ana-fronts as being quite distinct. Now it appears that pure kata-cold fronts and pure ana-cold fronts are the extremes of a continuous spectrum of structures, as sketched in Fig. 18. The early development of a cloud head may perhaps be interpreted in terms of a progression from (a) to (c) in Fig. 18 due to the upward extrusion of boundary-layer air. This process may be initiated by the approaching region of high potential vorticity originating at upper levels.

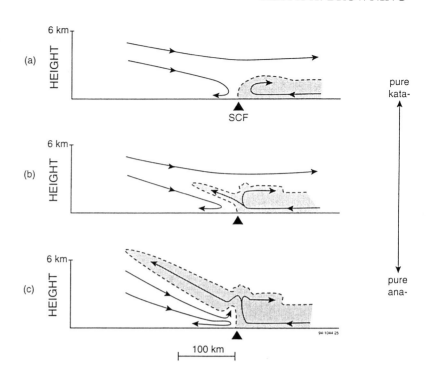

FIG. 18. Schematic cross-sectional representations of the spectrum of cold-front types, ranging from a pure kata-cold front to a pure ana-cold front. Arrows show transverse flow relative to the front. Stippled shading bounded by dashed contour represents boundary-layer or extruded boundary-layer air.

6. Outbreaks of Convection Associated with the Dry Intrusion

Section 4 described how the dry intrusion, as it advances ahead of the center of a developing cyclone, produces a dry slot detectable in the satellite imagery. This region is characterized by low-θ_w air overrunning a shallow moist zone characterized by high θ_w, thereby leading to potential instability. The air, as it enters the region from the west, is usually convectively stable above the boundary layer because of the adiabatic warming of the dry-intrusion air during its earlier descent, but within the dry slot it begins to ascend, along with the moist higher-θ_w air beneath, and eventually the potential instability may be realized in the form of convection. This is especially likely in association with a rapidly developing cyclone where the lifting is strong and/or, over certain continental (oceanic) areas in summer (winter), when there is substantial input of heat at the surface. The purpose of this section is to examine the location and nature of the resulting convection.

Figure 19 is an adaptation of the results of a composite study by Carr and Millard (1985) showing the location of severe convection-related weather events in relation to the pattern of cloud associated with a mature cyclone. A key

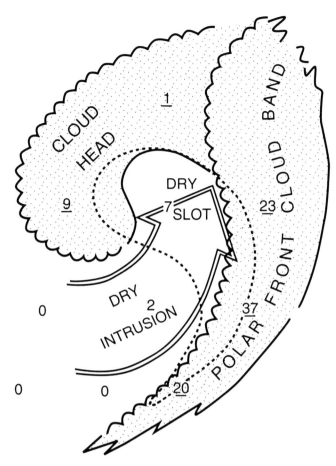

FIG. 19. Occurrence of convection-related severe weather events in relation to the cloud pattern associated with a cyclone. The numbers represent percentage of events in the central United States occurring within subzones in the main cloud areas (underlined numbers) or developing within subzones of the dry slot (numbers not underlined) (from Carr and Millard 1985). Most of the underlined events occurred near the inner cloud boundary and so we have added the dashed line to show the main region of strong convection.

feature of their study was the identification of the dry intrusion (arrow in Fig. 19) that was responsible for the satellite-detected dry slot within a large comma-shaped cloud system. We interpret the overall comma cloud here as being the combination of cloud head and polar-front cloud band as described above, but a clear distinction between these two cloud features does not always exist. The numbers plotted in Fig. 19 represent the percentage occurrence of severe weather events averaged over ten large zones. According to Carr and Millard, the high number of events associated with the polar-front cloud band occurred mainly close to the boundary with the dry intrusion. Thus, our estimate of the area most prone to severe convection is shown by the dashed envelope in Fig. 19. About 7 percent of the convection occurred within the dry slot itself; the rest occurred at its boundaries. Many convective lines were found to form in the dry slot and then to move eastward into (or to redefine the western edge of) the polar-front cloud band as they became severe.

A visualization by Hobbs et al. (1990) of the outbreak of such a line of deep convection ahead of a dry intrusion is shown in Fig. 20. The situation depicted here is essentially the same overrunning situation that is represented in the split-front model (Browning and Monk 1982) but the surface feature is referred to as a surface trough rather than a surface cold front. These surface features are often a source of confusion: in northwestern Europe they tend to have a small temperature drop and a somewhat larger drop in dew point but local conditions can lead to a temperature rise behind the surface front/trough.

The study summarized in Fig. 19 was of deep convection from the surface. Often in northwestern Europe convection associated with the dry intrusion is limited to a shallow mid-tropospheric layer. Figure 21 depicts a sounding close to a line of convective clouds along an upper cold front at the

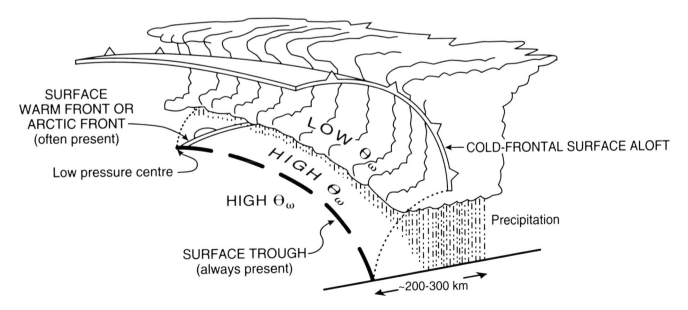

FIG. 20. Schematic of a cold front aloft with a line of deep cumulonimbus situated ahead of a surface trough (Hobbs et al. 1990).

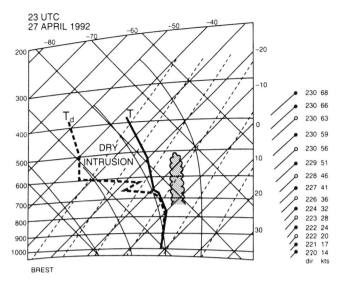

FIG. 21. Sounding at the southern end of a line of showers at an upper cold front, showing the convective instability due to the overrunning dry intrusion and the vertical extent of the observed convection (adapted from Browning et al. 1995).

leading edge of a dry intrusion. It shows that the convection was associated with the buoyant ascent of air parcels from the 800–700 mb layer. Midlevel convection frequently occurs where the dry intrusion overruns the bent-back front or warm front, in which case it can be associated with mesoscale gravity-wave activity. This is often observed in the United States (Uccellini and Koch 1987).

Convection breaking out within the dry slot has a tendency to form in lines. Occasionally, especially in the central United States, these become intense squall lines. In maritime parts of northwestern Europe the convective cloud lines are usually of more moderate intensity. The tendency for the dry-slot convection to form in parallel lines may be due to forcing by preexisting multiple vortex sheets (shear layers) within the dry-intrusion air, similar to those shown in Section 5 to be responsible for multiple bandedness in the vicinity of the surface cold front. Similar interpretations may apply to the multiple bandedness and hyperfine baroclinic zones observed aloft in the main frontal cloud band by Kreitzberg (1968) and many others since.

7. Concluding Remarks

A theme that emerges from this review is the major role played by the dry intrusion. The dry intrusion consists of air with low wet-bulb potential temperature (θ_w) brought down from near tropopause level. Some of this air, originating just beneath the tropopause, has low potential vorticity and some of it, originating just above the tropopause, has high PV. The effect of the high-PV air, especially where it overruns a low-level baroclinic zone, is to trigger (or enhance) cyclogenesis

and ascent. The pockets of low-PV air may lead to other forms of instability. The effect of the overrunning of low-θ_w air is to develop potential instability that is realized by the ascent.

There is evidence of significant substructure within the dry-intrusion air. Small regions of locally higher PV in the dry intrusion organize the cloud and precipitation on the mesoscale. Progress in understanding these processes requires a combination of further observational, modeling, and theoretical studies. There is currently a dearth of observations on the mesoscale, especially in the dry air where at present the only effective method of resolving the fine-scale dynamic and thermodynamic structure is to release closely spaced sondes from an aircraft.

The importance of upper-level PV anomalies for cyclogenesis is already well established: what is still a matter of debate is the relative importance of upper-level and lower-level factors to the evolution of the weather systems. There is also evidence (e.g., Browning and Reynolds 1994; Browning and Golding 1995) that severe surface winds and precipitation events associated with convection are closely related to the descent of mesoscale dry-intrusion features with high PV and low θ_w. If further research shows that this association has some generality, it will signify an opportunity for improving operational short period forecasts of hazardous weather. To realize such an opportunity, additional observations would be needed, but this is an era when all the pressures are towards reducing such costs. However, it may be possible to develop a focused and cost-effective observing strategy for meeting the requirement. As pointed out by Emanuel et al. (1995), the strategy of targeting extra observations may become attractive if adjoint techniques fulfill their promise in revealing, in near real time, those parts of the atmosphere that are particularly susceptible to initial error and would thus benefit from additional observations.

Acknowledgments

I am grateful to Mike Bader, Sid Clough, Roger Grant, Tim Hewson, and Martin Young for helpful comments on an earlier draft.

References

Bader, M. J., G. S. Forbes, J. R. Grant, R. B. E. Lilley, and A. J. Waters (Eds.) 1995: *Images in Weather Forecasting*. Cambridge University Press. 499 pp.

Bergeron, T., 1937: On the physics of fronts. *Bull. Amer. Meteor. Soc.*, **18**, 265–275.

Böttger, H., M. Eckardt, and U. Katergiannakis, 1975: Forecasting extratropical storms with hurricane intensity using satellite information. *J. Appl. Meteor.*, **14**, 1259–1265.

Browning, K. A., 1990: Organization of clouds and precipitation in extratropical cyclones. *Extratropical Cyclones*, Erik Palmén Memorial Volume, C. W. Newton and E. O. Holopainen, Eds., Amer. Meteor. Soc., 129–153.

—— 1994: Life cycle of a frontal cyclone. *Met. Apps.*, **1**, 233–235.

—— 1995: On the nature of the mesoscale circulations at a kata-cold front. *Tellus*, **47A**, 911–919.

——, and G. A. Monk, 1982: A simple model for the synoptic analysis of cold fronts. *Quart. J. Roy. Meteor. Soc.*, **108**, 435–452.

——, and F. F. Hill, 1985: Mesoscale analysis of a polar trough intersecting with a polar front. *Quart. J. Roy. Meteor. Soc.*, **111**, 445–462.

—— and R. Reynolds, 1994: Diagnostic study of a narrow cold-frontal rainband and severe winds associated with a stratospheric intrusion. *Quart. J. Roy. Meteor. Soc.*, **120**, 235–257.

—— and N. M. Roberts, 1994: Structure of a frontal cyclone. *Quart. J. Roy. Meteor. Soc.*, **120**, 1535–1557.

——, and B. W. Golding, 1995: Mesoscale aspects of a dry intrusion within a vigorous cyclone. *Quart. J. Roy. Meteor. Soc.*, **121**, 463–493.

——, S. A. Clough, C. S. A. Davitt, N. M. Roberts, T. D. Hewson, and P. G. W. Healey, 1995: Observations of the mesoscale substructure in the cold air of a developing frontal cyclone. *Quart. J. Roy. Meteor. Soc.*, **121**, 1229–1254.

Carbone, R. E., 1982: A severe frontal rainband. Part I: Stormwide hydrodynamic structure. *J. Atmos. Sci.*, **39**, 258–279.

Carlson, T. N., 1980: Airflow through midlatitude cyclones and the comma cloud pattern. *Mon. Wea. Rev.*, **108**, 1498–1509.

Carr, F. H., and J. P. Millard, 1985: A composite study of comma clouds and their association with severe weather over the Great Plains, *Mon. Wea. Rev.*, **113**, 370–387.

Danielsen, E. F., 1964: Project Springfield Report. Defense Atomic Support Agency, DASA 1517, 97 pp. [NTIS # AD-607980.]

Emanuel, K., and many others, 1995: Report of the first Prospectus Development Team of the U.S. Weather Research Program to NOAA and the NSF. *Bull. Amer. Meteor. Soc.*, **76**, 1194–1208.

Evans, S. M., D. Keyser, L. F. Bosart, and G. M. Lackmann, 1994: A satellite-derived classification scheme for rapid maritime cyclogenesis. *Mon. Wea. Rev.*, **122**, 1381–1416.

Harrold, T. W., 1973: Mechanisms influencing the distribution of precipitation within baroclinic disturbances. *Quart. J. Roy. Meteor. Soc.*, **99**, 232–251.

Hobbs, P. V., J. D. Locatelli, and J. E. Martin, 1990: Cold fronts aloft and the forecasting of precipitation and severe weather east of the Rocky Mountains. *Wea. Forecasting*, **5**, 613–626.

Hobbs, P. V., and P. O. G. Persson, 1982: The mesoscale and microscale structure and organization of clouds and precipitation in midlatitude cyclones. Part V: The substructure of narrow cold-frontal rainbands. *J. Atmos. Sci.*, **39**, 280–295.

Hoskins, B. J., M. E. McIntyre, and A. W. Robertson, 1985: On the use and significance of isentropic potential vorticity maps. *Quart. J. Roy. Meteor. Soc.*, **111**, 877–946.

Kreitzberg, C. W., 1968: The mesoscale wind field in an occlusion. *J. Appl. Meteor.*, **7**, 53–67.

Kuo, Y.-H., R. J. Reed, and S. Low-Nam, 1992: Thermal structure and airflow in a model simulation of an occluded marine cyclone, *Mon. Wea. Rev.*, **120**, 2280–2297.

Letestu, A.-C., 1994: Case study analysis of warm conveyor belt and low level jet phenomena. Ph.D. thesis, University of Reading, 210 pp.

McGinnigle, J. B., M. V. Young, and M. J. Bader, 1988: The development of instant occlusions in the North Atlantic. *Meteor. Mag.*, **117**, 325–341.

Mansfield, D., 1994: The use of potential vorticity in forecasting cyclones: operational aspects. *The Life Cycles of Extratropical Cyclones, Vol. III*, Bergen Symposium, 27 June–1 July 1994, 326–331.

Mass, C. F., and D.,M. Schultz, 1993: The structure and evolution of a simulated midlatitude cyclone over land. *Mon. Wea. Rev.*, **121**, 889–917.

Monk, G. A., and M. J. Bader, 1988: Satellite images showing the development of the storm of 15–16 October 1987. *Weather*, **43**, 130–135.

Reed, R. J., 1955: A study of a characteristic type of upper level frontogenesis. *J. Meteor.*, **12**, 226–237.

Rodgers, D. M., M. J. Magnano, and J. H. Arns, 1985: Mesoscale convective complexes over the US and during 1983. *Mon. Wea. Rev.*, **113**, 888–901.

Shapiro, M. A., and D. Keyser, 1990: Fronts, jet streams and the tropopause. *Extratropical Cyclones, The Erik Palmén Memorial Volume*, C. W. Newton and E. O. Holopainen, Eds., Amer. Meteor. Soc., 167–191.

Shutts, G. J., 1990: Dynamical aspects of the October storm, 1987: A study of a successful fine-mesh simulation. *Quart. J. Roy. Meteor. Soc.*, **116**, 1315–1347.

Uccellini, L. W., and S. E. Koch, 1987: The synoptic setting and possible source mechanisms for mesoscale gravity wave events. *Mon. Wea. Rev.*, **115**, 721–729.

——, D. Keyser, K. F. Brill, and C. H. Wash, 1985: The Presidents' Day cyclone of 18–19 February 1979: Influence of upstream trough amplification and associated tropopause folding on rapid cyclogenesis. *Mon. Wea. Rev.*, **113**, 962–988.

Weldon, R. B., 1977: An oceanic cyclogenesis—its cloud pattern and interpretation. NWS/NESS Satellite Applications Information Note 77/7, 11 pp.

Wernli, H., 1995: Lagrangian perspective of extratropical cyclogenesis. Ph.D. Dissertation No. 11016, Swiss Federal Inst. of Tech., Zürich, 157 pp.

——, and A. M. Rossa, 1994: Identification of structures within extratropical cyclones based upon objective criteria. *The Life Cycles of Extratropical Cyclones, Vol. II*, Bergen Symposium, 27 June–1 July 1994, 548–553.

Young, M. V., 1989: Investigation of a cyclogenesis event, 26–29 July 1988, using satellite imagery and numerical model diagnostics. *Meteor. Mag.*, **118**, 185–196.

——, 1994: A classification scheme for cyclone life-cycles: applications in analysis and short-period forecasting. *The Life Cycles of Extratropical Cyclones, Vol. III*, Bergen Symposium, 27 June–1 July 1994, 380–385.

——, G. A. Monk and K. A. Browning, 1987: Interpretation of satellite imagery of a rapidly deepening cyclone. *Quart. J. Roy. Meteor. Soc.*, **113**, 1089–1115.

Zillman, J. W., and P. G. Price, 1972: On the thermal structure of mature Southern Ocean cyclones. *Aust. Meteor. Mag.*, **20**, 34–48.

Dynamics of Mesoscale Structure Associated with Extratropical Cyclones

ALAN J. THORPE

Department of Meteorology, University of Reading, Reading, United Kingdom

1. Introduction

Extratropical cyclones exhibit a wide variety of subsynoptic-scale organization, which is of importance to weather forecasting as well as being of interest from an academic standpoint. The key mesoscale feature is the presence of fronts. Most, but not all, of the mesoscale structure of extratropical cyclones is associated with the frontal zones. Various mesoscale structures occuring in association with extratropical cyclones can be listed as:

> frontal wave cyclones
> frontal rain and snowbands
> line convection including segmentation
> comma clouds
> tropopause folds

The mathematical theory of frontogenesis (Hoskins 1982) gives a consistent description of the dynamics of two-dimensional frontal zones in which geostrophic balance is imposed on the along-front wind. This semigeostrophic theory filters all inertia-gravity waves and their related instabilities. Despite these restrictions it is able to describe many fundamental dynamical aspects of fronts such as the process of frontogenesis and cross-frontal circulation. The theory predicts discontinuous fronts forming in a finite time and it remains an area of active research interest to understand the physical processes responsible for halting such a frontal collapse. The processes that have been discussed in this context include unbalanced motion (e.g., surface friction, gravity waves, and Kelvin-Helmholtz instability); three-dimensional eddies; and removal of the large-scale frontogenetic flow.

Some of the mesoscale features mentioned above cannot be described by the semigeostrophic theory due to the filtering mentioned previously. Others, such as frontal wave cyclones, are three-dimensional structures, which are clearly not described in the two-dimensional theory but are of a sufficiently large horizontal scale that a balanced flow assumption may be applicable. It is the aim of this review to discuss recent ideas concerning the dynamics of two of these phenomena: frontal wave cyclones and frontal rainbands. These are, respectively, examples of a three-dimensional feature which can probably be described by semigeostrophic theory, and a two-dimensional phenomenon, which cannot be so represented. This review is not a historical survey and so it has to be stressed that recent ideas have built on earlier papers. These include those on the stability of flows typical of fronts such as the pioneering work of Kotschin (1932), Eliasen (1960), and Fjortoft (1950) as well as many papers on fronts by Arnt Eliassen.

It is clear that for the majority of oceanic storm-track cyclones there is a significant role played by cloud processes. Progress in extending the baroclinic instability model of cyclogenesis to include these processes has been made. The Eady model in its archetypal form assumes dry frictionless dynamics. The archetypal moist baroclinic instability model involves ascending air taken to be saturated and furthermore neutral to slantwise ascent (see Section 4.2) and descending air being unsaturated. Such a theory of moist cyclogenesis has been developed by Emanuel et al. (1987). In Fig. 1 the growth rate curve for the Eady problem is shown with various stabilities to slantwise ascent shown including both the classical dry case and the fully moist neutral case. This shows the increase of growth rate and decrease of horizontal scale of the maximum growing unstable cyclone wave due to moist processes. In addition, idealized frontogenesis models have been made more realistic by including latent heat release in this way. The development of mesoscale structure can therefore be discussed within the context of fully moist cyclone and frontal dynamical models.

From a theoretical viewpoint an important framework in which to describe the dynamics of these systems is via what

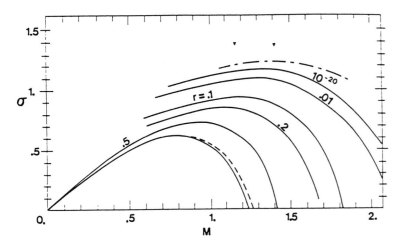

FIG. 1. Growth rate curves shown for the Eady problem with various stabilities to slantwise ascent shown by the r parameter: $r = 1$ is dry and $r = 0$ is neutral to saturated slantwise ascent (taken from Joly and Thorpe 1989). The growth rate is in units of 0.67 days and the wavenumber is in units of (415 km^{-1}).

$$\frac{DPV_e}{Dt} = 0 \text{ if } \boldsymbol{F} = 0 \text{ and } \frac{D\theta_e}{Dt} = 0$$

where θ_e is the equivalent potential temperature and $PV_e = \frac{1}{\rho}\nabla \cdot (\zeta\theta_e)$ is the equivalent potential vorticity.

These equations imply that for frictionless flow in mesoscale systems PV is conserved only in unsaturated air whereas PV_e is also conserved in saturated motion. The lack of conservation of PV in clouds is still significant, however, because it is easily seen that latent heat release will be associated with a dipole of PV anomalies with positive below and negative above the zone of maximum heating. These anomalies will remain in the atmosphere after the latent heating has ceased and attain a degree of balance. For long-lived systems such as fronts, this balance will apply during the lifetime of the clouds themselves. The role of friction, such as that which primarily occurs at the Earth's surface, is to act to produce local PV anomalies. As an illustration, a simple horizontal frictional force in the atmosphere would produce a local horizontally oriented PV dipole. Further discussion of the action of friction in modifying PV at fronts can be found in Cooper et al. (1992). It is interesting to note that often at the surface front both the latent heating and friction act to produce a positive potential vorticity anomaly.

The PV anomalies generated by these physical processes are thought to be important in the dynamics of the cyclone. From a dynamical as well as an observational viewpoint it is crucial, using PV-thinking, to be able to isolate the contribution to the flow and temperature from individual anomalies. This allows insight to be derived as to the relative contributions from latent heat, friction, upper-level forcing, etc. The process by which this can be done is referred to as "attribution" by Bishop and Thorpe (1994c). It relies on the mathematical tools of piecewise PV inversion, as described by Davis and Emanuel (1991), and piecewise omega inversion, as described by Clough et al. (1996). These tools allow the geostrophic and ageostrophic motion attributable to individual features, such as PV anomalies or zones of (omega equation) forcing, to be found. Attribution is an important process in determining the reasons for the development and evolution of extratropical cyclones. It is especially appropriate for mesoscale features, which retain a degree of (nonlinear) balance.

has become known as potential vorticity (PV)–thinking. Here it is pointed out that PV-thinking is not just applicable to geostrophically balanced flows on the synoptic scale. It is now known that even for smaller scale systems such as convection description of the dynamics using PV provides important insight (see, e.g., Raymond 1992). It has been argued that PV-thinking, even in its traditional use for geostrophically-balanced flows, does not give any new insight over-and-above that available from previously used principles not involving PV. There is no doubt that many processes in the physical world can be described equally well using alternative formulations. But the objective is to find the most succinct concept or set of concepts that explain the essential features. For the phenomenon discussed here PV does indeed give this "maximally simple but sufficiently complete" description.

2. Theoretical Concepts

2.1 PV Thinking

For many applications of PV ideas on the mesoscale, then, the flow is neither dry adiabatic nor frictionless so that the PV is not conserved. The PV evolution equation is:

$$\frac{DPV}{Dt} = \frac{1}{\rho}\nabla \cdot (\zeta\dot{\theta} + \theta\nabla \times \boldsymbol{F})$$

where $PV = \frac{1}{\rho}\nabla \cdot (\zeta\theta)$, $\dot{\theta} = \frac{D\theta}{Dt}$ and $\boldsymbol{F}$ is the frictional force. A simplification arises if the flow is saturated adiabatic and frictionless, i.e.,

2.2 Instability

The potential vorticity is also involved in certain instability criteria. These criteria arise by consideration of a balanced stationary basic flow in which an infinitesimal perturbation is implanted. The tendency of the perturbation to grow in time, with constant shape, can then be found mathematically. It is important to remember that the stability of the flow is, in the classical analysis, dependent only on the properties of the basic state. Such stability analyses reveal which aspect of the basic flow indicates its stability. In the case of the two mesoscale substructures to be described here, that property of the basic flow is most succinctly described in terms of its potential vorticity. In other words with a knowledge of the potential vorticity of the flow one can assess its susceptibility to perturbation. This indicates the fundamentality of the PV viewpoint. For a general geostrophically-balanced basic state, instability depends on either the existence of extrema of PV or on the absolute sign of PV relative to the sign of the Coriolis parameter. We postpone discussion of these criteria to the following sections.

In addition to instability criteria, the PV viewpoint gives considerable insight into the interaction of finite amplitude disturbances. For example, the essential dynamics of baroclinic instability can be described by the the vertical interaction of two (Rossby) edge waves occurring on the tropopause and at the surface; see Hoskins et al. (1985) for further details. Similarly, barotropic growth depends on the horizontal interaction of edge waves, as discussed in more detail later in this chapter.

3. Frontal Wave Cyclones

3.1 Observations

As is well-known, the Bergen school formulated a concept of a semipermanent polar front along which cyclone families would form and propagate. From a theoretical viewpoint, a fundamental aspect of this model is the existence of a special environment in which the cyclones form. The observational basis for the Bergen picture was at that time limited. Theoretical ideas subsequent to the development of this model moved attention away from the polar front concept. Baroclinic instability, in its archetypal form, involves an instability of the broad baroclinic climatological zonal flow. The developing cyclones are of synoptic scale and produce, during their life-cycle, zones of stretching and shearing deformation leading to frontogenesis. One can question whether the environment in which observed cyclones develop corresponds better to the Bergen polar front model or to the broad zonal current of classical baroclinic instability.

It is the contention here that frontal zones spawn secondary wave cyclones, which at their conception are typically of a much smaller horizontal scale than the parent cyclone. It is proposed here that these secondary mesoscale systems are the frontal cyclones described so eloquently by Bjerknes and colleagues.

It is worth asking whether there is good observational evidence for these secondaries and by inference for the

Norwegian picture, given improvements in the observational base. Ayrault et al. (1995) give results from a study of an objectively determined evaluation of very-high frequency variability in the Atlantic storm-track. This is in an attempt to differentiate storms into larger longer-lived storms which explain variability on the 2- to 5-day time-scale and the smaller short-lived ones, which occur in the 0.5- to 2-day frequency range. It is this latter category that involves the secondary waves having a horizontal scale of 1000 km or less.

Here we describe the subjective evidence from operational surface charts. Current practice in drawing surface charts over oceanic regions is, in this context, worth commenting on. The bench forecaster has many data available as he or she constructs a surface analysis including NWP first guess fields, satellite imagery, satellite soundings, merchant ship surface and upper-air, air-reps, buoys, etc. This represents an inadequate database but does mean that the forecaster will include structures only if evidence exists.

During the FRONTS 92 experiment, an analysis was performed by Tim Hewson (JCMM, University of Reading) of the occurrence on surface charts of secondary wave cyclones. In keeping with the theoretical requirement to evaluate the environment in which these systems develop and to maintain consistency with standard nomenclature, a classification scheme was used that identifies five types of secondary wave cyclones according to the synoptic environment at the time at which they are first detectable; see the hypothetical surface analysis in Fig. 2.

The charts used were for March, April, and May 1992 at 12 Z in the north Atlantic sector. Evidence of the frontal wave was required on two consecutive charts, and the intermediate 6-hourly charts were checked for consistency. The definition of a frontal wave adopted concerned the local structure of the front: moving along the front a local maximum must be reached in the vorticity and/or the curvature of the vector defining the along-front direction.

It was found that on average one frontal wave appeared somewhere in the region per day and that 50 percent of these were developmental. The number of events of each of the five types in the 3-month period and the percentage that were developmental, were:

   i. cold air trough (15, 67%)
   ii. cold front wave (25, 48%)
  iii. warm front wave (9, 11%)
  iv. col wave (16, 63%)
   v. warm sector wave (9, 11%)

It was found that col waves tended to develop into more intense systems than cold front waves. Furthermore, development was generally more pronounced ahead of a confluent upper trough than ahead of a diffluent upper trough. Col waves account for many major storms in the northeast Atlantic, such as on 16 October 1987, 11 April 1989, 21 January

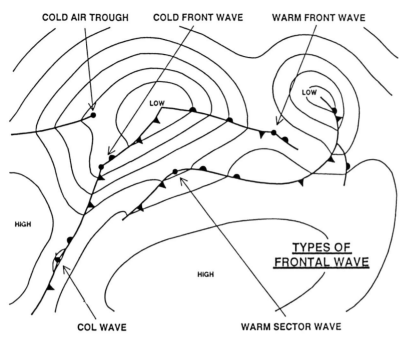

COLD AIR TROUGH   COLD FRONT WAVE   WARM FRONT WAVE

LOW

LOW

HIGH

TYPES OF
FRONTAL WAVE

HIGH

COL WAVE   WARM SECTOR WAVE

FIG. 2. Schematic surface analysis showing the various categories of frontal wave cyclones found by Hewson.

1990, and 2 February 1990. The IOP3 case of FRONTS 92 (Hewson 1993) on 27 April 1992 also falls into this category.

In conclusion there is strong evidence for a class of cyclones that develop on pre-existing tight frontal zones. We refer to these as frontal waves or frontal wave cyclones and often they occur as a secondary development on the fronts formed within a parent cyclone. Such frontal waves consequently occur at the end of the storm track. They are often of a short horizontal scale, putting them in the mesoscale category. They can be developmental but often are not. From a theoretical viewpoint certain critical questions arise from the observational evidence:

- Do these cyclones develop from the same or different mechanisms from the parent cyclones?
- Why are some frontal wave cyclones developmental and others are not?
- What is the importance of upper- versus lower-level forcing?

3.2 Development Mechanisms

The various scenarios for the development of cyclonic storms in the extratropics can be summarized by the interaction of so-called edge waves. These edge waves can exist at a potential vorticity gradient or at a potential temperature gradient as measured along a material surface. Such a material surface might be an isentropic surface or an iso-PV surface, such as the tropopause, or the Earth's surface. Ordinarily an edge wave is neutral, involving no development. A synoptic example might be an upper-level propagating short-wave trough.

Baroclinic and/or barotropic instability involves the mutual interaction of two edge waves allowing wave growth to occur. The term instability is being used here in its loosest sense to mean an increase in amplitude; it is not necessary to make any supposition about constant shape–exponential growth as in the normal mode. Arguments about normal mode versus initial value growth confuse the fact that edge wave interaction is the common factor.

The classic baroclinic instability model of Eady encapsulates the mutual interaction of an edge wave on the tropopause thermal gradient and an edge wave on a surface baroclinic zone. In general, the interaction between two edge waves will occur only if they are close enough to one another. For horizontal interaction this means that they must have a separation less than or of order the horizontal wavelength of the wave. The same applies in the vertical but with the vertical distance having been scaled with N/f. If the effective static stability is low, due to latent heat release, then this effective height separation is considerably reduced.

The archetype for barotropic interaction features two edge waves on the edges of a horizontally oriented PV strip. This has been discussed in the context of a horizontal-shear forced front by Joly and Thorpe (1990). In an equivalent way, horizontal interaction can occur at the ground if there is a extremum of potential temperature (see Schär and Davies 1990).

Here we will explore the vertical and horizontal interaction mechanisms with particular reference to mature fronts as the basic flow environment for the development. Consider a mature frontal zone in which there has been a period of frontogenesis leading to a tight frontal gradient. The role of moisture and friction is to produce a lower-tropospheric strip of high potential vorticity lying along the front. Observational evidence for this strip comes from the FRONTS 87 campaign; see Thorpe and Clough (1991) and Fig. 3 shown here. Theoretical evidence for its existence can be found in Thorpe and Emanuel (1985) (moist processes) and Cooper et al. (1992) (frictional processes).

Given the geostrophic flow and temperature of such a front, one can perform a traditional stability analysis. This requires, for normal modes to be found, that the frontogenesis be assumed to have ceased. This perhaps becomes a plausible assumption for a mature front, but in Section 3.4 we also consider the role of active frontogenesis. In Joly and Thorpe (1990) the basic state chosen was taken from a front produced at a certain stage during a moist two-dimensional Eady wave development (as previously described by Emanuel et al. 1987) assuming that the development has rapidly ceased. The growth rate curve is shown in Fig. 4 with both semigeostrophic

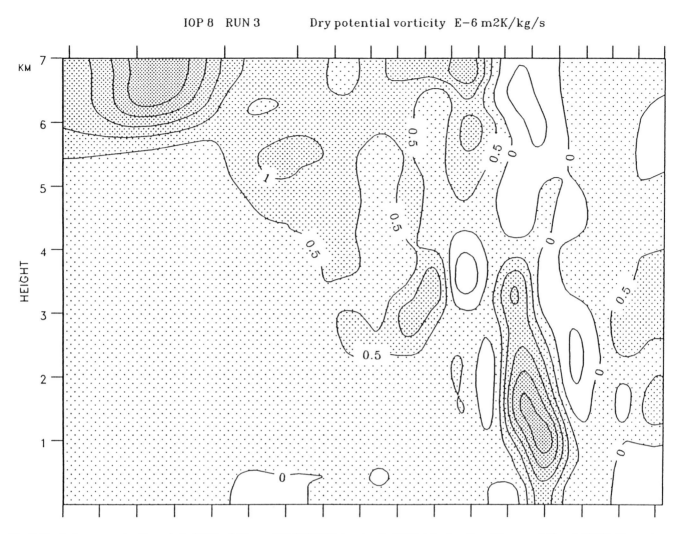

IOP 8   RUN 3   Dry potential vorticity E-6 m2K/kg/s

FIG. 3. Vertical cross section through the IOP8 cold front of FRONTS 87 showing the PV distribution; from Thorpe and Clough (1991). The contour interval is 0.5 PVU; ticks on the horizontal axis are 50 km apart and those on the vertical axis are 1 km apart. The surface cold front is immediately to the west of the lower tropospheric PV maximum indicated by the heavy stippling.

and primitive equation results being shown. The significant increase in the growth rate and (slight) movement of the peak to smaller wavelengths using primitive equations compared to using the geostrophic momentum approximation is discussed in Malardel (1994) and Malardel et al. (1997).

This study showed that a new family of frontal waves exist at wavelengths of order 800 km associated not with upper jets but with lower-level jets. These waves have structure that is more surface confined than classical baroclinic waves. The new modes derive a significant part, but not all, of their energy from the frontal horizontal shear via barotropic conversion. If waves have a greater horizontal scale, then they also use the baroclinic conversion. The key feature of these waves is that they depend on the presence at the front of anomalies of potential vorticity (or equivalently potential temperature). This is, in turn, the result of the effects of diabatic processes in the past history of the front.

The theory predicts that the horizontal scale of the waves

is of order 5 times the width of the frontal PV strip. Given that this width is dictated by the stage in the evolution of the parent front it can assume a wide range of values. For example for a frontal PV band of width 10 km, an along-front horizontal scale of the frontal waves of 50 km would be implied. Observations imply that such frontal waves do indeed have a wide range of scales, which is to some extent consistent with these predictions.

3.3   Nonlinear Structure

The low-level instability of the frontal potential vorticity strip has been studied into the nonlinear regime by Malardel et al. (1993). They found that there is a vortex roll-up, as shown in Fig. 5. The initial line of PV becomes distorted into a series of more or less circular vortices.

There is only a small pressure deepening associated with such development. This is an extremely significant aspect,

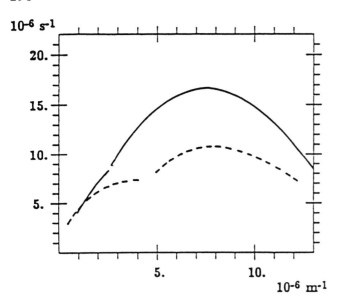

FIG. 4. Growth rate as a function of the along-front wavenumber for a frontal flow exhibiting a low-level potential vorticity maximum. The dashed curve uses semigeostrophic equations and is taken from Joly and Thorpe (1990). The maximum growing wave has a horizontal scale of 800 km with a growth rate of just less than 1 day. The solid line uses primitive equations and is taken from Malardel (1994).

showing that although such instability can account for waves at fronts it cannot typically account for significant development in the pressure field. This finding is consistent with the primarily barotropic nature of the instability. Many frontal wave cyclones are not developmental (about 50% from the subjective climatology quoted earlier) so this finding is relevant. Forecasters are probably not used to looking for such vortices because of their weak signature in the pressure field.

3.4   Upper-Level Forcing

PV ideas suggest that to allow cyclones to draw on the ample supply of potential energy stored at fronts there needs to be significant vertical interaction. This can occur only if the horizontal scale of the wave, L, is sufficiently long; i.e., $L \geq Nh/f$, where h is the vertical separation between the edge waves given, for example, by the distance between the tropopause and the surface. Note that N becomes the effective static stability in regions of saturation. Thorncroft and Hoskins (1990) have given an example in an idealized baroclinic life-cycle simulation of the interaction of an upper-level PV anomaly, associated with a tropopause trough, with a surface front. The resulting isolated frontal wave cyclone achieved a pressure drop of order 10 mb.

This result is significant and shows that one might be able to propose a simple differentiation of frontal wave cyclones into the following two categories:

- type D—developmental, requiring upper-level PV anomalies interacting with a surface front; large space (> 1000 km) and time (> 1 day) scales
- type ND—non-developmental, requiring a lower tropospheric frontal PV anomaly lying along the front; short space (< 1000 km) and time (< 1 day) scales

The term "developmental" is being used here to refer only to the pressure-deepening—type ND can, for example, involve significant increases in across-front wind as the vortices form. The term "ordinary" frontal wave has been coined by Malardel et al. (1993) for type ND. The above constitutes a hypothesis that can be tested from appropriate field observations as well as providing a forecasting rule of some significance.

Extension of these ideas to arbitrary initial-value type perturbations has been studied and reported in Joly (1995). There the point is made that, using the adjoint technique, nonmodal growth is sensitive to the norm chosen for maximization during the growth. It remains unclear at this stage whether these various initial perturbations, which maximize an aspect of the wave development such as kinetic energy growth, are realistic. By this we mean that at least in terms of PV structure, then, one can imagine that these disturbances may arise from mesoscale structures such as convective systems where a PV signature with a particular structure would be expected. On the other hand, disturbances resulting from "observational errors" might be expected to have an arbitrary PV structure. The intriging possibility of targetting observations in the storm-track where there is a known sensitivity to such growing disturbances is being investigated.

3.5   Role of Deformation

For frontal waves growing at a pre-existing front an important question is what effect any larger-scale frontogenetic forcing has on the wave. Frontogenesis (or frontolysis) is due to shearing and/or stretching deformation and, of course, leads to an intensification of the front. For theoretical analysis of frontal waves, this means that the frontal flow, whose stability is being examined, is not stationary in time. Therefore simple normal mode-type instability calculations are not applicable.

In Joly and Thorpe (1991), the case of pure shearing deformation was studied in the context of the Eady basic state. They found that such frontogenesis actually enhanced the instability although much less so than the inclusion of latent heating. The technique used was a version of the so-called linear tangent-resolvent matrix method.

On the other hand, stretching deformation might be expected to have a deleterious effect on incipient waves because of factors like the compression of waves in the across-front direction and stretching in the along-front

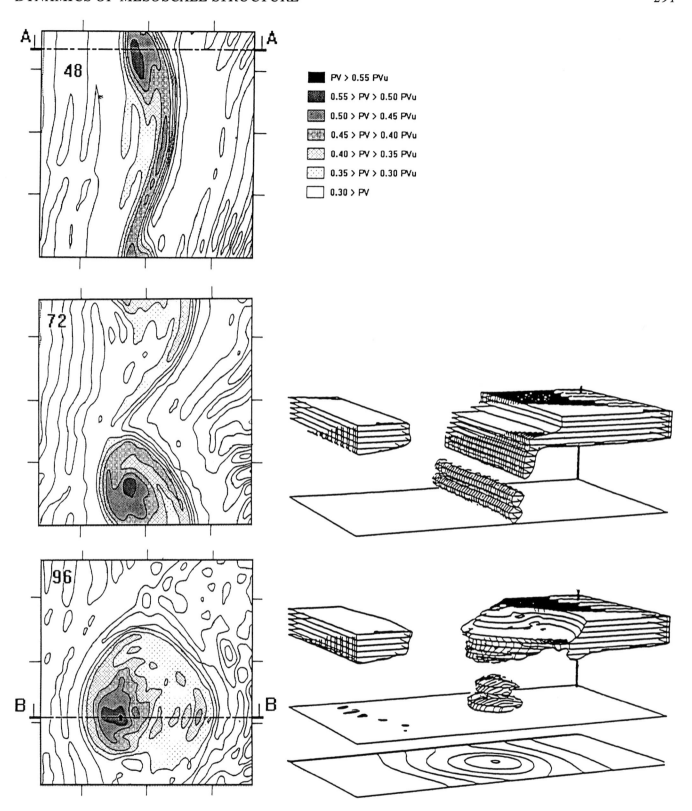

FIG. 5. (a) Evolution of the surface potential vorticity associated with frontal wave instability; from Malardel et al. (1993). (b) A three-dimensional view of the potential vorticity before and after the wave has developed, taken from Malardel (1994). Also plotted is the surface pressure distribution. These figures show the distortion of the initial two-dimensional PV strip into vortices.

direction. Also the cross-frontal ageostrophic circulation plays a significant role in this respect. Therefore, in this case there are no constant shape normal mode solutions possible. This problem has recently been studied by Bishop and Thorpe (1994a, b). They considered a basic state representing a two-dimensional moist front and found that a deformation rate in excess of $f/4$ prevents wave amplification. For strain rates greater than 0.6×10^{-5} s^{-1} the model predicts that wave slope amplification greater than a factor of about 3 is impossible. Hence the theory predicts how small deformation rates must be before frontal wave instability can lead to significant development.

There are two further corollaries from this work. The first is that we need to find the stretching deformation rate at actual fronts in order to assess their propensity for wave development. This is not straightforward and an attempt to do this using the ideas of attribution described in Bishop and Thorpe (1994c) are being made currently; see Renfrew et al. (1997). This involves partitioning the flow into the part due to the front itself and the part due to the frontal environment. This is being done using a Green's function approach. The second is that such deformation acts to increase the "potential" for wave growth while the frontogenesis proceeds. By this we mean that the vorticity and potential vorticity of the frontal zone is increasing in intensity and so that using the linear instability criterion (of course, derived in the absence of the deformation) the front is becoming more unstable. This potential instability can be released later, conceivably explosively, when the deformation ceases. The deformation prevents the waves but also acts to increase the likelihood of instability by narrowing and intensifying the frontal potential vorticity strip. Such potential (dynamical) instability is akin to potential convective instability wherein the atmosphere becomes unstable if a capping inversion can be removed. In fact one can argue that this is the only way the flow can become unstable as without the deformation as soon as the slightest instability is present the unstable circulations would remove it.

4. Frontal Rainbands

4.1 Conditional Symmetric Instability

Observations indicate that frontal precipitation is often organized into a series of bands parallel to fronts, see Matejka et al. (1980) for details. The nature of frontal cloud processes has been studied for some time dating back at least as far as Kleinschmidt's early research (see Thorpe 1993) with ideas of the possibility of an analogue to convective instability. Frontal precipitation is only convective in origin in narrow regions near the surface front and elsewhere the observed vertical stability and low precipitation rates are consistent with a nonconvective origin.

Ascent at fronts is shown in the theory of frontogenesis to be a consequence of frontogenesis itself (see Eliassen (1962) and subsequent research summarized in Hoskins (1982)). Condensation of water vapor in the frontal ascent will naturally lead to precipitation and this will, for example, appear in NWP models as so-called large-scale or dynamic precipitation. So one can ask what is the requirement for further explanation of frontal rain and snowfall. The main answer to this, historically, has been the requirement to explain the banded structure of the clouds as opposed to a broader region of nearly uniform ascent.

It has also been found recently from high-resolution frontal observations that certain structures exist in the wind and temperature fields that do not seem to be accounted for in the simplest theories of frontogenesis. An example of this is the observation that the frontal ascent is often characterized by absolute momentum (m) and equivalent potential temperature surfaces being parallel. The absolute momentum is proportional to the absolute angular momentum on an f-plane and is defined as $m = v + fx$ where v is the along-front wind and x is the across-front coordinate.

The theory of conditional symmetric instability (CSI) dates from the work of Bennetts and Hoskins (1979), although various incomplete descriptions of the mechanisms involved had been given earlier by Solberg, Kleinschmidt, and Sawyer. The theory describes an instability of a two-dimensional geostrophic flow to perturbations oriented along that flow direction. There are various equivalent forms for the criterion that has to be satisfied by the basic state to exhibit the instability:

$$fPV_e < 0, \quad Ri < \frac{f}{\zeta}$$

where ζ is the vorticity measured on a moist isentropic surface and Ri is the Richardson number based on the θ_e vertical gradient and the vertical shear of the basic flow. Note that the term CSI is reserved for motions that are moist convectively stable $\frac{\partial \theta_e}{\partial z} > 0$ and inertially stable $f\zeta > 0$ but that have $fPV_e < 0$.

Parcel theory provides another way to assess instability by assuming that the environment is unaffected as the parcel moves through it. Also the direction must be specified along which the parcel is lifted. The parcel criterion for unstable displacement can be written in a general form involving the anomalies produced between the parcel and its environment as the parcel moves away from its original location:

at neutral buoyancy: $fm'x' > 0$
along momentum surface: $g\theta'z' > 0$

where the prime indicates a difference between the parcel and its environment, and (x', z') is the parcel displacement in the x–z plane. The linear theory suggests that typically parcel ascent is close to neutral bouyancy, that is, along the sloping θ_e surface. These criteria are satisfied if θ_e surfaces are more steeply inclined than momentum surfaces. Note that if a cloud base and top have been established then the amount of so-called slantwise convective available potential energy is independent of the parcel path (see Emanuel 1983).

Evidence for the existence of CSI at fronts is now available from, for example, Emanuel (1988), Thorpe and Clough (1991), and Lagouvardos et al. (1993). Observations often show evidence of momentum anomalies giving a buckled appearance to the momentum surfaces; these are taken as evidence of CSI as CSI roll circulations have been shown to act in this way. In Fig. 6 examples of cross sections from FRONTS 87 from Thorpe and Clough (1991) are given showing the θ_e structure and the accompanying buckled m-surfaces.

4.2 Frontal Rainband Circulations

The rainband circulations deduced within the context of the linear theory are in the form of a series of rolls with alternating zones of ascent and descent sloping almost parallel to the basic state θ_e surfaces. The evolution of these circulations into the nonlinear regime has been described by using numerical simulations (see Thorpe and Rotunno 1989; Persson and Warner 1991; Innocentini and Neto 1992). For rainbands confined to a certain vertical layer, the rolls that involve flow, at the top of the ascent zone, across the basic state isentropes from cold to warm air, are suppressed due to the inertial resistance implied by crossing the basic state momentum surfaces. Therefore at a front the preferred roll circulations of rainbands will be in the thermally direct sense. It is plausible to suppose that such circulations will adjust the atmosphere over a period of time to a near-neutral state such as is commonly observed at fronts. This state involves momentum and θ_e surfaces becoming closely parallel to each other or alternatively $PV_e \approx 0$.

4.3 Three-Dimensional Structure

An important aspect of the practical application of the theory of CSI to fronts is that observations show that rainbands are distinctly finite in length. This means that we need to assess the effects of the three-dimensional structure necessarily imposed by this finite length. This has been addressed by Jones and Thorpe (1992), who show that the (two-dimensional) theory of CSI may hold as long as the length of the rainband is longer than about twice the across-front wavelength of the rolls. For shorter bands the CSI growth is substantially reduced. They also found that the bands tend to be rotated anticyclonically relative to the front, depending on the importance of viscous effects; such an orientation is often observed.

4.4 Role in Frontogenesis and Cyclogenesis

Another important factor is to determine the role of CSI on frontogenesis and cyclogenesis. This has beeen studied in the context of full numerical modeling studies, discussed in the next section, as well in more general terms. The semigeostrophic theory of frontogenesis has been extended to include the effects of latent heat release occurring in an atmosphere with effectively zero stability to slantwise ascent, that is, for $PV_e \approx 0$. This has been reported for deformation frontogenesis by Thorpe and Emanuel (1985) and for horizontal shear frontogenesis, such as occurs in an Eady wave, by Emanuel et al. (1987). These results show horizontal scale contraction of the ascent zone and an increased growth rate of the front.

From a physical viewpoint this shows that the frontal environment is an ideal one for the feedback between latent heat release and frontogenesis. The parcel displacements favorable for CSI are aligned along the sloping front and so are exactly those produced by the cross-frontal circulation. In other words, not only is the atmosphere unstable but also the dynamics of the front move parcels in just the right direction for releasing the potential energy.

A further question to be addressed is the role of snow sublimation and rain evaporation from the precipitation bands. This has recently been considered by Parker and Thorpe (1995). They find that the effect on the geostrophic flow and frontogenesis is minimal but there is a strong coupling to the ageostrophic motion leading to an elevated jet in the cross-frontal flow beneath the frontal surface; this has been observed in the FRONTS 87 experiment (see Fig. 7) and in radar data.

4.5 Modeling Studies

The use of high-resolution cloud models to describe frontal circulations in two-dimensions has taken place recently. Examples include Hsie and Anthes (1984), Knight and Hobbs (1988), Benard et al. (1992a, b), and Redelsperger and Lafore (1994). New features have emerged such as the important role of the frictional boundary layer in the potential vorticity budget of frontal zones. Alternative ideas about rainband dynamics are emerging involving the role of gravity waves organizing convection. Such studies are approaching the complete three-dimensional description of the cyclone with high-resolution models including zones embedded at fronts resolving the mesoscale detail. It remains for appropriate dynamical concepts to be developed to provide an understanding of such modeling results.

5. Discussion

The mesoscale substructures described here—namely frontal wave cyclones and frontal rainbands—have been subject

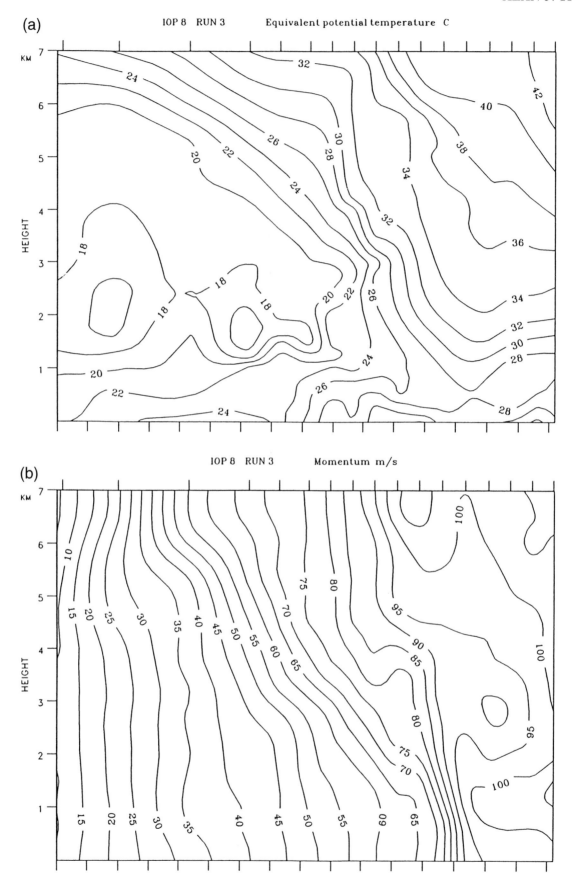

FIG. 6. (a) A vertical cross section from Thorpe and Clough (1991) from the FRONTS 87 dataset showing the frontal θ_e structure. (b) As in (a) but for the momentum surfaces showing evidence of buckling attributable to active CSI.

IOP 7   RUN 3   u-component of velocity m/s

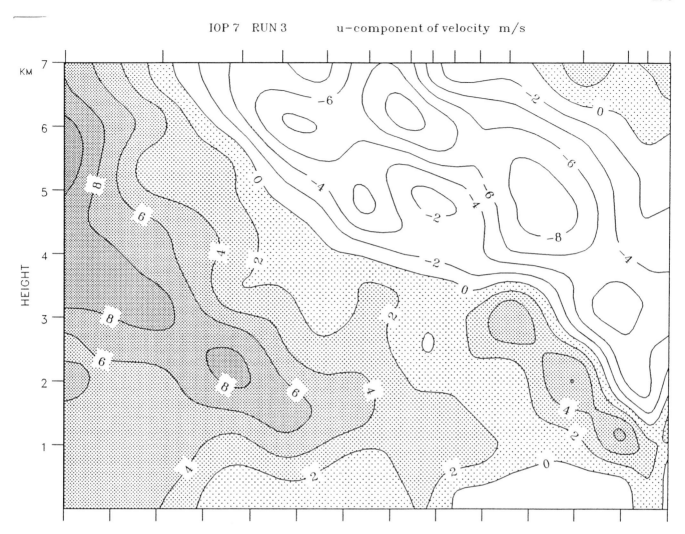

FIG. 7. A vertical cross section taken from FRONTS 87 of the cross-frontal flow showing the existence of strong flow toward the front immediately below the frontal surface. This is taken as evidence of the acceleration due to cooling from snow sublimation.

to theoretical study in the last few years. Hypotheses are emerging that need to be tested with high-resolution field observations. There have, of course, been many such campaigns already focusing in the general area of midlatitude cyclones. However, there are several aspects for which almost no detailed mesoscale observations exist. These include the structure of the tropopause with particular reference to the potential vorticity. Also the data that are available tend to involve nonsynchronous observations, which can be analyzed only via a space-time conversion. Therefore, there is a need for simultaneous observation of a large mesoscale area in the cloudy zones of such cyclones. It is hoped that programs such as the Fronts and Atlantic Storm Track Experiment (FASTEX), will sucessfully address these questions. The motivation is not only to improve understanding but also to provide input to weather forecasting and climate issues. The secondary wave cyclones are responsible for much of the problems in forecasting for northwest Europe at the end of the Atlantic storm-track. They are also associated with major zones of cloudiness, which impact strongly via radiative processes onto the climate system.

From a theoretical viewpoint it is clear that the pioneering research of the Bergen school and also of the German-speaking community early in this century has been the springboard to the understanding of midlatitude cyclones. Many of the components of the current picture were in place at that time but the modern era has allowed these pieces to be fitted together into a more complete and rigorous framework.

Acknowledgments

I would like to acknowledge the help from and collaboration with the following colleagues on the research summarized in this article: Craig Bishop, Sid Clough, Iain Cooper, Kerry Emanuel, Tim Hewson, Alain Joly, Sarah Jones, Sylvie Malardel, Doug Parker, and Rich Rotunno.

References

Ayrault, F., F. Lalaurette, A. Joly, and C. Loo, 1995: North-Atlantic ultra high frequency variability: An introductory survey, *Tellus*, **47A**, 671–696

Benard, P., J.-L. Redelsperger, and J.-P. Lafore, 1992a: Nonhydrostatic simulation of frontogenesis in a moist atmosphere. Part I: General description and narrow rainbands. *J. Atmos. Sci.*, **49**, 2200–2217.

——, ——, and ——, 1992b: Nonhydrostatic simulation of frontogenesis in a moist atmosphere. Part II: Moist potential vorticity budget and wide rainbands. *J. Atmos. Sci.*, **49**, 2218–2235.

Bennetts, D. A., and B. J. Hoskins, 1979: Conditional symmetric instability—a possible explanation for frontal rainbands. *Quart. J. Roy. Meteorol. Soc.*, **105**, 945–962.

Bishop, C. H., and A. J. Thorpe, 1994a: Frontal wave stability during moist deformation frontogenesis. Part 1: Linear wave stability. *J. Atmos. Sci.*, **51**, 852–873.

——, and ——, 1994b: Frontal wave stability during moist deformation frontogenesis. Part 2: The suppression of non-linear development. *J. Atmos. Sci.*, **51**, 874–888.

——, and ——, 1994c: Potential vorticity and the electrostatics analogy: Quasi-geostrophic theory. *Quart. J. Roy. Meteor. Soc.*, **120**, 713–731.

Clough, S. A., C. S. A. Davitt, and A. J. Thorpe, 1996: Attribution concepts applied to the omega equation. *Quart. J. Roy. Meteor. Soc.*, **122**, 1943–1962.

Cooper, I. M., A. J. Thorpe, and C. H. Bishop, 1992: The role of diffusive effects on potential vorticity in fronts. *Quart. J. Roy. Meteor. Soc.*, **118**, 629–647.

Davis, C. A., and K. A. Emanuel, 1991: Potential vorticity diagnostics of cyclogenesis. *Mon. Wea. Rev.*, **119**, 1929–1953.

Eliasen, E., 1960: *On the Development of Frontal Waves*. Det Danske Meteorologiske Institut. Meddelelser No. 13, pp. 107.

Eliassen, A., 1962: On the vertical circulation in frontal zones. *Geofys. Publ.*, **24**, No. 4, 147–160.

Emanuel, K. A., 1983: The Lagrangian parcel dynamics of moist symmetric instability. *J. Atmos. Sci.*, **40**, 2368–2376.

——, 1988: Observational evidence of slantwise adjustment. *Mon. Wea. Rev.*, **116**, 1805–1816.

——, Fantini M, and Thorpe A J, 1987: Baroclinic instability in an environment of small stability to slantwise moist convection, Part 1: Two-dimensional models. *J. Atmos. Sci.*, **44**, 1559–1573.

Fjortoft R, 1950: Application of integral theorems in deriving criteria of stability for laminar flows and for the baroclinic vortex. *Geofys. Publ.*, **17**, No. 6, 55.

Hewson, T., 1993: The FRONTS 92 Experiment: A Quicklook Atlas, *JCMM Internal Report No. 15*, available from the Joint Centre for Mesoscale Meteorology, University of Reading.

Hoskins, B. J., 1982: The mathematical theory of frontogenesis. *Ann. Rev. Fluid Mech.*, **14**, 131–151.

——, M. E. McIntyre, and A. W. Robertson, 1985: On the use and interpretation of isentropic potential velocity maps. *Quart. J. Roy. Meteor. Soc.*, **111**, 877–946.

Hsie, E. Y., and R. A. Anthes, 1984: Simulation of frontogenesis in a moist atmosphere using alternative parameterizations of condensation and precipitation. *J. Atmos. Sci.*, **41**, 2701–2716.

Innocentini, V., and C. Neto, 1992: A numerical study of the role of humidity in the updraft driven by moist slantwise convection. *J. Atmos. Sci.*, **49**, 1092–1106.

Joly, A., 1995: The stability of steady fronts and the adjoint method: Non-modal frontal waves. *J. Atmos. Sci.*, **52**, 3082–3108.

——, and A. J. Thorpe, 1989: Warm and occluded fronts in two-dimensional moist baroclinic instability *Quart. J. Roy. Meteor. Soc.*, **115**, 513–534.

——, and ——, 1990: Frontal instability generated by tropospheric potential vorticity anomalies. *Quart. J. Roy. Meteor. Soc.*, **116**, 525–560.

——, and ——, 1991: The stability of time-dependent flows: An application to fronts in developing baroclinic waves. *J. Atmos. Sci.*, **48**, 163–182.

Jones, S. C., and A. J. Thorpe, 1992: The three-dimensional nature of "symmetric" instability. *Quart. J. Roy. Meteor. Soc.*, **118**, 227–258.

Knight, D. J., and P. V. Hobbs, 1988: The mesoscale and microscale structure and organization of clouds and precipitation in mid-latitude cyclones. Part XV: A numerical modelling study of frontogenesis and cold-frontal rainbands. *J. Atmos. Sci.*, **45**, 915–930.

Kotschin, N., 1932: Uber die Stabilitat von Margulesschen Diskontinuitatsflachen. *Beitr. z. Physik d. freien Atmosphare*, **18**, 129–164.

Lagouvardos, D., Y. Lemaitre, and G. Scialom, 1993: Dynamical structure of a wide cold-frontal cloudband observed during FRONTS 87. *Quart. J. Roy. Meteor. Soc.*, **119**, 1291–1319.

Malardel, S., 1994: Ordinary secondary frontal waves. PhD thesis, University of Reading, 224 pp.

——, A. Joly, F. Courbet, Ph. Courtier, 1993: Non-linear evolution of ordinary frontal waves induced by low-level potential vorticity anomalies. *Quart. J. Roy. Meteor. Soc.*, **119**, 681–713.

——, A. J. Thorpe, and A. Joly, 1997: Consequences of the geostrophic momentum approximation on barotropic instability. *J. Atmos. Sci.*, **54**, 103–112.

Matejka, T., R. A. Houze, and P. V. Hobbs, 1980: Microphysics and dynamics of clouds associated with mesoscale rainbands in extratropical cyclones. *Quart. J. Roy. Meteor. Soc.*, **106**, 29–56.

Parker, D. J., and A. J. Thorpe, 1995: The role of snow sublimation at fronts. *Quart. J. Roy. Meteor. Soc.*, **121**, 763–782.

Perrson, P. O. G., and T. T. Warner, 1991: Model generation of spurious gravity waves due to inconsistency of the vertical and horizontal resolution. *Mon. Wea. Rev.*, **119**, 917–935.

Raymond, D. J., 1992: Nonlinear balance and potential-vorticity thinking at large Rossby number. *Quart. J. Roy. Meteor. Soc.*, **118**, 987–1015.

Redelsperger, J.-L., and J.-P. Lafore, 1994: Non-hydrostatic simulations of a cold front observed during the FRONTS 87 experiment. *Quart. J. Roy. Meteor. Soc.*, **120**, 519–556.

Renfrew, I. A., A. J. Thorpe, and C. H. Bishop, 1997: The role of the environmental flow in the development of secondary frontal cyclones. *Quart. J. Roy. Meteor. Sci.*, **123**, 1653–1676.

Schar, C., and H. C. Davies, 1990: An instability of mature cold fronts. *J. Atmos. Sci.*, **47**, 929–950.

Thorncroft, C. D., and B. J. Hoskins, 1990: Frontal cyclogenesis. *J. Atmos. Sci.*, **47**, 2317–2336.

Thorpe, A. J., 1993: An appreciation of the meteorological research of Ernst Kleinschmidt. *Meteorol. Zeitschrift, N.F.*, **2**, 3–12.

——, and S. A. Clough, 1991: Mesoscale dynamics of cold fronts: Structures described by dropsoundings in FRONTS 87. *Quart. J. Roy. Meteor. Soc.*, **117**, 903–941.

——, and K. A. Emanuel, 1985: Frontogenesis in the presence of small stability to slantwise convection. *J. Atmos. Sci.*, **42**, 1809–1824.

——, and R. Rotunno, 1989: Non-linear aspects of symmetric instability. *J. Atmos. Sci.*, **46**, 1285–1299.

Numerical Simulations of Mesoscale Substructures and Physical Processes within Extratropical Cyclones

THOR ERIK NORDENG

Norwegian Meteorological Institute, Oslo, Norway

1. Introduction

Nineteenth century and early twentieth century meteorologists were clearly aware of mesoscale structures within cyclones. Automatically recorded weather parameters were available and clearly visualized mesoscale structures. But it was the Bergen meteorologists who started to use the mesoscale structures of the available data as an integrated and necessary part of their analyses of extratropical cyclones. The scale of a typical extratropical cyclone may be several thousands of kilometers, but the Bergen school recognized and utilized that the extreme weather of the cyclone and its most dynamically active regions were confined to certain lines of discontinuity—the fronts.

Today high resolution numerical weather prediction models are daily verifying that the conceptual model of cyclone development and three-dimensional structure discovered in Bergen, Norway, 75 years ago (Bjerknes 1919; Bjerknes and Solberg 1922) are indeed still valid. Results from observational studies with dedicated research aircrafts packed with sophisticated instruments support this as well. According to the Bergen school conceptual model, cyclones develop as unstable waves on the polar front. Today we know that several factors may influence the spin up of the cyclone and that cyclones may form in various environments. Polar front cyclones are however, rather common, and fronts play a major role in most (if not all) extratropical cyclone developments.

In this chapter we focus on numerical simulations of mesoscale substructures within extratropical cyclones and how physical processes interact with the flow to create these substructures. Vilhelm Bjerknes had already pointed out in 1904 (Bjerknes 1904) the physical and mathematical bases for what we today recognize as numerical weather prediction: (1) analysis to obtain a three-dimensional state of the atmosphere, and (2) integration of the governing equations forward in time. The prospects and advantages of using numerical weather prediction models in atmospheric research was reviewed by Keyser and Uccellini (1987). Such models have frequently been used to understand the dynamics of extratropical cyclones and polar lows (e.g., Keyser et. al. 1978; Kuo and Reed 1988; Kuo et al. 1991a; Nordeng 1990; Nordeng and Rasmussen 1992). Recently, very high resolution models, which even take nonhydrostatic effects into account, have been constructed for the same purpose (Dudhia 1993). Recent studies have pointed at mesoscale substructures within extratropical cyclones and their possible role for the development of the cyclones (Neiman et al. 1993; Neiman and Shapiro 1993; Bond and Shapiro 1991, Grønås 1995). These substructures are largely a result of diabatic processes, which tend to reduce the scale of the most active parts of cyclones and henceforth play an interactive role in their dynamics (Kuo et al. 1991 b). There are, however, mesoscale cyclones in the atmosphere that in many respects looks like their larger-scale counterparts and seem to have a full life cycle of their own imbedded within larger-scale systems. To this category belong, for example, polar lows and comma clouds.

Numerical simulations of frontal structures in general are dealt with in Section 2. We concentrate on processes that are relevant for extratropical cyclones. In Section 3 we look at large-scale extratropical cyclones and in particular those that undergo explosive growth. We concentrate on the effect of diabatic processes and show how diabatic processes interact with the flow to create mesoscale structures. As numerical simulation of extratropical cyclones in general is covered elsewhere in this volume, this topic will here be covered in the context of demonstrating the similarities between large-scale and other extratropical cyclone developments, which, at a first glance, do not seem to fit into the Norwegian framework. These mesoscale cyclones are dealt with in Section 4. The chapter ends with a discussion in Section 5.

2. Frontal Structures and Dynamics

2.1 On the Creation of Low-Level Fronts and Baroclinic Zones

One of the main discoveries of the Bergen school was that the discontinuity surfaces within a cyclone (the fronts) are an integral part of the cyclone and play an important role in the cyclone development and dynamics. The conceptual model of cyclone development (Bjerknes 1919; Bjerknes and Solberg 1922) was at first viewed with skepticism among leading meteorologists at the time, but later became universally accepted.

Numerical simulations of the life cycle of cyclones developing, for example, along a broad baroclinic zone indeed show that fronts establish as in the conceptual cyclone model of the Bergen school (e.g., Simmons, this volume). The scale of the fronts may be several hundreds of kilometers in the along-front direction while the transverse direction is truly of the mesoscale. Fronts develop under the influence of frontogenetical forcing (confluent motion as suggested by Bergeron 1928) and shear. Edelmann (1963) was probably the first to show numerically that fronts intensify in developing cyclones. Today, numerical simulations of polar front developments, which closely follow the conceptual models

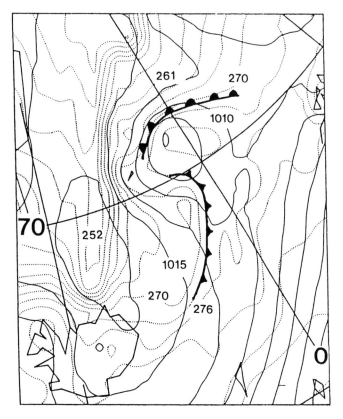

Fig. 1 Mean sea level pressure (solid lines, contour intervals 2.5 hPa) and equivalent potential temperature (dotted lines, contour interval 3 K) after 12 h simulation from 0000 UTC 26 January 1988 (Nordeng 1990).

of the Bergen school, may be found daily in all weather forecasting offices. Fronts may also form due to differential heating. This is particularly important in polar regions where there are large contrasts between land and ocean or in connection with ocean currents.

Arctic fronts were also introduced by the Bergen school meteorologists. In Nordeng (1990) the arctic front developed along the ice-edge east of Greenland and was subsequently advected into the open sea as a polar low developed. The ice-edge–based front later redeveloped. Shapiro and Fedor (1989) and Shapiro et al. (1989) observed arctic fronts with strong low level baroclinicity. The fronts were created along the ice-edge and were later advected into the open sea. Cyclones may form on these fronts as they do on the polar front. However, due to the lower tropopause and weaker static stability they become smaller than their polar front counterparts. Figure 1 shows a small-scale cyclone forming at the stationary front along the ice edge east of Greenland as an upper level (in this case upper level is actually at 700 hPa!) trough approaches from Greenland. During the Polar Lows Project conducted at the Norwegian Meteorological Institute (Lystad 1986) it was found that the North Atlantic Ocean had a large number of polar low developments. Geographically, this area consists of open relatively warm water (almost to 80°N west of Spitzbergen) due to the Gulf Stream. In the very north, the open water cuts into the ice with a characteristic V-shape. It is by no means strange that fronts easily form, taking the contrast between the open sea and the very cold surroundings into account. It is, however, frequently observed that polar lows tend to develop on a very special type of a frontal system. Figure 2 shows geopotential heights and temperatures at 1000 hPa for two cases from Grønås et al. (1987). In both cases there are troughs west of Spitzbergen extending northward from the main synoptic scale lows over northern Norway, and there is a characteristic wedge of warm air almost in phase with the trough.

Grønås et al. (1987) speculated whether the combined effect of diabatic heating from the warm open ocean and flow distortion from the mountains on Spitzbergen could cause such a flow pattern. The warm air anomaly may subsequently be advected southward. In Nordeng (1987) it was shown that the system had strong shearing frontogenesis at its leading edge and a somewhat weaker confluence at its rear part. As the warm air moves southward polar lows are frequently observed to develop. The relevance of a surface temperature anomaly for the invertibility principle of retrieving the flow from potential vorticity was discussed by Hoskins et al. (1985) and Thorpe (1986). A surface temperature anomaly will act in a similar way as a potential vorticity anomaly making it possible for transient vorticity anomalies aloft to develop a cyclone through coupling and mutual intensification. A cross section through the warm wedge (Fig. 3) reveals the very special nature of the frontal system. Its leading part looks like a backward-moving cold front whereas its rear part

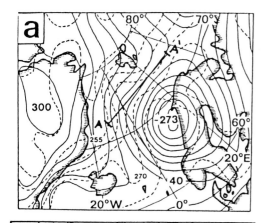

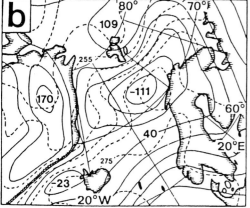

FIG. 2. ECMWF 1000 hPa height analyses (solid lines, contour interval 40m) and potential temperature (dashed lines, contour interval 5 K). (a) 1200 UTC 29 February 1984. (b) 1200 UTC 5 March 1984 (Grønås et al. 1987).

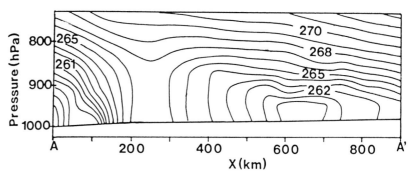

FIG. 3. Potential temperature after 6 hours of integration starting from 1200 UTC 29 February 1984 (i.e., from the field in Fig. 2a). Position of cross section is shown in Fig. 2a (Nordeng 1987).

2.2 Transverse Circulations in Frontal Zones

The Sawyer-Eliassen equation (Sawyer 1956; Eliassen 1962) describes the ageostrophic transverse circulation in frontal zones. The circulation may be viewed as a compensating circulation set up to restore thermal wind balance. In this equation potential vorticity plays the role of a stability parameter so that for low values of potential vorticity the vertical velocity may become large. The mathematical nature of the elliptic operator that determines the stream function of the problem is such that the flow will be elongated in the direction of the isentropic surfaces. (See Eliassen 1990 for a thorough review on transverse circulations in frontal zones). When latent heat is released in the ascending branch of the circulation, the stability (mathematically measured in terms of moist potential vorticity) becomes small and may even vanish, and the effect is a strong response in a small area. The flow will follow the most unstable direction, which is along the isotherms (θ_e). Eliassen (1990) showed the results of Todsen (1964), who was probably the first to demonstrate this numerically for real data. Thorpe and Emanuel (1985) assumed saturation and constant low values of moist potential vorticity in the ascending branch of the circulation around a frontal zone in the Eady wave problem integrated with a semigeostrophic model. The response to a confluent forcing was that condensation heating was speeding up the development with concentrated ascending motion as a result. The importance of low values of moist potential vorticity for cyclone developments is discussed further in Section 3.

If moist potential vorticity becomes negative inside clouds, the motion is symmetrically unstable. Growing transverse circulations may then be found even without forcing. Persson and Warner (1993) demonstrated the nature of these free symmetric modes by using a two-dimensional primitive equation model. Of particular importance for modelers is that the resolution of the model had to be in the order of 15 km in the horizontal direction and 170 m in the vertical to resolve the most unstable modes. This means that it may have to be parameterized in coarser resolution models (Nordeng 1987).

Bennets and Hoskins (1979) suggested that prefrontal rainbands were caused by symmetric instability. Cold frontal rainbands have been studied by Hobbs and collaborators in a series of papers. In Knight and Hobbs (1988) they showed that the first cold frontal rainband developed in a region of negative moist potential vorticity. Other bands developed behind the first and intensified as they moved into the region of negative moist potential vorticity. In numerical simulations of large-scale extratropical cyclones it is frequently

is less distinct and looks like a weak cold front. Nordeng (1987) showed that the leading part of the system was symmetrically unstable. In fact, the leading part of the frontal system looks very similar to the bent-back warm front (Shapiro and Keyser 1990) observed in connection with strong oceanic extratropical cyclones (e.g., Neiman and Shapiro 1993). In the present case warm air is enclosed between masses of colder air. In the large-scale cyclone of Neiman and Shapiro a seclusion process enclosed warm air in the core of the cyclone.

The importance of fronts and warm air seclusion is discussed further in Section 3. In Section 4 we show another example of warm air being trapped between colder air masses and that this had an impressive effect of the spin-up of a mesoscale cyclone.

found that the air within the strongest ascending motion in connection with the warm front is neutral or even unstable with respect to symmetric instability (e.g., Kuo and Reed 1988; Kuo et al. 1991 b), but this is also found in numerical simulations of smaller-scale cyclones such as polar lows (Nordeng and Rasmussen 1992). It should be mentioned that Knight and Hobbs' (1988) work was two-dimensional and the along-front variation was neglected. More recently, Jones and Thorpe (1992) discussed the three-dimensional nature of symmetric instability and showed that a flow with negative potential vorticity is unstable not only in the two-dimensional case.

The transverse circulation as described here should not be confused with the Lagrangian circulation of individual particles. The transverse circulation is merely the transverse component of the three-dimensional motion. This has relevance to the conveyor belt concept. See, for example, the chapter by Browning in this volume, or Browning (1990).

2.3 Along-Frontal Variability

There are also mesoscale structures in the along-front direction. A phenomenon truly on the mesoscale is the so-called frontal fracture stage discussed by Shapiro and Keyser (1990). Based upon results from idealized (Eady-wave type) as well as real data numerical simulations and observational studies, Shapiro and Keyser (1990) suggested a modification of the classical Bergen school cyclone model. They noticed that in the early stages of some cyclone developments there was no clear connection between the cold front and the warm front (a "missing" triple point) and named this the "frontal fracture stage." Eliassen (1990), in describing his personal experience

as a synoptician, noted that the inner part of the cold front was not always discernible from the data. More recently, Hines and Mechoso (1993) showed some results indicating that this is a phenomenon for oceanic cyclones. In idealized simulations with midlatitude jet-streams as initial conditions they found cyclogenesis in accordance with the Shapiro and Keyser model only in the case where the surface drag was small with "oceanic values." For oceanic values of surface drag there was a region of frontolysis between the two strong frontogenesis regions in connection with the cold and the warm fronts (Fig. 4). It is interesting that similar structures may even be found for small scale cyclones. There is clearly a "fractured" frontal system in connection with the polar low in Fig. 1. The fronts in Fig. 1 were drawn subjectively from the contours of equivalent potential temperature at the lowest model level.

That there are considerable variations in, for example, rainfall along a cold front is something everyone has experienced and is a result of the convective nature of the flow. A thorough review of observations, theories, and numerical simulations may be found in Browning (1990). Dudhia (1993) has investigated the three-dimensional structure of a cold front by using a nonhydrostatic model with a horizontal resolution of 6.7 km. He managed to simulate several of the mesoscale structures often observed in connection with these type of fronts, such as the facts that (1) the front was split into line-elements (e.g., James and Browning 1979; see Fig. 5); (2) that there was prefrontal rain in connection with an upper cold front (Browning and Monk 1982); and (3) there was a strong low-level jet ahead of the front at the surface. The ascending motion in connection with the front consisted of convective updrafts imbedded in a more gently sloping

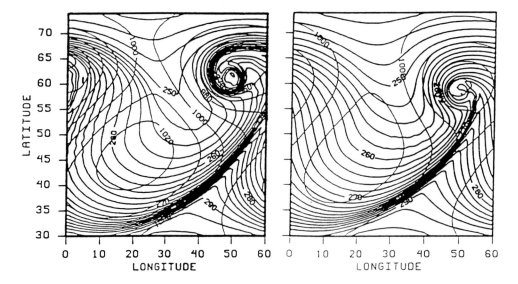

FIG. 4. Longitude-latitude contour plot of the model's lowest-level potential temperature (K, thick lines) and surface pressure (mb, thin lines) for ocean drag (left panel) and land drag (right panel) after 5 days of integration (Hines and Mechoso 1993).

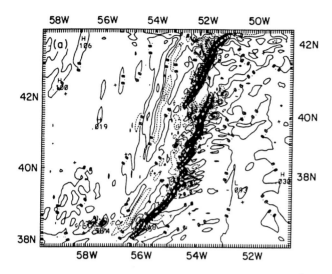

FIG. 5. Vertical velocity $s = 0.865$ at contour intervals of 0.2 m s^{-1} from a numerical simulation of a cold front (Dudhia 1993).

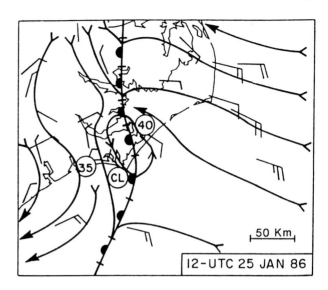

FIG. 6. Analyzed surface streamlines for eastern North Carolina for 1200 UTC 25 January 1986. One full wind barb corresponds to 5 m s^{-1}. The numbers refer to Portable Automated Mesonet (PAM) sites (Doyle and Warner 1993).

conveyor belt of ascending air at neutral stability (Dudhia 1993). A similar structure was observed by Neiman et al. (1993) in connection with the ascending motion when the warm conveyor belt overrides the cold conveyor belt in the warm front. Sensitivity studies revealed that by far the most important diabatic process was release of latent heat. Without latent heat the ascent in connection with the front became broad and gentle without the intense convection. A similar experiment was performed by Doyle and Warner (1993) using the same model on a front on the east coast of the United States (5 km resolution). They also managed to simulate several observed aspects of the front, including mesoscale vortices that developed along the front (Figs. 6 and 7).

3. Mesoscale and Physical Aspects of Large-Scale Extratropical Cyclones

3.1 Surface Processes

We know that there must be some differences between cyclogenesis over land and over water, since the largest deepening rates are found for cyclones over the oceans (one exception is cited in Uccellini 1990). One obvious candidate for the differences is, of course, friction. We have already mentioned that the "frontal fracture stage" of Shapiro and Keyser (1990) may be result of the smaller surface drag over oceans than over land. The results of Hines and Mechoso (1993) indicate that this may be one reason for the bent back warm front as well (see Fig. 4). Thorncroft et al. (1993) show, however, that cyclone life cycles according to the Shapiro–Keyser model (i.e., with the bent-back warm front) are connected to planetary-scale flows with small meridional barotropic shear as well. The role of friction is by no means

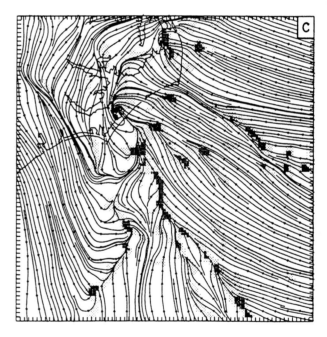

FIG. 7. Simulated surface layer stream lines valid at 1200 UTC 25 January 1986. Areas with surface layer convergence exceeding 10 s have dark shading (Doyle and Warner 1993).

simple. It slows down surface winds, but it also enhances surface convergence. A sensitivity study by Anthes and Keyser (1979) showed that the intensity of a cyclone over the United States was remarkably weakened in a simulation with surface friction as compared to a frictionless simulation. The total precipitation was, however, roughly the same, probably due to larger surface convergence. Theoretical work (e.g., Jansen 1989; Nordeng 1991) and observations (e.g., Geernaert et al. 1986) suggests that the surface drag coefficient is a function of the so-called "wave-age" parameter ($c_p/u*$), where c_p is the phase speed of the ocean waves. $c_p/u*$ is typically small for "young sea," not in equilibrium with the wind, whereas it is large in the ocean swell regime. Doyle (1995) tested these ideas with coupled ocean/wave atmosphere mesoscale model simulations of cyclogenesis, and found in particular that roughness effects associated with ocean waves modulate the deepening rate during rapid cyclogenesis and enhance the cyclone filling process. In addition, he found that regions of young windsea act to locally enhance the low-level frontogenesis, convergence, and rainfall.

Kuo et al. (1991a) investigated the role of surface fluxes of heat and moisture for seven explosive cyclones, but could not find any major effect on the storm development. However, surface fluxes may have been important in the preconditioning of the air prior to the rapid development. Both positive effects (e.g., Bosart 1981; Uccellini et al. 1987) and negative effects (e.g., Reed and Simmons 1991; Kuo and Low-Nam 1990) are reported in the literature. It should be mentioned that there is still controversy over whether the surface exchange coefficients for heat and moisture are constants or if they increase with wind speed as the surface drag coefficient does. In spite of these problems, it seems, however, that state-of-the-art numerical models are capable of reproducing intense cyclogenesis. Dell'Osso (1992) and Donall et al. (1991) have reported good agreement between their numerical simulations of the ERICA IOP4 storm and the analysis of Neiman et al. (1993). The numerical simulations clearly show a bent-back warm front and a secluded pocket of warm air in agreement with observations and the conceptual model of Shapiro and Keyser (1990). In this case the low deepened by 62 hPa in 24 hours, a deepening rate that was matched by the numerical models.

3.2 The Role of Latent Heat

In Section 2 the role of released latent heat from condensation for the circulation in frontal zones was discussed. It was shown that the major effect of latent heat is to intensify the circulation and to concentrate the circulation on a smaller scale. Since fronts are an integral part of cyclones it is therefore no surprise that latent heat has a large effect on the intensification of cyclones as well.

Eliassen (1984) demonstrated the three-dimensional nature of the ascending motion in frontal zones. Taking x to be the direction along the isotherms with the colder air to the left, y to be the direction normal to the isotherms and the vertical direction parallel to the three-dimensional absolute vorticity vector, the elliptic operator for the vertical velocity ω may be written as,

$$L(\omega) = \left(\sigma \frac{\partial^2}{\partial x^2} + \sigma_1 \frac{\partial^2}{\partial y^2} + f\eta \frac{\delta^2}{\delta p^2} \right) \omega$$

where $\sigma = -(\alpha/\theta)(\partial\theta/\partial p)$ is the (standard) static stability parameter and $\sigma_1 = -(\alpha/\theta)(\delta\theta/\delta p)$ is the static stability as measured taking derivatives along the direction of the vorticity vector. η is the (standard) absolute vorticity. Since σ_1 tends to be small when there is ascending motion, and particularly when release of latent heat is taken into account, one would expect that the principal axes of the operator be elongated in the direction of the isotherms, that is, parallel to the front and shrunk in the direction normal to the front. In addition, the ascending motion will follow the third principal axis in the direction of the absolute vorticity vector. This effect is commonly found in weather maps (precipitation areas are elongated along fronts) and shows up in numerical simulations as well. Figure 8 is a cross-section through a deepening cyclone taken from Kuo and Reed (1988). The relative short cross frontal extension and the tilting structure of the ascending motion along absolute momentum surfaces are evident.

Theoretically, the growth rate for the most unstable normal mode in the Eady wave problem doubles if the ascending motion takes place under neutral stratification (e.g., slantwise neutrality) (Emanuel et al. 1987; Joly and Thorpe 1987). Recently, Cao and Cho (1995) showed by using a numerical model that negative moist potential vorticity first appears in the warm sector near the north end of the cold frontal zone at the development stage and then intensifies in the bent-back warm front at the mature stage. After the cyclone matures, the negative moist potential vorticity region moves toward the warm core. Pedersen (1963) was probably the first to demonstrate scale contraction and larger intensity when latent heat is released in a numerical simulation of an extratropical cyclone. He used a simplified quasigeostrophic model and parameterized the release of latent heat by assuming that the lapse rate is saturated-adiabatic in regions of condensation. Kuo et al. (1991b) used the quasigeostrophic omega equation and results from a full primitive equation model simulation to demonstrate that the effect of released latent heat on numerical simulation of the QEII storm was to establish a strong interaction between the baroclinic and diabatic processes. The rapid cyclogenesis was strongly related to moist frontogenesis at the warm front and the effect of latent heat was to modify the frontal structure to reinforce the adiabatic secondary circulation (Fig. 9). Large values of potential vorticity were found at low levels within the frontal cloud band. It was absent in the adiabatic simulation, showing that it was created by the release of latent heat. In addition, a tongue of warm air was found in the frontal region.

Similar results were found by Ohm (1993) in her simulation of the New Year Storm of 1992. This storm is considered one of the largest natural disasters in Norway in modern

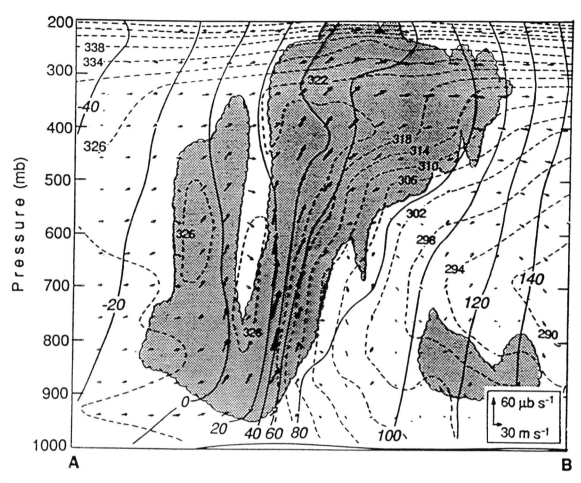

FIG. 8. Cross section of absolute momentum (solid, m s^{-1}) and equivalent potential temperature (dashed). Arrows give vertical and horizontal motion in the plane of the cross section according to the attached scale. Shading denotes cloud mass (Kuo and Reed 1988).

times. The observed deepening rate was exceptional. It deepened 38 hPa in the 12 hours between 0600 and 1800 UTC on 31 December 1991 and a total of 54 hPa in the 24 hours between 0600 UTC 31 December 1991 and 0600 UTC 1 January 1992. Maximum observed mean wind was 46 m s^{-1} at two lighthouses. The strongest winds were observed when the bent-back warm front hit the coast. (The Bergen school used the name bent-back occlusion rather than bent-back warm front, cf. Bergeron 1980). Bjerknes (1930) described how the bent-back occlusion sometimes can appear as a secondary cold front. Ohm simulated the storm with the Norwegian Limited Area Model and the Sundqvist scheme (see, e.g., Sundqvist et al., 1989) for parameterization of condensation and cloud processes. This storm has also been simulated by Breivik et al. (1992) and by Grønås (1995). Figure 10 shows pressure and wind relative to the storm in the 290 K isentropic surface. The warm air approaches the cyclone from south parallel to the cold front rising cyclonically around and over the bent-back warm front, while cold air is spiraling in toward the center of the cyclone behind the cold front. There is clearly a wedge of warm air extending

toward the center of the cyclone along the bent-back warm front in the same manner as observed in connection with the ERICA IOP4 storm (Neiman et al. 1993) and simulated for the QEII storm (Shapiro and Keyser 1990). According to Grønås (1995) the storm first developed in a classical way as an upper air anomaly moved in over a region with large surface baroclinicity. Release of latent heat particularly in connection with the bent-back warm front created potential vorticity at low altitudes, which eventually coupled and phase locked with the potential vorticity aloft. The bent-back warm front was detectable up to 500 hPa, where it coupled with an upper level front in connection with a tropopause fold. This is in many respects similar to the results for the ERICA IOP4 storm (Neiman et al. 1993). For this case, as for other strongly deepening cyclones, the release of latent heat was crucial for simulating the correct intensity and structure of the storm. This is nicely shown in Fig. 11 (Shapiro et al. 1991) from a simulation of the ERICA IOP4 storm. In the dry simulation (top panel) the upper level forcing passes smoothly over the surface center. In the moist simulation, however (bottom panel), release of latent heat at low levels in connec-

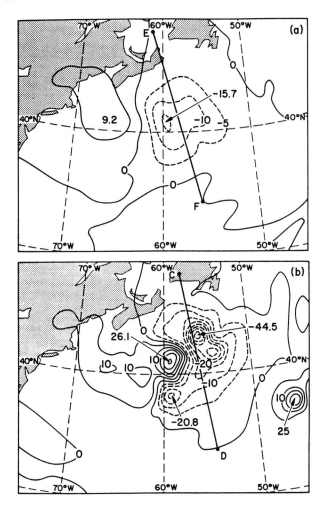

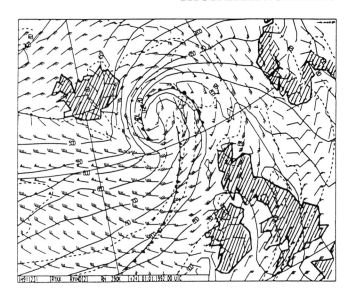

FIG. 10. Winds relative to the low-pressure system, pressure (hPa, solid contours) and relative humidity (dashed contours) at isentropic surface 290 K. Valid at 0000 UTC 1 January 1992 (24 h simulation) (Ohm 1993).

FIG. 9. Diagnosed vertical motion (ω, in hPa s^{-1}) from the quasigeostrophic omega equation at 700 hPa based on the quasigeostrophic baroclinic forcing, i.e., no contribution from diabatic effects (a) from the adiabatic primitive equation experiment, and (b) from the full physics primitive equation experiment. Twelve hour forecasts valid at 0000 UTC 10 September 1978 (Kuo et al. 1991).

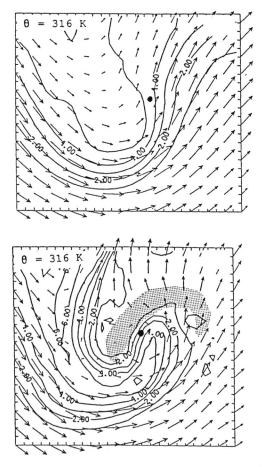

tion with the bent-back warm front has created potential vorticity at low levels, but at the same time destroyed potential vorticity aloft. The movement of the upper level potential vorticity is hence slowed down and allowed to interact with the lower anomaly for a longer time. The special hooked shape of the upper anomaly will in addition maximize vertical velocity forcing. Their simulation also showed the seclusion process, in which relatively warm air was cut off from the surrounding colder air in the cyclone center.

Grønås (1995) described the evolution of the New Year Storm of 1992 as a two-stage process. When the warm air is being secluded by cold air, the cyclone intensifies through a mesoscale lower tropospheric cyclogenesis connected to the warm anomaly of the secluded air. The surface pressure decrease connected to the seclusion low was ~30 hPa within 12 hours and it had a horizontal scale of ~500 km.

FIG. 11. Eighteen-hour adiabatic simulation (bottom panel) and diabatic simulation (top panel) from 0000 UTC 4 January 1989 of IPV (10^{-5} K s^{-1} mb^{-1}) at isentropic surface 316 K. Area with large values of released latent heat is shaded. Dots denote the center of the cyclone at the surface (Shapiro et al. 1991).

4. Nonclassical Mesoscale Cyclone Developments

That small scale lows may regenerate on the occlusion was known among the Bergen school meteorologists. Bergeron (1954) described the process as a special kind of nonfrontal thermodynamic, hurricane-like disturbances of small dimensions (less than 500 km) and often having surface pressure less than 980 hPa. Mesoscale vortices may be seen as small-scale perturbation vortices in the center of a parent cyclone as (e.g., in Bond and Shapiro 1991), quite often they develop in the rear part of an existing larger scale cyclone (as e.g., described in Nordeng and Rasmussen 1992) and take over as the main circulation center, but sometimes they seem to take a life of their own, developing into polar lows or comma clouds while the parent cyclone continues its own life cycle. Neiman and Shapiro (1993) in fact compare the inner core of the ERICA IOP4 storm with a polar low, while Grell and Donall Grell (1994) actually show by using a multiscale nonhydrostatic model system (highest horizontal resolution in innermost nested grid was 2.8 km) that mesovortices developed along the bent-back warm front of the ERICA IOP4 storm. The most intense vortex had the character of a warm core hurricane and moved slowly along the bent-back warm front into cold polar air before eventually becoming the main cyclone center.

4.1 Polar Low Developments

A frequent birth area for polar lows seems to be at the edge between the relatively calm inner part of a synoptic scale cyclone and the much stronger (cold) northerly flow further out. This area is optimal for growth of disturbances because of differential thermal advection, which quickly generates a surface-based baroclinic zone (front). This front may subsequently interact with an approaching potential vorticity anomaly aloft, creating a polar low. Examples of this may be found in Nordeng et al. (1989) and Nordeng and Rasmussen (1992). In the latter, in a particular area, there is an upper-level forcing, as seen from the distribution of IPV (Fig. 12). At this particular place the mesoscale low develops. It quickly takes over as the main disturbance in the area and moves southward toward the coast of northern Norway.

This low developed in a reversed shear flow, that is, the thermal wind had a direction opposite to the movement of the cyclone. The larger-scale IPV anomaly in Fig. 12 (centered on latitude 20°E) moved in the same direction as the surface low. The superimposed anomaly that triggered the development is seen as a wave along the eastern side of the larger-scale anomaly. This smaller-scale anomaly rotated cyclonically around this larger-scale anomaly, opposite to the movement of the surface low, and quickly lost its influence on the polar low development.

Although a "triggering" mechanism for which a mobile upper air IPV anomaly interacts with an area of large

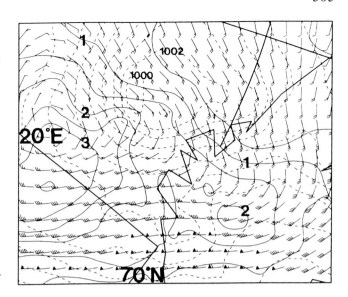

FIG. 12. Solid lines are potential vorticity at isentropic surface 278 K from 1200 UTC 26 February 1987 in units of PVU (1 PVU = 10^{-6} m^2 s^{-1} K kg^{-1}). Also shown are horizontal winds at the 278 K isentropic surface valid at the same time, a flag is 25 m s^{-1}, a full barb is 5 m s^{-1} and a half barb is 2.5 m s^{-1}. Dotted lines are surface pressure (hPa) valid 6 hours later (Nordeng and Rasmussen 1992).

baroclinicity or temperature anomaly at the surface seems to be prevalent for polar low developments, the scale and intensity of polar lows may be explained by traditional normal mode methods as well (e.g., Duncan 1978; Reed and Duncan 1987). Duncan (1978) demonstrated the structure of a reversed shear polar low. Haugen (1985), in his idealized experiment using an adiabatic primitive equation model on a baroclinic reversed shear flow in a channel, showed that a disturbance developed much in line with observed polar lows on the time scale of 24 hours. An interesting result was that even in his dry simulation an area of conditional instability in the region of strongest ascent was created showing that the adiabatic part as well as the diabatic part of the dynamics may cooperate in making the configuration optimal for further growth. Another interesting result was that the shape of the warm air and the surface pressure trough was very similar to the warm wedge frontal system (Fig. 3).

Montgomery and Farrell (1992) simulated a cyclone development as an initial-value problem imposing mobile upper-level and lower-level potential vorticity anomalies with a three-dimensional nonlinear geostrophic momentum model that incorporated moist processes. They showed that polar low developments may be viewed as a two-stage process where after an initial period of interaction between the upper and lower vorticity anomaly (called induced self-development), a further development is possible (called diabatic destabilization), which is associated with production of low-level potential vorticity by diabatic processes. In fact, the diabatic destabilization mechanism is robust enough to be able to sustain growth even without an upper-level anomaly. In the case studied by Nordeng and Rasmussen (1992) and in one of the cases in Nordeng (1990) an upper level triggering anomaly was found, but no external forcing was found in the subsequent development. Grønås et al. (1994) described a small-scale low development over Germany that apparently was not triggered by an upper-level anomaly but intensified purely by diabatic processes creating low-level potential vorticity, which again induces a secondary circulation enhancing the diabatic processes.

Nordeng and Rasmussen (1992) did not discuss the source of the upper-air triggering disturbance. Zick (1983) showed by using satellite-retrieved winds that a comma cloud developed as an upper-air vorticity maximum actually was advected around a 500 hPa low and subsequently triggered a comma cloud development after a full turn.

4.2 Preconditioning of the Environments

Polar lows seem to develop entirely over sea. That they quickly lose their energy and die over land is not very difficult to understand, because of stronger friction and disappearance of latent and sensible heat supply. However, that they are rarely (if not at all) observed to develop over land is more difficult to understand. If release of baroclinic instability triggered by an upper-level disturbance is the main driving mechanism, this should happen frequently over land. In order to understand this, we may have to understand the differences in preconditioning the environments over land and sea.

First of all, the lower tropospheric air is much warmer over sea than over land (in winter). Anomalies in sea surface temperature will create air temperature anomalies in the lower atmosphere as well. Several examples of this have been presented in this chapter. These regions will subsequently be exposed as birth areas for polar lows through transient upper-air disturbances. Section 2.1 gives several examples of this process.

There are, however, other processes that make the precondition different between land and sea. Although not as strong, the polar low described by Nordeng and Rasmussen (1992) and other polar lows as well (e.g., Nordeng et al. 1989; Grønås et al. 1987; Bond and Shapiro 1991) seem to develop in an area corresponding to the position of the bent-back warm front or along the bent-back occlusion. Neiman et al. (1993) compared the center of a synoptic-scale extratropical cyclone with a polar low development. According to the results of Hines and Mechoso (1993), the bent-back warm front is typically a phenomenon appearing over sea.

4.3 An Example of a Strong Mesoscale Vortex Development

We have just mentioned that it has been observed that a vorticity anomaly was advected around an upper-level low and eventually triggered a comma cloud development after it had made a full turn (Zick 1983). A physical mechanism somewhat similar to that was found to apply for a strong disturbance taking place in the Bay of Biscay in the beginning of 1994. Figure 13 shows a satellite picture for 1200 UTC 6 January 1994 with superimposed surface wind observations (ships). The data coverage was good and the central pressure could be estimated to be around 970 hPa (976 hPa was observed by two ships on either side of the low in the area of strongest winds). The Norwegian Meteorological Institute mesoscale limited-area model (see Nordeng and Rasmussen 1992, for references to the model), initiated with ECMWF analysis 12 hours before the low appeared on the satellite pictures, is clearly able to describe the development of the mesoscale low (Fig. 14). The model was run with a horizontal resolution of 10 km and with 30 layers. The simulation shows a tight inner core surrounded by strong surface winds. The modeled central pressure is 976 hPa, that is, roughly 5 hPa higher than the estimated central pressure. The surface pressure gradient is, however, as observed because there is a large scale bias in the simulation. The scale of the vortex is only 250 km, based on the width of the cloud shield. The scale of the tight inner core is, however, less than 150 km, that is, an even smaller scale than the polar low observed by Shapiro et al. (1987). There are, however, several similarities with that low. The pattern of maximum wind is approximately the same (35 ms^{-1} at 300 m height) and maximum relative vorticity is $21 \cdot 10^{-4}$ s^{-1}. Sensible heat flux from the ocean is smaller than in the polar low (max 250 W m^{-2}), whereas latent heat flux is of same size (600 W m^{-2}).

In Fig. 15 we show potential vorticity at the 295 K isentropic surface at 1200 UTC 6 January 1994 from the ECMWF model integration started at 0000 UTC 6 January 1994 (i.e., a 12-h forecast) and for the run with the mesoscale model. The IPV anomaly is related to a strong 500 hPa low, which crossed the Atlantic just southwest of a synoptic-scale surface low. The surface low reached its mature stage west of Ireland at 1200 UTC 5 January 1994. Close to the time when the mesoscale cyclone started its explosive growth, the IPV anomaly was advected over the region of strongest surface baroclinicity. Trajectories at ≈295 K reveal that air parcels crossed the Atlantic in one day from 1200 UTC 4 January to 1200 UTC 5 January, and that they actually rotated within the upper-level cyclone scale before contributing to the growth of the mesoscale cyclone. Diabatic effects played a crucial role in developing the cyclone. With latent heat turned off in the model, we obtain only a modest trough much in line with the large-scale ECMWF model simulation. The low-level potential vorticity (at isentropic surface 286 K) was of the same size as the upper-level potential vorticity anomaly (7 PVU at isentropic surface 286 K and 5 PVU at isentropic surface 295 K), showing the importance of diabatic processes.

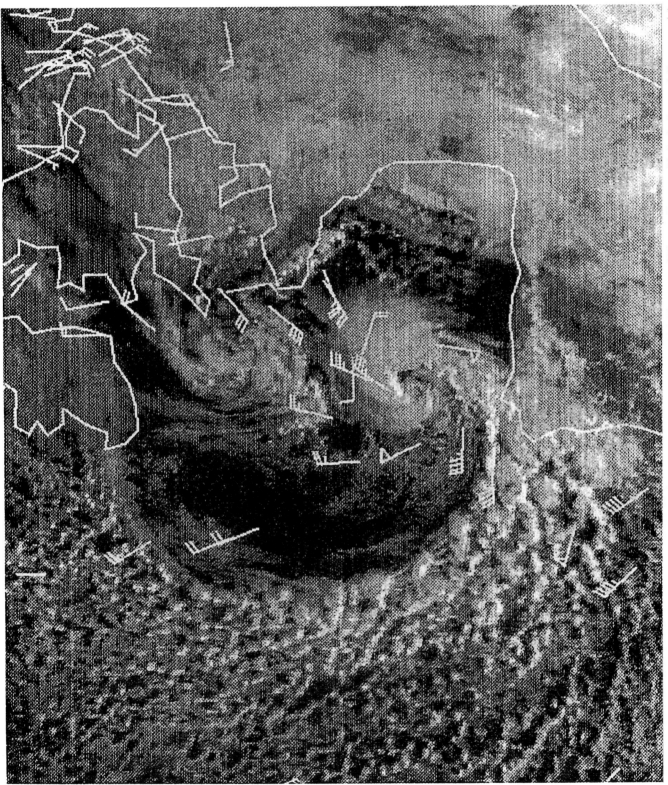

FIG. 13. Meteosat 4 (visible) and surface wind observations (ships), 1200 UTC 6 January 1994. The coast of Western Europe with Spain, Portugal, France, and the British Isles are outlined.

Fig. 14. Twelve-hour simulation valid at 1200 UT 6 January 1994 showing simulated cloud cover (white) and mean sea-level pressure in contour internals of 1 hPa (red). The coast of Western Europe is outlined as a thick line. The ocean has a bluish color and green is cloudfree land areas.

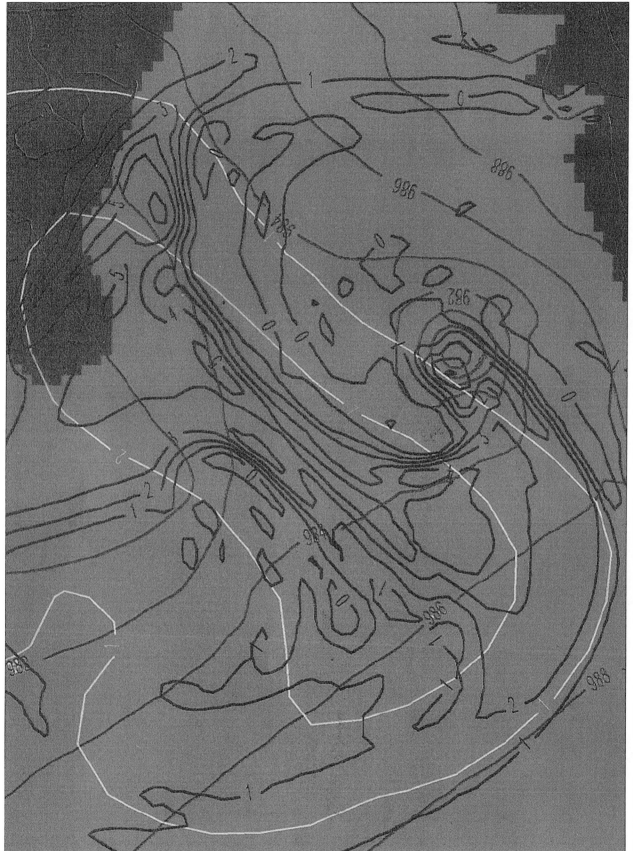

Fig. 15. IPV (in units of PVU) at isentropic surface 295 K, 1200 UTC 6 January 1994, obtained from the ECMWF operational model (white lines) and from the mesoscale model (black lines). Mean sea-level pressure at contour intervals of 2 hPa is shown as red lines (12 h simulations). The ocean is blue and land areas are green.

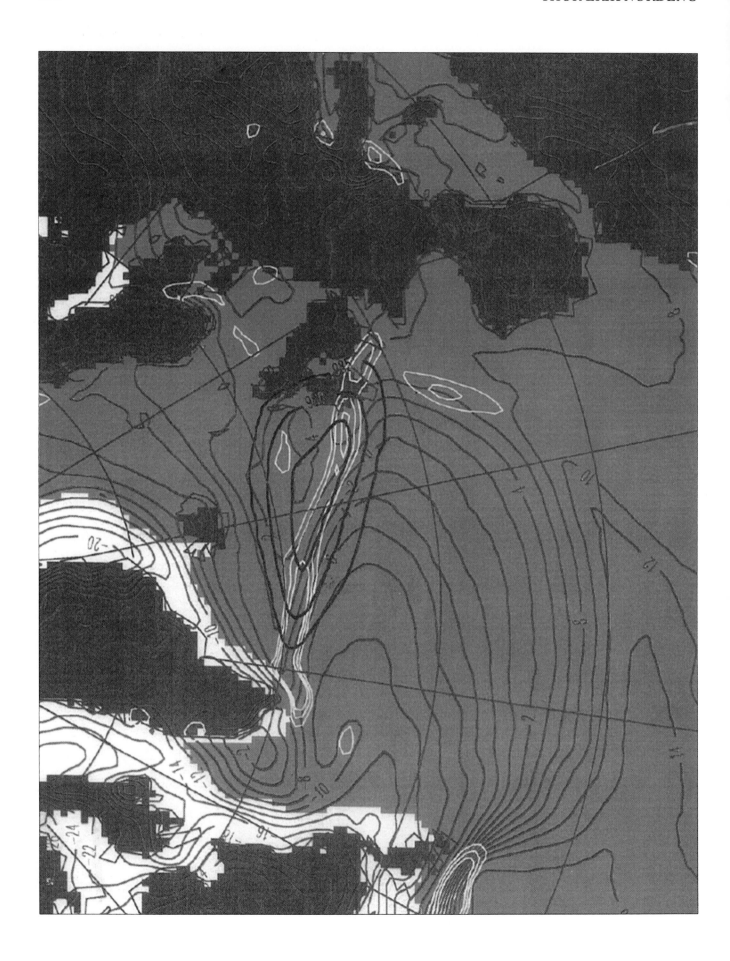

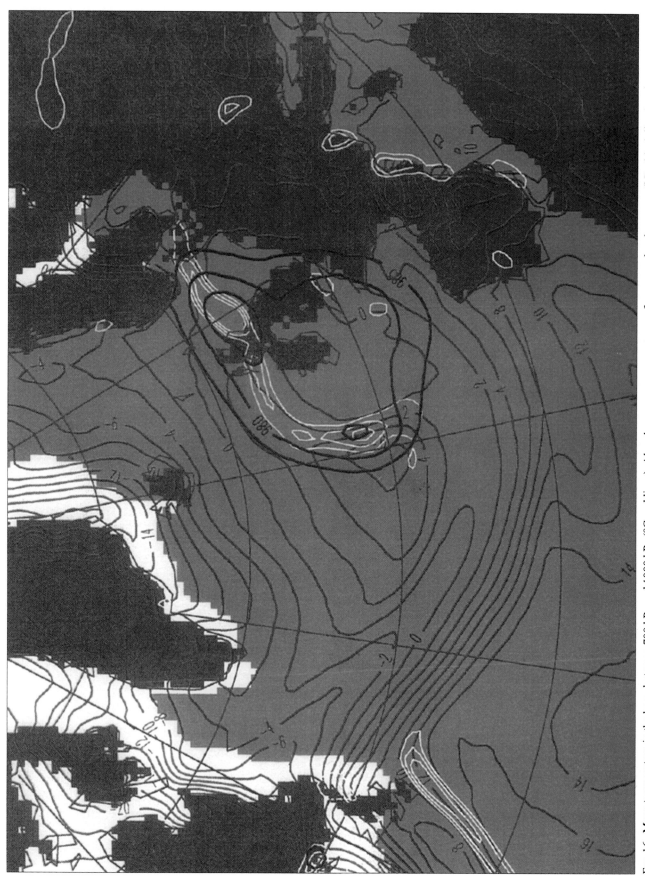

FIG. 16. Mean temperature in the layer between 700 hPa and 1000 hPa (°C, red lines). Also shown are some contours of mean sea level pressure (hPa, black lines) and some contours of relative vorticity at 1000 hPa (s⁻¹, white lines); (top) valid at 0000 UTC 5 January 1994, (bottom) valid at 0000 UTC 6 January 1994. The ocean is blue, land areas are green, and sea-ice is white.

FIG. 17. Wind relative to the low-pressure system (every second grid point is plotted, i.e., they are 20 km apart) and pressure (contour intervals 25 hPa, white lines) at isentropic surface 288 K. Valid at 1200 UTC 6 January 1994 (12 h simulation). The fronts are drawn subjectively from contours of equivalent potential temperature. The ocean is blue and land areas are green.

This example shows neatly how a mesoscale low is developed from a mesoscale anomaly within a synoptic-scale low. The development has many aspects similar to polar lows, and may indeed be classified as such. It has, in particular, many similarities with the very strong polar low development observed by Shapiro et al. (1987). To the author's knowledge, this is the first time that a mesoscale low with such a scale and intensity has been simulated numerically. However, to fully understand the dynamics involved, we must look at the synoptic-scale fields at low levels as well. The satellite picture (Fig. 13) shows that there is an area with excessive convection from the rear of the cyclone extending backward into the Atlantic. Particularly impressive is the sharp edge between the deepest convective clouds and an area of no convection. At 0000 UTC 5 January warm air was squeezed between two cold air outbreaks, one from the Norwegian Sea, and the other from Canada. A wedge of warm air extends from Britain to the south tip of Greenland (Fig. 16, top). Later this warm air was advected southward by the wind from a stationary synoptic-scale low west of Britain, enwinding colder air at the center of the low (Fig. 16, bottom). The mesoscale low developed where the temperature contrasts were largest (strongest low-level baroclinicity) on the interface between the cold air at the center of the synoptic-scale low and the much warmer air around it. The large-scale ECMWF model forecast a disturbance at the surface as well, but only as a weak trough, very much in line with a simulation with the mesoscale model with latent heating turned off.

The low deepened at a position relative to the large-scale low almost exactly as in Nordeng and Rasmussen (1992). There were, however, important differences. This low had cold air on its left side (relative to the direction of movement), it was a "normal" shear flow (thermal wind in the same direction as the moving cyclone), and the upper level forcing was in phase with the low (same direction as the low).

5. Discussion

Both sophisticated numerical models and simplified balanced models (semigeostrophic models) have extensively been used to understand the dynamics of the atmosphere. From these simulations it has emerged that even large-scale cyclones' mesoscale structure play a crucial part in their dynamics. Mesoscale structures become particularly important when diabatic processes are taken into account, because moist processes tend to reduce the scale of the phenomenon and concentrate the most active dynamical part of the flow on the small scales.

Cyclones seem to form within a large range of scales, and mesoscale vortices may form within larger-scale cyclones. It is amazing how similar cyclones are. Satellite pictures show more or less the same comma shaped cloud shield for all cyclones regardless of scale. Numerical modeling and observational studies indeed verify this, suggesting that there are no fundamental differences between cyclones on various scales. This is nicely depicted in Fig. 15. The larger-scale IPV structure from the ECMWF model has been deformed into the shape of a "hammer-head" by interaction with the diabatically enhanced flow of the mesoscale vortex. This shape is very similar to the shape of the IPV forcing of the IOP4 storm (Fig. 11). The horizontal scales are, however, quite different (ratio 1:10); that is, the same ratio as between the scales of the surface lows. In addition, the flow relative to the cyclone is very similar to the flow in large-scale explosive cyclones; for example, compare Fig. 17 with Fig. 10. The fronts in Fig. 17 have been drawn subjectively based on low-level equivalent potential temperature.

The preconditioning of the environment seems to be important for rapid cyclogenesis. All mesoscale cyclones reported here intensified in regions of low-level baroclinicity. This preconditioning often takes place as a seclusion of warm air by frontogenetical deformation. Large-scale explosive cyclones also tend to develop where there is large baroclinicity at the surface (e.g., in connection with ocean currents). The role of a warm air anomaly at the center of a cyclone, either preconditioned, or "self-made," as seen in large-scale cyclones (warm air seclusion), remains to be understood. At this stage it is worth remembering that a warm surface-based temperature anomaly plays mathematically the same role as an upper-level potential vorticity anomaly for the "invertibility principle," indicating a preference for (further) growth.

Acknowledgments

I am grateful to all those colleagues who have made this work possible and particularly to Graeme Kelly at ECMWF, who helped me to retrieve satellite pictures and to construct Fig. 13. Comments from and discussions with Graeme Kelly, Lorenzo Dell'Osso, and Martin Miller are gratefully acknowledged. Most of this work was performed while the author was working at the European Centre for Medium Range Weather Forecasts (ECMWF) in Reading, Great Britain.

References

Anthes, R. A., and D. Keyser, 1979: Test of a fine-mesh model over Europe and the United States. *Mon. Wea. Rev.,* **107,** 963–984.

Bennets, D. A., and B. J. Hoskins, 1979: Conditional symmetric instability—a possible explanation for frontal rainbands. *Quart. J. Roy. Meteor. Soc.,* **105,** 945–962.

Bergeron, T., 1928: Uber die dreidimensional verknupfende Wetteranalyse. *Geof. Publ.,* **5,** No. 6, 1–111.

——, 1954: Reviews of tropical hurricanes. *Quart. J. Roy. Meteor. Soc.,* **80,** 131–164.

——, 1980: Synoptic meteorology: An historic review. *Pure Appl. Geophys.,* **119,** 443–473.

Bjerknes, J., 1919: On the structures of moving cyclones. *Geofys. Publ.,* **1**, 1–8.

——, 1930: Practical examples of polar-front analysis over the British Isles. *Geophys. Mem.,* **50**, Meteorological Office, London.

—— and H. Solberg, 1922: Life Cycle of Cyclones and the Polar Front Theory of Atmospheric Circulation. *Geofys. Publ.* **3**, No.1.

Bjerknes, V., 1904: Das Problem der Wettervorhersage, betrachtet vom Standpunkte der Mechanik und Physik, *Met. Z.* **19**, 97.

Bond, N. A., and M. A. Shapiro, 1991: Polar lows over the Gulf of Alaska in conditions of reverse shear. *Mon. Wea. Rev.,* **119**, 551–572.

Bosart, L. F., 1981: The President's Day snowstorm of 18-19 February 1979: A subsynoptic-scale event. *Mon. Wea. Rev.,* **109**, 1542–1566.

Breivik, L. A., J. E. Kristjansson, K. H. Midtbø, B. Røsting, and J. Sunde, 1992: Simulation of the 1 January 1992 North Atlantic storm. Technical Report 99, The Norwegian Metorological Institute, Oslo, Norway, 69 pp.

Browning, K. A., 1990: Organization of clouds and precipitation in extratropical cyclones. *Extratropical Cyclones. The Erik Palmén Memorial Volume.* C. Newton and E. O. Holopainen, Eds., Amer. Meteor. Soc., 129–153.

—— and G. A. Monk, 1982: A simple model for the synoptic analysis of cold fronts. *Quart. J. Roy. Meteor. Soc.,* **108**, 435–452.

Cao, Z., and H-R. Cho, 1995: Generation of moist potential vorticity in extratropical cyclones. *J. Atmos. Sci.,* **52**, 3263–3281.

Dell'Osso, L., 1992: Extreme oceanic cyclogenesis. In ECMWF Seminar Proceedings: *Validation of Models over Europe,* Vol.1, 1992, 183–210.

Duncan, C. N., 1978: Baroclinic instability in a reversed shearflow. *Met. Mag.,* **107**, 17–23.

Doyle, J. D., 1995: Coupled ocean wave/atmosphere mesoscale model simulations of cyclogenesis. *Tellus,* **47A**, No. 5:1, 766–778.

—— and T. T. Warner, 1993: Nonhydrostatic simulations of coastal mesobeta scale vortices and frontogenesis. *Mon. Wea. Rev.,* **121**, 3371–3392.

Dudhia, J., 1993: A nonhydrostatic version of the Penn State–NCAR mesoscale model: Validation tests and simulation of an Atlantic cyclone and cold front. *Mon. Wea. Rev.,* **121**, 1493–1513.

Donall, E. G., M. A. Shapiro, and P. J. Neiman, 1991: Frontogenesis in a rapidly intensifying extratropical cyclone. *First International Symposium on Winter Storms,* New Orleans, 14–18 January 1991. Amer. Meteor. Soc., 393–397.

Edelmann, W., 1963: On the behavior of disturbances in a baroclinic channel. Summary Report Research in Objective Weather Forecasting, Part F, Contract AF61 (052)-373. Deut. Wetterd., Offenbach/Main, 35p.

Eliassen, A., 1962: On the vertical circulation in frontal zones. *Geofys. Publ.,* **24**, 147–160.

——, 1984: Geostrophy. *Quart. J. Roy. Meteor. Soc.,* **110**, 1–12.

——, 1990: Transverse circulations in frontal zones. *Extratropical Cyclones. The Erik Palmén Memorial Volume.* C. Newton and E.O. Holopainen, Eds., Amer. Meteor. Soc., 155–165.

Emanuel, K. A., M. Fantini, and A. J. Thorpe, 1987: Baroclinic instability in an environment of small stability to slantwise moist convection. Part 1: Two-dimensional models. *J. Atmos. Sci.,* **44**, 1559–1573.

Geernaert, G. L., K. B. Katsaros, and K. Richter, 1986: Variation of the drag coefficient and its dependence on the sea state. *J. Geophys. Res.,* **91**, 7667–7679.

Grell, G., and E. Donall Grell, 1994: A multiscale simulation of an oceanic cyclone. *Proceedings. The Life Cycle of Extratropical Cyclones.* M. A. Shapiro and S. Grønås, Eds., Geophysical Institute, University of Bergen, Norway, Vol III, 83–88.

Grønås, S., 1995: The seclusion intensification of the New Year's storm 1992. *Tellus,* **47A**, No 5:1, 733-746.

——, A. Foss, and M. Lystad, 1987: Numerical simulations of polar lows in the Norwegian Sea. *Tellus,* **39A**, 334–354.

——, N. G. Kvamstø, and E. Raustein, 1994: Numerical simulation of the northern Germany storm of 27–28 August 1989. *Tellus,* **46A**, 635–650.

Haugen, J. E., 1985: Numerical simulations with an idealized model. The Polar Lows Project. Technical Report No.11, Norwegian Meteorological Institute, Oslo, Norway. (In Norwegian.)

Hines, K. M., and C. R. Mechoso, 1993: Influence of surface drag on the evolution of fronts. *Mon. Wea. Rev.,* **121**, 1152–1175.

Hoskins, B. J., M. E. McIntyre, and A. W. Robertson, 1985: On the use and significance of isentropic potential vorticity maps. *Quart. J. Roy. Meteor. Soc.,* **111**, 877–946.

James, P. K., and K. A. Browning, 1979: Mesoscale structure of line convection at surface cold fronts. *Quart. J. Roy. Meteor. Soc.,* **105**, 371–382.

Jansen, P. A. E. M., 1989: Wave-induced stress and the drag of air flow over sea waves. *J. Phys. Oceanogr.,* **19**, 745–754.

Joly, A., and A. J. Thorpe, 1987: Warm and occluded fronts in two-dimensional moist baroclinic instability. *Quart. J. Roy. Meteor. Soc.,* **115**, 513–534.

Jones, S. C., and A. J. Thorpe, 1992: The three-dimensional nature of symmetric instability. *Quart. J. Roy. Meteor. Soc.,* **118**, 227–258.

Keyser, D., and L. W. Uccellini, 1987: Regional models: Emerging research tools for synoptic meteorologists. *Bull. Amer. Meteor. Soc.,* **68**, 306–320.

——, M. A. Shapiro, and D. J. Perkey, 1978: An examination of frontal structure in a fine-mesh primitive equation model for numerical weather prediction. *Mon. Wea. Rev.,* **106**, 1112–1124.

Knight, D. J., and P. V. Hobbs, 1988: The mesoscale and microscale structure and organization of clouds and precipitation in midlatitude cyclones. Part XV: A numerical modeling study of frontogenesis and cold-frontal rainbands. *J. Atmos. Sci.,* **45**, 915–930.

Kuo, Y.-H., and S. Low-Nam, 1990: Prediction of nine explosive cyclones over the western Atlantic Ocean with a regional model. *Mon. Wea. Rev.,* **118**, 3–25.

——, and R. J. Reed, 1988: Numerical simulation of an explosively deepening cyclone in the eastern Pacific. *Mon. Wea. Rev.,* **116**, 2081–2105.

——, ——, and S. Low-Nam, 1991a: Effects of surface energy fluxes during the early development and rapid intensification stages of seven explosive cyclones in the Western Atlantic. *Mon. Wea. Rev.,* **119**, 457–476.

——, M. A. Shapiro, and E. Donall, 1991b: Interaction between baroclinic and diabatic processes in a numerical simulation of a rapidly developing marine cyclone. *Mon. Wea. Rev.,* **119**, 368–384.

Lystad, M., 1986 (Ed.): Polar lows in the Norwegian, Greenland and Barents Seas. Final Report, Polar Lows Project. The Norwegian Meteorological Institute, Oslo, Norway, 196 pp.

Montgomery, M. T., and B. F. Farrel, 1992: Polar low dynamics. *J. Atmos. Sci.,* **49**, 2484–2505.

Neiman, P. J., and M. A. Shapiro, 1993: The life cycle of an extratropical marine cyclone. Part 1: Frontal-cyclone evolution and thermodynamic air-sea interaction. *Mon. Wea. Rev.,* **121**, 2153–2176.

——, ——, and L. S. Fedor, 1993: The life cycle of an Extratropical Marine Cyclone. Part II: Mesoscale structures and diagnostics. *Mon. Wea. Rev.*, **121**, 2177–2199.

Nordeng, T. E., 1987: The effect of vertical and slantwise convection on the simulation of polar lows. *Tellus*, **39A**, 354–375.

——, 1990: A model-based diagnostic study of the development and maintenance mechanism of two polar lows. *Tellus*, 42A, 92–108.

——, 1991: On the wave age dependent drag coefficient and roughness length at sea. *J. Geophys. Res.*, **96**, 7167–7174.

—— and E. Rasmussen, 1992: A most beautiful polar low. A case study of a polar low development in the Bear Island region. *Tellus*, **44A**, 81–99.

——, A. Foss, S. Grønås, M. Lystad, and K.H. Midtbø, 1989: On the role of resolution and physical parameterization for numerical simulations of polar lows. *Polar and Arctic Lows*. A. Deepak Publishing, 217–232.

Ohm, C., 1993: Numerical simulation of the New Year's Storm 1992. Master's thesis, Geophysical Institute, University of Bergen, Norway.

Pedersen, K., 1963: On qualitative precipitation forecasting with a quasi-geostrophic model. *Geofys. Publ.*, **25**, 1–25.

Persson, P. O. G., and T. T. Warner, 1993: Nonlinear hydrostatic conditional symmetric instability: Implications for numerical weather prediction. *Mon. Wea. Rev.*, **121**, 1821–1833.

Reed, R. J., and C. N. Duncan, 1987: Baroclinic instability as a mechanism for the serial development of polar lows: A case study. *Tellus*, **39A**, 376–384.

——, and A. J. Simmons, 1991: An explosive deepening cyclone over the North Atlantic that was unaffected by concurrent surface fluxes. *Wea. Forecasting*, **6**, 117–122.

Sawyer, J. S., 1956: The vertical circulation at meteorological fronts and its relation to frontogenesis. *Proc. Roy. Soc. London.* **A234**, 346–362.

Shapiro, M. A., and L. S. Fedor, 1989: A case study of an ice-edge boundary layer front over the Barents Sea. *Polar and Arctic Lows*. A. Deepak Publishing, 257–278.

——, and D. A. Keyser, 1990: Fronts, jet streams and the tropopause.

Extratropical Cyclones. The Erik Palmén Memorial Volume. C. Newton and E. O. Holopainen, Eds., Amer. Meteor. Soc., 167–191.

——, E. G. Donall, P. J. Neiman, L. S. Fedor, and F. Gonzalez, 1991: Refinements in the conceptual models of extratropical cyclones. Preprints, *First International Winter Storm Symposium*, New Orleans, Amer. Meteor. Soc., 6–13.

——, L. S. Fedor, and T. Hampel, 1987: Research aircraft measurements of a polar low over The Norwegian Sea. *Tellus*, **39A**, 272–306.

——, T. Hampel, and L. S. Fedor, 1989: Research aircraft observations of an arctic front over the Barents Sea. *Polar and Arctic Lows*. A. Deepak Publishing, 279–290.

Sundqvist, H., E. Berge, and J. E. Kristjansson, 1989: Condensation and cloud parameterization studies with a mesoscale numerical weather prediction model. *Mon. Wea. Rev.*, **117**, 1641–1657.

Thorncroft, C. D., B. J. Hoskins, and M. E. McIntyre, 1993: Two paradigms of baroclinic-wave life-cycle behavior. *Quart. J. Roy. Meteor. Soc.*, **510**, 17–55.

Thorpe, A. J., 1986: Synoptic scale disturbances with circular symmetry. *Mon. Wea. Rev.*, **114**, 1384–1389.

——, and K. A. Emanuel, 1985: Frontogenesis in the presence of small stability to slantwise convection. *J. Atmos. Sci.*, **42**, 1809–1824.

Todsen, M., 1964: A Study of Vertical Circulations in a Cold Front. Part IV, Final Report, Contr. No. AF 61 (052)-525.

Uccellini, L.W., 1990: Processes contributing to the rapid development of extratropical cyclones. *Extratropical Cyclones. The Erik Palmén Memorial Volume.* C. Newton and E. O. Holopainen, Eds., Amer. Meteor. Soc., 81–105.

——, R. A. Petersen, K. F. Brill, P. J. Kocín, and J.J. Tuccillo, 1987: Synergistic interactions between an upper-level jet streak and diabatic processes that influence the development of a low-level jet and a secondary coastal cyclone. *Mon. Wea. Rev.*, **115**, 2227–2261.

Zick, C., 1983: Method and results of an analysis of comma cloud developments by means of vorticity fields from upper tropospheric satellite winds. *Meteor. Rundsch*, **36**, 69–84.

Advances in Forecasting Extratropical Cyclogenesis at the National Meteorological Center

LOUIS W. UCCELLINI

Office of Meteorology, National Weather Service, Silver Spring, MD, USA

PAUL J. KOCIN

Hydrometeorological Prediction Center, National Center for Environmental Prediction, National Weather Service, Camp Springs, MD, USA

JOSEPH M. SIENKIEWICZ

Marine Prediction Center, National Center for Environmental Prediction, National Weather Service, Camp Springs, MD, USA

Abstract

The National Meteorological Center[1] represents one of the major meteorological forecast centers in the world and is the central component of a "forecaster–machine" mix that characterizes the forecast process in the United States. In this chapter, the lineage from the concepts of Vilhelm Bjerknes and the Bergen school of the late 1910s to the creation of the National Meteorological Center in 1960 is traced. Furthermore, the improvements made in forecasting cyclogenesis at the National Meteorological Center are examined, with emphasis placed on (1) the advancements in numerical prediction, and (2) the role of the forecasters who have become skillful at understanding the strengths and weaknesses of numerical prediction models and applying them in forecasting on a day-to-day basis.

By all subjective and objective measures, tremendous advances have been made in predicting explosive cyclogenesis, especially since the middle 1980s. Not only have short-range predictions increased in accuracy, but the range of "useful" predictions of major storm events has been extended out to five days in advance. Rapid and secondary cyclogenesis are now routinely predicted, even for the "data-void" regions over the oceans with false alarm rates now less than 10% for day-4 forecasts.

These advancements are a direct result of (1) the introduction of numerical models with higher resolution and more sophisticated treatment of physical processes, (2) the improvements in model-based data assimilation schemes which rely on an increasingly asynoptic dataset covering the entire globe, and (3) a better trained work force capable of working with the many existing numerical models and associated model output statistics, recognizing the strengths and weaknesses of each to produce consistent and increasingly accurate weather forecasts.

Given the improvements in forecasting extratropical cyclogenesis on a routine basis and predicting major storm events marked by rapid cyclogenesis four and five days in advance, one can only conclude that the original vision of forecasting the weather through objective mathematical approaches that motivated the creators of the Bergen school has, in fact, been realized. This accomplishment is the result of the efforts of many meteorologists located throughout the world over the past 80 years and represents one of the major intellectual achievements of the twentieth century.

1. Introduction

The creation of the Bergen School by Vilhelm Bjerknes and his collaborators has had a profound influence on the understanding of the baroclinic nature of cyclones and fronts. Papers presented at the 1994 Symposium on the Life Cycles

---

[1]In 1995, the National Meteorological Center was restructured and is now known as the National Centers for Environmental Prediction (NCEP).

of Extratropical Cyclones[2] and at the 1988 Palmen Memorial Symposium on Extratropical Cyclones (Erik Palmen Memorial Volume, 1990) illustrate the success of combining the basic physical and mathematical principles derived by Bjerknes with careful synoptic analysis developed at the Bergen school to provide a basis for our understanding of (1) the life cycle of cyclones, (2) the structure of fronts, (3) the air flow through these systems, and (4) the associated distribution of clouds and precipitation.

However, the increased understanding of cyclones and fronts was not the sole motivating force behind the Bergen school. Vilhelm Bjerknes's vision went beyond devising a circulation theorem that could explain the existing state of the atmosphere. He also promoted the idea that these same mathematical concepts could be used as a basis for weather forecasting. Thor Bergeron described the importance of forecasting as a motivating force for Bjerknes's efforts:

> During the period [early 1900s], Bjerknes got more and more convinced that dynamic meteorology had only one main, all embracing task: to predict future atmospheric states. This problem, he meant, should then not be treated by sheer empiricism, i.e., cataloguing and memorizing, of isobaric patterns and weather types, as in the current weather forecasting of those days, but as a mathematically well defined physical problem. He even thought of the mechanical integration of differential equations, i.e., numerical computations, when they were not analytically integrable. (Bergeron et al. 1962, page 14)

Thus, with the creation of the Leipzig school in Germany and then the Bergen school in Norway, Bjerknes sought to revolutionize weather forecasting and to make it a viable aid to agriculture and to fishermen and explorers who risked their lives on the high seas. He based this approach on: (1) expanded observations, (2) detailed analysis, and (3) the application of mathematically based physical principles to the study of convergence lines (fronts) and cyclones.

Realization of the Bergen school's vision for the application of dynamic principles to weather forecasting was delayed first by the reluctance of the established meteorological community to accept these new approaches and then by the events surrounding World War II. After World War II, however, scientists in the United States took up the challenge originally offered by Bjerknes. After emphasizing that Bjerknes "defined the aim of dynamic meteorology to be the prediction of the future state of the atmosphere" (p. 231), Charney (1950) reviewed the link between the accomplishments of the Leipzig and Bergen Schools in the early part of the century and the numerical prediction effort initiated in the United States after World War II. This linkage extends from

[2]"The Life Cycles of Extratropical Cyclones." An International Symposium. 27 June – 1 July 1994, Bergen, Norway. Volumes I, II, and III, published by the Geophysical Institute, University of Bergen, Bergen, Norway.

(1) J. Bjerknes and Solberg's basic analytic and synoptic analyses of fronts and cyclones and Richardson's attempt at numerical prediction in the 1920s and (2) their simple wave models promoted to explain cyclone evolution in the 1930s to (3) Bjerknes and Holmboe's application of these wave models to upper-level trough/ridge systems and Rossby's wave theory in the 1940s to (4) the evolution of the baroclinic wave theory (and associated set of equations) for which Charney (who was J. Bjerknes first Ph.D. student at UCLA) played such a major role and which established a firm basis for the successful application of numerical models that was to follow.

Thompson (1987) describes the revolutionary advances made by the Meteorology Group established under the direction of Jules Charney in the late 1940s and early 1950s. This group was established by Jon Von Neumann at the Institute of Advanced Studies at Princeton. Von Neumann was very active in the initial application of the newly emerging computer technologies and made a concerted effort to build upon the scientific potential of a high-speed "stored-program" computer. He decided the major challenge to address was weather prediction (Thompson 1987). The collaboration among the many scientists associated with the Meteorology Group (under Charney's direction) within the Electronic Computer Project (under Von Neumann's direction) led to the first computer-based 24-hour prediction of the 500-mb flow using a barotropic model developed by Charney and Fjortoft.

At the 75th Annual Meeting of the American Meteorological Society, George Cressman provided a review of this exciting era and related the camaraderie among meteorologists from all over Europe and the United States during their quest to develop mathematically based numerical forecasts as a "Meteorological Camelot." The advances in numerical weather prediction made by the Meteorology Group, most notably the development of baroclinic prediction models and the early application of these models to several major cyclone events (e.g., the 6 November 1953 case in the eastern United States), led directly to the decision to form an operational modeling group in 1954 called the Joint Numerical Weather Prediction Unit (JNWPU) (Shuman 1989). The creation of the JNWPU was the forerunner to the formation of the National Meteorological Center (NMC) in 1960. The NMC brought together those tasked with developing numerical models with a large group of forecasters responsible for providing national guidance products for use by local forecast offices throughout the country.

With the creation of the NMC in 1960, the United States embarked on transforming the forecast process from one of observations and subjective forecasting to a "forecaster-machine" mix that today involves:

1. more extensive global observations based on a wide variety of in situ and remotely sensed measurements

2. expanded use of model-based data assimilation systems (Dey 1989)

3. the application of global and regional numerical models (Bonner 1989)

4. extensive use of model output statistics (Glahn and Lowry, 1972; Carter et al. 1989)

5. forecaster interpretation and comparison of numerical model forecasts to create national guidance products (Corfidi and Comba 1989)

6. the production of local warnings and local forecasts at Weather Forecast Offices (WFO) that are disseminated to the public and other Federal agencies

7. the production of hydrologic and flood forecasts by River Forecast Centers (Stallings and Wenzel 1995)

8. the coordination of forecasts and warnings with a diverse user community, especially those involved with emergency management.

The purpose of this chapter is to review the performance of the "forecaster–machine" mix at the NMC with a specific emphasis on the ability to predict cyclogenesis. The advances in the model forecasts of cyclones are reviewed in Section 2. The role of the forecasters in the Meteorological Operations Division (MOD)[3] in the NMC is discussed in Section 3. The performance of the recently created Marine Forecast Branch in predicting explosive cyclogenesis over the Pacific Ocean is discussed in Section 4. In Section 5, several recent forecast achievements are described, including the unprecedented forecast effort put forth for the March 1993 "Superstorm" over the eastern third of the United States. Prospects for the future of NMC's forecast approaches are discussed in Section 6.

2. NMC Model Performance in Cyclone Forecasting

The creation of the JNWPU (and then the NMC) was based on the initial success of the Meteorology Group at Princeton in simulating the large-scale aspects of major cyclone events and was driven by the requirements related to the operational production of numerical weather forecasts (Shuman 1989). As advances were made in devising and applying numerical schemes on bigger and faster computers and as numerical models evolved from a one-layer barotropic model to a six-level primitive equation model, there was a steady improvement in the 500 mb forecasts to levels well above the skill exhibited by subjective forecasts before the advent of numerical prediction (Fig. 1). As noted in Fig. 1, the increasing accuracy in the 500 mb height forecasts over a 35-year period

---

[3]In 1995, the Meteorological Operations Division was renamed the Hydrometeorological Prediction Center.

represents a steady trend and is a direct result of improvements in the numerical models, model-based data assimilation systems, and the increasing power of the computers that could be applied to the integration of the full set of predictive equations.

In contrast to the steady improvements in the prediction of 500 mb height fields shown in Fig. 1, the impact of the early numerical models on predicting surface cyclogenesis was far less encouraging. Shuman (1989) states: "In the beginning, numerical predictions could not compete with those produced manually. They had several serious flaws, among them over-prediction of cyclone development. Far too many cyclones were predicted to deepen into storms" (p. 287). The optimism reached by the occasional correct prediction of a major cyclone event (e.g., Brown and Olson 1978) was more than offset by the general inability of the early hemispheric and regional models to predict surface cyclogenesis with acceptable reliability one to two days in advance throughout the 1960s, 1970s and early 1980s. Even with the introduction of the Limited Area Fine Mesh Model (LFM) (Gerrity 1977), the prediction of rapid cyclogenesis did not markedly improve. After his first review of the performance of operational models in predicting cyclogenesis, Sanders (1986) concludes that the operational models essentially captured the "baroclinic nature of cyclogenesis but the intensity of the response to the baroclinic forcing remains intractable" (p. 2207).

The inability of the earliest hemispheric and regional numerical models to provide reliable forecast guidance for rapid cyclogenesis, especially those that occurred near coastlines or over the oceans, caused many problems for the forecast community and even led Ramage (1976) to call for an end to the application of numerical models for cyclone prediction by urging that the meteorological community "abandon research that uses weather sequences in a computer as bases for deduction about the real atmosphere" (p. 9). Yet, the experience gained by identifying and examining the shortcomings of the early operational models combined with the rapid advances made by the application of research models to the study of synoptic and mesoscale systems (see e.g., Keyser and Uccellini 1987) contributed to an acceleration in the introduction of new operational models and model-based analysis systems beginning in the early to middle 1980s (Bonner 1989).

Perhaps the most important advancements at the NMC during this era were the introduction of (1) the Regional Analysis and Forecast System (RAFS) on 27 March 1985 (see Hoke et al. 1989) based on the Nested Grid Model (NGM) devised by Norman A. Phillips, and (2) the global Medium-Range Forecast (MRF) model in April 1985 (Kanamitsu 1989). These developments were based on improved physics, increased resolution, improved data assimilation and resulting initial analyses, and improved numerics, which were all designed to take maximum advantage of the

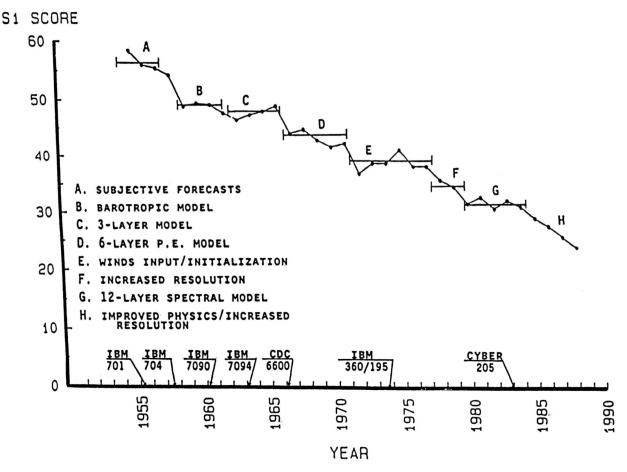

Fig. 1. Record of S_1 score (Teweles and Wobus 1954) for 36-h predictions of geopotential height at 500 mb. The S_1 score is roughly a measure of normalized rms vector error of geopotential height gradient. The area of verification is North America and adjacent waters. To calibrate the S_1 score in terms of practical skill, a forecast with a score of 20 is virtually perfect, and one with a score of 70 is worthless, e.g., skill (percent) $= 2 \times (70 - S_1)$ (Shuman 1989). Since 1988, the S1 score has continued to decrease to a level just above 20, indicating increasing skill associated with the introduction of higher resolution models now run on the Cray C90 coimputer.

new powerful computers utilized by the NMC. The notable advancements in the operational global observing network, data assimilation schemes and numerical forecast models in the early and middle 1980s were a direct result of the Global Weather Experiment planned and carried out in the 1960s and 1970s, culminating with the First GARP Global Experiment (FGGE) in 1978 and 1979 (see, e.g., Johnson 1986). The impact of the model advances at the NMC in 1985 on the forecasts of cyclogenesis was almost immediate.

Just one year after Sanders published his review of the performance of the LFM and termed the problem of predicting rapid oceanic cyclogenesis as "intractable," he began publishing a series of papers in which he tracked the positive trend in cyclone prediction. Sanders (1987) emphasizes that the NGM and the MRF showed distinct improvements in forecasting rapid cyclogenesis, with the NGM showing useful skill in predicting the extreme deepening rates associated with explosive cyclogenesis out to 48 hours and the MRF out to 60 hours, while noting decreasing skill in the later time

periods for each model. He also noted that the models showed more skill in the Atlantic than the Pacific Ocean basins, suggestive of the importance of the dense North American data network, along with the improved resolution and physical parameterizations, as necessary components for advancing the numerical prediction of cyclones. He concluded that "These results alter the nature of the problem of research on explosive cyclogenesis from one of discovering a missing ingredient to one of improving the performance and extending the range of predictability" (p. 322).

A series of papers by Junker et al. (1989), Sanders and Auciello (1989), Mullen and Smith (1990), Sanders (1992), Grumm et al. (1992), Smith and Mullen (1993), Oravec and Grumm (1993) and Grumm (1993) all point to the increasing skill of the NMC operational global and regional models over the 10 years in predicting explosive cyclogenesis, with the global models generally outperforming the regional models in predicting cyclones. This skill is measured by the increases in the Probability of Detection (POD) of major cyclogenesis.

Sanders and Auciello (1989) found a POD of 72% for the 0- to 24-h NGM and 42% for the 36-to-60 h Aviation run of the MRF (known as the AVN). Grumm (1993) found that the POD had risen to 83% at 48 h and to 50% at 72 h for the AVN. The False Alarm Rate (FAR) also decreased from the 17% at 24 h for the NGM and 34% at 60 hours for the AVN (Sanders and Auciello 1989) to a FAR that now remains below 10% for all time periods out to 72 h in the AVN (Grumm 1993).

A dramatic increase in skill has also been documented in the 3- to 5-day range. The long-term Mean Sea-Level Pressure (MSLP) anomaly correlation score for the entire globe is now as high on day 5 as it was for day 3 only 15 years ago (Stokols et al. 1991; Fig. 2). Sanders (1992) and more recently Bedrick et al. (1994) emphasize that the global models all show "useful skill" in the 3- to 5-day range in predicting rapid cyclogenesis, especially for the western Atlantic. Grumm (1993) concurs with Sanders's finding and concludes his 15 month review of the AVN by noting that the global models can now predict well-developed cyclone circulations at any forecast length and that "if the AVN forecasts a strong cyclone 3 days in advance, it is likely to happen." From these results, it is readily apparent that numerical models (1) have provided highly accurate forecasts of rapid cyclogenesis out to three days in advance, and (2) have extended the range of useful prediction of major cyclone events to day 5.

3. The Role of the Forecasters in the NMC

Shuman's (1989) review article on the development and application of operational numerical models at the NMC emphasizes the importance of collocating the modelers who develop and apply numerical prediction techniques with the forecasters in the Meteorological Operations Division who interpret the model solutions and issue daily national weather forecasts. The day-to-day feedback offered by this collaboration no doubt contributed to the remarkable improvements in forecast skill from 1958 to the present as documented by Shuman (1989) and Bonner (1989).

The MOD of the NMC is the primary national analysis and forecast center of the NWS which provides many guidance products to weather forecasters throughout the United States and around the World (Corfidi and Comba 1989). The MOD forecasters combine information from the NMC, ECMWF, and UKMET global models to produce "Medium-Range" (3- to 5-day) forecasts and combines output from several regional models to produce "Short-Range" (1- to 2-day) forecasts. The MOD products are designed to meet the requirements of general weather, aviation, and marine forecasting. Given the day-to-day uncertainties of numerical model forecasts and the intermodel differences, the subjective evaluations and interpretations made by MOD forecasters are an essential component of the forecast process that attempt to unify and coordinate the local forecasts released to the general public and other users (McPherson and Olson 1991).

Tracton (1993) reviews the performance of forecasters relative to the MRF and ECMWF model forecasts in the Medium Range. He showed that the forecaster's skill as measured by the standardized anomaly correlation score for the MSLP (Murphy and Epstein 1989) is generally 2 to 3 points higher than the best model on any given day. These results agree with those presented by Stokols et al. (1991), which are shown in Fig. 3 and illustrate that the forecaster's

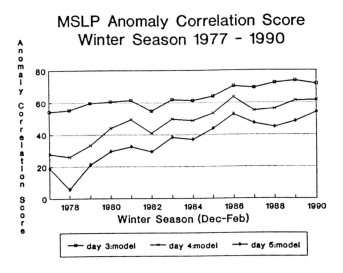

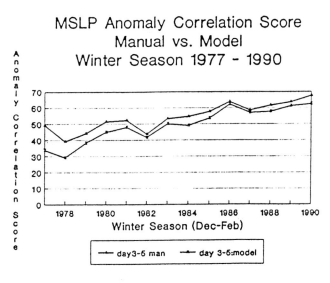

Fig. 2. Long-term Mean Sea-Level Pressure (MSLP) Anomaly Correlation Score for days 3, 4, and 5 for the Global Spectral Model during the period 1977–1990 (Stokols et al. 1991).

Fig. 3. Long-term Mean Sea-Level Pressure (MSLP) Anomaly Correlation Score for days 3–5: model vs. manual for the period 1977–1990 (Stokols et al. 1991).

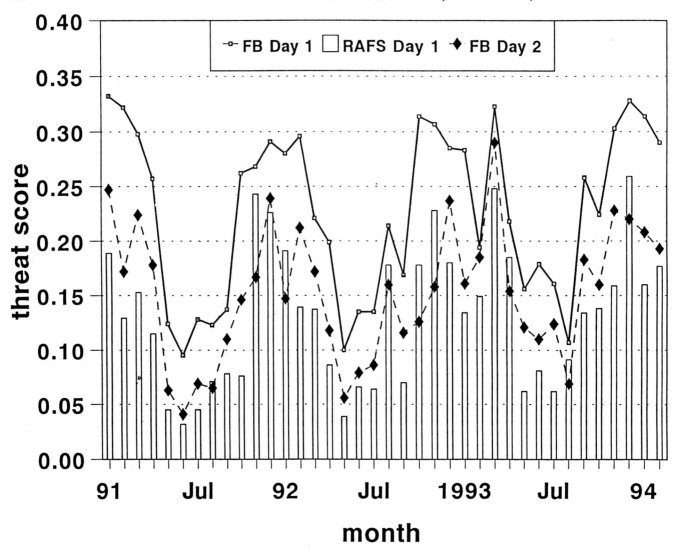

FIG. 4. Monthly threat scores for 24-hour 1" quantitative precipitation forecasts. Comparison between Day 1 (0-h to 24-h) and Day 2 (24-h to 48-h) manual total precipitation forecasts and the Day 1 (0- to 24-hour) NGM (RAFS) forecasts for the period January 1991 through January 1994 (Olson et al., 1995).

skill level has increased with the increasing skill of the models. Tracton (1993) also emphasizes that the forecaster's scores are more consistent from day to day than any of the models despite the "large differences among the model solutions" on any given day and "considerable temporal variability" in numerical model predictions on a day-to-day basis (p. 152). These comparisons reflect the human forecaster's ability to select the "most appropriate 'model of the day' in formulating their predictions" (p. 152) to produce surface forecasts that are generally more accurate than any given model.

The forecasters have also been able to make consistent improvements upon the model's quantitative precipitation forecasts in the 1- to 2-day range (Fig. 4). The forecasters day-1 threat scores are always significantly higher than the models for all categories (Olson et al. 1995). Furthermore, the day-2 forecaster scores are generally higher than the day-1

model scores for the higher precipitation categories (see e.g., Fig. 4). The forecasters and models exhibit the most skill in the winter season, given the relative success of the forecaster-machine mix to predict the baroclinic systems and associated weather patterns compared to the difficulties the models have in predicting summer convective systems (Funk 1991).

The positive trend in forecasting skill (Fig. 3) reflects an increasing level of confidence among the forecasters in recognizing model strengths (e.g., height and wind prediction) and weaknesses (e.g., quantitative precipitation forecasts) as a function of various flow regimes and their ability to make the correct adjustments to the model predictions. The forecasters have become particularly adept at extending the range of prediction for major cyclone events to 3 to 5 days in advance (Stokols et al. 1991; Uccellini et al. 1995). An illustration of this skill as it applies to predicting rapid oceanic cyclogenesis is discussed in the next section.

4. Forecasting Oceanic Cyclones

4.1. Marine Forecasting at NMC

In the summer of 1989, the newly formed Satellite Marine Section (SMS) of MOD began issuing High Seas marine warnings and forecasts (up to 36 hours) for the western North Atlantic. Traditionally, this function had been done at the local forecast office level but was centralized as an early step in the modernization of the NWS. The MOD was no longer solely responsible for guidance products designed to be further interpreted by the meteorological community, but was now actively issuing wind warnings and sea-state forecasts that are released directly to the marine community, including ships on the high seas. Similarly, the High-Seas warning and forecast functions for the central and eastern North Pacific and Bering Sea were transferred to SMS in March 1991 from the San Francisco Weather Service Forecast Office. These high-seas text bulletins fulfill the U.S. requirement for providing marine warnings for the safety of vessels at sea in accordance with international agreement.

In spring 1993, the SMS was upgraded to the Marine Forecast Branch (MFB)[4] solidifying NWS and NMC commitment to providing accurate warnings and forecasts of oceanic cyclones in support of safety at sea. MFB operational forecasters have worked directly with the developers of numerical guidance products within NMC's Ocean Products Center (Rao 1989) and other NMC components to improve the quality of the forecasts. In addition, MFB forecasters have also responded to the requests and needs of ocean mariners (Chesneau 1994), resulting in an enhanced set of products for a high-frequency radiofacsimile transmission for both the North Pacific and North Atlantic. This expansion was implemented in January 1994 and provides the marine community a graphical depiction and associated watches and warnings out to day 4 for each ocean basin. The product suites consist of: surface analyses; 24-, 48-, and 96- hour surface forecasts; 500 mb analyses and forecasts; sea-state analyses and forecasts; and oceanographic analyses.

In preparing analyses and forecast products, the MFB meteorologists work with a variety of atmospheric numerical models (AVN run of the NMC Global Spectral Model, NGM, ETA, MRF, UKMET, ECMWF, and U.S. Navy NOGAPS) and ocean wave models developed by the NWS and the U.S. Navy. By knowing the strengths and weaknesses of these models, the MFB forecasters routinely forecast explosive cyclogenesis, secondary development, low latitude cut-offs, and conversion of tropical storms to extratropical cyclones out to day 4 for each ocean basin.

---

[4]In 1995, the Marine Forecast Branch was transferred from MOD into the newly formed Marine Prediction Center within NCEP.

4.2 Skill in Day 4 Forecasts of Significant Ocean Cyclones

To illustrate the performance of MFB forecasts, a comparison was made between forecast (manual and MRF day 4 forecasts) and observed (NMC 0000 UTC Northern Hemisphere (NH) analyses) positions and central pressures for significant cyclones over the North Pacific for the 5-month period 1 November 1992 to 31 March 1993. For the purpose of this study, a significant cyclone was defined as any cyclone that reached a minimum central pressure of less than or equal to 980 mb on the NMC Northern Hemisphere manual surface analyses valid at 0000, 0600, 1200, and 1800 UTC. No reanalysis of NMC NH surface analyses was done, although in the event of a rare error, some interpolation was done to determine central pressure.

Since mariners on ocean transit are most concerned with the significant cyclones, comparisons were made between forecast and observed cyclones at the 0000 UTC synoptic time as close as possible to the minimum central pressure of each identified cyclone. This coincides with the synoptic observing time with the maximum number of ship observations available for the North Pacific (Sienkiewicz 1992). Therefore, for each cyclone only one comparison between forecast (manual and MRF) and observed central pressure and position was made. The intent was to determine if the significant events are being forecast four days in advance by both manual and model forecasts. It was also assumed that the satellite-estimated position of the cyclone center for deep-ocean storms is most accurate near the time of minimum central pressure. The forecast domain was the eastern and central North Pacific east of 160°E and north of 30°N.

4.2.1 Mean pressure errors

During the study period, 87 cyclones met the selection criteria. Figure 5a shows the distribution of pressure error for both the manual and MRF forecasts, and Fig. 5b shows the cumulative frequency distribution of central pressure for both observed and matching forecast cyclones. Observed central pressures at verification time (0000 UTC) varied from 943 mb (2 cyclones) to 998 mb with approximately 25% of the cyclones 965 mb or deeper. For the 87 observed cyclones, 79 were forecast manually, with 80 forecast by the MRF. The difference in the number of matching forecasts between the manual product and the MRF was due to one cyclone center being forecast west of the area used to verify the forecasts.

For this sample, the average life cycle for all 87 cyclones was 87 hours, or a little more than 3½ days. For the 7 cyclones not forecast by the MRF, the average life cycle was approximately 2 days. Therefore, cyclones not forecasted by the MRF or forecast with the least accuracy on average had significantly shorter life cycles than those forecast well four to five days in advance.

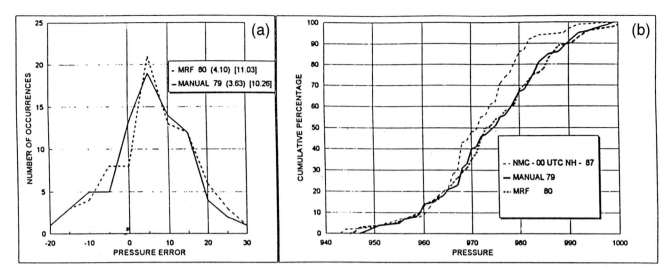

FIG. 5. (a) Frequency distribution of mean pressure errors for manual (solid) and MRF (dashed) day-4 forecasts for all cyclones observed to reach a minimum pressure of 980 mb or less. The number of forecast cyclones, mean pressure error (in parentheses) and RMS (brackets) for both manual and MRF forecasts are shown in the key. (b) Cumulative percent frequency distribution of central pressures for all cyclones to have reached a minimum pressure of 980 mb or less. Observed cyclones from NMC NH analyses are shown by a dashed line, Manual forecasts are solid, and MRF forecasts are shown by a dotted line.

The frequency distribution of sea-level pressure error (Fig. 5a) illustrates the tendency for both manual and MRF forecasts to underforecast the intensity of the significant cyclones at their time of maximum strength. The manual forecasts show improvement over the MRF in the error range from −5 to +5 mb and again at pressure errors from +15 to +28 mb. There is virtually no difference between MRF and manual forecasts of the outlyers at errors of +30 and −20 mb.

In Fig 5b. it is significant to see that the manual and MRF forecasts parallel the observed distribution for cyclones below 965 mb then begin to gradually lose ground above 965 mb to approximately 978 mb, then more rapidly above 978 mb. Manual forecasts are better than the MRF specifically in the 968- to 973-mb range and again above 984 mb. This is most likely due to MFB forecasters systematically deepening the MRF forecast pressures in these ranges. This distribution also suggests that forecasters may, in fact, hesitate to deepen the MRF forecast pressures below 960 mb to any significant degree.

The manual mean and RMS pressure errors only slightly bettered the 96-hour MRF forecasts. This is consistent with the findings by Stokols et al. (1991) that the trend in recent years has been for manual forecasts of the MSLP to improve upon the model forecasts with only slightly more skill. According to Stokols et al. (1991) this gap has continued to narrow as models have continued to improve.

Cyclones were further stratified into a subset consisting of all those that reached a minimum pressure of 965 mb or less. The frequency distribution of pressure error for this stratum is shown in Fig. 6a. Again, the tendency of the MRF

to forecast positive pressure errors is clearly evident. The manual forecasts do show improvement over the MRF between the −10 and +25 mb pressure error and, in particular, for pressure error equal to 0. Two outlyers at −15 and one at +30 mb pressure error are slightly worse than MRF forecasts.

The cumulative percent frequency distribution for this category is shown in Fig. 6b. Thirty observed cyclones met this criteria, with 28 and 29 forecast by the manual and MRF, respectively. The distributions of forecasts and observed central pressures parallel each other to approximately 960 mb, then diverge with manual forecasts consistently 1 to 2 mb less than the MRF. Again, this suggests that forecasters are bettering the MRF MSLP forecasts for cyclones observed that have central pressure greater than 960 mb. However, the distribution shows that both manual and MRF forecasts fail to forecast a noticeable number of cyclones observed to have a central MSLP of less than 970 mb.

Mean pressure errors and RMS for this category of cyclones were slightly larger than for all cyclones to have reached a minimum central pressure less than or equal to 980 mb. A total of 28 cyclones had mean deepening rates of 1 Bergeron[5] (Sanders and Gyakum 1980) or greater. This is considerably different from the results discussed by Sanders (1992) and suggests 1992–1993 was significantly more active over the North Pacific than the period used in the Sanders study.

---

[5]One Bergeron is defined as a 24-hour deepening rate for a cyclone as a function of latitude; where 1 Bergeron is equivalent to 14 mb at 30°N increasing to 24 mb at 60°N.

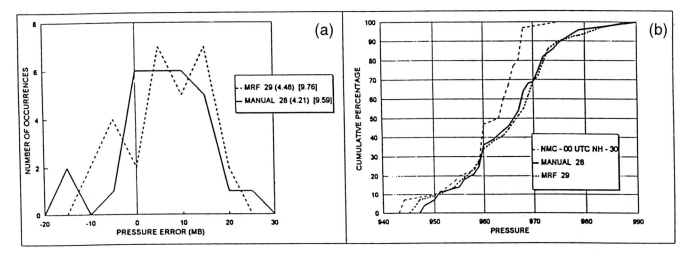

FIG. 6. As in Fig. 5 except for all cyclones observed to reach a minimum pressure of 965 mb or less.

TABLE 1

Number of forecast cyclones for manual, MRF, UKMET, and ECMWF day-4 forecasts.
Included are Mean Position Errors (MPE) in km, RMS, and Mean Vector Errors (MVE) (deg, km)
for all cyclones observed to reach a central pressure of 980 mb or less and 965 mb or less.

	Manual		MRF		UKMET		ECMWF	
	≤980 mb	≤965 mb	≤980 mb	≤965 mb	≤980 mb	≤965 mb	≤980 mb	≤965 mb
N	79	28	80	29	66	22	68	23
MPE	527	461	526	439	663	505	540	513
RMS	615	517	610	484	656	576	604	563
MVE	−35330	−41115	−26256	−34588	−3297	−95153	−184147	−146131

4.2.2 Position errors

Mean Position Errors (MPEs), RMS, and Mean Vector Errors (MVEs) were calculated for all forecast cyclones for the manual, MRF, UKMET, and ECMWF for both classifications of cyclones and are shown in Table 1 (ECMWF 96-hour forecasts of cyclones were compared to NMC NH analyses valid at 1200 UTC). Unfortunately, the UKMET and ECMWF forecasts were not available for the entire forecast domain for the period 1 November 1992 to 17 December 1992. This in part accounts for the lower numbers of forecast cyclones. The manual MPEs were comparable to the MRF forecasts, showing no increased skill over the MRF model forecast positions. However, both manual and MRF forecasts were better in the mean than both UKMET and ECMWF forecasts for this sample. The MPEs for all forecasts for the first classification of storms are in excellent agreement with the results of Sanders (1992). For the stratum

of stronger cyclones, the MPEs were significantly smaller, indicating that *the positions of the strongest cyclones with a central MSLP less than 965 mb were forecast more accurately than for all cyclones in this sample that had a MSLP of 980 mb or less.*

Mean Vector Errors (MVE) varied widely with the ECMWF forecasts having the largest southerly MVE. This is consistent with NMC operational forecasters observation that the high resolution T213 ECMWF model tends to displace surface cyclones too far south in the Pacific domain as compared to the MRF and UKMET models. Interestingly, for the stronger classification of cyclones, the magnitude of MVEs grew significantly with all forecasts with the exception of the ECMWF. In addition, MVEs gained an increasing easterly component for the stronger cyclones with the exception of the MRF, which gained northerly component. This suggests that the forecasts of deep ocean storms that are less than 965 mb (with the exception of the MRF) tend to be too fast with the zonal component of motion. It also suggests that the MRF may be moving surface lows too far north into the cold air.

Interestingly, the mean track speed of motion (13.5 ms^{-1}) for all the cyclones was significantly faster than the 6 ms^{-1} value discussed by Sanders (1992). Track speeds of the stronger category of cyclones during the deepening phase was 17 ms^{-1}. This is comparable to the mean track speeds Sanders (1992) discussed for the Atlantic cases. Sanders (1992) also suggests that for the Pacific cases the growth rate of position error was a little more than 10% of the track speed and that for the Atlantic cases was a little less than 10% of the track speed. Thus, it was surmised that although position errors were larger in the Atlantic than Pacific, the Pacific position forecasts were relatively poorer. Although the growth rate of position error was not calculated for this study, the significantly faster track speeds with comparable MPEs to Sanders (1992) suggests that this may not be the case. The fact that the MPEs decreased for the stronger classification of storms for all forecasts, despite the large mean track speeds, is quite encouraging. This suggests that even in a data-sparse region such as the North Pacific, the occurrence and position of intense, fast-moving oceanic storms can be forecast in the day-4 time range.

The false alarm rate was also computed for the manual forecasts and was considered to be cyclones forecast to be less than 980 mb but failed to reach that depth. Negative pressure errors associated with cyclones that did eventually deepen and with filling cyclones were not considered to be false alarms. A total of 7 false alarms for the manual forecasts were observed during this period. In general, these were cases for which the manual forecast portrayed a single strong cyclone with the NMC Northern Hemisphere analyses showing multiple centers with the energy associated with the storm system considerably more fragmented. For the 79 manual forecasts this yields a FAR of 8%. This is consistent with results by Grumm (1993) showing a FAR of less than 10% for the AVN 72-hour forecasts.

In summary, these results show the dependence of MFB forecasters on model forecasts in preparing manual day-4 forecasts of significant ocean cyclones in the North Pacific Ocean. MFB forecasters slightly bettered MRF forecasts of central pressure and equaled MRF mean position errors. In addition, all forecasts (MRF, UKMET, ECMWF and manual) showed significantly smaller Mean Position Errors for the stronger category of cyclones. Of the 87 significant cyclones observed, 91% and 92% were forecast by manual and MRF day-4 forecasts, respectively. The false alarm rate for the manual forecasts was 8%. For the stronger category of storms, 93% and 97% were forecast four days in advance by the manual and MRF forecasts. *Therefore, if a major cyclone is forecast by the Marine Forecast Branch for the relatively data-sparse North Pacific at day 4, it generally will occur,* *including explosively deepening primary cyclones and secondary cyclogenesis.*

5. Recent Examples of Cyclone Forecasts

The previous two sections provide substantial evidece for the overall improvements in forecasting extratropical cyclones, including the rapid development of these storms over the North Pacific Ocean. In this section, several cases of cyclone forecasts are presented to illustrate the progress associated with the entire forecast process and the challenges that still confront the forecast community.

5.1 The 12–14 March 1993 "Superstorm"

The 12–14 March 1993 cyclone was one of the most intense extratropical cyclones ever to affect the eastern half of the United States (Kocin et al. 1995). Heavy snow, high winds, coastal flooding, and tornadoes left dozens dead, caused more than $2 billion in property damage and produced the most widespread disruption of air travel in the history of aviation. The storm developed in the western Gulf of Mexico and tracked northeastward up the Eastern Seaboard of the United States, producing one of the most widespread areas of heavy snow for any East Coast storm (Fig. 7).

The cyclone developed along a stationary frontal boundary in the western Gulf of Mexico just prior to 1200 UTC 12 March (Fig. 8). In the following 24 hours, the cyclone deepened explosively over the Gulf of Mexico, with a central pressure of 984 mb south of New Orleans, Louisiana, at 0000 UTC 13 March and a pressure of 971 mb over eastern Georgia by 1200 UTC 13 March. The rapid development of the storm was accompanied by coastal flooding across western Florida as the cold front approached from the west, and by the development of a very broad area of heavy snow extending northeastward from Alabama, northern Georgia, northwestern Florida along the Appalachian Mountains to the Northeast United States. The storm continued to move northeastward and reached its lowest pressure of 960 mb over the Chesapeake Bay by 0000 UTC 14 March and then began to fill slowly as it continued northeastward into New England by 1200 UTC 14 March (Kocin et al. 1995).

Many earlier East Coast snowstorms are notorious for their lack of predictability, as far back as the March Blizzard of 1888 (Kocin 1983) to a more contemporary failure such as the Presidents' Day Storm of February 1979 (Bosart 1981; Uccellini et al. 1984). The poor forecasts for these coastal snowstorms have contributed to a public perception that these weather systems are either unpredictable and/or the forecasters are unable to forecast snowstorms, a perception that has survived even the good forecasts.

A remarkable aspect of the 12–14 March 1993 storm was

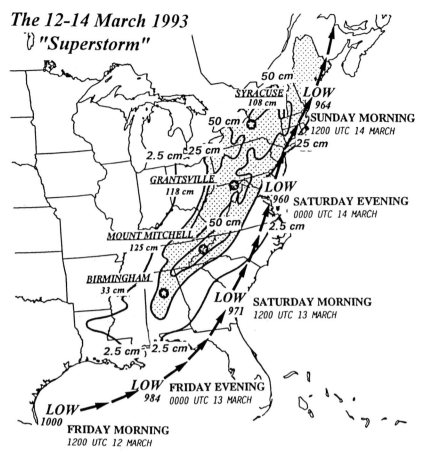

FIG. 7. Storm track with 12-hour positions, central sea-level pressure (mb) and total snowfall (cm) with selected amounts.

the general performance by the forecast community that alerted the public to the potential of a dangerous winter storm of "historic proportions" two to three days in advance (Uccellini et al. 1995). The overall success of the forecast performance can be related, in part, to the performance of the operational global numerical models that produced forecasts of a major cyclone event five to six days in advance. The 5-day, 4-day, and 3-day forecasts produced by the MOD forecasters and verifying at 1200 UTC 13 and 14 March (Fig. 9) show the forecaster-derived positions and intensities of the surface cyclone. These forecasts were derived following an examination of MRF, ECMWF, and UKMET.

At 1200 UTC 13 March, the actual analysis shows a 971 mb cyclone over eastern Georgia (Fig. 8). The 5-, 4- and 3-day model forecasts verifying at that time show that the three models had widely varying positions of the cyclone along the East Coast. Forecasters initially looked at a dual-cyclone system at day 5, and then focused on one primary cyclone over the southeastern United States at days 4 and 3. By the 3-day forecast, forecasters decided on a cyclone position over Georgia close to the position generated from the MRF since the MRF displayed more run-to-run consistency than either

the ECMWF or UKMET models for this case in placing the storm over the southeastern United States. While forecasters selected a position that was very close to the verifying location, the storm was actually much more intense than forecast at this stage of its evolution.

At 1200 UTC 14 March, the actual storm had moved northeastward along the East Coast and was located near Portland, Maine, with a central pressure of approximately 964 mb (see Fig. 8). The forecasts generated by the medium-range numerical models were consistent in highlighting a major storm in the eastern United States. However, despite run-to-run model differences in both the ECMWF and UKMET models in forecasting the cyclone location at days 4 and 3 (note the reversal in the models solutions of an inland versus an oceanic track of the surface low in Fig. 9), forecasters leaned toward the more consistent MRF in keeping the storm close to the East Coast, albeit too far south of the verifying position and slightly too weak. Despite run-to-run model differences, forecaster consensus resulted in a consistent forecast of a major cyclone that would move from the Gulf of Mexico into Georgia and then northeastward along the Atlantic coast.

As the time of the actual event approached, the forecasts provided relatively consistent locations of the storm as it was developing. For example, the short-range forecasts valid at 1200 UTC 13 March provided consistent locations of the storm center over southern Georgia. However, the 48-h and 36-h manual forecasts (Fig. 10) produced central sea-level pressures of the storm system over Georgia that were much higher than observed (971 mb) as forecasters dealt with model predictions that underpredicted the intensity of the cyclone in the Gulf of Mexico even one to two days in advance (Uccellini et al. 1995). As the verifying time drew near, the short-range models began to converge on a solution of a rapidly deepening cyclone over the Gulf of Mexico. The manual forecasts at 24 h and 12 h reflect the forecasters' belief that the storm emerging from the Gulf of Mexico would in fact be very intense.

Thus, while the longer-range forecasts provided the consistency that led to forecaster confidence that a major storm was likely to occur, the short-range models were able to provide more detailed guidance on the intensification of the storm. Nevertheless, the ability of the forecasters to select the more accurate scenario from a wide range of solutions was an important factor in the successful forecast released to the general public, which is consistent with other

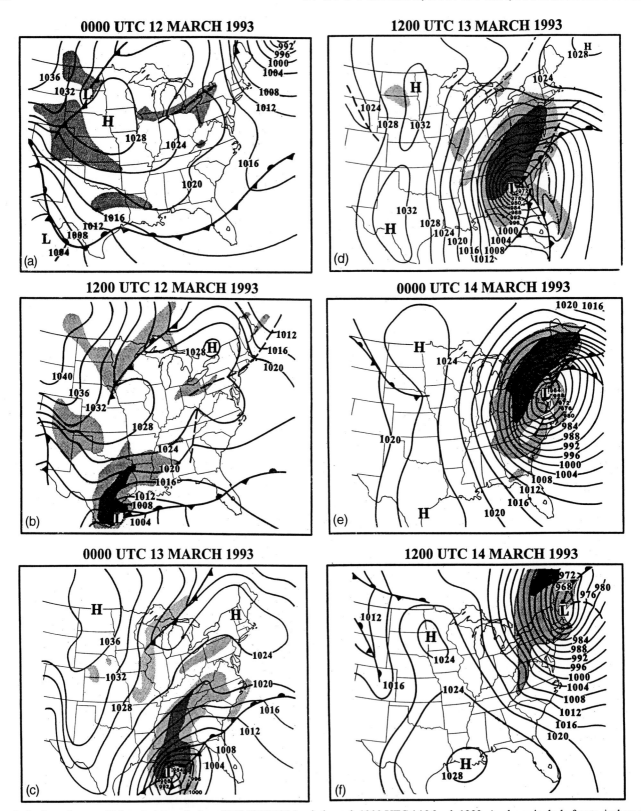

FIG. 8. Surface charts at 12-hourly intervals from 0000 UTC 12 March through 1200 UTC 14 March 1993. Analyses include fronts, isobars (4 mb increments) and precipitation (shading; dark shading for heavy precipitation).

FIG. 9 (facing page). The 5-day, 4-day, and 3-day manual surface fronts and isobar forecasts issued by the MOD Weather Forecast Branch based on the 132-, 108-, and 84-h MRF and UKMET model forecasts and the 144-, 120-, and 96-h ECMWF forecasts. The forecasts are valid at 1200 UTC 13 March (left) and 1200 UTC 14 March (right). The U, E, and M notations refer to UKMET, ECMWF, and MRF forecast positions of the surface cyclone, with the last two digits representing central pressure (E94 = 994 mb from the ECMWF). Solid, dashed, and dot-dashed lines on the right-hand side refer to forecast tracks of the surface cyclone from their respective models.

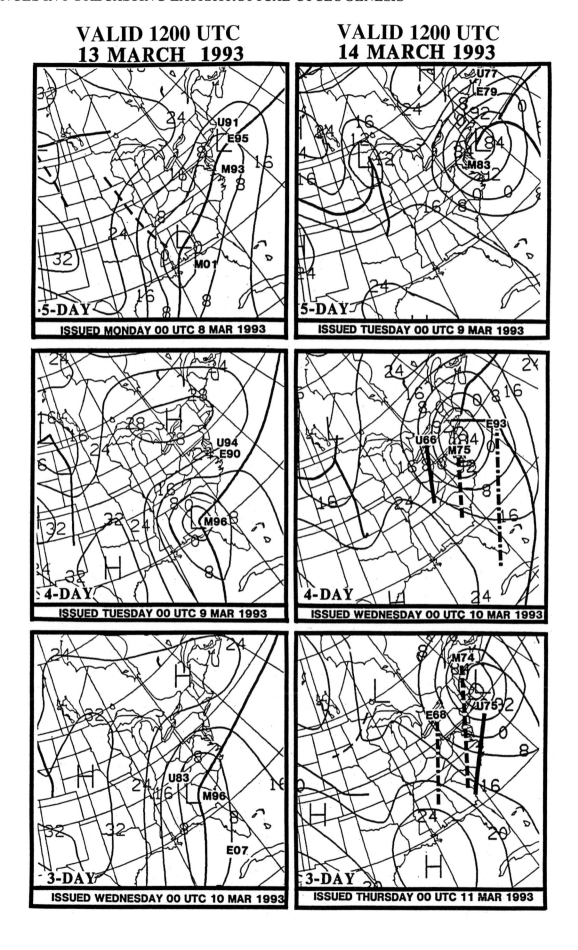

VALID 1200 UTC 13 MARCH 1993

VALID 1200 UTC 14 MARCH 1993

5-DAY — ISSUED MONDAY 00 UTC 8 MAR 1993

5-DAY — ISSUED TUESDAY 00 UTC 9 MAR 1993

4-DAY — ISSUED TUESDAY 00 UTC 9 MAR 1993

4-DAY — ISSUED WEDNESDAY 00 UTC 10 MAR 1993

3-DAY — ISSUED WEDNESDAY 00 UTC 10 MAR 1993

3-DAY — ISSUED THURSDAY 00 UTC 11 MAR 1993

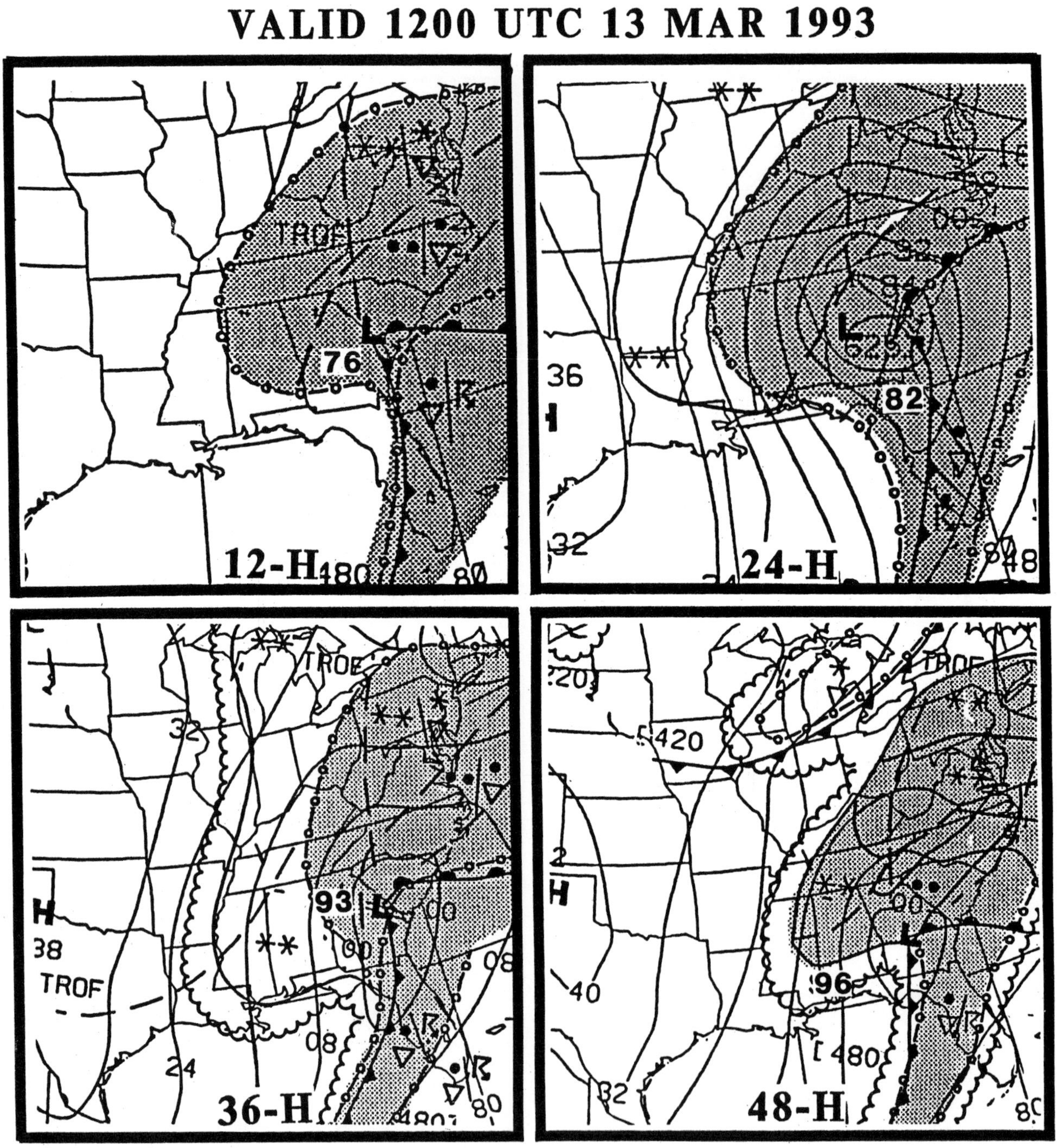

VALID 1200 UTC 13 MAR 1993

FIG. 10. The 12-, 24-, 36- and 48-h manual surface forecast issued by the MOD Weather Forecast Branch valid at 1200 UTC 13 March 1993. The 2-digit values near the surface cyclone represent the forecast central pressure of the cyclone.

recent measures of forecaster performance, as presented by Stokols et al. (1991) and Tracton (1993) and discussed in Section 3.

The storm yielded unprecedented snowfall amounts from the southeastern states through the Appalachian Mountains (Fig. 7). Fifty or more cm of snow was common along much of the Appalachian Mountains. Forecaster consensus of a storm track from eastern Georgia through the Chesapeake Bay to New York state (only the forecast track from the

Chesapeake Bay into eastern New York was west of the actual track) was the critical factor that led to the following forecasts: (1) the axis of heaviest snow was forecast to occur along the spine of the Appalachian Mountains; (2) the major metropolitan areas were located within the gradient of moderate to heavy amounts of snow (snow was forecast to change over to ice pellets and rain from the Washington, D.C., area to Philadelphia, New York, and Boston); (3) the 16-inch (40 cm) forecast issued at 0130 UTC 13 March was unprec-

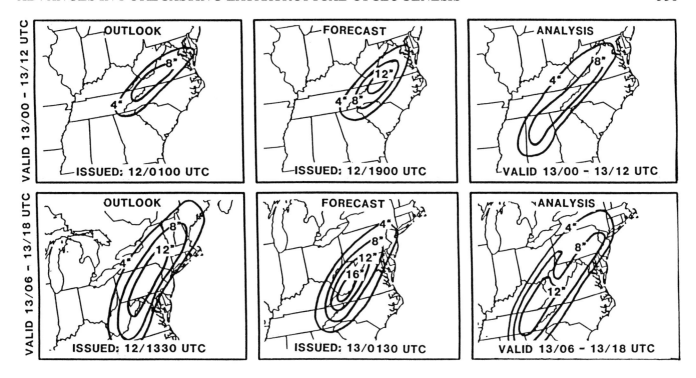

FIG. 11. Twelve-hourly snowfall forecasts and analyses valid between 0000 UTC and 1200 UTC 13 March (top) and 0600 and 1800 UTC 13 March (bottom) issued by the MOD Weather Forecast Branch on 12–13 March 1993.

edented and came within 1 inch of verifying; and (4) the total predicted snowfall exceeding 24 inches (60 cm) over West Virginia was unprecedented (Fig. 11).

As must be expected, the forecasts were not perfect, with snowfall underforecast in portions of Alabama, Kentucky, Ohio, and western Pennsylvania. In addition, the initial development of the cyclone in the Gulf of Mexico was considerably underforecast, resulting in unexpected flooding and destruction, particularly in Florida. This aspect of the storm evolution is already the subject of numerous research projects involving modeling centers throughout the United States (see e.g., Huo et al., 1995). Even with this deficiency, the overall consistency and accuracy of the forecast with respect to (1) the lead-time for a major cyclone to affect the East Coast and (2) the timing, areal distribution and amounts of snow led to widespread praise for the numerical guidance and the general forecast community for their efforts in forecasting the March 1993 "Superstorm" days in advance.

5.2   Marine Cyclones

5.2.1   Atlantic case of 18 February 1994

On the other end of the spectrum, the case of 18 February 1994 was a relatively innocuous case of Atlantic cyclogenesis of concern primarily to mariners. However, it illustrates that explosive cyclogenesis can be predicted with remarkable accuracy over the "data-void" oceans as discussed in section 4.

The marine cyclone originated as a weak upper trough/ surface cyclone that passed across the northeastern United States on 15–16 February 1994 (not shown). While only producing light, scattered snows across New York and New England, a cyclone developed off the New England coast early on 16 February with an analyzed central pressure of 1011 mb at 40°N 64°W, and moved east over the western to central Atlantic in the following 24 to 48 hours and began to deepen rapidly. By 1200 UTC 18 February, the cyclone had deepened to 978 mb in the central Atlantic at 46°N 36°W (Fig. 12; bottom).

The 96-h forecast produced by forecasters at the Marine Forecast Branch verifying at 1200 UTC 18 February was made at 1200 UTC 14 February, relying heavily on the MRF, ECMWF, and UKMET models. Four days prior to the event, the trough responsible for the upcoming cyclone event over the Atlantic was located along the Montana–North Dakota border and adjacent Canadian provinces. The 96-h surface forecast is shown in Fig. 12 (top). Note that the 4-day forecast position is nearly coincident with the analyzed position and the central pressure forecast is 4 mb higher than observed. While the frontal position is forecast too fast near the cyclone, the overall forecast pattern, with (1) the cyclone located south of a separate dual-center cyclone between Greenland and Iceland, and (2) an associated front moving across Ireland, is generally accurate. The corresponding 48-h forecast verifying at the same time (Fig. 12; middle) is consistent with the earlier forecast and the analysis.

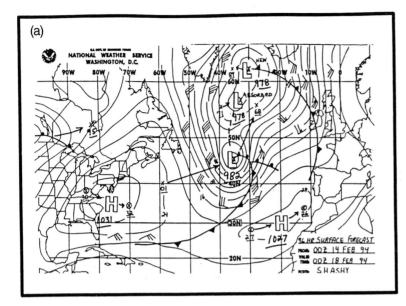

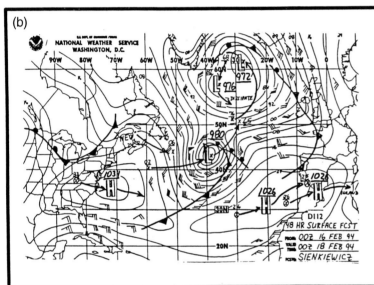

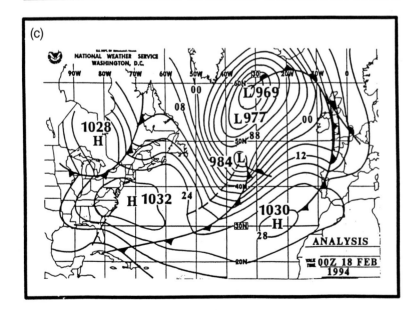

FIG. 12 (left). MFB manual 96-hour forecast (a), 48-hour forecast (b), and verifying NMC operational surface analysis (c) for 0000 UTC 18 February 1994.

5.2.2   *Pacific cases of 12 November 1992 and 29 November 1992*

Included are two examples of forecasts of Pacific cyclogenesis during November 1992. Four-day (or 84-hour) forecasts prepared by MFB meteorologists are compared to the corresponding MRF forecast and verifying NMC 0000 UTC Northern Hemisphere surface analyses.

The first example shown in Fig. 13 is from 0000 UTC 12 November 1992. Two significant cyclones were observed and forecast 4 days in advance over the North Pacific. The depth of both cyclones was slightly overforecast at the verifying time. However, the structures within the cyclones were well predicted in the manual forecast. This is particularly evident for the occluded front associated with the cyclone near 46N 159W, with the triple point displaced well to the northeast of the center. Secondly, the cold frontal position over the Bering Sea and Aleutian chain matches the analyzed position quite well. In addition, the wind field associated with each cyclone was well forecast four days in advance. The cyclone to the east of the Kamchatka Peninsula did in fact continue to deepen to 960 mb by 1200 UTC 12 November. This is an example in which forecasters correctly followed model guidance and enhanced the model forecast fields to produce an accurate depiction of frontal features and the wind field.

The second example in Fig. 14 is from 0000 UTC 29 November 1992 and depicts a mature cyclone over southwestern Alaska. This was a secondary cyclone that developed to the south of a 940-mb storm over the central Bering Sea and moved rapidly into southwestern Alaska. In this case the manual forecast shows a significantly weaker cyclone than forecast by either the 96-hour UKMET or MRF. Also shown in (b) are the MRF (circle), UKMET (triangle), and ECMWF (square) 96-h (120-h for ECMWF) forecast positions and central pressures (967, 938, and 946 mb, respectively).

This is an excellent illustration of the problems faced with forecasters in dealing with both secondary development and filling cyclones. The 960-mb center observed over the central Bering Sea was in fact the former 940-mb storm. The manual forecast shows only troughing in this

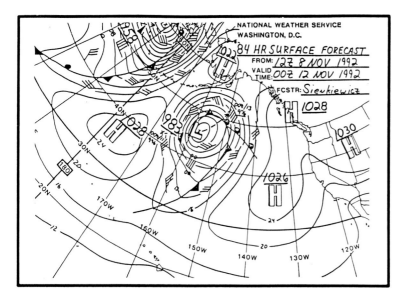

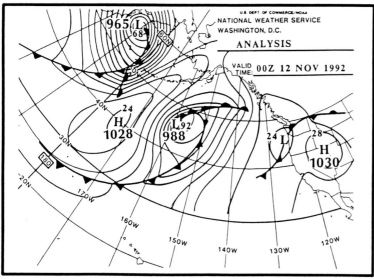

Fig. 13. MFB manual 84-hour forecast (top) and verifying NMC operational manual surface analysis (bottom) for 0000 UTC 12 November 1994.

vicinity with a sole 967-mb center over southwest Alaska. In this case, the forecaster was attempting to correct for a well-known model deficiency associated with the AVN being too slow to increase the minimum MSLP for filling/weakening cyclones (Grumm 1993). However, the forecaster overcompensated by filling the central Bering Sea low too quickly, but still predicted the secondary development. As compared to the NMC analysis, the general position of features, the eastward progression of the front over the Gulf of Alaska and the extent of gale force winds southward to approximately 43°N was forecast fairly well. Although this was not a perfect forecast, it still provided useful information concerning the wind field to the mariner associated with an intense, but weakening, storm.

6. Summary

The revolutionary approach to synoptic analysis and the development of a mathematical basis for weather forecasting initiated at the Bergen school in the early part of this century provided a foundation for the numerical modeling started in the United States after World War II. The initial success of these models in the early 1950s in simulating the large-scale aspects of cyclogenesis led to the creation of the NMC in 1960 and the application of numerical models in day-to-day weather forecasting, all within a ten-year period. With the establishment of the NMC and the associated collocation of modelers and forecasters came the complete overhaul of the forecast process which is now marked by a forecaster–machine mix. Today, the forecast process is heavily dependent on numerical prediction. Nevertheless, the success or failure of this process still relies on forecasters who depend on their understanding of (1) synoptic and mesoscale meteorology and (2) the strengths and weaknesses of the operational models to make the needed adjustments to the model solutions to produce accurate 5-day weather forecasts.

The introduction of new models and associated model output statistical packages, the application of model-based analysis and initialization schemes, and the increased experience of forecasters in using the models have all led to a tremendous improvement in the forecast of cyclogenesis and the associated weather patterns. After nearly 30 years of model applications and development, improvements in the prediction of rapid cyclogenesis came on quite suddenly in the middle 1980s with (1) the introduction of higher-resolution models with more sophisticated treatment of the physical processes that contribute to extratropical cyclogenesis, (2) better data assimilation and model initialization schemes based on an expanded global observational network, and (3) the application of the numerical forecast systems on much more powerful computers. These advancements were a direct result of the Global Weather Experiment, which involved researchers and operational meteorologists from all over the world. With the continued progress in numerical analysis and forecast systems since the 1980s, explosive cyclogenesis (related to primary and secondary development) can now be predicted with reliable accuracy and decreasing false alarm rates three, four, and in some cases five days in advance, even for the large ocean areas surrounding the United States.

It is clear that the operational analysis and modeling systems are quite capable of simulating the scale-interactive, nonlinear processes that contribute to rapid cyclogenesis and

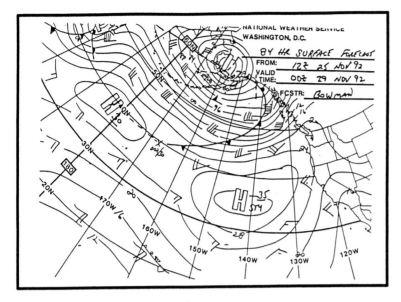

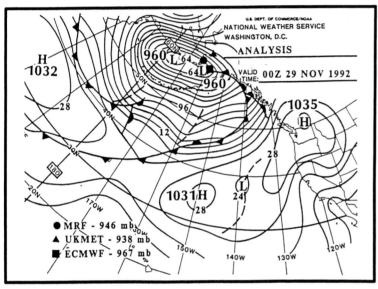

FIG. 14. MFB manual 84-hour forecast (top) and verifying NMC operational manual surface analysis (bottom) for 0000 UTC 29 November 1992.

The success of the implementation of the fore-caster–machine mix at the NMC in improving forecasts of major weather events, such as the major oceanic storms over the Atlantic and Pacific Oceans, "Superstorm of March 1993," and more recently, the consistent medium-to-short-range forecasts of the 6–8 January 1996 "Blizzard of '96," is tempered by the knowledge that the forecasts are not perfect and that much can be done to improve upon current operational prediction systems.

Several major efforts are under way to improve the forecasts and extend our practical limits of predictability toward current theoretical limits. One effort involves the development and implementation of new modeling systems such as the Meso-ETA (Black 1994) and the development of a coupled ocean-atmospheric general circulation model (Ji et al. 1994). Both of these models are designed to capture more effectively the interaction among the various physical and dynamical processes that contribute to rapid cyclogenesis and have significant impact on the prediction of these storms in both the 1- to 2-day range (Meso-ETA) and for seven days and beyond with improved global models.

The prediction of major cyclone events in the extended range will also be facilitated by the recent implementation of the ensemble approach to numerical prediction at the NMC described by Tracton and Kalnay (1993). Currently, the NMC model ensemble includes 46 different runs of the MRF, providing a probabilistic forecast approach to determining a most likely solution from the wide range of possibilities offered by the models on a daily basis. The ensemble predictions are essentially an application of a stochastic approach to forecasting, in contrast to the use of one model run in a deterministic fashion, which should have its greatest impact in the medium-range predictions.

Last but not least, the NMC has taken on a major initiative to develop and implement computer workstations for forecast desks in order to facilitate access to the four-dimensional grids of model data and to provide a basis for the visualization of model forecasts in a real-time mode. This effort represents an important step in providing the forecaster with the tools needed to understand better the interaction of the physical processes contributing to rapidly changing weather situations. Furthermore, the improved access to the model fields will provide insights into the model simulations, which will likely assist the forecasters' decision-making process in selecting the "model of the day" and making the necessary adjustments.

Given the continuing improvements in the operational models and the enhanced access to the model simulations through the introduction of powerful workstations to the

also to secondary development of cyclones. Furthermore, forecasters have developed sufficient knowledge to work with the models and make weather forecasts and issue storm warnings well into the 3-, 4-, and 5-day range with an increasing level of confidence. By all measures, the probability of predicting major cyclone events continues to rise for all forecast periods out to five days while the false alarm rate continues to decrease. Given the performance of the forecaster-machine mix at the NMC over the past five to ten years and the recent creation of the Marine Forecast Branch to provide storm warnings for the high seas, one can only conclude that the original vision that motivated those who fostered the Bergen school has in fact been realized, and the impact on forecasting rapid cyclogenesis and related weather systems has been remarkable.

forecast floor, we expect the forecaster–machine mix to build upon the general upward trend in forecast skill and recent dramatic improvements in forecasting rapid surface cyclogenesis. It is clear that the goal of V. Bjerknes and his collaborators to improve the forecasts of major storm systems over the ocean through the application of mathematical principles to weather forecasting is being realized some 80 years later. This accomplishment represents one of the major intellectual achievements of the twentieth century. The challenge now is to predict these major storm events with even more reliability and extend the range of predictability of rapid cyclogenesis to its theoretical limits.

References

Bedrick, M., A. J. Cristaldi III, S. J. Colucci, and D. S. Wilds, 1994: An analysis of the accuracy of 120-h predictions by the National Meteorological Center's Medium-Range Forecast Model. *Wea. Forecasting*, 9, 3–20.

Bergeron, T., O. Devik, and C. L. Godske, 1962: Vilhelm Bjerknes March 14, 1862–April 19, 1951. *Geofys. Publ.*, 24, 7–32.

Black, T., 1994: The new NMC mesoscale ETA model: Description and forecast examples. *Wea. Forecasting*, 9, 265–278.

Bonner, W. D., 1989: NMC Overview: Recent Progress and Future Plans. *Wea. Forecasting*, 4, 275–285.

Bosart, L. F., 1981: The Presidents' Day snowstorm of 18–19 February 1979: A subsynoptic-scale event. *Mon. Wea. Rev.*, 109, 1542–1566.

Brown, W. E., and D. A. Olson, 1978: Performance of NMC in forecasting a record-breaking winter storm, 6–7 February 1978. *Bull. Amer. Meteor. Soc.*, 59, 562–575.

Carter, G. M., J. P. Dallavalle, and H. R. Glahn, 1989: Statistical forecasts based on the National Meteorological Center's Numerical Weather Prediction System. *Wea. and Forecasting*, 4, 401–412.

Charney, J. G., 1950: Progress in Dynamic Meteorology. *Bull. Amer. Meteor. Soc.*, 31, 231–236.

Chesneau, L.S., 1994: Aboard the Ike. *Mar. Wea. Log*, 38, 8–14.

Corfidi, S. F., and K. E. Comba, 1989: The Meteorological Operations Division of the National Meteorological Center. *Wea. Forecasting*, 4, 343–366.

Dey, C. H., 1989: The Evolution of Objective Analysis Methodology at the National Meteorological Center. *Wea. Forecasting*, 4, 297–312.

Extratropical Cyclones. The Erik Palmén Memorial Volume, 1990: Amer. Meteor. Soc., 262 pp.

Funk, T. E., 1991: Forecasting techniques utilized by the Forecast Branch of the National Meteorological Center during a major convective rainfall event. *Wea. Forecasting*, 6, 548–564.

Gerrity, J., 1977: The LFM Model–1976: A documentation. NOAA Technical Memorandum. NWS NMC-60 (NTIS PB-279-419), 68 pp.

Glahn, H. R., and D. A. Lowry, 1972: The use of model output statistics (MOS) in objective weather forecasting. *J. Appl. Meteor.*, 11, 1203–1211.

Grumm, R. H., 1993: Characteristics of surface cyclone forecasts in the aviation run of the global spectral model. *Wea. Forecasting*, 8, 87–112.

——————, R. J. Oravec, and A. L. Siebers, 1992: Systematic model forecast errors of surface cyclones in NMC's Nested-Grid Model, December 1988 through November 1990. *Wea. Forecasting*, 7, 65–87.

Hoke, J. E., N. A. Phillips, G. J. DiMego, J. J. Tuccillo, and J. G. Sela, 1989: The Regional Analysis and Forecast System of the National Meteorological Center. *Wea. Forecasting*, 4, 323–334.

Huo, Zonghui, D.-L Zhang, J. Gyakum, and A. Staniforth, 1995: A diagnostic analysis of the Superstorm of March 1993. *Mon. Wea. Rev.*, 123, 1740–1761.

Ji, M., A. Kumar, and A. Leetmaa, 1994: A multi-season climate forecast system at the National Meteorological Center. *Bull. Amer. Meteor. Soc.*, 75, 569–577.

Johnson, D.R., 1986: Summary of the Proceedings of the First National Workshop on the Global Weather Experiment. *Bull. Amer. Meteor. Soc.*, 67, 1135–1143.

Junker, N. W., J. E. Hoke, and R. H. Grumm, 1989: Performance of NMC's Regional Models. *Wea. Forecasting*, 4, 368–390.

Kanamitsu, M. 1989: Description of the NMC Global Data Assimilation and Forecast System. *Wea. Forecasting*, 4, 335–342.

Keyser, D., and L. W. Uccellini, 1987: Regional models: Emerging research tools for synoptic meteorologists. *Bull. Amer. Meteor. Soc.*, 68, 306–320.

Kocin, P. J., 1983: An analysis of the "Blizzard of '88." *Bull Amer. Meteor. Soc.*, 64, 1258–1272.

——————, P. N. Schumacher, R. F. Morales, Jr., and L. W. Uccellini, 1995: Overview of the 12–14 March 1993 Superstorm. *Bull. Amer. Meteor. Soc.*, 76, 165–182.

McPherson, R. D. and D. A. Olson, 1991: Reply. *Wea. Forecasting*, 6, 572.

Mullen, S. L., and B. B. Smith, 1990: An analysis of sea-level cyclone errors in NMC's Nested Grid Model (NGM) during the 1987–1988 winter season. *Wea. Forecasting*, 5, 433–447.

Murphy, A. H., and E. S. Epstein, 1989: Skill scores and correlation coefficients in model verification. *Mon. Wea. Rev.*, 117, 572–581.

Olson, D. A., N. W. Junker, and B. Korty, 1995: An evaluation of three decades of quantitative precipitation forecasting at the Meteorological Operations Division of the National Meteorological Center. *Wea. Forecasting*, 10. 498–511.

Oravec, R. J., and R. H. Grumm, 1993: The prediction of rapidly-deepening cyclones by NMC's Nested-Grid Model in Winter 1989 through Autumn 1991. *Wea. Forecasting*, 8, 248–270.

Ramage, C. S., 1976: Prognosis for weather forecasting. *Bull. Amer. Meteor. Soc.*, 57, 4–10.

Rao, D. B., 1989: A Review of the Program of the Ocean Products Center. *Wea. Forecasting*, 4, 427–443.

Sanders, F., 1986: Explosive cyclogenesis over the west central North Atlantic Ocean, 1981–1984. Part II: Evaluation of LFM model performance. *Mon. Wea. Rev.*, 114, 2207–2218.

——————, 1987: Skill of NMC operational dynamical models in prediction of explosive cyclogenesis. *Wea. Forecasting*, 2, 322–336.

——————, 1992: Skill of operational dynamical models in cyclone prediction out to five-days range during ERICA. *Wea. Forecasting*, 7, 3–25.

——————, and E. P. Auciello, 1989: Skill in prediction of explosive cyclogenesis over the western North Atlantic Ocean, 1987/1988: A forecast checklist and NMC dynamical models. *Wea. Forecasting*, 4, 157–172.

——————, and J. R. Gyakum, 1980: Synoptic-dynamic climatology of the "bomb." *Mon Wea. Rev.* 108, 1590–1606.

Shuman, F. G., 1989: History of numerical weather prediction at the National Meteorological Center. *Wea. Forecasting*, 4, 286–296.

Sienkiewicz, J.M., 1992: The observation. *Mar. Wea. Log*, 36, 4–12.

Smith, B. B., and S. L. Mullen, 1993: An evaluation of sea-level cyclone forecasts produced by NMC's Nested-Grid Model and Global Spectral Model. *Wea. Forecasting*, **8**, 37–56.

Stallings, E. A., and L. A. Wenzel, 1995: Organization of the River and Flood Program in the National Weather Service. *Wea. Forecasting,* **10**, 457–464.

Stokols, P. M., J. P. Gerrity, and P. J. Kocin, 1991: Improvements at NMC in numerical weather prediction and their effect on winter storm forecasts. *Preprints: First International Symposium on Winter Storms;* Amer. Meteor. Soc., 15–19.

Teweles, S., and H. Wobus, 1954: Verification of prognostic charts. *Bull. Amer. Meteor. Soc.*, **35**, 455–463.

Thompson, P., 1987: The maturing of the science. *Bull. Amer. Meteor. Soc.*, **68**, 631–637.

Tracton, M. S., 1993: On the skill and utility of NMC's medium-range central guidance. *Wea. Forecasting*, **8**, 147–153.

——, and E. Kalnay, 1993: Operational ensemble prediction at the National Meteorological Center: Practical aspects. *Wea. Forecasting*, **8**, 379–398.

Uccellini, L. W., P. J. Kocin, R. A. Petersen, C. H. Wash, and K. F. Brill, 1984: The Presidents' Day cyclone of 18–19 February 1979: Synoptic overview and analysis of the subtropical jet streak influencing the precyclogenetic period. *Mon. Wea. Rev.*, **115**, 2227–2261.

——, P. J. Kocin, P. M. Stokols, and R. A. Dorr, 1995: Forecasting the Superstorm of 12–14 March 1993. *Bull. Amer. Meteor. Soc.*, **76**, 183–199.

Numerical Weather Prediction: A Vision of the Future, Updated Still Further

MICHAEL E. McINTYRE

Centre for Atmospheric Science at the Department of Applied Mathematics and Theoretical Physics,* Cambridge, United Kingdom

1. Introduction

It is a great pleasure to be reminded of the Bergen Symposium, which I enjoyed so much, and to take the opportunity to update my own small, somewhat personal contribution (McIntyre 1994) culminating in a vision of the future of numerical weather prediction. One reason to update it and to reemphasize the points made at Bergen is that the most important, and perhaps the most controversial, of those points still seem to me likely to hold good at the future time envisaged—despite my being forced by today's politics to go a bit further into the future. More of that later; but let me just say for now that I shall go one decade further, twenty years into the Millennium.

The controversial points in question are the points about ergonomics. That is, they are the points about the potential for humans to interact usefully and efficiently with computer based systems, to whatever extent the latter become artificially intelligent in one sense or another—the potential for optimizing the person–machine synergies on which operational weather prediction and the responsible use of same will continue to depend (e.g., Tennekes 1988). It still seems to me, after sampling the most recent literature, that while seeking to exploit developments in artificial intelligence as fully as possible we should also be skeptical of the claims that tend to be made about such developments, most especially the claim that artificial intelligence will become, over the next several decades, absolutely superior to human intelligence (e.g., Warwick 1997). That claim seems to be based on a textbook model of the human brain (neuron–synapse networks as simple logic circuits) that is outdated today, and many orders

of magnitude too simple, leading to gross underestimates of what would be involved in simulating or matching, let alone surpassing, human brain function in every respect. Indeed, hard evidence is now accumulating to confirm what the principles of molecular biology have long suggested, including what Jacques Monod (1971) called "microscopic cybernetics," the arbitrary computational powers of protein molecules. Microscopic, more aptly nanometric, cybernetics suggests—and the new evidence is beginning to say it more directly—that textbook neurons and synapses represent only the tip of a hypermassively parallel computational iceberg, of which we are very far indeed from possessing an adequate conceptual, let alone detailed computational, model.[1]

Of course electronic computers are much better than we are at certain things; and what is going to be done with various kinds of computers in the next few decades—electronic, photonic, and no doubt biomolecular—will surely be marvelous in itself. It would equally be a mistake to underestimate that. Indeed, my vision of the future assumes that computers and humans will together be doing some very marvelous things indeed.

So let me proceed as before. I believe that the points to be discussed are very fundamental, and I make no apology for repeating them. As I said at Bergen, they are fundamental from two quite different viewpoints. They are fundamental not only from a fluid dynamical viewpoint—and I shall mention some recent advances in understanding, for instance, Lighthill radiation, Hamiltonian balanced models, and potential vorticity inversion—but fundamental, also, from the viewpoint of how human perception works, and human cognition, especially as regards the visual–cognitive

---

*The Centre for Atmospheric Science is a joint initiative of the Department of Chemistry and the Department of Applied Mathematics and Theoretical Physics.

[1] Keys to the burgeoning literature may be found at our web site, URLs <http://www.atmos-dynamics.damtp.cam.ac.uk/> and <ftp://ftp.damtp.cam.ac.uk/pub/papers/mem/lucidity.ps>.

system. Notwithstanding what I just said about neurons, synapses, and protein molecules, we have considerable knowledge of at least the typical general properties of perception and cognition. That knowledge, which again points toward hypermassive parallelism and has far-reaching human as well as technical importance, comes from general biological reasoning and from analogies with simpler biological systems whose molecular-scale functioning is relatively well understood. It also comes from the wealth of perceptual phenomena, many of them counterintuitive yet easily observable but all pointing toward the same simple principles, and many of which have been studied in detail by psychologists, neurologists, and scientists working in related fields.[2]

For present purposes the implication of all this is still, it seems to me, an extraordinary potential for effective, efficient person–machine interaction in future weather prediction operations, and, one might add, in public liaison as well, including a significant contribution from human perceptual–cognitive skills that machines will not be able to match. It would be rash to predict exactly how the person–machine division and interface—the division of labor and responsibility—will be organized. But it is strongly arguable that, for the foreseeable future, there always will be some such division and that it will come much closer to optimal as the state of the art matures, a nontrivial process that is barely beginning. Any such progress will require not only that we continue to bring powerful computing and data handling technologies fully to bear on those processes that can be automated—including ensemble forecasting and four-dimensional variational data assimilation, both of which now seem to be coming of age in an exciting way (see, e.g., within a burgeoning literature, Molteni et al. 1996, Palmer 1996, Thépaut et al. 1996)—but also that we continue to move toward exploiting fully, in particular, the potentialities of the human visual system, that most powerful of data interfaces between computers and humans. For clear biological reasons, connected with the survival of species, the human visual system has, for instance, a "four dimensional intelligence" that even in terms of raw computing power still dwarfs, by many orders of magnitude, the power of today's largest electronic supercomputers.

And there must surely be great scope for exploiting what is perhaps the most remarkable human cognitive skill of all, forms of which we deploy every day with hardly a moment's thought. This is the ability, after appropriate learning, to interrelate intuitively, to simultaneously take account of, and to act swiftly on, very disparate kinds of information, if the information can be presented in an appropriate form and in a sufficiently familiar context. The successful development of this kind of cognitive skill is sometimes called "getting a feel

for the problem." We all have some "feel" for what happens, and why, if someone or something upsets a bucket of water on a table, despite the formidable complexity of the sequence of events when viewed in detail. We can make quick judgments of the consequences and their dependence on the configuration of neighboring objects, sometimes even quick enough to help with damage limitation. More specialized such skills, like flying aircraft safely, or playing tennis, with attention to a variety of sensory inputs, depend on other kinds of familiarity. These require more specialized learning on top of the range of childhood experiences. The potential for developing analogous skills on the human side of weather prediction is something that we surely tend to underestimate, if only because that potential, and the degree of skill attainable, and how to attain it, have hardly begun to be explored. This will need pioneering by talented individuals. Given the right background conditions—compare J. S. Bach's establishment of the equal tempered musical keyboard—there could yet be a Paganini, a Rachmaninov, or a Charlie Parker of operational weather prediction or, if you prefer to switch metaphors, a Chuck Yeager or a Martina Navratilova.

2. Research Background

Several existing lines of research in the atmospheric sciences, some of them with a long history, have already given us some slight hint of the potential for person–machine efficiency. Much of this research has been oriented to synoptic and mesoscale tropospheric phenomena (e.g., within today's vast literature, Hibbard et al. 1989; Reed et al. 1992, 1994; Demirtas and Thorpe 1997; and many papers of the Bergen Symposium), and some of the underlying ideas go back all the way to the original air mass concepts developed at Bergen and to the dynamical properties of potential vorticity discovered by Rossby (1936, 1940) and Ertel (1942). There has also been relevant research in, and crossfertilization from, adjacent fields. One such field is stratospheric meteorology and chemistry, where the research leading to our present understanding was stimulated first by the greater observational accessibility and dynamical simplicity of the stratosphere and second, more recently, by scientific and public concern about the ozone layer and its continuing depletion.

My own professional interest comes largely from trying to understand the fluid dynamics of the stratosphere, and dates back to attempts, in the early 1980s, to imagine what isentropic distributions of Rossby–Ertel potential vorticity (PV) might look like in the wintertime middle stratosphere and how they might help one to understand the preconditioning of the northern stratosphere for midwinter sudden warmings. This was followed by the intense excitement of being involved in seeing, and helping to make sense of, some of the earliest observational estimates of middle stratospheric

---

[2]Keys to the literature, and some demonstrations of perceptual phenomena, may be found on the same web and ftp sites (footnote 1).

PV distributions from satellite data. The results became an important landmark because the estimated PV distributions turned out to be of better quality than conventional wisdom said was possible with the satellite data. Although we were seeing only a "blurred view of reality" through "knobbly glass" (McIntyre and Palmer 1984), it was still a good enough view, when combined with theoretical insight—including insight from the set of idealized models labeled "Rossby wave critical layer theory," and insight into what is now called "PV invertibility" and its scale dependence—to catalyze important advances in our understanding of the real stratosphere both from a dynamical and, as it turned out, also from a chemical viewpoint.

Today, thanks to a wealth of new data on the stratosphere, the metaphorical glass has already become far less knobbly, as suggested by Figure 1a. There is a wealth of new data, and clever new data analytic methods that exploit the kinematics of advection, vividly suggesting the potential for "four dimensional intelligence" in data analysis and quality control. The research continues apace and I cannot do it justice here: a random and inadequate sample of the most closely relevant papers might include those by Schoeberl et al. (1992), Randel et al. (1993), Waters et al. (1993), Waugh (1993), Chen et al. (1994), Manney et al. (1994), Strahan et al. (1994), Sutton et al. (1994), Fisher and Lary (1995), Lahoz et al. (1996), Nakamura (1996), O'Sullivan and Chen (1996), Lee et al. (1997), and Schoeberl et al. (1997), among many others. Not only cleverness with computing and data analysis, but also with remote sensing techniques and other observational techniques, has been crucially important here.

I should mention also that, aside from the conceptual background of wave–mean interaction theory (going back to Lord Rayleigh), of which the Rossby wave critical layer theory is just a particular case, some of the dynamical ideas that adumbrated the relatively clear view we have today were discussed in a typically perceptive early contribution by Huw Davies (1981), on the nature of stratospheric warmings. More of the history is surveyed in my (1982) review and in a more recent review (1993a) written for an audience from the theoretical mechanics community, emphasizing how wave–mean and PV-related concepts have displaced what the admirable Doctor de Bono would call the "intermediate impossible" or "crazy ideas" stage of trying to understand atmospheric circulations, invoking things like "negative viscosity"—now clearly understandable in terms of less crazy ideas, like Rossby wave propagation.

This stratospheric research, exciting though it has been for me personally, and continues to be, is only one of several lines of research that have reminded us of a key point about untapped potential in operational weather prediction. The point, which now seems to be widely appreciated—and indeed is coming close now to new forms of operational exploitation (A. Hollingsworth, personal communication

1997–98)—is that it makes good sense scientifically and data analytically to consider, either implicitly or explicitly, PV and chemical tracers together. Whether or not you find it insightful to think of the dynamics in terms of PV, the bottom line is that there is much more information in chemical and dynamical fields taken together than in either taken separately.

This point has cropped up again and again in atmospheric research, and has been another point of crossfertilization between our understanding of the troposphere and the stratosphere, including conditions near the tropopause. It emerges from many studies using data from remote sensors like the water vapor imagers, and the famous Total Ozone Mapping Spectrometer, and goes back to much earlier work by pioneers like Starr and Neiburger (1940), Reed and Danielsen (1959), and Danielsen (1968)—natural developments, in turn, of the Bergen air mass concepts and the advocacy of isentropic analysis by Napier Shaw (1930). Figures 1a–c, taken with kind permission from Appenzeller et al. (1996), beautifully illustrate the point with a three-way snapshot of typical midlatitude, near-tropopause phenomena. Figure 1a shows a hypothetical chemically inert tracer advected on the 320K isentropic surface over a 4-day period. The pattern was computed from analyzed winds by means of an accurate, sophisticated "contour advection" technique, representing a significant "technology transfer to the real world" from so called "academic" theoretical fluid dynamical work (Dritschel 1989). This particular idea for technology transfer was a beautiful idea independently thought of and implemented by Norton (1994) on the one hand, and by Waugh and Plumb (1994) on the other. In Figure 1a (see caption for details), colors represent nominally stratospheric or ex-stratospheric air, and white tropospheric. Figures 1b,c show respectively a corresponding water vapor image and isentropic map of PV. The resemblance among the three is obvious at a glance, and tells us, I believe, something significant. The three maps show three quite different variables—yet human cognition, even without the help of animation, or forward radiation models, can instantly discern that the three pictures are all showing aspects of the same structure. Furthermore, we know that the resemblance is significant dynamically as well as morphologically, for the reasons to be recalled in Sections 4 to 6.

In a recent effort to get an idea of the potential suggested by such cases, Demirtas and Thorpe (1997) have shown that other cases can be found that, by contrast, lack the kind of resemblance that is so conspicuous in Figures 1a–c, and that this can be symptomatic of highly significant analysis errors. In the cases they present, a surprisingly crude and simple correction procedure turns out to be enough to improve some forecasts decisively. It was simply, in effect, to move PV contours—by implication, reshaping the ridge–trough structure of the tropopause—to give better qualitative agreement

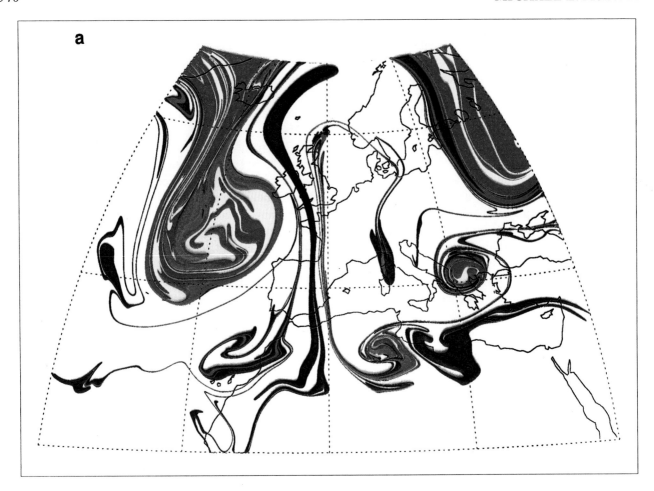

Figure 1 (above and facing page). **(a)** Hypothetical passive tracer on the 320 K isentropic surface for 14 May 1992 at 1200Z, calculated by the *contour advection* technique (Norton 1994; Waugh and Plumb 1994) using winds from the operational analyses of the European Centre for Medium Range Weather Forecasts (ECMWF), initialized four days earlier to coincide with the PV (potential vorticity) contours whose values are 1, 2, 3, 4, and 5, in the standard PV units of $10^{-6}\,m^2\,s^{-1}\,K\,kg^{-1}$, of a smoothed isentropic map of PV from the same operational analyses. Values from 1 unit upwards are colored rainbow-wise from dark blue to red and represent nominally stratospheric air. **(b)** Meteosat 5.7–7.1 μm water vapor image at approximately the same time as in (a). **(c)** Isentropic map of PV for the same time as in (a) and from the same operational analyses. The dotted contours near the tropopause have values 0.5, 1.0, 1.5 PV units and the solid contours 2, 3, 4, . . . PV units (Appenzeller et al. 1996).

with a water vapor image as judged subjectively. Then PV inversion was used to generate the corresponding "bogus data." The subjective judgments involved were nontrivial. It seems plain that ozone can be used in a similar way; and there have even been first attempts, involving chemical modeling, at using information not only from relatively inert but also from relatively *reactive* chemicals, which are sensitive to solar zenith angle (Austin 1992; Fisher and Lary 1995). Much of this will no doubt be automated and made much more objective, reliable, and accurate, as soon as the state of the art of 4D variational assimilation becomes sufficiently developed. But my bet is that human vision will remain potentially very important for grasping how well any such assimilation is performing, for "getting a feel for the problem" and hence insight into what to do or say if—or rather when—things go wrong, when the moisture field and all the other dynamical

and chemical fields fail to fit together neatly in spacetime, within the analysis–forecast system.

3. Cyclogenesis and Rossby Wave Propagation

There have been still other remarkable crossfertilizations between stratospheric and tropospheric research, yet again highlighting the dynamical relevance of isentropic distributions of PV and hence the dynamical information implicit in chemical tracer fields. To a dynamicist, of course, the stratosphere and troposphere are inseparable parts of the same system; and the extratropical tropopause, with its strong isentropic gradients of PV, is hardly a passive "ceiling" but much more an erodable, highly deformable, dynamically active "Rossby-elastic wall" (McIntyre and Palmer 1984; Holton et al. 1995), dynamically more like the edge of the

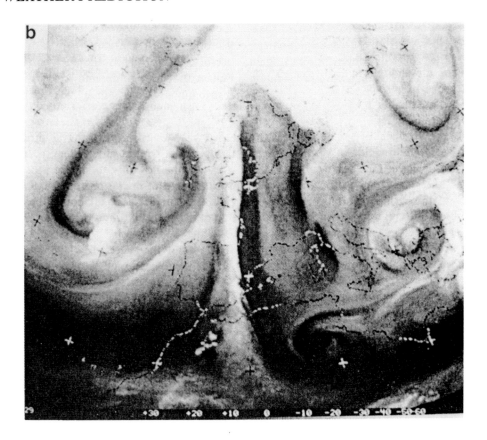

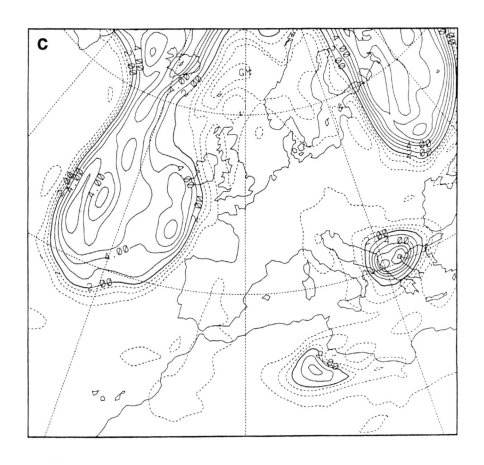

stratospheric polar vortex. Its sideways "Rossby-wave elasticity" obviously has an effect on downstream development, for instance, as well as on local development, once again bringing in the notions of advection and wave propagation simultaneously.

Indeed, to a dynamicist the phrase "tropospheric" weather system sounds self-contradictory. High PV stratospheric air is well known to be crucial to midlatitude cyclogenesis, as many papers in the Bergen Symposium reminded us. Nevertheless, it is mainly the lower stratosphere that has direct importance for cyclogenesis, the middle stratosphere being relatively remote from so-called tropospheric weather and having stronger conceptual than direct relevance. Here, again speaking from my own experience, I am thinking especially of a line of research from which another—this time really surprising—insight emerged about tropospheric cyclogenesis (Edmon et al. 1980; see also Dunkerton et al. 1981; Hoskins et al. 1985; Held and Hoskins 1985; Thorncroft et al. 1993; Magnusdottir and Haynes 1996).

The surprise was that, contrary to what some textbooks still say, the major part of the growth of eddy kinetic energy seen in the typical Simmons–Hoskins baroclinic wave life cycle is due to upward Rossby wave propagation and Doppler shifting, not to baroclinic instability.

The upward Rossby wave propagation is dynamically indistinguishable from the Charney–Drazin type of propagation traditionally thought of as relevant, on a larger scale, to the wintertime middle stratosphere. Less than half of the growth of eddy kinetic energy in the life cycle could be attributed to baroclinic instability. The initial baroclinic instability saturates early in the cycle (e.g., Thorncroft et al. 1993, Figure 4); and the subsequent peaking of the eddy kinetic energy is associated with a more or less distinct second saturation event involving an extreme, irreversible deformation of the Rossby-elastic tropopause, and dynamically much the same as the breaking of planetary-scale Rossby waves in the wintertime middle stratosphere (McIntyre and Palmer 1983–85) and indeed significantly like the idealized form of breaking described by the Rossby wave critical layer theory. When the importance of upward Rossby wave propagation in the baroclinic life cycle was first discovered by Edmon et al., the surprise was total. I don't think anyone had the slightest inkling that stratospheric planetary-scale Rossby wave propagation had anything to teach us about such a highly nonlinear phenomenon as tropospheric cyclogenesis.

4. Why PV?

This question is still asked. I think it is prompted by the mathematical truism that one dependent variable is as good as another if they carry the same information. Let us, therefore, recall the fluid dynamical reasons why the PV calls for

special attention, or, more precisely, summarize why it is useful to make a particular choice of dependent variables, useful to focus prognostic attention on near-surface distributions of potential temperature (PT) and interior isentropic distributions of PV—which latter, once upon a time, I rashly called "IPV distributions." The presentation follows that in McIntyre (1993b) and builds on lecture material I have used for many years. (Why was "IPV distributions" rash? I have to take the blame here. I thought I had come up with a neat mnemonic acronym but failed to remember something else about human perception, the phenomenon of "stray adjective" or "disabled toilet syndrome"—see footnote 4 on page 345. This is a potentially confusing manifestation of our power of lateral thinking. The "I" for "isentropic" was meant to apply to "distributions," "maps," "gradients," "fluxes," etc., and not to "PV.")

The important idea, going back at least as far as Charney (1948) and Kleinschmidt (1950–1)—see also the review by Hoskins et al. (1985), hereafter HMR—is to recognize that certain aspects of the PV–PT fields can be regarded as controlling, in a certain sense, the dynamical evolution. More precisely, there is an "invertibility principle"—so familiar to theoreticians that it is hardly ever mentioned explicitly—saying that a certain subset of the information in the PV–PT field can be used to diagnose everything about the other dynamical fields, apart from any inertia–gravity oscillations that may be present including equatorial Kelvin waves. This diagnostic process is what we now call PV inversion. A precise statement will be given in the next section. The advantages of this viewpoint include the following:

1. The evolution of the PV–PT field incorporates the effects of advection in the conceptually simplest way possible. This is a powerful advantage, since there is no escape from considering the advective nonlinearity somehow. As every meteorologist knows, advection is fundamental to practically all weather developments, whatever else is going on.

2. The PV–PT viewpoint recognizes, makes explicit, and keeps conceptually separate those aspects of the dynamics that are intrinsically nonlocal. The nonlocal or action-at-a-distance aspects are all incorporated into the idea of "PV inversion."

3. The PV–PT viewpoint makes precise the basis, extent, and limitations of the partial analogy with two dimensional (2D) nondivergent barotropic vortex dynamics.[3] Barotropic dynamics is included as a particular case characterized by a particular inversion operator, a simple inverse Laplacian.

---

[3] See also the supplement to HMR (*Quart. J. Roy. Meteorol. Soc.*, **113**, 402), regarding "equivalent barotropic" dynamics.

Thus the PV–PT viewpoint shows, for instance, why certain classical aerodynamical phenomena such as 2D shear instability, vorticity shedding, vortex rollup, merging, and vortex core isolation all seem to have baroclinic, "layerwise-2D" counterparts in synoptic and mesoscale atmosphere-ocean dynamics, even down to the vortex streets behind oceanic islands and, of more concern to weather forecasters, the larger scale shedding of PV anomalies from, for instance, the European Alps. According to the recent work of Aebischer and Schär (1998), this last contributes significantly to Alpine lee cyclogenesis. Again, vortex core isolation, as in smoke rings, is a central theme in today's stratospheric research because of its importance to ozone hole chemistry (references cited in Section 2 above). Vortex core isolation is related to Rossby-wave elasticity—another phenomenon most easily understood from the PV–PT viewpoint—as well as to the horizontal shear just outside the vortex edge (Juckes and McIntyre 1987). It is probably important also, I suspect, to tropical cyclone maintenance (McIntyre 1993b) because it can protect the moist updraft from drier incursions, keeping the mature tropical cyclone thermodynamically efficient. Whether anything similar happens in extratropical explosive marine cyclogenesis is an open question, but an interesting one.

Point (1) is of central importance whether or not the motion is frictionless and adiabatic. Of course, many significant weather developments do depend on fast upper air motions that are, to a first approximation, frictionless and adiabatic, so that both PV and PT are materially conserved, i.e., simply advected. Then features in the PV–PT fields often become sharp edged and frontlike because of the strong deformation rates in the large-scale wind field, with their well-known tendency to create steep gradients in the distributions of materially conserved quantities, as in so-called "chaotic advection"—the advected quantities rapidly developing spatial scales far smaller than those of the advecting wind fields, and doing so, in most cases, exponentially rapidly. Figure 1a well illustrates this ubiquitously important process, all the fine scales and steep gradients shown there having developed advectively over just 4 days. The figure also reminds us of one of the forecaster's nightmares: a high PV air mass drawn out into a long sheet or filament, and eluding operational analyses, can sometimes, if large-scale deformation rates slacken off, undergo vortex rollup and turn into a compact cyclone like, for instance, the cyclone over the Balkan Peninsula near the right-hand edge of the figure.

Synopticians have long been familiar, of course, with the zoology of all the various advectively produced structures, including jets, shear lines, tropical upper tropospheric troughs, and so on. The PV–PT viewpoint when linked to the classical insights about advection and deformation, including those of the Bergen school about surface frontogenesis, provide by far the simplest explanation of why such structures occur so commonly. In a nutshell, one has the steepening of low-altitude temperature gradients and of high-altitude PV gradients.

The PV–PT viewpoint enables one to regard advection as a quasi-horizontal, quasi-2D process. This is another powerful conceptual simplification—and, incidentally, not at all dependent on the familiar but crude approximations of quasigeostrophic theory, a point that is underlined by Figure 2 on page 349. Advection for this purpose can be considered quasi-2D despite its really being 3D. It is quasi-horizontal and quasi-2D in the sense of referring to horizontally projected motion along isentropic, that is to say constant PT, surfaces, and to horizontally projected motion near the Earth's surface—meaning, in practice, just above the boundary layer (HMR)—with no need to refer explicitly to vertical motion despite the latter's dynamical importance in many cases.

When material PV and PT conservation fails, as happens for instance in moist convection, the evolution of the PV–PT field can still be described as resulting from dynamically local effects only: advective, diabatic, frictional, and so on. Moreover—and very surprisingly—the quasi-two-dimensionality persists, in a certain sense, even when diabatic heating, for instance latent heating, is taking air parcels across isentropic surfaces. To see this, one need only adopt a viewpoint that is arguably a natural consequence of taking seriously the analogy between PV and chemicals (Section 5, item 7).

The resulting view of the dynamical evolution is fundamentally simpler than any view that refers directly to the primitive equations. The primitive equations intimately combine the local frictional and diabatic effects with *three dimensional* advection and dynamically *nonlocal* interactions. The nonlocal interactions are mediated by the pressure and buoyancy fields, constrained by the requirements of mass conservation, hydrostatic balance, and other forms of balance (see Section 6 below). So although the primitive equations are useful in showing how Newton's second law of motion is satisfied, they impede understanding by intertwining all the different aspects of the stratified, rotating fluid dynamics—3D and quasi-2D, balanced and unbalanced, local and nonlocal, prognostic and diagnostic.

The importance of dynamically nonlocal interactions, incidentally, is one of the reasons why studying the local balance of terms in a single equation, while sometimes useful when seen as part of a wider picture, can often give a totally misleading impression of causal linkages and the workings of the dynamics (e.g., Holton et al. 1995).

5. PV and Its Invertibility

The standard meteorological definition of PV for a three-dimensional, baroclinic, hydrostatic atmosphere,

$$P = (f + \hat{\mathbf{z}} \cdot \nabla_\theta \times \mathbf{V})(-g \partial \theta / \partial p) \qquad (1)$$

is essentially that proposed by Rossby in 1940 as a natural generalization of the single layer "shallow water" PV he had discovered in 1936 (Rossby 1936, 1940). In the expression on the right, f is the Coriolis parameter, $\hat{z}$ is a unit vertical vector, V is the horizontal wind vector, θ is the PT or any function of PT alone, such as specific entropy, p is pressure altitude, and g is the gravity acceleration. The subscript θ indicates differentiation along an isentropic surface, so that although the quantity $\hat{z} \cdot \nabla_\theta \times V$ looks like relative vorticity, it is quite different, in principle, from the ordinary relative vorticity. It is more appropriately called Rossby's "isentropic vorticity" to keep the distinction clear, as explained in HMR. As is well known, Ertel's (1942) more general formula reduces to (1) in the shallow-geometry, hydrostatic case.

A convenient SI unit for the PV is $10^{-6} \, m^2 \, s^{-1} \, K \, kg^{-1}$; it could perhaps be called the microrossby, but since Ertel has an independent claim I shall settle for calling it the PV unit or PVU, as was done in HMR. It is a convenient unit since, with the convention of using the PT as the materially conserved thermodynamic variable, θ in the formula (1), PV values $P < 1$ usually imply that we are looking at tropospheric air, and $P > 2$ extratropical stratospheric air. One of the exceptions to this is the case of a mature tropical cyclone, where lower tropospheric PV values well over 10 PVU may be encountered, in dramatic contrast with typical surrounding values, of order 10^{-1} PVU. Extratropical marine cyclogenesis is another exception, again with substantial contrasts in lower tropospheric PV values (e.g., Reed et al. 1992; Davis et al. 1993).

One way of stating the invertibility principle is that proposed in HMR. It begins by assuming that the mass under each isentropic surface is specified, or some equivalent information giving the static stability of a suitable reference state, just as is done in the theory of available potential energy. The principle asserts that, given this information, together with the global distribution of the PV on each isentropic surface, and of the PT at the lower boundary, one can deduce, diagnostically, all the other dynamical fields such as winds, temperatures, geopotential heights, local static stabilities, and the adiabatic part of the vertical motion, to the extent that, and to the accuracy with which, the motion can be regarded as balanced (again see Section 6 below).

The diagnostic nature of PV inversion means, of course, that the prognostic aspects of the problem have been entirely confined to describing the advective and nonadvective evolution of the PV and PT. This is part of the conceptual separation referred to in Section 4. The following additional points can be made:

1.   Just as in the simpler case of the inverse Laplacian, in order to carry out an inversion we must solve the diagnostic problem globally, with proper attention to boundary conditions. This is an inescapable consequence of "nonlocalness". The inversion problem is an elliptic problem partially analo-

gous to membrane deformation problems or problems in electrostatics (as discussed in HMR and in my more recent reviews), with the geopotential qualitatively like the electrostatic potential and the wind vector approximately at right angles to the electrostatic field.

2.   The principle, in the form just stated, helps to explain why isentropic gradients of PV and surface gradients of PT keep on turning up as key factors in theoretical studies of barotropic and baroclinic instabilities, large-scale waves, vortices, so-called "geostrophic" turbulence, and other phenomena involving balanced motion. In the electrostatic analogy, the interior PV and the surface PT are on the same footing: they have comparable roles, and comparable status as, so to speak, free electric charges subject to advection.

3.   For practical purposes the phrases "at the surface" and "at the lower boundary," in connection with PT distributions and gradients, will usually mean just above the planetary boundary layer, as already mentioned.

4.   The principle works only for "dry inversion," e.g., when the PV and PT fields refer to the dry and not to the moist equivalent PT, say θ_e. The latter, and its associated moist PV, say P_e, are more useful for the purposes of parameterizing upright and slantwise moist convection, for instance by setting $P_e = 0$ in the eye wall of a tropical cyclone (e.g., Emanuel 1991) or in the updraft of an intense extratropical marine cyclone. The P_e and θ_e fields are not useful for inversion, for mathematical reasons (lack of ellipticity).

5.   The invertibility principle, as stated here, carefully avoids any prior commitment as to the best balance condition under which to carry out the inversion. Indeed a strong reason for elevating it to the status of a "principle" is to focus attention on the idea that the balance and invertibility concepts need not be tied to any particular set of approximations, filtered equations, or explicit formulae, and to leave open the possibility that more accurate ways of quantifying balance and invertibility may yet be found—an outlook already being vindicated, as it turns out, at least for shallow water models as illustrated in Figure 2 below.

6.   The statement that vertical motion can be deduced is related of course to the omega equation principle. A simple illustration, the "vacuum cleaner effect," was given in HMR Section 4 (see also the Appendix of HMR); it shows how upper air PV advection can induce large-scale, adiabatic upward motion just ahead of it, in practice often triggering moist convection in the manner familiar to synopticians and forecasters studying what is called Petterssen type B cyclogenesis.

7.   PV distributions and their possible evolution, and the associated transports and budgets of PV—more precisely, of the transportable "charge" or signed quasi-chemical "substance," PVS say, whose mixing ratio or amount per unit mass is the PV—are constrained by two exact, general theorems that hold even in the presence of diabatic heating

and frictional or external forces. They are simple, almost trivial, to prove by standard vector calculus (e.g., Haynes and McIntyre 1990), but have nontrivial significance. Specifying a PV distribution that violated the implied constraints would presumably lead to failure of any attempt at inversion; see also the caveat in HMR Section 3, equations (17b)ff. The first theorem says that PVS, considered as an additive, conservable, transportable quasi-charge or quasi-substance, is *indestructible*, like real electric charge—one can have pair production and mutual annihilation, but no net charge creation—except where isentropic surfaces intersect a boundary such as Earth's surface. The second theorem says that PV distributions behave as if isentropic surfaces were *impermeable* to the "charged particles" of PVS, even in the presence of diabatic heating. The mathematics permits us to think of an isentropic surface in a stably stratified atmosphere as acting like a *semipermeable membrane*, allowing mass to cross it but not the notional particles of PVS.

It is this "impermeability" or "semipermeability," then, that allows the quasi-2D representation of advection to be retained, very surprisingly, even when mass and real chemical substances are crossing isentropic surfaces diabatically. The PVS particles are to be imagined as moving strictly on the isentropic surfaces, regardless of how those surfaces themselves move. In this picture, values of the PV, which have the nature of chemical mixing ratios, can of course change; but they can change only through the PVS being transported, diluted, or concentrated in various ways.[4] Both the "indestructibility theorem" and the "impermeability theorem" are straightforward consequences of the way in which the PV is constructed mathematically, equation (1) above and its nonhydrostatic generalization due to Ertel. This has the significant further consequence that the theorems apply not only to the exact Rossby–Ertel PV that could be constructed from exact wind and PT fields if one knew them (which is

never the case in practice) but also, exactly, to any "coarse grain PV" constructed from observational datasets for the wind and PT fields (Haynes and McIntyre 1990; Keyser and Rotunno 1990).

In the interesting example of explosive marine cyclogenesis, the highly concentrated low altitude PV anomalies that are characteristic of the phenomenon are "concentrated" in a literal sense, if one describes what goes on in terms of PVS. According to that description, one can picture such PV anomalies as having resulted from a strongly convergent inflow of PVS particles trapped on dry-isentropic surfaces and hence, inevitably, concentrated near the convergence maximum. The PVS is concentrated there in just the way a real chemical can't be (because the molecules of the real chemical aren't, of course, trapped on the dry isentropes but are carried across them in the cyclone's upward vertical motion).

6. Limitations on Balance and PV Invertibility

Over the past decade or so there has been interesting work on PV inversion at accuracies comparable to that of the gradient wind and Bolin–Charney balance approximations, a level of accuracy that is often, in practice, comparable to data errors and data assimilation errors (e.g., Thorpe 1985; Raymond 1992; Davis 1992; Davis et al. 1993). I shall not go into this except to express my continuing interest and to say that I believe that the development of robust and efficient inversion algorithms having this kind of accuracy will sooner or later be seen to be of the greatest practical importance—not, of course, as a means of timestepping the numerical weather models themselves, which are most efficiently based on the primitive equations, but rather as a means of generating "bogus data" in the manner of Demirtas and Thorpe (1997), and generally for "getting a feel for the problem" in research as well as in operational mode. Because inversion poses a nonlinear elliptic problem, massively parallel computation should help future developments here.

I would like to say just a little, however, about some recent and theoretically important advances in our understanding of PV invertibility and its *ultimate* limitations—the limitations on how accurate it can be in principle, the limitations beyond which we cannot go no matter how clever we are computationally. These limitations are, of course, bound up with the ultimate limitations on the concepts of "balance" and "slow manifold."

Some of these limitations have become clearer from studies of a phenomenon characteristic of the deep tropics, asymmetric inertial instability (Ciesielski et al. 1989; O'Sullivan and Hitchman 1992; Dunkerton 1993; Clark and Haynes 1994). If we add moist processes (Bennetts and Hoskins 1979) we similarly encounter what might have to be

---

[4]At this point I should remind the reader that there has been an unfortunate and dangerous confusion of terminology in the literature. Words like "source" have been used in different senses for PV and chemicals—even, sometimes, in the same discussion with PV and chemicals considered together. If the mixing ratio of an inert chemical like helium were to change through dilution, I don't think we would want to speak of a "sink" of helium. Yet the parallel situation for PVS is often spoken of in terms of a "sink for PV." Further discussions of the various incompatible meanings of words like "source," "sink"—and "transport" as well—are given in my 1992 review, Section 11, and under STRAY ADJECTIVE in the file lucidity-supplem.tex on the web site already noted. (So powerful can be the straying of adjectives that, very confusingly, through a kind of linguistic–cognitive domino effect, the word "substance" has become, in the present context, almost a synonym for "per unit volume." More precisely, you will find the term "PV substance" used to mean the PV times the mass density ρ, even though "water substance," for instance, is never used to mean the corresponding thing, water mixing ratio times ρ. The domino effect well illustrates the danger of dismissing all such problems as "mere semantics.")

called "conditional *asymmetric symmetric* instability" (just as we have to speak nowadays of the "*variable* solar *constant*"). Perhaps it should be "asymmetric slantwise conditional instability." In any case, its importance in intense extratropical as well as tropical cyclogenesis seems to be in little doubt, and its velocity fields are, of course, another kind of unbalanced motion inaccessible to PV inversion as usually understood.

When and where such dry or moist instabilities are unimportant, there remains another, entirely different, more subtle, and ever present limitation on PV invertibility, which has no necessary connection with any instability and which might be summarized in the phrase "fuzziness of the slow quasi-manifold." It manifests itself in various subtle ways, including

1. the characteristic "schizophrenia" or "velocity splitting" of filtered equation models, (usually in the form of two coexisting, unequal velocity fields, there being no such thing as "the" velocity field of the model; this has recently been shown to be inescapable for a large and important class of Hamiltonian balanced models, whether constructed by filtering or other means (Allen and Holm 1996; McIntyre and Roulstone 1997)),
2. the mathematical phenomenon called Poincaré's "homoclinic tangle," and
3. the physical phenomenon described by the Lighthill theory of aerodynamic sound emission generalized to the rotating frame (Ford et al. 1998).

This last physical phenomenon might be described as the weak but nonvanishing "spontaneous-adjustment emission" of inertia–gravity waves by unsteady, 2D or layerwise-2D vortical motion. The wave emission mechanism is sometimes referred to, illogically, as "geostrophic" adjustment even though—as a consideration of circular vortices immediately suggests—it need not take the system toward geostrophic balance. (This has long been a source of confusion, I suspect, about the significance of "ageostrophic" winds.)

The phrase *spontaneous* adjustment is also intended to distinguish the phenomenon in question from the quite different phenomenon of Rossby or initial condition adjustment. The latter is simply due to the initial conditions being out of balance. The essential character of spontaneous-adjustment emission is its remarkable weakness. It is often far weaker, in circumstances of interest, than you would deduce from any simple order-of-magnitude or scaling analysis, from the values of Froude and Rossby numbers and so on. It is this remarkable weakness that gives rise to a correspondingly remarkable accuracy in PV inversion, or, rather, possible accuracy, a possibility that can be realized if one is clever enough computationally.

It was James Lighthill's great contribution, in the simpler but fundamentally similar aeroacoustic or sound-emission problem (Lighthill 1952), to show why the emission is often far weaker than you would think from a standard order-of-magnitude or scaling analysis. He achieved the required conceptual clarification despite having no clues from numerical modeling, and despite the dauntingly nonlinear yet delicate nature of the problem. The Lighthill theory tells us, in effect, that spontaneous-adjustment emission is often weak because destructive interference among the emitted waves is often important. This is not something that you can notice from scaling analysis. Recent work by my former students Warwick Norton and Rupert Ford, and by other colleagues including Jim McWilliams and Lorenzo Polvani, has added to our detailed knowledge of how all this works in a rotating system and, in particular, has given us some detailed quantitative checks on the Lighthill theory and its range of validity, which for order-of-magnitude purposes turns out to be astonishingly wide (Ford 1994; Polvani et al. 1994).

In a rotating system, Coriolis effects tend to enhance still further the possibilities for accurate balance and PV invertibility, because they tend to weaken still further the spontaneous-adjustment emission. One can say, following Lighthill, that in the acoustic or pure gravity wave problem one has a mismatch of spatial scales at small Mach or Froude number. With Coriolis effects, one has in addition a frequency mismatch at small Rossby number. Both things, in combination with the destructive interference, weaken the spontaneous-adjustment emission. The weakening due to the frequency mismatch at small Rossby number was already observed and noted in numerical experiments by Errico (1982).

Even more remarkably, when one goes to larger Froude and Rossby numbers the spontaneous-adjustment emission tends to remain weak by any practical criterion, and *weaker* than the Lighthill theory might appear to predict without detailed consideration of the vortical motion (Ford 1994). The reason is that the unsteadiness of the vortical motion, necessary to excite spontaneous adjustment, tends to become weaker as a result of the same parameter changes. This in turn is because inversion operators have a *short range* character under these conditions, or "small deformation radius," related to the slow propagation speeds of the inertia–gravity waves that mediate the apparent action-at-a-distance (McIntyre 1992, p. 345).

If the spontaneous-adjustment emission were zero, then we could have a true slow manifold within phase space, in the strict mathematical sense of the term "manifold." States on this (invariant) manifold would correspond to what Edward Lorenz once called "superbalance," implying exact and unique PV invertibility and exactly zero inertia–gravity wave activity, given suitable initialization. The above-mentioned and other evidence is overwhelming, however, in favor of there being no such thing as a strict slow manifold, as originally argued via another approach by Warn (1997, q.v.) and by Warn and Ménard (1986). Spontaneous-adjustment emission turns what might otherwise be a manifold into a much fuzzier mathematical object in phase space. That object almost certainly has the character of a thin chaotic or "stochastic" layer, hence the term "slow quasi-manifold." This means that PV inversion itself must have a corresponding fuzziness, or inherent approximateness, representing an ultimate limitation on its possible accuracy.

Though rigorous proof seems beyond present day mathematical resources, for realistic fluid models with their infinite dimensional phase spaces, it is worth recalling that fuzzy mathematical objects of the kind in question have long been known to dynamical systems experts studying low-order Hamiltonian systems like perturbed pendulums and other weakly coupled oscillators. Experts on these topics, who have studied multitudes of cases and developed some general theoretical perspectives, assure me that stochastic layers seem to be generic—that is to say typical of nearly all such systems—as expected intuitively from Liouville's theorem and considerations of the infinite time available to disturb what is called a homoclinic orbit. This is the relevance of Poincaré's "homoclinic tangle"; for some relevant recent work see Bokhove and Shepherd (1996 and references therein). But the good news is that, under the usual parameter conditions, the atmosphere's slow quasi-manifold, though fuzzy, appears likely to be thin enough to be, for practical purposes, almost a manifold, at least the parts of it corresponding to motions whose Froude number is not too large (vertical scale not too small). The upshot is that PV inversion, despite being inherently approximate and inherently nonunique, can nevertheless, with enough computational ingenuity, be made astonishingly accurate—again provided that the vertical scale is not too small—and certainly far more accurate than any simple filtered theory or scaling analysis could ever suggest.

Figure 2 presents an example, taken from McIntyre and Norton (1990), showing direct evidence for such accuracy, far greater than any of the standard filtered theories would suggest and, by implication, confirming the weakness of spontaneous-adjustment emission in at least some circumstances. Like most of the fundamental theoretical work to date, the computations have been done, so far, only for the simplest fluid-dynamical system for which the balance and inversion concepts are nontrivial, namely, a shallow water system. This example is from the PhD thesis work of my former student Warwick Norton (1988), full details of which we are hoping to publish soon in a long delayed joint paper (McIntyre and Norton 1998). The computation was done with a hemispherical pseudospectral shallow water model at fairly high resolution, triangular truncation T106. A carefully initialized primitive equation integration produced the geopotential height, wind, and divergence fields shown in the top two panels, representing a complicated, unsteady vortical motion somewhat like an exaggerated 300-hPa flow in the

real atmosphere, with a strongly meandering midlatitude jet. There is a gigantic blocking anticyclone, and three major troughs. Froude and Rossby numbers reach values of the order of 0.5 in parts of the jet; Rossby numbers are of course infinite at the equator. Such values can hardly be thought of as small, and we may expect that the standard filtered theories of balanced motion will not be accurate. Nevertheless, in this and a range of similar cases, it proved possible to reconstruct the height, wind, and divergence fields by PV inversion— i.e., knowing only the mass of the fluid layer and the PV distribution, to reconstruct all the other fields from that information alone —and to do so with astonishing accuracy. Here of course the relevant PV is the shallow water PV of Rossby (1936), namely absolute vorticity divided by layer depth. In this example, the PV distribution is shown by contours and grayscale in the middle two panels; and the reconstruction of the other fields from it, the output of an accurate PV inversion algorithm, is shown in the bottom two panels of Figure 2. In comparing the original primitive equation fields in the top two panels to the reconstructions in the bottom two panels, you have to look carefully to see the differences: by implication, the invertibility principle is at least this good in this particular example.

To get results with this sort of accuracy, a rather elaborate inversion algorithm had to be used; the equations are formidable looking and take about half a page to write out, and are non-trivial to solve numerically (Norton 1988). The computations are like a high-order extension of nonlinear initialization in the manner of Hinkelmann (1969). In other words, we are dealing with a fairly complicated nonlinear elliptic problem, about which little is known in any mathematically rigorous sense; however, a powerful check on the numerical procedures, and on the accuracy of the invertibility principle itself, came from tests of cumulative accuracy in 10-day experiments (not shown here) running the primitive equations in parallel with the balanced model defined by advecting the PV and using the inversion algorithm. These 10-day experiments again showed astonishing accuracy. Similar results, indeed somewhat more accurate still, as it turned out, were obtained using the nonlinear normal mode approach. All these results are in stark contrast with those that would be obtained using standard quasi-geostrophic inversion, such inversion being hopelessly *inaccurate* under the parameter conditions of Figure 2. In the case of Figure 2, a quasigeostrophic inversion (not shown) produced wind speeds in error by factors of order 2 even well away from the equator.

7. Implications for Weather Prediction

It has been said that "there is nothing so practical as a good theory." One could say that the theoretical developments I have been discussing underline the fact that the dynamics of weather developments has as strongly advective a character as does the moist and dry thermodynamics. One has moist advection, warm advection, and PV advection. This does not capture everything about real weather developments, but it surely captures an important part of them. To say that the dynamics has a strongly advective character is perfectly compatible, incidentally, with classic perceptions about the role of Rossby wave propagation, as I have already remarked in connection with downstream and local development and the "Rossby-elastic tropopause." The Rossby wave propagation mechanism is, indeed, partly advective—one could say "sideways advective"—as explained for instance in the reviews cited, including HMR Section 6. Ordinary textbook plane polarized waves on stretched strings are "sideways advective" in a partly analogous sense: the material of the string is carried sideways, back and forth, whenever we excite any wave motion. Of course, to complete the intuitive picture for Rossby waves, and to include for instance such things as group velocity (HMR Section 6c), one must obviously, as always, consider advection and PV inversion together.

If we now bring this view of the dynamics and thermodynamics together with an appreciation of the nature of human perception, especially human vision and visual cognition—the "4D intelligence" I have referred to — then the practical implications for weather prediction become conspicuous. What follows is a slight update of a discussion (1988) that I published some years ago in the Royal Meteorological Society's popular magazine "*Weather*," in response to an article by Tennekes (1988), which also appeared in *Weather* and which I read with great interest.

Tennekes gave what I saw as a thoughtful, provocative and timely warning of what may be at stake; and I wanted to add something to his discussion of that vexed and critical question, the future role of the human forecaster. My remarks were based on earlier lectures I had given to the ECMWF Seminar of 1987. I suspect that the same remarks—Tennekes' remarks as well as mine—are still worth making, and that the issues are going to stay alive for quite a few years to come.

Tennekes suggested that the automation of analysis and forecasting has been taking us into a situation where already, for some purposes, "the added value of human skills has become marginal," and that commercial pressures, and commercial thinking, will lead to pressure to "bypass the forecasters altogether." Are we, he asks, "going to leave it to private enterprise to redefine the added value of human skills?" Against this, Tennekes pleaded that "only human professionals can bear the responsibility for life and property in emergencies."

I agree; and I believe that the case can be further strengthened. The peculiar way in which the atmosphere works argues strongly in itself, as I have been hinting, for an enhanced and not a diminished role, sooner or later, for human professional skills.

Consider the most basic aspect of the forecaster's responsibility, the quality control, and the quality *assessment*, of the analysis–forecast process on which everything else depends. In view of the massive data throughput and the requirement to "work at lightning speed," as Tennekes put it, the tasks involved are daunting. Clearly they do indeed call for as much automation as possible, wherever appropriate. However, there are some things that humans can do faster, and better, than any existing or foreseeable machine. It may be useful to step sideways and look at these first in another context.

A motorist driving in city traffic is continually keeping track of, and extrapolating the positions of, a variety of independently moving objects most of which are rapidly changing their apparent size and shape, some in very complicated ways. Some of the objects intermittently obscure others from the field of view, and dirt on the windscreen may impose further complications. A skilled driver seeing a pedestrian about to emerge from behind a parked vehicle some way down the street may "instinctively" begin to slacken speed within a few tenths of a second. It is possible that only the pedestrian's feet were visible at first, their characteristic motion showing beneath the parked vehicle. "So what?" you may say. "These are familiar facts, but what is the relevance to weather prediction?" Be patient a moment longer.

In performing these data processing tasks that we blithely call "instinctive" or "intuitive," the human brain and its visual peripherals are performing almost unimaginably prodigious computational feats, involving massive data throughput and much more, all taking place at more than "lightning speed" by the standards of any electronic computer. Furthermore, the data in my example, carried by the light entering the driver's eyes, on which safe driving depends, are gappy, noisy data. These are among the reasons why, for instance, computers don't yet drive taxis. The fact that we are not conscious of all this computational activity does not mean that it is not taking place. It may even be quite literally unimaginable, in any complete and detailed sense. As Marvin Minsky has emphasized, an ability to grasp intuitively how one's own brain works is biologically speaking an unlikely sort of mental ability. We quite naturally tend to underestimate what is involved. In watching young video games addicts at play, we tend to be impressed by the fast-moving action on the color displays, though what is truly remarkable is the human, not the electronic, side of the person–machine interface.

The actual computing power that the human brain deploys on visual data processing and interpretation is hardly a well-known quantity. But I well recall hearing in 1979 a fascinating radio interview in which John Maddox talked to David Marr, then an outstanding young thinker among the new generation of researchers on vision and artificial intelligence. Marr made the point that even the earliest (and evolutionarily most ancient) computational processes in the human visual system—the sort of thing we have in common with frogs and other creatures we call primitive—already

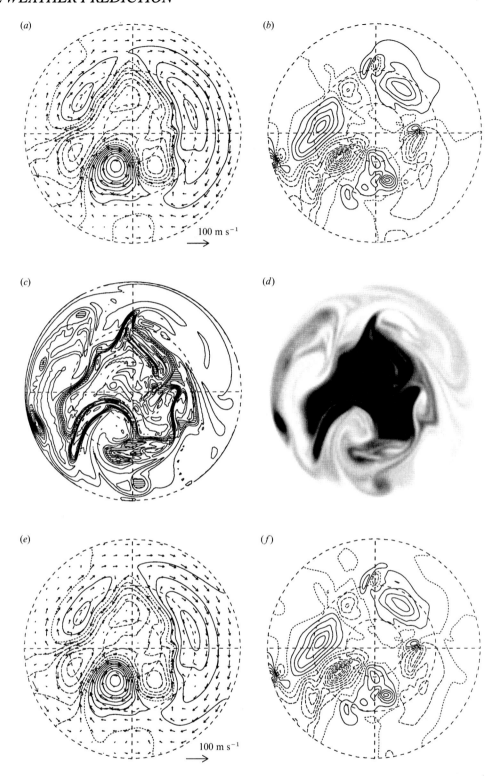

Figure 2. Demonstration of highly accurate PV inversion in a shallow water system, the simplest dynamical system for which the balance (slow quasi-manifold) and inversion concepts are nontrivial. Quasigeostrophic inversion would be grossly inaccurate in this example, if only because there are substantial layer depth variations, with fractional departures 0.4 or more of the area-mean depth, 2 km. The domain is the hemisphere and the map projection is polar stereographic. Solid contours show positive values, dashed contours negative values, and dotted contours zero. (a) Wind field from the primitive equation solution, arrows scaled by the 100 m s^{-1} standard arrow shown, and geopotential height, contoured at intervals of 0.1 km, defined as departure of layer depth from its area mean 2 km. (b) (horizontal) divergence from the primitive equation solution, contoured at intervals of 0.6×10^{-6} s^{-1}. (c,d) Grayscale and contour maps of the shallow water PV from the primitive equation solution, i.e., absolute vorticity divided by local layer depth; the contour interval is 1×10^{-8} m^{-1} s^{-1} and the hatching in the contour plot highlights values lying between 4 and 6 of these units. The grayscale representation of the same information is monotonic from light to dark, from zero to a maximum value of 1×10^{-7} m^{-1} s^{-1}. (e,f) As in (a,b) but reconstructed from the PV alone using an accurate nonlinear PV inversion algorithm (McIntyre and Norton 1990, 1998).

appear to need several orders of magnitude more computing power than a "fast general purpose computer." When one considers the vastly more complicated functioning of the entire visual cortex, there seems little doubt that the total equivalent computing power must be staggering. The advantage of three-dimensional miniaturization down to cellular and molecular scales overwhelms the disadvantage of what might loosely be called a slower "cycle time." There is also the robust and highly developed "programming"—here the distinction between "software" and "hardware" would be pushing the analogy too far—the cumulative result of hundreds of millions of years of natural selection.

It is dangerous to guess what human technology might look like even a decade or two from now, but my own guess is that the human eye–brain system will continue to make even our biggest supercomputers, our most sophisticated artificial-intelligence software, and our most massive experiments on artificial neural networks look feeble for some decades to come, when it comes to processing, interpreting, and interacting at high speed with appropriate kinds of visual data, and doing so with what we call "understanding," or cognitive grasp. On the other hand, independent commercial pressures to develop electronic computers to do what they are good at, including animated graphics all the way through to what is charmingly called "virtual reality" (as with "asymmetric symmetric instability" and the "variable solar constant") will produce mindboggling possibilities for the person–machine interface.

You will have no difficulty in seeing the point I am coming to. Provided that the interface with the forecaster is appropriately designed, the prodigious computing power—and cognitive–perceptual power—of the human brain and its peripherals could be harnessed far more efficiently to some very considerable electronic or photonic computing power and through this to the quality control and quality assessment of weather forecasts under operational pressures—just as it is already harnessed to other high speed, "four dimensional" problems under operational pressures such as playing world-class football, flying military aircraft, or for that matter driving in heavy traffic. It is in such situations, involving the visual detection of coherent motion, that the eye–brain's "4D intelligence" is most conspicuously powerful. One of the computational tasks to which our visual systems are most exquisitely adapted is that of following moving, coherent features, even when the features continually change their shape and intensity, and even in the presence of noise, and of data gaps—as my traffic example was meant to suggest. This prompts the question, why should we not use, among other things, this "4D intelligence" in weather prediction?

In order to exploit the implied potential to the full, it will be crucial to present information to the duty forecaster's eye–brain system in an appropriate form. To use "4D intelligence" to the full we need to use animation, and we need to find, if possible, ways of presenting the evolution of the atmosphere in terms of moving, visibly coherent features. It is an extraordinary piece of good fortune, therefore, that so much of the dynamical and thermodynamical behavior of the atmosphere can be described in terms of quantities that often behave advectively. And when they do not behave advectively, as with the low-altitude PV changes involved in explosive marine cyclogenesis or tropical cyclogenesis, then this fact calls for the forecaster's attention, the more so since it means that the numerical prediction model is invoking so-called "physics" parametrizations that may be less soundly based than the model's dynamics. We therefore need animated displays that make it instantly obvious whether the behavior is advective or not. Such displays are possible: if a moving vehicle hits a brick wall, or suddenly doubles in size, or suddenly appears from thin air, so to speak, it is a very visible thing.

One of the promising graphical choices for the near future seems to be the tropopause "isostrophic PT maps" used at Reading, of which Thorncroft et al. (1993) give examples both from the real atmosphere and from life cycle experiments. These are maps of PT on the 2 PVU surface, approximately marking the extratropical tropopause and reflecting its character as an "erodable, highly deformable, dynamically active Rossby-elastic wall." The tropopause could be marked even more accurately if we were to redefine the PV using some suitable function of PT in place of the PT itself, θ in the Rossby–Ertel formula (1); it is merely a lucky accident that the conventional definition of the PV, using θ rather than an alternative thermodynamic function such as specific entropy, or some other function of θ, happens to be a good approximate extratropical tropopause marker. A single tropopause map has the advantage, with present-day graphical methods, of summarizing most of the upper air information that would otherwise require a whole stack of isentropic maps of PV.

It is possible of course that, some years from now, the "virtual reality" techniques will get so clever that it becomes perfectly easy to look at a whole stack of maps anyway, or to fly oneself very quickly, supermanlike, around some kind of 3D representation of significant features such as the tropopause and other regions of steep isentropic gradients of PV. Whatever visualization methods turn out to be best, it seems to me that experience with the typical patterns of behavior of such features, once they are made highly visible, cannot fail to lead to sharpened insights into the problems of data analysis and quality control; and as the automated system itself becomes more consistently "four dimensional" and thence "advectively intelligent" there will be an increasing potential, as hinted at by Figure 1a, for improvement in the accuracy and clarity of any graphical representation of the PV–PT fields, greatly enhancing their visual intelligibility as well as the accuracy of the numerical forecasts that use them.

8. A Vision, or Dream, of the Future

I should like to conclude, then, by sharing, in a lighthearted vein, the vision—or perhaps I should not risk sounding too pretentious, and just say *dream*, or just plain *fantasy*—about future weather prediction operations. I hope it will be taken in the spirit of the real visionaries, the great pioneers Vilhelm Bjerknes and Lewis Fry Richardson. The original dream was something that Glenn Shutts and I indulged in during a brief conversation at the UK Meteorological Office, probably sometime in 1986 or 1987. Today's politico-economic conditions make the whole thing even more dreamlike, in some ways, than it did in the 1980s; and, as I said earlier, I am going to push the time of the dream a further decade into the future. But, wearing my optimist's hat, I shall also dare to make it part of a larger dream in which, through practical necessity, the chaotic pendulum of politics and economics has swung toward a new attractor basin representing a degree of sanity for the longer term, including saner uses of our new "psychological nuclear energy," the supranational power of market forces (von Weizsäcker et al. 1997). Though there is hardly room for complacency I actually think that, despite everything, there is a tenable optimist's theory of the future, meaning the future of many things besides weather prediction.

The year is 2020. The international demand for good one week deterministic–stochastic weather forecasts has led to the establishment of a greatly enhanced global observing network, and to an unprecedented concentration of electrophotonic and biomolecular computing power and human skills at the great new International Centers for weather and climate prediction. The observing network includes not only sophisticated remote sensors but also a large fleet of ultra-light, ultra-cheap, remotely controlled miniature aircraft or "aerosondes" that can be flexibly deployed in forecast-sensitive regions as necessary, including, for instance the centers of tropical cyclones.[5]

An added incentive has been the excess greenhouse warming, now conspicuous and exacerbated by a slight upward fluctuation in solar irradiance. These changes were accompanied by an increase in the frequency and severity of moisture-related weather phenomena, not just tropical cyclones of increasing power but also destructive storms in higher latitudes, due to maritime explosive cyclogenesis—the latter one of the fastest evolving forms of hazardous weather and one of the trickiest challenges to the weather forecaster. The changes, moreover, had all been predicted

fairly convincingly, and a few years in advance of their occurrence, by climate models, solar physics models, and mesoscale models. All of these models had reached a state of refinement undreamt of, by serious scientists at least, before the Millennium, and the climate models had had notable successes in reproducing Pleistocene as well as present climate regimes. Development of the solar models had been greatly stimulated by the improved helioseismic data that began to be available in the 1990s. In terms of their perceived impact on human societies, weather and climate prediction had become increasingly important everywhere, including the low-lying coastal areas where the effects of sea level rise had become unequivocal and where further population growth, only just coming under control, had been accommodated by new amphibian dwellinghouse technologies.

Weather and climate prediction had been made far more socially effective. This was not only through new communication skills and technologies but also through one of the spinoffs from those skills and technologies—the new cooperation and mutual understanding between scientists, social scientists, and the intelligent lay public that was built up in the aftermath of the damaging "Science Wars" of the 1990s and the accompanying "Little Dark Age" of commercially amplified superstitious belief. The effort to ward off a greater Dark Age had begun with attention to what was called the problem of the public understanding of science, followed by recognition that public respect for the scientific ideal, for rational thinking and respect for evidence—and a wider appreciation that such questions are distinct from questions about personal faith and spiritual health—were even more important than understanding particular bits of science. This in turn had led scientists to develop a publicly declared code of professional conduct that helped to reduce arrogance and theologizing. Although these various efforts had seemed quite futile at first—not least (Gelbspan 1997) in the dark days before the Kyoto Climate Convention of 1997, with the might of the fossil fuel industry solidly behind arrogance and theologizing—the scientists' efforts finally began to have a big effect when, to everyone's astonishment, the international mass newsmedia swung behind them.

This happened when, after fully establishing their supranational power base, the international newsmedia and Internet moguls had begun to realize that their practice of promoting scientifically irresponsible mass journalism, though profitable in the short term, was a bit like playing with matches in a planetary-scale explosives factory. They saw that they were playing games, on a global scale, with exceedingly powerful human instincts—which instincts and their bioclimatological origins were by then better understood scientifically[6]—and they saw that such games could lead in the end to catastrophic

---

[5]There are already precursors to such a development: see, e.g., Lighthill (1993) on the Aerosonde Project for the improvement of tropical cyclone forecasts: current designs call for a 12 kg, 3 meter wingspan aircraft capable of in situ and surface pressure measurements using radar altimetry. See also, e.g., Palmer et al. (1998) on singular vectors as a guide to deployment of observations.

[6]Indeed, new understanding of these matters was already emerging in the 1990s (as discussed for instance in lucidity3.ps; see footnote 1).

social destabilization and widespread totalitarian repression, even without the extra pressure from climate change. This in turn, they saw, would cause the destruction of their own power base, dependent as it was, and is, on the continued existence of the free market democracies and the Internet. The growth of technologically sophisticated kamikaze terrorism, and apocalyptic social balkanization or sectarianization—not only in the poorer countries but also, during the 1990s, in the affluent USA—was one of the signs to which the international newsmedia eventually paid heed, it being recognized that, with biotechnological advances, the terrorism might even become quite literally a global-scale phenomenon, even without climate refugees. The newsmedia also saw that they were the only political force able to stand up to the fossil fuel industry and its suicidal zealotry.

Scientists, for their part, having seen more clearly the damage that had been done to science through its deliberate or inadvertent promotion as the Answer to Everything and the Way to the Mind of God, had developed as part of their professional code a publicly declared "humility principle," in turn leading to a better public understanding of the nature of scientific uncertainty. The upshot had been a renewed respect for science and a popular revival of the notion that good science, rather than astrology, witchcraft, or head-in-sand zealotry, would be our best eyes and ears on an uncertain future, and an indispensable aid to managing medical, environmental, and societal risk.

The same professional principle of humility and recognition of uncertainty—indeed the development of an improved, simple *language* of uncertainty—had helped to increase respect for weather forecasting in particular. Public weather bulletins, early in the Millennium, were already routinely incorporating well-explained, semiquantitative estimates of forecast uncertainty. It was widely believed that these largely successful, and widely publicized, estimates of forecast uncertainty had been the single most important influence toward building public confidence in the scientific respectability of weather prediction—helped of course by improvements in actual forecasting skill, and by the increasingly informative animated-graphical techniques used in public presentations, in broadcasts, and on the Internet. (Some early versions of those techniques had actually been seen, from time to time, in the British Broadcasting Corporation's television weather bulletins as long ago as the 1980s, some of which dared to be as informative as possible, going against the then prevailing "look of the thing" or "fool the public" culture. Even such simple but effective devices as cutting straight from an animated rainfall radar map into an animation, at the same speed, of the predicted forecast rainfall pattern—much fuzzier of course—instantly conveyed to an interested lay person a useful intuitive idea of predictable and unpredictable spatial scales, and some inkling of the overall complexity of the problem.)

Today, in the year 2020, the global observing network has long been largely automated—including even aerosonde flight plans, whose bureaucracy had been successfully computerized—and the network performs far more consistently than in the 1980s. But data assimilation and forecasting procedures have deliberately not been quite fully automated. Artificial intelligence systems, though very powerful, are still far from being directly competitive with the human brain in certain respects—particularly the visual–cognitive system of a gifted individual who has had the full childhood educational and virtual-reality games experience, and who has been given the incentive of a highly respected professional status after surviving rigorous training and selection procedures. This continuing supremacy of certain aspects of human intelligence is now better understood, even though it had surprised some artificial-intelligence researchers who had based their predictions of absolute machine supremacy on textbook models of neurons and synapses that were, in fact, outdated well before the Millennium.

The data and forecasting quality control systems at the new International Centers are among the many twenty-first-century information systems in which the efficiency of person–machine interaction has been fully developed and exploited, with a finesse undreamt of before the Millennium. The basic mode of operation is, of course, simple in principle: the visual inspection and manipulation of animated thermodynamical and dynamical fields—not only to inform the forecasters about the evolution of the system's model atmosphere and its response to data assimilation, but also to facilitate extremely rapid interactive repairs to the model fields. Techniques such as variable-speed animation within virtual-reality environments have long been taken for granted: the duty forecaster is, in a sense, thoroughly immersed in the four dimensional weather. Also taken for granted is the succinctness and high visibility with which advected quantities, especially PV, PT, water vapor, and other chemical mixing ratios, convey large amounts of dynamical and thermodynamical information in an intuitively assimilable form, while fully recognizing, of course, the importance of large-scale wave propagation effects.

On some occasions, a forecaster carries out the repairs directly on the PV–PT and moisture fields, for instance molding the shapes of missing sharp edges or shear lines whose likely presence in an upper-air PV–PT distribution might be a good guess from experience. The system maintains the appropriate integral constraints on the potential vorticity field (respecting the internal indestructibility of "PV substance"), and a system of audio signals instantly informs the forecaster whether, and in what way, he or she is reducing or increasing the stress between the 4D analysis and the observational constraints. The latter include satellite brightness temperatures in various spectral regions, giving information not only on temperatures but also on advected quantities such as water vapor, ozone, and many other trace chemicals. There is also a warning signal when the limitations of the balance and potential vorticity invertibility concepts are approached, and an ability to diagnose any spontaneous-adjustment emission of inertia–gravity waves that results, as well as emission from other sources such as cumulonimbus activity, and to assess the consequences for mesoscale developments and for aviation fuel savings in the lower stratosphere.

Even in the year 2020, there are still fairly severe limitations on the size of the ensemble of initial conditions that can be used to help assess forecast uncertainty. So although a basic ensemble of a few thousand forecasts is always run automatically, for which the initial conditions are varied objectively using forms of singular vector analysis, there is also provision for a practically unlimited number of special forecast runs, based on the forecasters' subjective assessment of the most sensitive locations for varying the initial conditions, helped by the locations of features in displays of advected quantities and an array of other diagnostics. This subjective or semisubjective assessment of sensitive locations is regarded as one of the most important parts of the duty forecaster's responsibility, if only because it is the smallest scales that are most conspicuous to human perception but least well handled by the automatic analysis–forecast process. Human awareness of sensitive locations is considered to be an essential part of the background to the judgments that have to be made when, as happens increasingly often these days, there is a risk of hazardous weather.

As a typical manifestation of the more enlightened uses of market forces, forecasters are provided with strong incentives to develop their subjective skills—even though there will always be some element of luck to add to the excitement. For instance, one member of the ensemble of forecasts always takes the fully automated objective analysis as initial condition. After verification the system logs the improvement in skill, if any, in the final forecast, resulting from the forecaster's repair work, if any, and subjective assessment of ensemble properties. This is added to the forecaster's personal score and thence bonus payment. A particularly high score is earned by finding initial conditions for a special forecast run that proves to have a higher rate of divergence from the main run than a typical member of the automatic ensemble, for a given root mean square initial difference. It is becoming something of a legend that a select few in the forecasting profession have begun to develop an almost uncanny level of skill, like the legendary "top guns" or ace fighter pilots; and such individuals are coming to be known as *ace forecasters*. A few of these individuals have, in newsmedia interviews, described their subjective experience as something like merging with, or becoming part of, the weather, or not just being in the eye of the storm but becoming the eye of the storm, as one of them put it. And the urge to aspire to such skill and to join the ranks of internationally famous ace forecasters is intense, even though only a select few have so far earned that status. It is just one small but

significant part of the drama that has contributed to the new prestige, glamour, and human interest of the atmospheric sciences.

Acknowledgments

Many colleagues have kindly shared their knowledge, ideas, historical recollections, and unpublished work over the years, including John Allen, Rainer Bleck, Onno Bokhove, Lance Bosart, Keith Browning, Oliver Bühler, Jule Charney, Mike Cullen, Ed Danielsen, Chris Davis, Huw Davies, David Dritschel, Franco Einaudi, Arnt Eliassen, Kerry Emanuel, Mike Fisher, Rupert Ford, Bill Grose, Peter Haynes, Raymond Hide, Tony Hollingsworth, Darryl Holm, Jim Holton, Brian Hoskins, Ian James, Martin Juckes, Steve Koch, Dan Keyser, Ed Lorenz, Bob Lunnon, Robert MacKay, Jerry Mahlman, Taroh Matsuno, Jim McWilliams, Geoff Monk, Mike Montgomery, Phil Morrison, Mikhail Nezlin, John Norbury, Warwick Norton, Alan O'Neill, Tim Palmer, Anders Persson, Norman Phillips, Ray Pierrehumbert, Alan Plumb, Dave Raymond, Dick Reed, Peter Rhines, Rich Rotunno, Ian Roulstone, Rick Salmon, Prashant Sardeshmukh, Wayne Schubert, Mike Sewell, Mel Shapiro, Ted Shepherd, Glenn Shutts, Adrian Simmons, Chris Snyder, Susan Solomon, George Sutyrin, Henk Tennekes, Alan Thorpe, Jürgen Theiss, Joe Tribbia, Adrian Tuck, Louis Uccellini, Tom Warn, Darryn Waugh, Jeff Whitaker, Geoff Vallis, Martin Young, and Vladimir Zeitlin. Work at Cambridge received support in part from the Natural Environment Research Council, through the British Antarctic Survey and through the UK Universities' Global Atmospheric Modelling Programme, from the Science and Engineering Research Council, subsequently Engineering and Physical Sciences Research Council, through research grants and through their generous award of a Senior Research Fellowship.

References

Aebischer, U., and C. Schär, 1998: Low-level potential vorticity and cyclogenesis to the lee of the Alps. *J. Atmos. Sci.*, **55**, 186–207.

Allen, J. S., and D. D. Holm, 1996: Extended-geostrophic Hamiltonian models for rotating shallow water motion. *Physica D*, **98**, 229–248.

Appenzeller, C., H. C. Davies, and W. A. Norton, 1996: Fragmentation of stratospheric intrusions. *J. Geophys. Res.*, **101**, 1435–1456.

Austin, J., 1992: Towards the four-dimensional assimilation of stratospheric chemical constituents. *J. Geophys. Res.*, **97**, 2569–2588.

Bennetts, D. A., and B. J. Hoskins, 1979: Conditional symmetric instability—a possible explanation for frontal rainbands. *Quart. J. Roy. Meteor. Soc.*, **105**, 945–962.

Bokhove, O., and T. G. Shepherd, 1996: On Hamiltonian balanced dynamics and the slowest invariant manifold. *J. Atmos. Sci.*, **53**, 276–297.

Buizza, R., and T. N. Palmer, 1993: The singular-vector structure of the atmospheric global circulation. *J. Atmos. Sci.*, **52**, 1434–1456.

Charney, J. G., 1948: On the scale of atmospheric motions. *Geofysiske Publ.*, **17**(2), 3–17.

Chen, P., J. R. Holton, A. O'Neill, and R. Swinbank, 1994: Isentropic mass exchange between the Tropics and Extratropics in the stratosphere. *J. Atmos. Sci.*, **51**, 3006–3018.

Ciesielski, P. E., D. E. Stevens, R. H. Johnson, and K. R. Dean, 1989: Observational evidence for asymmetric inertial instability. *J. Atmos. Sci.*, **46**, 817–831.

Clark, P. D., and P. H. Haynes, 1994: Inertial instability of an asymmetric low-latitude flow. *Quart. J. Roy. Meteor. Soc.*, **122**, 151–182.

Danielsen, E. F., 1968: Stratospheric-tropospheric exchange based on radioactivity, ozone and potential vorticity. *J. Atmos. Sci.*, **25**, 502–518.

Davies, H. C., 1981: An interpretation of sudden warmings in terms of potential vorticity. *J. Atmos. Sci.*, **38**, 427–445.

Davis, C. A., 1992: Piecewise potential vorticity inversion. *J. Atmos. Sci.*, **49**, 1397–1411.

——, M. T. Stoelinga, and Y.-H. Kuo, 1993: The integrated effect of condensation in numerical simulations of extratropical cyclogenesis. *Mon. Wea. Rev.*, **121**, 2309–2330.

Demirtas, M., and A. J. Thorpe, 1997: Sensitivity of short-range weather forecasts to local potential-vorticity modifications. *Mon. Wea. Rev.*, in press.

Dritschel, D. G., 1989: Contour dynamics and contour surgery: numerical algorithms for extended, high-resolution modelling of vortex dynamics in two-dimensional, inviscid, incompressible flows. *Computer Phys. Rep.*, **10**, 78–146.

Dunkerton, T. J., 1993: Inertial instability of nonparallel flow on an equatorial beta-plane. *J. Atmos. Sci.*, **50**, 2744–2758.

——, C.-P. F. Hsu, and M. E. McIntyre, 1981: Some Eulerian and Lagrangian diagnostics for a model stratospheric warming. *J. Atmos. Sci.*, **38**, 819–843.

Edmon, H. J., B. J. Hoskins, and M. E. McIntyre, 1980: Eliassen-Palm cross-sections for the troposphere. *J. Atmos. Sci.*, **37**, 2600–2616. (Also Corrigendum, *J. Atmos. Sci.*, **38**, 1115, especially second last item.)

Emanuel, K. A., 1991: The theory of hurricanes. *Ann. Rev. Fluid Mech.*, **23**, 179–196.

Errico, R. M., 1982: Normal mode initialization and the generation of gravity waves by quasi-geostrophic forcing. *J. Atmos. Sci.*, **39**, 573–586.

Ertel, H., 1942: Ein Neuer hydrodynamischer Wirbelsatz. *Met. Z.*, **59**, 271–281.

Fisher, M., and D. J. Lary, 1995: Lagrangian four-dimensional variational data assimilation of chemical species. *Quart. J. Roy. Meteor. Soc.*, **121**, 1681–1704.

Ford, R., 1994: Gravity wave radiation from vortex trains in rotating shallow water. *J. Fluid Mech.*, **281**, 81–118.

——, M. E. McIntyre, and W. A. Norton, 1998: Balance and the slow quasi-manifold: some explicit results. *J. Atmos. Sci.*, submitted.

Gelbspan, R., 1997: *The Heat Is On: The High Stakes Battle over Earth's Threatened Climate.* Addison-Wesley, 278 pp.

Haynes, P. H., and M. E. McIntyre, 1990: On the conservation and impermeability theorems for potential vorticity. *J. Atmos. Sci.*, **47**, 2021–2031.

Hibbard, W., et al., 1989: Application of the 4-D McIDAS to a model diagnostic study of the Presidents' Day cyclone. *Bull. Amer. Meteor. Soc.*, **70**, 1394–1403.

Hinkelmann, K. H., 1969: Primitive equations. In: *Lectures in Numerical Short-range Weather Prediction. Regional Training Seminar*, Moscow. World Met. Org. No. 297, pp. 306–375.

Holton, J. R., P. H. Haynes, M. E. McIntyre, A. R. Douglass, R. B. Rood, and L. Pfister, 1995: Stratosphere–troposphere exchange. *Rev. Geophys.*, **33**, 403–439.

Hoskins, B. J., M. E. McIntyre, and A. W. Robertson, 1985 [HMR]: On the use and significance of isentropic potential-vorticity maps. *Quart. J. Roy. Meteor. Soc.*, **111**, 877–946. Also **113**, 402–404. [*Note:* The title should have been "On the use and significance of isentropic maps of potential vorticity."]

Juckes, M. N., and M. E. McIntyre, 1987: A high resolution, one-layer model of breaking planetary waves in the stratosphere. *Nature*, **328**, 590–596.

Keyser, D., and R. Rotunno, 1990: On the formation of potential-vorticity anomalies in upper-level jet-front systems. *Mon. Wea. Rev.*, **118**, 1914–1921.

Kleinschmidt, E., 1950–1: Über Aufbau und Entstehung von Zyklonen (1–3 Teil). *Met. Runds.*, **3**, 1–6, 54–61; **4**, 89–96.

Lahoz, W. A., A. O'Neill, A. Heaps, V. D. Pope, R. Swinbank, R. S. Harwood, L. Froidevaux, W. G. Read, J. W. Waters, and G. E. Peckham, 1996: Vortex dynamics and the evolution of water vapour in the stratosphere of the southern hemisphere. *Quart. J. Roy. Meteor. Soc.*, **122**, 423–450.

Lee, A. M., G. D. Carver, M. P. Chipperfield, and J. A. Pyle, 1997: Three-dimensional chemical forecasting: a methodology. *J. Geophys. Res.*, **102**, 3905–3919. [Special issue on the ASHOE airborne experiment.]

Lighthill, M. J., 1952: On sound generated aerodynamically. *Proc. Roy. Soc. London*, **A 211**, 564–587. Also *Collected Papers*, Vol. III, ed. M. Y. Hussaini. Oxford, University Press.

Lighthill, J., 1993: Final recommendations of the Symposium. In: *Proc. ICSU/WMO Internat. Symp. on Tropical Cyclone Disasters*, Beijing, ed. J. Lighthill, Z. Zheng, G. Holland, K. Emanuel; Beijing, Peking University Press, 582–587.

Magnusdottir, Gudrun, and P. H. Haynes, 1996: Application of wave-activity diagnostics to baroclinic-wave life cycles. *J. Atmos. Sci.*, **53**, 2317–2353.

Manney, G. L., R. W. Zurek, A. O'Neill, and R. Swinbank, 1994: On the motion of air through the stratospheric polar vortex. *J. Atmos. Sci.*, **51**, 2973–2994.

McIntyre, M. E., 1982: How well do we understand the dynamics of stratospheric warmings? *J. Meteor. Soc. Japan, Special Centennial Issue*, **60**, 37–65. [Note: "latitude" should be "altitude" on page 39a, line 2.]

——, 1988: Numerical weather prediction: a vision of the future. *Weather* (Roy. Meteorol. Soc.), **43**, 294–298.

——, 1992: Atmospheric dynamics: Some fundamentals, with observational implications. *Proc. Internat. School Phys. "Enrico Fermi", CXV Course*, J. C. Gille, G. Visconti, Eds. (ISBN 0-444-89896-4), North-Holland, 313–386. [A list of updates and minor corrigenda is available via <ftp://ftp.damtp.cam.ac.uk/pub/papers/mem>; get mcintyre.ps. Also accessible via <http://www.atmos-dynamics.damtp.cam.ac.uk/>.]

——, 1993a: On the role of wave propagation and wave breaking in atmosphere–ocean dynamics. *Theoretical and Applied Mechanics 1992*, S. Bodner, J. Singer, A. Solan, and Z. Hashin, Eds. (Sectional Lecture, Proc. XVIII Int. Congr. Theor. Appl. Mech., Haifa.), Elsevier, 281–304.

——, 1993b: Isentropic distributions of potential vorticity and their relevance to tropical cyclone dynamics. *Proc. ICSU/WMO Internat. Symp. on Tropical Cyclone Disasters*, Beijing, J. Lighthill, Z. Zheng, G. Holland, K. Emanuel, Eds., Peking University Press, 143–156.

——, 1994: Numerical weather prediction: An updated vision of the future. *The Life Cycles of Extratropical Cyclones, Proceedings of an International Symposium at the University of Bergen, Norway, 27 June – 1 July 1994*, S. Grønås and M. Shapiro, Eds., ISBN 82-419-0144-5. Vol. I, Invited Papers, 275–286.

——, and W. A. Norton, 1998: Potential-vorticity inversion on a hemisphere. *J. Atmos. Sci.*, resubmitted.

——, and T. N. Palmer, 1983: Breaking planetary waves in the stratosphere. *Nature*, **305**, 593–600.

——, and ——, 1984: The "surf zone" in the stratosphere. *J. Atmos. Terr. Phys.*, **46**, 825–849.

——, and ——, 1985: A note on the general concept of wave breaking for Rossby and gravity waves. *Pure Appl. Geophys.*, **123**, 964–975.

——, and I. Roulstone, 1998: Hamiltonian balanced models: slow manifolds, constraints and velocity splitting. *J. Fluid Mech.*, in revision.

Molteni, F., R. Buizza, T. N. Palmer, and T. Petroliagis, 1996: The ECMWF ensemble prediction system: Methodology and validation. *Quart. J. Roy. Meteor. Soc.*, **122**, 73–119.

Monod, J., 1971: *Chance and Necessity*, A. Wainhouse, Transl., Collins, 187 pp.

Nakamura, N., 1996: Two-dimensional mixing, edge formation, and permeability diagnosed in an area coordinate. *J. Atmos. Sci.*, **53**, 1524–1537.

Norton, W. A., 1988: Balance and potential vorticity inversion in atmospheric dynamics. Ph.D. Thesis, University of Cambridge, 167 pp.

——, 1994: Breaking Rossby waves in a model stratosphere diagnosed by a vortex-following coordinate system and a technique for advecting material contours. *J. Atmos. Sci.*, **51**, 654–673.

O'Sullivan, D. J., and M. H. Hitchman, 1992: Inertial instability and Rossby wave breaking in a numerical model. *J. Atmos. Sci.*, **49**, 991–1002.

——, and P. Chen, 1996: Modeling the QBO's influence on isentropic tracer transport in the subtropics. *J. Geophys. Res.*, **101**, 6811–6821.

Palmer, T. N., 1996: Predictability of the atmosphere and oceans: from days to decades. *Decadal Climate Variability* (NATO Advanced Study Institute, Les Houches, 1995), D. Anderson and J. Willebrand, Eds., Series 1, Global Environmental Change, Vol. 44, Springer.

——, R. Gelaro, J. Barkmeijer, and R. Buizza, 1998: Singular vectors, metrics, and adaptive observations. *J. Atmos. Sci.*, **55**, 633–653.

Polvani, L. M., J. C. McWilliams, M. A. Spall, and R. Ford, 1994: The coherent structures of shallow water turbulence: deformation-radius effects, cyclone/anticyclone asymmetry and gravity-wave generation. *Chaos*, **4**, 177–186 and 427–430. [Special Volume on geophysical flows; pp. 427–430 are colour plates.]

Rabier, F., E. Klinker, P. Courtier, and A. Hollingsworth, 1996: Sensitivity of forecast errors to initial conditions. *Quart. J. Roy. Meteor. Soc.*, **122**, 121–150.

Randel, W. J., J. C. Gille, A. E. Roche, J. B. Kumer, J. L. Mergenthaler, J. W. Waters, E. F. Fishbein, and W. A. Lahoz, 1993: Stratospheric transport from the tropics to middle latitudes by planetary-wave mixing. *Nature*, **365**, 533–535.

Raymond, D. J., 1992: Nonlinear balance and potential-vorticity thinking at large Rossby number. *Quart. J. Roy. Meteor. Soc.*, **118**, 987–1015.

Reed, R. J., and E. F. Danielsen, 1959: Fronts in the vicinity of the tropopause. *Arch. Met. Geophys. Biokl.*, **A 11**, 1–17.

——, M. T. Stoelinga, and Y.-H. Kuo, 1992: A model-aided study of the origin and evolution of the anomalously high potential vorticity in the inner region of a rapidly deepening marine cyclone. *Mon. Wea. Rev.*, **120**, 893–913.

——, Y.-H. Kuo, and S. Low-Nam, 1994: An adiabatic simulation of the ERICA IOP 4 storm: An example of quasi-ideal frontal cyclone development. *Mon. Wea. Rev.*, **122**, 2688–2708.

Rossby, C. G., 1936: Dynamics of steady ocean currents in the light of experimental fluid mechanics. *Mass. Inst. of Technology and Woods Hole Ocean. Inst. Papers in Physical Oceanography and Meteorology*, **5**(1), 1–43.

——, 1940: Planetary flow patterns in the atmosphere. *Quart. J. Roy. Meteor. Soc.*, **66**(Suppl.), 68–97.

Schoeberl, M. R., L. R. Lait, P. A. Newman, and J. E. Rosenfield, 1992: The structure of the polar vortex. *J. Geophys. Res.*, **97**, 7859–7882. [Polar Ozone Special Issue, no. D8]

——, A. E. Roche, J. M. Russell III, D. Ortland, P. B. Hays, and J. W. Waters, 1997: An estimation of the dynamical isolation of the tropical lower stratosphere using UARS wind and trace gas observations of the quasi-biennial oscillation. *Geophys. Res. Lett.*, **24**, 53–56.

Shaw, Sir Napier, 1930: *Manual of Meteorology. Vol. III: The Physical Processes of Weather.* Cambridge, University Press.

Starr, V. P., and M. Neiburger, 1940: Potential vorticity as a conservative property. *J. Marine Res.*, **3**, 202–210.

Strahan, S. E., and J. D. Mahlman, 1994: Evaluation of the SKYHI general circulation model using aircraft N_2O measurements. 1. Polar winter stratospheric meteorology and tracer morphology. *J. Geophys. Res.*, **99**, 10305–10318.

Sutton, R. T., H. Maclean, R. Swinbank, A. O'Neill, and F. W. Taylor, 1994: High-resolution stratospheric tracer fields estimated from satellite observations using Lagrangian trajectory calculations. *J. Atmos. Sci.*, **51**, 2995–3005.

Tennekes, H., 1988: Numerical weather prediction: Illusions of security, tales of imperfection. *Weather* (Roy. Meteorol. Soc.), **43**(4), 165–170.

Thépaut, J.-N., P. Courtier, G. Belaud, and G. Lemaître, 1996: Dynamical structure functions in a four-dimensional variational assimilation. A case study. *Quart. J. Roy. Meteor. Soc.*, **122**, 535–561.

Thorncroft, C. D., B. J. Hoskins, and M. E. McIntyre, 1993: Two paradigms of baroclinic-wave life-cycle behaviour. *Quart. J. Roy. Meteor. Soc.*, **119**, 17–55.

Thorpe, A. J., 1985: Diagnosis of balanced vortex structure using potential vorticity. *J. Atmos. Sci.*, **42**, 397–406.

Warn, T., 1997: Nonlinear balance and quasigeostrophic sets. *Atmos.-Ocean*, **35**, 135–145. [*Note:* This pioneering paper was written in 1983 but rejected by the journal to which it was then submitted. It seems to have been the first to recognize that the "slow quasi-manifold" of the parent dynamics is in some sense fuzzy.]

——, and R. Ménard, 1986: Nonlinear balance and gravity–inertial wave saturation in a simple atmospheric model. *Tellus*, **38A**, 285–294.

Warwick, K., 1997: *March of the Machines: Why the New Race of Robots Will Rule the World.* Random House (Century Books), 263 pp.

Waters, J. W., L. Froidevaux, W. G. Read, G. L. Manney, L. S. Elson, D. A. Flower, R. F. Jarnot, and R. S. Harwood, 1993: Stratospheric ClO and ozone from the Microwave Limb Sounder on the Upper Atmosphere Research Satellite. *Nature*, **362**, 597–602.

Waugh, D. W., 1993: Subtropical stratospheric mixing linked to disturbances in the polar vortices. *Nature*, **365**, 535–537.

Waugh, D. W., and R. A. Plumb, 1994: Contour advection with surgery: A technique for investigating fine scale structure in tracer transport. *J. Atmos. Sci.*, **51**, 530–540.

von Weizsäcker, E., A. B. Lovins, and L. H. Lovins, 1997: *Factor Four: Doubling Wealth, Halving Resource Use—The New Report to the Club of Rome.* Earthscan Publications, 322 pp.

Index

WE REMEMBER BERGEN

A photographic recollection of our friends
Bergen, Norway, 27 June–1 July 1994

This collection of photographs was taken during the lecture sessions and at the special social events of the 1994 International Symposium on the Life Cycles of Extratropical Cyclones, held in Bergen, Norway. The contributing photographers are Carlye Calvin (album photo editor), Nadine Lindzen, Paul Schultz, and Melvyn Shapiro. The special social events included the evening fjord cruise on the Norwegian sailing ship "Statsraad Lehmkuhl," the reception hosted by city of Bergen at the historic 12th century "Haakonshallen" City Hall of Bergen, and the tour of Norwegian composer Edvardt Greig's home "Trollhaugen" and a piano recital featuring Greig's compositions in the adjacent concert hall designed by architect Peter Helland-Hansen. Following the recital, the participants and their guests continued their festivities at the symposium banquet at the old hotel at Solstrand, a hotel where Vilhelm and later Jack Bjerknes hosted banquets at international meetings.

Memories captured in a blink of the camera's eye

Melvyn Shapiro

Arnt Eliassen

Chester Newton

Brian Hoskins

Thor Erik Nordeng

Mel Shapiro

Dan Keyser

Keith Browning

Lennart Bengtsson

Don Johnson

Phil Smith

Rich Rotunno

Kerry Emanuel

Nick Bond

R. Saravanan

Isaac Held

Richard Anthes

Fred Sanders

Sigbjørn Grønås

Marty Ralph and Chris Snyder

Randy Dole

Warren Blier

Ola Persson

Paul Kocin

Claus-Peter Hoinka and Fedor Mesinger

Bill Kuo and Christoph Schar

Dave Schultz and Jon Martin

Huw Davies

Simo Jarvenoja and Juhanni Rinne

Mankin Mak

Mike Wallace and Dick Reed

Eero Holopainen, Louis Uccellini, Hans Volkert and Alan Thorpe

Elmer Raustein and Anna Sandvik

Bob Serafin

Stan Benjamin and Bill Hooke

Georg Grell and Qin Xu

Gudrun Magnusdottir, Erik Rasmussen and Sigbjørn Grønås

Lorenzo Dell'Osso and Peter Lynch

Bob Gall and Howie Bluestein

Paula McCaslin

Alan Thorpe

Michael McIntyre

Brian Farrell and Petros Ioannou

Dick Lindzen and John Green

Jon E. Kristjansson and Gudrun Magnusdottir

Dick Reed and Bill Cotton

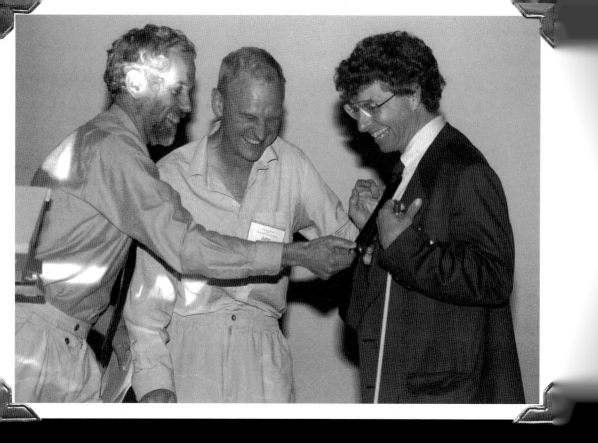

Brian Hoskins, John Green and Brian Farrell

Evelyn Mazur and Dorothea Johnson

Manfred Kurz

Tony Hollingsworth, Nils Gunnar Kvamsto and Adrian Simmons

Hans Volkert

Mike Wallace

Dave Parsons and Qin Xu

Michael McIntyre and Mel Shapiro

Hilding Sundqvist

Edvard Grieg's Home

Edvard Grieg's Home

Volkmar Wirth, Bill Hooke, and Phil Smith

Banquet at Solstrand

Dick Reed, Nancy Sanders, Joan Reed and Fred Sanders

Mr. and Mrs. Troll

Warren and Mary Washington

Bob and Betsy Serafin

Stephanie and Dick Dirks

Jackie and John Snook

Lise Levy and Egil Stokke

Chris and Mary Ann Davis

Isidoro and Beatrice Orlanski

Jina and Roger Wakimoto

Sandy Rush

Bob Serafin and Ned Ostenso

Ken Spengler

Steve Tracton

Christian Zick

Lance Bosart

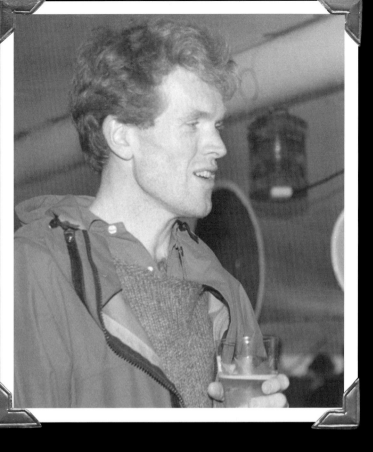

Froda Flatøy

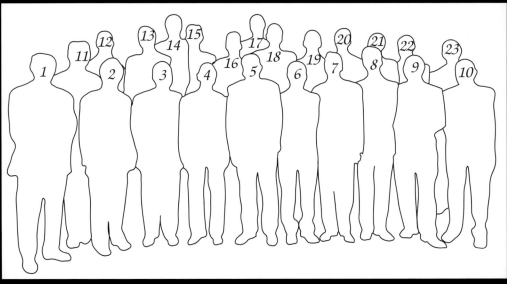

1- Isaac Held

2- Brian Farrell

3 - Dick Lindzen

4 - Dick Reed

5 - Arnt Elliason

6 - Dan Keyser

7 - Mel Shapiro

9 - Tony Hollingsworth

10 - Fred Sanders

11 - Mike Wallace

12 - John Green

13 - Lance Bosart

14 - Rick Anthes

15 - Chester Newton

17 - Lennart Bengtsson

18 - Adrian Simmons

19 - Louis Uccellini

20 - Eero Holopainen

21 - Huw Davies

22 - Brian Hoskins

23 - Sigbjørn Grønås